Convergence of One-Parameter Operator Semigroups

This book presents a detailed and contemporary account of the classical theory of convergence of semigroups and its more recent developments treating the case where the limit semigroup, in contrast to the approximating semigroups, acts merely on a subspace of the original Banach space (this is the case, for example, with singular perturbations). The author demonstrates the far-reaching applications of this theory using real examples from various branches of pure and applied mathematics, with a particular emphasis on mathematical biology. These examples also serve as short, nontechnical introductions to biological concepts involved, allowing readers to develop intuitions underlying mathematical results.

The book may serve as a useful reference, containing a significant number of new results ranging from the analysis of fish populations to signaling pathways in living cells. It comprises many short chapters, which allows readers to pick and choose those topics most relevant to them, and it contains 160 end-of-chapter exercises so that readers can test their understanding of the material as they go along.

ADAM BOBROWSKI is a professor and Chairman of the Department of Mathematics at Lublin University of Technology, Poland. He has authored over 50 scientific papers and two books *Functional Analysis for Probability and Stochastic Processes* and *An Operator Semigroup in Mathematical Genetics*.

NEW MATHEMATICAL MONOGRAPHS

All the titles listed below can be obtained from good booksellers or from Cambridge University Press. For a complete series listing, visit www.cambridge.org/mathematics.

1. M. Cabanes and M. Enguehard *Representation Theory of Finite Reductive Groups*
2. J. B. Garnett and D. E. Marshall *Harmonic Measure*
3. P. Cohn *Free Ideal Rings and Localization in General Rings*
4. E. Bombieri and W. Gubler *Heights in Diophantine Geometry*
5. Y. J. Ionin and M. S. Shrikhande *Combinatorics of Symmetric Designs*
6. S. Berhanu, P. D. Cordaro and J. Hounie *An Introduction to Involutive Structures*
7. A. Shlapentokh *Hilbert's Tenth Problem*
8. G. Michler *Theory of Finite Simple Groups I*
9. A. Baker and G. Wüstholz *Logarithmic Forms and Diophantine Geometry*
10. P. Kronheimer and T. Mrowka *Monopoles and Three-Manifolds*
11. B. Bekka, P. de la Harpe and A. Valette *Kazhdan's Property (T)*
12. J. Neisendorfer *Algebraic Methods in Unstable Homotopy Theory*
13. M. Grandis *Directed Algebraic Topology*
14. G. Michler *Theory of Finite Simple Groups II*
15. R. Schertz *Complex Multiplication*
16. S. Bloch *Lectures on Algebraic Cycles (2nd Edition)*
17. B. Conrad, O. Gabber and G. Prasad *Pseudo-Reductive Groups*
18. T. Downarowicz *Entropy in Dynamical Systems*
19. C. Simpson *Homotopy Theory of Higher Categories*
20. E. Fricain and J. Mashreghi *The Theory of H(b) Spaces I*
21. E. Fricain and J. Mashreghi *The Theory of H(b) Spaces II*
22. J. Goubault-Larrecq *Non-Hausdorff Topology and Domain Theory*
23. J. Śniatycki *Differential Geometry of Singular Spaces and Reduction of Symmetry*
24. E. Riehl *Categorical Homotopy Theory*
25. B. A. Munson and I. Volić *Cubical Homotopy Theory*
26. B. Conrad, O. Gabber and G. Prasad *Pseudo-Reductive Groups (2nd Edition)*
27. J. Heinonen, P. Koskela, N. Shanmugalingam and J. T. Tyson *Sobolev Spaces on Metric Measure Spaces*
28. Y.-G. Oh *Symplectic Topology and Floer Homology I*
29. Y.-G. Oh *Symplectic Topology and Floer Homology II*
30. A. Bobrowski *Convergence of One-Parameter Operator Semigroups*

Convergence of One-Parameter Operator Semigroups

In Models of Mathematical Biology and Elsewhere

ADAM BOBROWSKI
Lublin University of Technology, Poland

CAMBRIDGE
UNIVERSITY PRESS

CAMBRIDGE
UNIVERSITY PRESS

University Printing House, Cambridge CB2 8BS, United Kingdom

Cambridge University Press is part of the University of Cambridge.

It furthers the University's mission by disseminating knowledge in the pursuit of education, learning and research at the highest international levels of excellence.

www.cambridge.org
Information on this title: www.cambridge.org/9781107137431

First published 2016

A catalogue record for this publication is available from the British Library

ISBN 978-1-107-13743-1 Hardback

To the most enthusiastic drummer, boxer, and eater ever – my son Marek

Contents

Preface

Now lift up your eyes, and look from the place where you are, northward and southward and eastward and westward; for all the land that you see I will give to you and to your seed forever.

(Gen. 13:14b–15. Holy Bible. Recovery Version. © 2003 Living Stream Ministry, Anaheim, California.)

The question of convergence and the related Trotter–Kato–Neveu–Sova–Kurtz Theorem lie in the center of the theory of one-parameter semigroups of operators. In fact, many key results, including the fundamental Hille–Yosida Theorem, are proved by approximating a semigroup involved by a family of semigroups that are easier to handle. On the other hand, the most elegant proof of the Trotter–Kato–Neveu–Sova–Kurtz Theorem is obtained by applying the Hille–Yosida Theorem to a certain, coordinate-wise acting operator in the space of convergent sequences [163, 211, 229]. Hence, all of the basic books devoted to the theory, except, of course, of the classics [115] and [180] published before the Trotter–Kato–Neveu–Sova–Kurtz Theorem was known, treat the subject at least roughly (see, e.g., [9, 101, 128, 163, 284, 334, 353]). Other books, like [132], take advantage of the theory and show its beautiful and far-reaching applications.

This book aims to present the classic theory of convergence of semigroups and also to discuss its applicability to models of mathematical biology and other branches of mathematics. At the same time, it is an attempt at expressing my long fascination with phenomena accompanying such convergence, and my continuous amazement at the variety of examples, especially of the stochastic type, found in the literature.

The main difficulty in applying the theory to the models mentioned here is that the Trotter–Kato–Neveu–Sova–Kurtz-type theorems characterize convergence that is almost uniform in time $t \in [0, \infty)$, whereas in the models one often encounters a situation in which convergence is not uniform in the neighborhood of $t = 0$. Therefore, this book splits naturally into two parts. In the first part, I discuss the models in which the classical theory applies. In the second part, I present methods developed for treating convergence that is not uniform near $t = 0$, such as singular perturbations. Therefore, in particular, the book is an outgrowth of Tosio Kato's project [199] to incorporate singular perturbations in the theory of convergence of semigroups of operators.

As already stated, this book is equally devoted to the general theory and to examples, which are selected from various branches of applied and pure mathematics with an emphasis on models of mathematical biology. These examples serve to exemplify applications of the main theorems, but at the same time are meant to be nontechnical, short introductions to related biological models. Deprived of these examples, the book loses more than half of its worth (if it possesses any). However, for readers who want to know the structure of the book, the main theoretical chapters in Part I are:

- Chapter 2, where the basic convergence theorem is presented
- Chapter 7, where the relations between the limit pseudoresolvent and the extended limit are presented
- Chapter 8, devoted to the characterization of the regularity subspace where the semigroups converge in a regular way
- Chapter 14, in which the question of convergence outside of the regularity space is discussed
- Chapter 15, in which the Hasegawa condition is introduced
- Chapter 18, which is devoted to discrete-time approximations
- Chapter 24, which is devoted to the case in which the approximating semigroups act in different spaces than the limit semigroup
- Chapter 25, in which the stability condition is discussed

In Part II, the most important theoretical chapters include:

- Chapters 26, 28, and 29, in which basic examples and fundamental properties of irregular convergence are presented
- Chapter 31, providing tools for dealing with parabolic problems
- Chapter 42, presenting the important Kurtz's Theorem, which allows dealing with variety of convergence questions, including problems of hyperbolic type
- Chapters 54 and 55, giving some insight into alternative approaches

In Part III, devoted to convergence of cosine families in addition to basic theory (which is in many aspects parallel to the theory of convergence of semigroups), we reveal the stunning result that there is no singular perturbation theory for cosine families.

Perhaps it should be added here that in Part II, Chapters 33 through 40 (with the exception of Chapter 37) form an essay on singular limits involving fast diffusion, including the treatment of the homogenization theorem of Conway, Hoff, and Smoller discussed in Chapter 38.

This book suffers from the omission of important theorems and examples, most of which I am simply ignorant of. Of those I am aware of, I should mention the beautiful phenomenon ascribed to Bafico and Baldi [13]. The reason for not including it here is that it appears to be related to convergence in weaker topology, and I determined to restrict myself to strong topology of Banach spaces. Other examples include the Janas–Berezin approximation method [33, 189], state-space collapse in the theory of queues (see [295] and the references given there), models of adaptive dynamics [79, 113, 285] (which are in most interesting cases nonlinear), diffusion processes on an open book [152], and definitely many, many more. See, for example, the recent book by J. Banasiak and M. Lachowicz [25] written in the spirit of asymptotic analysis, and the paper by S. N. Evans and R. B. Sowers [133] – a rich source of examples of stochastic nature (I thank T. G. Kurtz and S. N. Evans for this reference). Fortunately, there are other, perhaps more apt accounts of the geography of the vast land of singular-perturbation theory, including [205, 263, 280, 311, 316, 333], to mention just a few that I have on my bookshelf. (On April 30, 2015, MathSciNet returned 2,404 matches to the "singular perturbation" query.) The other side of this coin is that the lion's share of the examples presented here have not been available in book form as of this writing.

I have benefited from delivering a series of lectures at the Institute of Mathematics of the Polish Academy of Sciences and teaching two courses at Maria Curie-Skłodowska University (UMCS) in Lublin, Poland, all based on the material in this book. Judging by the feedback that I have received, audiences have benefited, too, but you can never be sure of things like that. I would like to thank both institutions for allowing me to teach these courses, despite the danger of having a new generation of students interested in the theory of semigroups of operators. Financially, beginning in 2011 and until 2014, I was supported by the Polish Government, grant number 6081/B/H03/2011/40.

Moreover, I would like to express my thanks to Professor Jan Kisyński, my teacher and my friend, who introduced me to the world of convergence of semigroups and for years has been an invaluable "point of reference," and to my students, particularly to Radek Bogucki, without whom this book would have been completely different, and Adam Gregosiewicz (a victim of my course at UMCS), who solved many exercises – his solution manual will be available on the web in due time, "but you can never be sure of things like that." Special thanks are also due to Jacek Banasiak for stimulating discussions, hints for solving problems, and references leading to new and interesting ones. Moreover, I would like to acknowledge friendly support I have received from A. Bátkai, K. Bogdan, D. Mugnolo, W. Chojnacki, J. Goldstein, M. Kimmel, T. Komorowski, M. Lachowicz, H. Leszczyński, T. Lipniacki, A. Marciniak-Czochra, R. Rudnicki, Y. Tomilov, and D. Wrzosek, who in various ways contributed to my understanding of the subjects covered in the book. They have done it in the unfeigned hope that I will be able to comprehend some complex issues in mathematics and biology; if their hopes are shattered by the book, they should not be held responsible for all or any misrepresentations, as they were acting in good faith.

Last but not least, thanks are due to my family: I would like to thank my wife Beata for her constant encouragement and help, without which it would be very difficult to devote so much time to writing this book. Special thanks go also to my sons, Radek and Marek, for supplementing this monograph with unique pictures.[1,2]

[1] Approximately 10 years ago, after I had completed writing my first book [49], I heartily wanted to thank Roger Astley from Cambridge University Press for his help and encouragement. However, he did not allow me, by saying something to the effect that Cambridge University Press does not approve of such acknowledgments. By looking at other books recently, to my astonishment I discovered that this was a flat lie. Hence, I am thanking him now in a footnote, which is printed in footnote-size, lest he should notice.

[2] Hey, I can also thank my friends Chyl and Jarecki for, hm, being my friends.

1

Semigroups of Operators and Cosine Operator Functions

In this introductory chapter, I present the rudiments of the theory of strongly continuous semigroups of operators and the related theory of cosine operator functions (cosine families). These theories, especially the first one, being well known, and given the number of books devoted to the subject, I felt it unnecessary to present all the involved details here, but appropriate to recall basic facts. Experts may simply skip this chapter, and students familiar with the theory may test their fluency by checking whether they remember how to prove the facts presented here (and perhaps do the nonstandard exercises at the end). All of the remaining readers are asked to consult one or several of [9, 49, 132, 101, 103, 118, 122, 127, 128, 129, 163, 180, 193, 201, 215, 256, 284, 324, 334, 343, 353] or other relevant monographs. The real newcomers may want to begin by reading [259]. A unified proof of the Hille–Yosida and Sova–Da Prato–Giusti Theorems is given in Appendix A.

C_0 Semigroups

A C_0 semigroup or a strongly continuous semigroup in a Banach space $\mathbb{X}$ is a family $T = (T(t))_{t\geq 0}$ (written also as $\{T(t), t \geq 0\}$) of bounded linear operators such that:

(a) $T(t)T(s) = T(t+s), t, s \geq 0$,
(b) $T(0) = I_{\mathbb{X}}$,
(c) $\lim_{t\to 0} T(t)x = x, x \in \mathbb{X}$,

with the last limit in the strong topology in $\mathbb{X}$. If merely the first two conditions are satisfied the family is said to be a **semigroup**. The three properties together, combined with the Banach–Steinhaus uniform boundedness principle, imply that there exist $M \geq 1, \omega \in \mathbb{R}$ such that $\|T(t)\| \leq M\mathrm{e}^{\omega t}$. Moreover, the map

$[0, \infty) \ni t \mapsto T(t)x$ is continuous (in the strong topology) for all $x \in \mathbb{X}$. Hence, the Laplace transform $R_\lambda = \int_0^\infty e^{-\lambda t} T(t)\,dt$ of the semigroup is well defined for all $\lambda > \omega$, as a strong improper Riemann integral, and we have $\|R_\lambda\| \le \frac{M}{\lambda-\omega}$. Using the semigroup property, we check that:

$$R_\lambda^n = \int_0^\infty \cdots \int_0^\infty e^{-\lambda \sum_{i=1}^n t_i} T\left(\sum_{i=1}^n t_i\right) dt_1 \ldots dt_n,$$

which gives $\|R_\lambda^n\| \le M(\lambda - \omega)^{-n}, n \ge 1$.

The infinitesimal generator of $(T(t))_{t \ge 0}$ is defined as:

$$Ax = \lim_{t \to 0} \frac{1}{t}(T(t)x - x),$$

for those x for which the limit exists in the strong topology. Because all integrals of the form $\int_0^h T(t)x\,ds, x \in \mathbb{X}, h > 0$ belong to the domain $D(A)$ of A and $\lim_{h\to 0} \frac{1}{h} \int_0^h T(t)x\,ds = x$, A is densely defined. Furthermore, for $x \in D(A)$, $T(t)x = x + \int_0^t T(s)Ax\,ds$, and this implies that A is closed. Then, R_λ turns out to be the **resolvent** of A in the sense that $\lambda - A$ has a bounded left and right inverse: (a) $R_\lambda(\lambda - A)x = x$ for $x \in D(A)$, and (b) $R_\lambda x$ belongs to $D(A)$ for all $x \in \mathbb{X}$ and $(\lambda - A)R_\lambda x = x$ (in particular, the range of $\lambda - A$ is the whole of $\mathbb{X}$; this is the all-important **range condition**). Therefore, we write $R_\lambda = (\lambda - A)^{-1}$. We note that the equation:

$$\lambda x - Ax = y, \tag{1.1}$$

where $y \in \mathbb{X}$ is given and $x \in D(A)$ is to be found, is called the **resolvent equation** for A.

This implies in turn that $R_\lambda, \lambda > 0$ satisfies the **Hilbert Equation**:

$$R_\lambda - R_\mu = (\mu - \lambda)R_\mu R_\lambda, \qquad \lambda, \mu > 0, \tag{1.2}$$

which could also be obtained from the fact that R_λ is the Laplace transform of the semigroup.

The Hille–Yosida–Feller–Phillips–Miyadera Theorem states that the conditions on A mentioned here are not only necessary but also sufficient for A to be the generator of a strongly continuous semigroup.

Theorem 1.1 (Hille–Yosida–Feller–Phillips–Miyadera) *An operator A is the generator of a strongly continuous semigroup $\{T(t), t \ge 0\}$ with $\|T(t)\| \le Me^{\omega t}$ iff:*

- *A is closed and densely defined,*
- *for all $\lambda > \omega$, $\lambda - A$ has a continuous left and right inverse $(\lambda - A)^{-1}$,*
- *$\|(\lambda - A)^{-n}\| \le M(\lambda - \omega)^{-n}$, $n \ge 1, \lambda > \omega$.*

Using the Hilbert Equation and the estimate $\|(\lambda - A)^{-1}\| \le \frac{M}{\lambda-\omega}$, we may prove that the map $(\omega, \infty) \ni \lambda \to (\lambda - A)^{-1} \in \mathcal{L}(\mathbb{X})$ is differentiable with $\frac{\mathrm{d}}{\mathrm{d}\lambda}(\lambda - A)^{-1} = -(\lambda - A)^{-2}$. Hence, by induction:

$$\frac{\mathrm{d}^n}{\mathrm{d}\lambda^n}(\lambda - A)^{-1} = (-1)^n n!(\lambda - A)^{-(n+1)}, \qquad n \ge 1$$

so that the third condition may be equivalently expressed as:

- $\|\frac{\mathrm{d}^n}{\mathrm{d}\lambda^n}(\lambda - A)^{-1}\| \le \frac{Mn!}{(\lambda-\omega)^{n+1}}$, $n \ge 0, \lambda > \omega$.

We note that A generates a strongly continuous semigroup $\{T(t), t \ge 0\}$ iff $(\lambda - A)^{-1} = \int_0^\infty \mathrm{e}^{-\lambda t} T(t)\,\mathrm{d}t$; this shows in particular that an operator must not generate two different semigroups. We will write $(\mathrm{e}^{tA})_{t\ge0}$, $\{\mathrm{e}^{tA}, t \ge 0\}$ or $\{T_A(t), t \ge 0\}$ for the semigroup generated by A.

The Phillips Perturbation Theorem says that if A is the generator of a strongly continuous semigroup, then so is $A + B$ (with domain $\mathcal{D}(B) = \mathcal{D}(A)$) where B is a bounded operator. Moreover,

$$\mathrm{e}^{t(A+B)} = \sum_{n=0}^{\infty} S_n(t), \text{ where}$$

$$S_0(t) = \mathrm{e}^{tA}, S_{n+1}(t) = \int_0^t \mathrm{e}^{(t-s)A} B S_n(s)\,\mathrm{d}s, \qquad n \ge 0.$$

Cosine Operator Functions

A strongly continuous cosine operator function (or a cosine family) is a family $\{C(t), t \in \mathbb{R}\}$ of bounded linear operators satisfying:

(a) $2C(t)C(s) = C(t+s) + C(t-s), t, s \in \mathbb{R}$,
(b) $C(0) = I_{\mathbb{X}}$,
(c) $\lim_{t\to0} C(t)x = x, x \in \mathbb{X}$.

As in the case of semigroups, there exist $M \ge 1$ and $\omega \ge 0$ (note the restriction on ω; see Proposition 3.14.6 in [9] and our Exercise 61.2), such that $\|C(t)\| \le M\mathrm{e}^{\omega t}$. Also, $C(t) = C(-t)$ and $\{C(t), t \in \mathbb{R}\}$ is a strongly continuous family. Finally, its Laplace transform is uniquely determined by:

$$\lambda(\lambda^2 - A)^{-1} = \int_0^\infty \mathrm{e}^{-\lambda t} C(t)\,\mathrm{d}t, \lambda > \omega, \tag{1.3}$$

where A is the infinitesimal generator defined as:

$$Ax = \lim_{t\to0} \frac{2[C(t)x - x]}{t^2},$$

for those x for which this limit exists. Again, A is of necessity closed and densely defined. The analogue of the Hille–Yosida–Feller–Phillips–Miyadera Theorem (due to Sova [319], and Da Prato and Giusti [290]) is as follows.

Theorem 1.2 *An operator A is the generator of a strongly continuous cosine operator family $\{C(t), t \in \mathbb{R}\}$ such that $\|C(t)\| \le M e^{\omega t}$ iff:*

- *A is closed and densely defined,*
- *for all $\lambda > \omega$, $\lambda^2 - A$ has a continuous left and right inverse $(\lambda^2 - A)^{-1}$,*
- $\| \frac{d^n}{d\lambda^n}[\lambda(\lambda^2 - A)^{-1}] \| \le M n! (\lambda - \omega)^{-(n+1)}$, $n \ge 0, \lambda > \omega$.

Formula (1.3) shows that the generator uniquely determines the cosine family. I will write $\{\mathrm{Cos}_A(t), t \in \mathbb{R}\}$ for the cosine family generated by A. I note that there is a cosine family equivalent of the Phillips Perturbation Theorem (Corollaries 3.14.10 and 3.14.13 in [9]).

Semigroups and Cosines

As it was seen earlier, the theories of strongly continuous semigroups and of cosine families are quite analogous (see, however, [62] or Chapter 61). In Appendix A, I will show that both the Hille–Yosida–Feller–Phillips–Miyadera and the Sova–Da Prato–Giusti generation theorems may be obtained from the Hennig–Neubrander Representation Theorem for Laplace transform. Here, I discuss two other connections between these theories.

Suppose that $\{\mathrm{Cos}_A(t), t \in \mathbb{R}\}$ is a strongly continuous cosine family. Then the formula:

$$T(t) = \frac{1}{2\sqrt{\pi t}} \int_{-\infty}^{\infty} e^{-\frac{s^2}{4t}} \mathrm{Cos}_A(s)\, ds, \qquad t > 0 \tag{1.4}$$

defines a strongly continuous semigroup whose generator turns out to be A. In other words, generators of cosine families are of necessity generators of semigroups, even of holomorphic semigroups (see Chapter 15). The converse, however, is not true. There are generators of holomorphic semigroups that are not generators of cosine families. Relation (1.4) is known as the **Weierstrass Formula** [9, 139] or **subordination principle** [292]; see [154] for probabilistic aspects of this relationship.

The **Kisyński Theorem** establishes another connection between semigroups and cosines, as follows [9, 213]: Let again $\{\mathrm{Cos}_A(t), t \in \mathbb{R}\}$ be a strongly continuous cosine family in a Banach space $\mathbb{X}$ and let $\mathbb{X}_{\mathrm{Kis}}$ be the subset of $x \in \mathbb{X}$ such that $t \mapsto \mathrm{Cos}_A(t)x$ is continuously differentiable. Then, $\mathbb{X}_{\mathrm{Kis}}$ is a Banach space when equipped with the norm $\|x\|_{\mathrm{Kis}} = \|x\| + \sup_{t\in[0,1]} \|\frac{dC_A(t)x}{dt}\|$, and the operator $\mathcal{A}(y,x) = (x, Ay)$ with domain $\mathcal{D}(\mathcal{A})$ composed of (y,x) with $y \in \mathcal{D}(A)$, $x \in \mathbb{X}_{\mathrm{Kis}}$ generates a C_0 group in $\mathbb{X}_{\mathrm{Kis}} \times \mathbb{X}$ (i.e., both $\mathcal{A}$ and $-\mathcal{A}$ generate C_0 semigroups there), and we have:

$$\mathrm{e}^{t\mathcal{A}} = \begin{pmatrix} \mathrm{Cos}_A(t) & \int_0^t \mathrm{Cos}_A(s)\,\mathrm{d}s \\ \frac{\mathrm{d}\mathrm{Cos}_A(t)}{\mathrm{d}t} & \mathrm{Cos}_A(t) \end{pmatrix}, \qquad t \in \mathbb{R}. \tag{1.5}$$

The converse is also true: if A is a linear operator in a Banach space $\mathbb{X}$, if $\mathbb{X}_1$ is a Banach space such that:

$$\mathcal{D}(A) \subset \mathbb{X}_1 \subset \mathbb{X}, \tag{1.6}$$

and the topology in $\mathcal{D}(A)$ is not weaker than the topology in $\mathbb{X}_1$ and the topology in $\mathbb{X}_1$ is not weaker than the topology in $\mathbb{X}$ and $\mathcal{A}$ given here generates a C_0 group in $\mathbb{X}_1 \times \mathbb{X}$, then A generates a strongly continuous cosine family in $\mathbb{X}$. Moreover, $\mathbb{X}_1$ is isomorphic to $\mathbb{X}_{\mathrm{Kis}}$. The $\mathbb{X}_{\mathrm{Kis}}$ is often referred to as the **Kisyński space**. Interestingly, if condition (1.6) is omitted, $\mathbb{X}_{\mathrm{Kis}}$ is not uniquely determined; see [237].

CONVENTION Throughout this book, when speaking about abstract Banach spaces, I use x, y, z, and so on to denote their elements. However, when dealing with applications, which usually involve function spaces, for their elements I write f, g, h or ϕ, ψ, and so on and use x, y, and z to denote arguments of functions.

Exercise 1.1 Let A be the generator of a strongly continuous semigroup in a Banach space $\mathbb{X}$. Show that $\lim_{\lambda\to\infty} \lambda(\lambda - A)^{-1}x = x, x \in \mathbb{X}$ and that $\lim_{\lambda\to\infty} \lambda[\lambda(\lambda - A)^{-1}x - x] = Ax, x \in D(A)$. In fact, the latter condition holds iff $x \in D(A)$.

Exercise 1.2 Check that a bounded operator A generates the semigroup (actually, a group) given by $\mathrm{e}^{tA} = \sum_{n=0}^{\infty} \frac{A^n t^n}{n!}$ and the cosine family given by $\mathrm{Cos}_A(t) = \sum_{n=0}^{\infty} \frac{A^{2n}t^{2n}}{(2n)!}$. Verify the Weierstrass Formula in this case.

Exercise 1.3 Let $\mathbb{X}$ be a Banach space, $z \in \mathbb{X}$ be a fixed element, and $f \in \mathbb{X}^*$ be a fixed linear functional. Consider the bounded linear operator $Ax = f(x)z$. Prove that:

$$\mathrm{e}^{tA}x = \begin{cases} x + \frac{f(x)}{r}(\mathrm{e}^{rt} - 1)z, & r \neq 0, \\ x + tf(x)z, & r = 0, \end{cases} \qquad t \geq 0, x \in \mathbb{X},$$

where $r = f(z)$.

Exercise 1.4 Let $\mathbb{X}$ be a Banach space.

(a) Let $A \in \mathcal{L}(\mathbb{X})$ be a bounded linear operator satisfying $A^2 = A$. Show that $e^{tA} = I_{\mathbb{X}} + (e^t - 1)A$.
(b) Let $A \in \mathcal{L}(\mathbb{X})$ be a bounded linear operator satisfying $A^2 = 0$. Show that $e^{tA} = I_{\mathbb{X}} + tA$.
(c) Let $A \in \mathcal{L}(\mathbb{X})$ be a bounded linear operator satisfying $A^2 = -I_{\mathbb{X}}$. Show that $e^{tA} = (\cos t) I_{\mathbb{X}} + (\sin t) A$.
(d) Let $A \in \mathcal{L}(\mathbb{X})$ be a bounded linear operator satisfying $A^2 = I_{\mathbb{X}}$. Show that $e^{tA} = (\cosh t) I_{\mathbb{X}} + (\sinh t) A$.

Exercise 1.5 Let $\mathbb{X} = C_0[0, \infty)$ be the space of continuous functions on $\mathbb{R}^+$, vanishing at infinity, and let μ be a non-negative, continuous function on $\mathbb{R}^+$ such that $\lim_{x\to\infty} \mu(x)$ exists (but may be infinite).

(a) Show that the related, possibly unbounded, operator $M : f \mapsto -\mu f$ (with domain composed of f such that $\mu f \in \mathbb{X}$) is the generator of the contraction semigroup in $\mathbb{X}$ given by:

$$e^{Mt} f(x) = e^{-\mu(x)t} f(x), \qquad x, t \geq 0. \tag{1.7}$$

In the case in which $\lim_{x\to\infty} \mu(x) = \infty$, check that formula (1.7) does not define a strongly continuous semigroup in the space $C[0, \infty]$ of continuous functions with (finite) limits at ∞. *Hint:* take $f \equiv 1$.
(b) Let μ be positive. For $b \in \mathbb{X}$, let a bounded linear operator B in $\mathbb{X}$ be defined by $Bf(x) = f(0)b(x)$, $x \in \mathbb{R}^+$. By (a) and the Phillips Perturbation Theorem, $M + B$ is the generator of a strongly continuous semigroup. Under the simplifying assumption that $\mu(0) = b(0)$, check directly that:

$$e^{t(M+B)} f(x) = e^{-\mu(x)t}(f(x) - f(0)q(x)) + f(0)q(x), \tag{1.8}$$

where $q(x) = \frac{b(x)}{\mu(x)}$ so that $q \in \mathbb{X}$. ($M + B$ is one of the terms in the predual of the generator of the McKendrick semigroup; see Chapter 50. See also Exercise 19.3.) *Hint:* Note that $q \in D(M)$; a direct calculation shows that the generator of the semigroup defined by the right-hand side of (1.8) extends $M + B$.

Exercise 1.6
(a) Let $\mathbb{X} = C[0, \infty]$ be the space of continuous functions on $[0, \infty)$ with limits at infinity, equipped with the usual supremum norm. Check that:

$$T(t)f(x) = f(x + t)$$

defines a strongly continuous semigroup of operators in $\mathbb{X}$, that the domain of its generator A is composed of continuously differentiable functions f such that $f' \in \mathbb{X}$ (deduce that $\lim_{x\to\infty} f'(x) = 0$), and that $Af = f'$.

(b) Let $\mathbb{X}$ be the space $C[0, 1]$ of continuous functions on $[0, 1]$, equipped with the usual supremum norm. Check that:

$$S(t)f(x) = f(xe^{-t})$$

defines a strongly continuous semigroup of operators in $\mathbb{X}$, that the domain of its generator B is composed of functions f such that $f'(x)$ exists for all $x \in (0, 1]$ and $\lim_{x\to 0} xf'(x) = 0$. Also, $Bf(x) = -xf'(x)$, $x \in (0, 1]$ (and clearly $Bf(0) = 0$).

(c) Check that:

$$U(t)f(x) = f(1 - (1 - x)e^{-t})$$

defines a strongly continuous semigroup of operators in the same Banach space $C[0, 1]$, that the domain of its generator C is composed of functions f such that $f'(x)$ exists for all $x \in [0, 1)$ and $\lim_{x\to 1}(1 - x)f'(x) = 0$. Also, $Cf(x) = (1 - x)f'(x)$, $x \in [0, 1)$ (and clearly $Cf(1) = 1$).

Exercise 1.7 Two semigroups $(e^{tA})_{t\ge 0}$ and $(e^{tB})_{t\ge 0}$ defined in Banach spaces $\mathbb{X}$ and $\mathbb{Y}$, respectively, are said to be isomorphic (or similar) iff there is an isomorphism $I : \mathbb{X} \to \mathbb{Y}$ such that:

$$Ie^{tA} = e^{tB}I, \qquad t \ge 0.$$

(a) Check that $x \in D(A)$ iff $Ix \in D(B)$ and that $IAx = BIx$.

(b) Check that all three semigroups of the previous exercise are similar and use this fact to find an independent proof of the characterizations of generators of semigroups in points (b) and (c) in the previous exercise.

Exercise 1.8 Let $L^1(\mathbb{R}^+)$ be the space of (equivalence classes) of absolutely integrable functions on $\mathbb{R}^+$.

(a) Let $\{T(t), t \ge 0\}$ be the semigroup of shifts to the right:

$$T(t)\phi(x) = \begin{cases} \phi(x - t), & x \ge t \\ 0, & x < t, \end{cases} \qquad \phi \in L^1(\mathbb{R}^+).$$

Show that the domain of its generator is composed of absolutely continuous functions ϕ with $\phi(0) = 0$, i.e. such that $\phi(x) = \int_0^x \psi(y)\,dy$, $x \ge 0$ for some $\psi \in L^1(\mathbb{R}^+)$, and that $A\phi = -\phi' = -\psi$.

(b) Let $\mu \ge 0$ be a bounded, integrable function on $\mathbb{R}^+$. Show that $I \in \mathcal{L}(L^1(\mathbb{R}^+))$ given by $(I\phi)(x) = e^{\int_0^x \mu(y)\,dy}\phi(x)$ is an isomorphism of $L^1(\mathbb{R}^+)$

(with $\|I\| \le \mathrm{e}^{\int_0^\infty \mu(x)\,\mathrm{d}x}$ and $\|I^{-1}\| \le 1$) and check that the isomorphic image $T_\mu(t) = I^{-1}T(t)I$ of the semigroup of point (a) via I is

$$T_\mu(t)\phi(x) = \begin{cases} \mathrm{e}^{-\int_{x-t}^{x} \mu(y)\,\mathrm{d}y}\phi(x-t), & x \ge t \\ 0, & x < t, \end{cases} \qquad \phi \in L^1(\mathbb{R}^+). \tag{1.9}$$

(c) Let A_μ be the generator of this semigroup. Using the fact that the product of two absolutely continuous functions is absolutely continuous, show that $D(A_\mu) = D(A)$, and that $A_\mu\phi = A\phi - \mu\phi$, $\phi \in D(A_\mu)$. (See also Exercise 2.3.)

Exercise 1.9 Let A be the operator in l^2, the space of square-summable sequences $(x_n)_{n\ge1}$, given by $D(A) = \{(\xi_n)_{n\ge1} \in l^1, (n\xi_n)_{n\ge1} \in l^2\}$, $A(\xi_n)_{n\ge1} = -(n\xi_n)_{n\ge1}$.

(a) Show that A is a generator and that:

$$\mathrm{e}^{tA}\,(\xi_n)_{n\ge1} = \left(\mathrm{e}^{-nt}\xi_n\right)_{n\ge1}.$$

(b) Let $L^2(0,\pi)$ be the space of (classes of) square integrable functions on $(0, 2\pi)$. Check that $e_n(x) = \sqrt{\frac{\pi}{2}}\sin x$, $x \in (0,\pi)$, $n \ge 1$, is a complete, orthonormal system (see, e.g., [107]), and conclude that $L^2(0,\pi)$ is isometrically isomorphic to l^2 with isomorphism $I : L^2(0,\pi) \to l^2$ given by:

$$If = ((f, e_n))_{n\ge1},$$

where the inner parenthesis denotes the scalar product in $L^2(0,\pi)$.

(c) Let B be the operator in $L^2(0,\pi)$ with domain $D(B)$ composed of functions f that can be represented in the form:

$$f(x) = ax + \int_0^x \int_0^y h(z)\,\mathrm{d}z\,\mathrm{d}y$$

where $a = -\frac{1}{\pi}\int_0^\pi \int_0^y h(z)\,\mathrm{d}z\,\mathrm{d}y$, for some $h \in L^2(0,\pi)$, and given by:

$$Af = h.$$

Use (a) and (b) to show that B is a generator, and find the form of e^{tB}.

Exercise 1.10 Let $BUC(\mathbb{R})$ be the space of uniformly continuous functions on the real line, equipped with the supremum norm. Check that:

$$C(t)f(x) = \frac{1}{2}[f(x+t) + f(x-t)], \qquad t, x \in \mathbb{R},$$

defines a strongly continuous cosine family. Prove that its generator is given by $Af = f''$ with domain composed of twice continuously differentiable members of $BUC(\mathbb{R})$ such that $f'' \in BUC(\mathbb{R})$.

Exercise 1.11 Let $BUC(\mathbb{R}^+)$ be the space of uniformly continuous functions on the half-line $\mathbb{R}^+$ equipped with the supremum norm. For $f \in BUC(\mathbb{R}^+)$, let $f_e \in BUC(\mathbb{R})$ be its even extension to the whole line. Check that:

$$C(t)f(x) = \frac{1}{2}[f_e(x+t) + f_e(x-t)], \qquad t \in \mathbb{R}, x \in \mathbb{R}^+,$$

defines a strongly continuous cosine family. Prove that its generator is given by $Af = f''$ with domain composed of twice continuously differentiable members of $BUC(\mathbb{R}^+)$ such that $f'' \in BUC(\mathbb{R}^+)$, and $f'(0) = 0$.

Exercise 1.12 Let $C_p[0, 1]$ be the space of continuous functions f on $[0, 1]$ such that $f(0) = f(1)$. Let f_p be the periodic extension of a $f \in C_p[0, 1]$ given by $f_p(x + k) = f(x), x \in [0, 1]$ where k is any integer. Show that:

$$C(t)f(x) = \frac{1}{2}[f_p(x+t) + f_p(x-t)], \qquad t \in \mathbb{R}, x \in \mathbb{R}^+,$$

defines a strongly continuous cosine family and prove that its generator is $Af = f''$ with domain composed of twice continuously differentiable members of $C_p[0, 1]$ such that $f', f'' \in C_p[0, 1]$.

Exercise 1.13 Let $C[-\infty, \infty]$ be the space of continuous functions on $\mathbb{R}$ with limits at $+\infty$ and $-\infty$, and let $\{T(t), t \geq 0\}$ be operators defined by:

$$T(t)f(x) = \begin{cases} f(x + \sigma_1 t), & x \in [0, \infty), \\ f(\frac{\sigma_1}{\sigma_{-1}} x + \sigma_1 t), & x \in (-\sigma_{-1} t, 0), \\ f(x + \sigma_{-1} t), & x \in (-\infty, -\sigma_{-1} t], \end{cases}$$

where σ_1 and σ_{-1} are given positive constants.

(a) Show that this is a strongly continuous semigroup with generator A given by:

$$Af = \begin{cases} \sigma_1 f'(x), & x \geq 0 \\ \sigma_{-1} f'(x), & x < 0, \end{cases}$$

defined for all f satisfying the following conditions: (i) f is continuously differentiable for $x > 0$, with the derivative having limits at $+\infty$ and 0 – the latter limit is denoted $f'(0+)$. (ii) f is continuously differentiable for $x < 0$ with derivative having limits at $-\infty$ and 0 – the latter limit is denoted $f'(0-)$. (iii) We have $\sigma_{-1} f'(0-) = \sigma_1 f'(0+)$.

(b) Let $\alpha(x) = \sigma_{\operatorname{sgn} x} x$. Check that $I : f \mapsto f \circ \alpha$ is an isometric isomorphism of $C[-\infty, \infty]$. Redo point (a) by noting that $S(t) = IT(t)I^{-1}$ is the translation to the left: $S(t)f(x) = f(x + t), x \in \mathbb{R}, t \geq 0$.

(c) Prove that the semigroup may be extended to a group $\{T(t), t \in \mathbb{R}\}$, and find an explicit formula for $T(t), t < 0$.

(d) Prove that $C(t) = \frac{1}{2}(T(t) + T(-t)), t \in \mathbb{R}$ defines a cosine family and find its generator.

PART I

Regular Convergence

2

The First Convergence Theorem

We start with a lemma concerning the relationship between convergence of functions and their Laplace transforms. We assume that $f_n : \mathbb{R}^+ \to \mathbb{X}, n \geq 1$, where $\mathbb{X}$ is a Banach space, are continuous functions satisfying $\|f_n(t)\| \leq M, t \geq 0, n \geq 1$ for some $M > 0$, and denote by $\hat{f}_n(\lambda) = \int_0^\infty \mathrm{e}^{-\lambda t} f_n(t)\,\mathrm{d}t, \lambda > 0$ their Laplace transforms.

Lemma 2.1 *Functions f_n converge almost uniformly in $t \in \mathbb{R}^+$ iff they are equicontinuous in n and their Laplace transforms converge (pointwise).*

Here, by almost uniform convergence we mean the convergence that is uniform in any compact subinterval of $\mathbb{R}^+$, and equicontinuity in n means that for any $t \in \mathbb{R}^+$ and $\epsilon > 0$ there exists a $\delta > 0$ such that $|h| < \delta$ and $t + h \in \mathbb{R}^+$ imply $\|f_n(t+h) - f_n(t)\| < \epsilon$.

Proof Equicontinuity is known to be necessary for uniform convergence, and convergence of functions (even pointwise) implies convergence of resolvents by Lebesgue Dominated Convergence Theorem. Hence, it is the converse that is nontrivial (and useful).

The following argument proving the converse is due to J. Kisyński. Let $l^\infty(\mathbb{X})$ be the space of bounded sequences with values in $\mathbb{X}$, equipped with the norm $\|(x_n)_{n\geq 1}\| = \sup_n \|x_n\|$, and let $c(\mathbb{X})$ be its subspace of convergent sequences. By equicontinuity assumption, the function $F : \mathbb{R}^+ \to l^\infty(\mathbb{X})$ given by $F(t) = (f_n(t))_{n\geq 1}$ is continuous. Furthermore, $\int_0^\infty \mathrm{e}^{-\lambda t} F(t)\,\mathrm{d}t = (\hat{f}_n(\lambda))_{n\geq 1}, \lambda > 0$. If a bounded linear functional Ψ on $l^\infty(\mathbb{X})$ is orthogonal to $c(\mathbb{X})$, then $\int_0^\infty \mathrm{e}^{-\lambda t} \Psi(F(t))\,\mathrm{d}t = \Psi(\hat{f}_n(\lambda))_{n\geq 1} = 0, \lambda > 0$, implying $\Psi \circ F \equiv 0$, by continuity. Hence, F has values in $c(\mathbb{X})$ or, which is the same, $\lim_{n\to\infty} f_n(t)$ exists. Because the limit function must be continuous, the limit is almost uniform in nonnegative t. □

We come to our first and the simplest approximation theorem. Notably, its proof contains, in a germ form, all the main ideas to be developed later.

Theorem 2.2 *Assume* $(\mathrm{e}^{tA_n})_{t\geq 0}$, $n \geq 1$ *are strongly continuous semigroups in a Banach space* $\mathbb{X}$ *that are equibounded in the sense that there exists a constant* $M > 0$ *such that:*

$$\|\mathrm{e}^{tA_n}\| \leq M, \qquad t \geq 0, n \geq 1. \tag{2.1}$$

Let $(\mathrm{e}^{tA})_{t\geq 0}$ *be another strongly continuous semigroup in* $\mathbb{X}$ *such that* $\|\mathrm{e}^{tA}\| \leq M$. *The following are equivalent:*

a) $\lim_{n\to\infty} \mathrm{e}^{tA_n} = \mathrm{e}^{tA}$ *(strongly).*
b) $\lim_{n\to\infty}(\lambda - A_n)^{-1} = \lim_{n\to\infty}(\lambda - A)^{-1}$ *(strongly) for all* $\lambda > 0$.
c) $\lim_{n\to\infty}(\lambda - A_n)^{-1} = \lim_{n\to\infty}(\lambda - A)^{-1}$ *(strongly) for some* $\lambda > 0$.
d) For any $x \in D(A)$ *there exist* $x_n \in D(A_n)$ *such that* $\lim_{n\to\infty} x_n = x$ *and* $\lim_{n\to\infty} A_n x_n = Ax$.

Moreover, if these equivalent conditions are satisfied then the limit in a) is almost uniform in $t \in \mathbb{R}^+$.

Proof Implication a) $\Rightarrow$ b) is obvious by the Lebesgue Dominated Convergence Theorem, and b) $\Rightarrow$ c) is trivial. Assuming that c) is satisfied, we take an $x \in D(A)$ and find the $y \in \mathbb{X}$ with $x = (\lambda - A)^{-1}y$. Then, we define $x_n = (\lambda - A_n)^{-1}y$ to see that $x_n \in D(A_n)$, $\lim_{n\to\infty} x_n = (\lambda - A)^{-1}y = x$ and $\lim_{n\to\infty} A_n x_n = \lim_{n\to\infty}(\lambda(\lambda - A_n)^{-1}y - \lambda y) = \lambda(\lambda - A)^{-1}y - \lambda y = A(\lambda - A)^{-1}y = Ax$, showing d).

To see that d) implies b), for $\lambda > 0$ and $y \in X$ we find $x \in D(A)$ such that $y = \lambda x - Ax$. Then, vectors $y_n = \lambda x_n - A_n x_n$ where x_n is chosen as in d), satisfy $\|(\lambda - A_n)^{-1}(y_n - y)\| \leq M\lambda^{-1}\|y_n - y\|$, $\lim_{n\to\infty} y_n = y$ and $(\lambda - A_n)^{-1}y_n = x_n \to x = (\lambda - A)^{-1}y$. This implies $\lim_{n\to\infty}(\lambda - A_n)^{-1}y = \lim_{n\to\infty}(\lambda - A)^{-1}y$, showing b).

Finally, we show that b) and d) imply a). For $x \in D(A)$, let x_n be as in d) and define $f_n(t) = \mathrm{e}^{tA_n}x_n = x_n + \int_0^t \mathrm{e}^{sA_n}A_n x_n \,\mathrm{d}s$, $t \geq 0$. Then f_n are Lipschitz continuous with constant $M\sup_{n\geq 1}\|A_n x_n\|$, and $\lim_{n\to\infty}\int_0^\infty \mathrm{e}^{-\lambda t}\mathrm{e}^{tA_n}x_n \mathrm{d}t = \lim_{n\to\infty}(\lambda - A_n)^{-1}x_n = (\lambda - A)^{-1}x = \int_0^\infty \mathrm{e}^{-\lambda t}\mathrm{e}^{tA}x \,\mathrm{d}t$, $\lambda > 0$. Hence, by Lemma 2.1, $\lim_{n\to\infty} \mathrm{e}^{tA_n}x_n = \mathrm{e}^{tA}x$ almost uniformly in $t \in \mathbb{R}^+$. Since the semigroups are equibounded this implies $\lim_{n\to\infty} \mathrm{e}^{tA_n}x = \mathrm{e}^{tA}x$ (almost uniformly in $t \in \mathbb{R}^+$). Thus, a) follows by density argument. □

The equivalence of a), b), and c) is usually referred to as the (first) Trotter–Kato Theorem, the adjective "second" being reserved for the version where the existence of the limit semigroup is not assumed. The fourth condition and the related notion of **extended limit** of linear operators (see Chapter 7) was introduced independently by M. Sova and T. G. Kurtz [132, 229, 320].[1] For convenience, we will refer to all convergence results involving resolvents as the Trotter–Kato Theorem, and to those involving generalized limit as the Sova–Kurtz Theorem.

Using the latter theorem greatly simplifies proofs of convergence. For example, consider $\mathbb{X} = C[-\infty, \infty]$, the space of continuous functions on $\mathbb{R}$ with limits at plus and minus infinity, and the sequence of semigroups generated by operators $A_n = h_n^{-1}(T_n - I)$ where $T_n f(x) = \frac{1}{2}(f(x + \Delta_n) + f(x - \Delta_n))$, $x \in \mathbb{R}$, and h_n and Δ_n are positive numbers. If $\lim_{n\to\infty} h_n = 0$ and $\lim_{n\to\infty} \Delta_n = 0$ with $\lim_{n\to\infty} \frac{\Delta_n^2}{h_n} = 1$, then the semigroups converge to the Brownian motion semigroup generated by the half of the one-dimensional Laplacian $A = \frac{1}{2}\frac{\mathrm{d}^2}{\mathrm{d}x^2}$ with maximal domain $C^2[-\infty, \infty]$ in $\mathbb{X}$:

$$\mathrm{e}^{tA} f(x) = E f(x + w_t) = \frac{1}{\sqrt{2\pi t}} \int_{-\infty}^{\infty} \mathrm{e}^{-\frac{y^2}{2t}} f(x + y)\,\mathrm{d}y, \tag{2.2}$$

where $w_t, t \geq 0$ is a one-dimensional standard Brownian motion. Proving this via the Trotter–Kato Theorem requires finding the form of the resolvents $(\lambda - A_n)^{-1}$, which is possible but time-consuming, while the proof via the Sova–Kurtz version is immediate if we use Taylor's Theorem – see Exercise 2.2. On the other hand, it should be noted that finding $(x_n)_{n\geq 1}$ of point (d) may require some ingenuity, while having a closed form of $(\lambda - A_n)^{-1}$ makes the reasoning quite straightforward. The reader will find ample illustrations of these facts in the following chapters of this monograph.

Remark 2.3 Interestingly, there is no explicit assumption of equicontinuity of semigroups' trajectories in Theorem 2.2. For $x \in D(A)$ this assumption is implicit in d), and convergence of the trajectories for $x \in X$ is obtained by density argument.

In the three chapters that follow we present three examples of application of Theorem 2.2: first in the space of continuous functions, then in the space of integrable functions, and finally in a Hilbert space of square integrable functions.

[1] Interestingly, extended limit has been introduced yet again, and its basic properties established, by P. E. T. Jørgensen much later [191], under the name of **graph extension**; see also [294, p. 293], where – in studying the case of self-adjoint operators – the same notion is introduced and termed **strong graph limit**, and [199], where it was used in the context of singular perturbations. In fact, once discovered, the notion seems so natural that one wonders why some – excellent – books do not cover it.

Exercise 2.1 Let A be the generator of a semigroup in a Banach space $\mathbb{X}$ and let B be a bounded linear operator in $\mathbb{X}$. Show that for each $a \in \mathbb{R}$, $A + aB$ generates a semigroup and that $\lim_{a\to 0} \mathrm{e}^{t(A+aB)} = \mathrm{e}^{tA}$.

Exercise 2.2 Prove convergence of semigroups spoken of after Theorem 2.2 by showing convergence of their resolvents, and then by showing convergence of their generators. To this end show first that given $\lambda > 1$ and a doubly infinite $(y_n)_{n\in\mathbb{Z}}$ with limits in $\pm\infty$ there exists a unique $(x_n)_{n\in\mathbb{Z}}$ with the same properties such that $\lambda x_n - \frac{1}{2}(x_{n+1} + x_{n-1}) = y_n$. This $(x_n)_{n\in\mathbb{Z}}$ is given by:

$$x_n = \frac{1}{\sqrt{\lambda^2 - 1}} \sum_{i\in\mathbb{Z}} \left(\lambda - \sqrt{\lambda^2 - 1}\right)^{|i-n|} y_i.$$

Also, recall that:

$$\left(\lambda - \frac{1}{2}\frac{\mathrm{d}^2}{\mathrm{d}x^2}\right)^{-1} f(x) = \frac{1}{\sqrt{2\lambda}} \int_{-\infty}^{\infty} \mathrm{e}^{-\sqrt{2\lambda}|x-y|} f(y)\,\mathrm{d}y,\ \lambda > 0, x \in \mathbb{R}. \quad (2.3)$$

Exercise 2.3 Let A be as in Exercise 1.8 and let B given by $B\phi = \mu\phi$ be the bounded multiplication operator in $L^1(\mathbb{R}^+)$, where $\mu \ge 0$ is assumed to be bounded (but not necessarily integrable over $\mathbb{R}^+$). By the Phillips Perturbation Theorem, $A + B$ is a generator. Let $\mu_n(x) = \mathrm{e}^{-n^{-1}x}\mu(x)$ so that μ_n is bounded and integrable and let $A_n := A_{\mu_n}$ be the generators of the semigroups defined by (1.9) with μ replaced by μ_n. Use the Trotter–Kato Theorem in its Sova–Kurtz version to show that $\lim_{n\to\infty} \mathrm{e}^{tA_n} = \mathrm{e}^{tA_\mu}$. Conclude that (1.9) holds without the assumption that μ is integrable.

☞ Chapter's summary

We have covered the first convergence theorem: for a sequence of equibounded semigroups to converge to a given limit semigroup it is necessary and sufficient that their resolvents converge to the resolvent of the limit semigroup, and that happens iff the limit semigroup's generator is in a sense a limit of approximating generators. By nature, such convergence of semigroups is almost uniform in $t \in [0, \infty)$.

3

Continuous Dependence on Boundary Conditions

For an example of Theorem 2.2, consider the space $C[0, \infty]$ of continuous functions on $\mathbb{R}^+ = [0, \infty)$ with limits at infinity. Let A be the operator $Af = f''$ with domain composed of twice continuously differentiable functions with $f'' \in C[0, \infty]$, satisfying the boundary condition:

$$af''(0) - bf'(0) + cf(0) - d\int_{\mathbb{R}^+_*} f \, \mathrm{d}\mu = 0, \tag{3.1}$$

where μ is a probability measure on $\mathbb{R}^+_* = (0, \infty)$, and a, b, c, and d are given nonnegative constants with $c \geq d$ and $a + b > 0$.

We argue that A generates a Feller semigroup on $C[0, \infty]$ (comp. [144]). Let us recall that by definition a strongly continuous contraction semigroup in $C[0, \infty]$ is said to be a Feller semigroup iff it maps non-negative functions into non-negative functions [49, 132, 296]. A well-known necessary and sufficient condition for a densely defined operator to generate a Feller semigroup is that it satisfies the **positive maximum principle** (if the maximum of a $f \in D(A)$ is attained at $x \in [0, \infty]$, and $f(x) \geq 0$, then $Af(x) \leq 0$) and the **range condition** (see next page).

Our A is densely defined (see Exercise 3.2) and satisfies the positive maximum principle. To see this, assume a positive maximum of $f \in D(A)$ is attained at $x = 0$ (if the maximum is attained at $x > 0$ or at infinity the claim is obvious). Consider first the case where $a \neq 0$. Then $f''(0) = a^{-1}(bf'(0) - cf(0) + d\int_{\mathbb{R}^+_*} f \, \mathrm{d}\mu) \leq 0$ because $f'(0) \leq 0$ and $d\int_{\mathbb{R}^+_*} f \, \mathrm{d}\mu \leq cf(0)$ (here $f(0) \geq 0$ is used; in particular we may have $d = 0$). If $a = 0$, we have $f'(0) \leq 0$ and, on the other hand, $f'(0) = b^{-1}(cf(0) - d\int_{\mathbb{R}^+_*} f \, \mathrm{d}\mu) \geq 0$. Therefore, $f'(0) = 0$ and the even extension of f to the whole of $\mathbb{R}$ is twice continuously differentiable. Since its maximum is attained at $x = 0$, we have $f''(0) \leq 0$, as desired.

It remains to be checked that the range condition is satisfied: for all $\lambda > 0$ and $g \in C[0,\infty]$ there exists an $f \in D(A)$ such that:

$$\lambda f - f'' = g. \tag{3.2}$$

We look for f of the form:

$$f(x) = C\mathrm{e}^{\sqrt{\lambda}x} + D\mathrm{e}^{-\sqrt{\lambda}x} - \frac{1}{\sqrt{\lambda}}\int_0^x \sinh\sqrt{\lambda}(x-y)g(y)\,\mathrm{d}y, \tag{3.3}$$

where C and D are constants to be determined. Clearly, f is twice continuously differentiable and satisfies (3.2) but we need to make sure $f \in D(A)$. Because f is to have a finite limit at ∞ and since $\lim_{x\to\infty} \mathrm{e}^{-\sqrt{\lambda}x}(D + \frac{1}{2\sqrt{\lambda}}\int_0^x \mathrm{e}^{\sqrt{\lambda}y}g(y)\,\mathrm{d}y) = \frac{1}{2\lambda}\lim_{x\to\infty} g(x)$, we must have:

$$C = \frac{1}{2\sqrt{\lambda}}\int_0^\infty \mathrm{e}^{-\sqrt{\lambda}y}g(y)\,\mathrm{d}y, \tag{3.4}$$

that is,

$$f(x) = D\mathrm{e}^{-\sqrt{\lambda}x} + \frac{1}{2\sqrt{\lambda}}\int_0^\infty \mathrm{e}^{-\sqrt{\lambda}|x-y|}g(y)\,\mathrm{d}y =: D\mathrm{e}^{-\sqrt{\lambda}x} + g_\lambda(x). \tag{3.5}$$

Also, taking:

$$D = \frac{(b\sqrt{\lambda} - a\lambda - c)C + ag(0) + d\int_{\mathbb{R}_*^+} g_\lambda\,\mathrm{d}\mu}{a\lambda + b\sqrt{\lambda} + c - d\int_{\mathbb{R}_*^+}\mathrm{e}^{-\sqrt{\lambda}x}\mu(\,\mathrm{d}x)}, \tag{3.6}$$

we see that (3.1) is satisfied.

The D defined here has a noteworthy property: if a_n, b_n, c_n, d_n, and μ_n tend to a, b, c, d, and μ, respectively (the limit of measures in the weak* topology), then the corresponding constants D_n tend to D, provided, as we assumed, at least one of a and b is nonzero. In other words, D depends continuously on the boundary conditions. But this shows that so does $f = (\lambda - A)^{-1}g$ defined by (3.4)–(3.6). Hence, by Theorem 2.2 the semigroups generated by A depend continuously on the parameters appearing in (3.1). This is an example of a much more general principle (see [92, 94, 93] where, however, the problem is seen in the context of L^p spaces, compare also [267, Section 4] and Corollary 56.4). To quote [93]: when a, b, c, and d are changed, "the domain of the corresponding semigroup generator apparently changes, so we seem to have an unbounded, domain-changing perturbation, and the continuous dependence result is not at all obvious."

A few words concerning interpretation of boundary condition (3.1) are here in order. The semigroup generated by A is related to a stochastic process on $\mathbb{R}^+$; the form of the generator ($Af = f''$) tells us that, while away from the

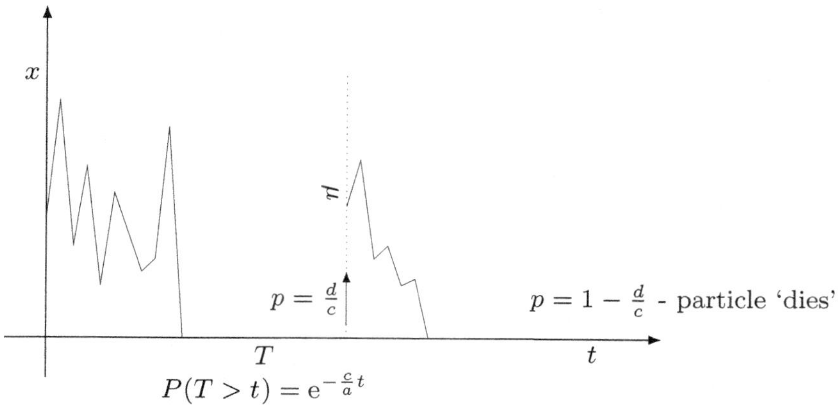

Figure 3.1 Elementary return Brownian motion (redrawn from [70])

boundary $x = 0$, the process behaves like a (re-scaled) Brownian motion. To specify the process completely, however, we need to provide rules of its behavior at the boundary, and this is the role of (3.1). Upon touching the boundary, the particle performing Brownian motion may be stopped there, reflected, or killed (i.e., removed from the space); it may also jump somewhere into $\mathbb{R}^+$. The coefficients a, b, c, and d may be thought of as describing relative frequencies of such events, and μ is the distribution of particle's position right after the jump. In particular, the case $a = 1, b = c = d = 0$ is the stopped (or: absorbing) Brownian motion (the particle reaching the boundary stays there forever), $a = c = d = 0, b = 1$ is the reflected Brownian motion whose paths are absolute values of paths of an unrestricted Brownian motion, and $a = b = d = 0, c = 1$ is the minimal Brownian motion (the particle reaching the boundary disappears).

Consider in more detail the case where $a \neq 0$ and $b = 0$:

$$af''(0) + cf(0) - d\int_{\mathbb{R}_*^+} f\,\mathrm{d}\mu = 0. \tag{3.7}$$

This is the case of **elementary return Brownian motion**, in which the process after reaching the boundary stays there for a random exponential time T with parameter c/a (see Figure 3.1):

$$P(T > t) = \mathrm{e}^{-\frac{c}{a}t}, \quad t \geq 0.$$

At time T, the process either terminates, with probability $1 - \frac{d}{c}$, or jumps, with probability $\frac{d}{c}$, to a random point in $\mathbb{R}_*^+$, the distribution after the jump being μ, and starts its movement afresh. (Interestingly, the possibility for a process

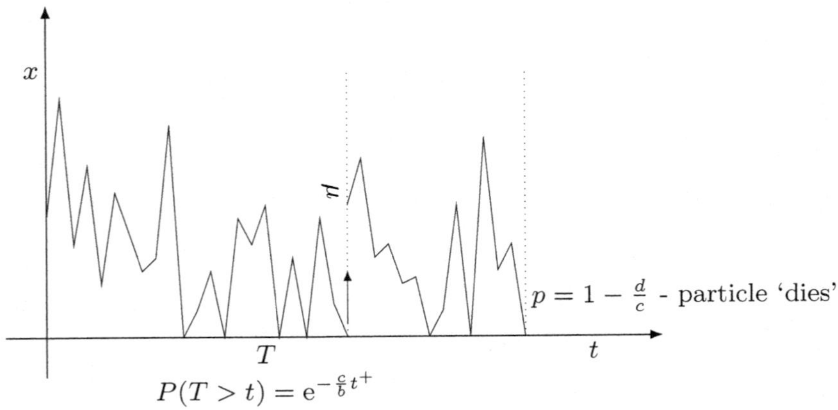

Figure 3.2 Elastic barrier (redrawn from [70])

to return from the boundary to $\mathbb{R}_*^+$ in this way was not suggested by physical considerations, but by the Wright–Fisher model with mutations, originating from population genetics, see [142]–[144], [134], and Chapter 17.)

The **elastic barrier** where $a = 0$ and $b \neq 0$:

$$bf'(0) = cf(0) - d \int_{\mathbb{R}_*^+} f \,\mathrm{d}\mu \tag{3.8}$$

may be described by analogy (see Figure 3.2). Here, after reaching the boundary the trajectory is reflected and the process continues in this fashion for a random time T. The times when the path touches $x = 0$ form a measurable subset of the time axis, and the Lebesgue measure of this set is zero. There is, nevertheless, a way to measure the time spent at the boundary, called the Lévy local time t^+; t^+ itself is a random process (on a separate probability space) increasing only when the Brownian path is at the boundary [188, 195, 196, 296]. As in the elementary return Brownian motion, at time T distributed according to (see [188] p. 45 or [195] p. 426):

$$P(T > t) = \mathrm{e}^{-\frac{c}{b}t^+}, \quad t \geq 0,$$

the process either terminates, with probability $1 - \frac{d}{c}$, or jumps, with probability $\frac{d}{c}$, to a random point in $\mathbb{R}_*^+$, the distribution after the jump being μ, and starts afresh.

For a detailed analytic and probabilistic treatment of boundary conditions see, for example, [144, 145, 247]. A very nice insight into the subject may also be gained by considering a characteristic operator of E. B. Dynkin – see [121, 122, 123, 124, 193]. Consult also [158], where a physical interpretation to very general boundary conditions is given, and reader-friendly and expertly written [324]. Here, we will merely find the most general form of boundary conditions for the one-dimensional Laplacian generating a Feller semigroup in $C[0,\infty]$ (comp. [247] p. 37); this part may be skipped on the first reading.

To this end, let A be the generator of a Feller semigroup in $C[0,\infty]$ with domain contained in $C^2[0,\infty]$ and such that $Af = f''$, $f \in D(A)$. Then, the solution to the resolvent equation for A must be of the form (3.3) with C specified in (3.4). Hence (comp. (4.3), later on):

$$
\begin{aligned}
(\lambda - A)^{-1} g(x) &= (C + D)\mathrm{e}^{-\sqrt{\lambda}x} \\
&\quad + \frac{1}{2\sqrt{\lambda}} \int_0^\infty [\mathrm{e}^{-\sqrt{\lambda}|x-y|} - \mathrm{e}^{-\sqrt{\lambda}(x+y)}] g(y)\, \mathrm{d}y.
\end{aligned}
$$

Here, $C = C_\lambda$ and $D = D_\lambda$ are in fact linear functionals indexed by $\lambda > 0$; we write $E_\lambda := \lambda(C_\lambda + D_\lambda)$. It is an exercise to check that E_λ is positive (compare Lemma 4.2, later on, note that $(\lambda - A)^{-1} g \geq 0$ provided $g \geq 0$) and $E_\lambda(e) \leq 1$, where $e = 1_{[0,\infty]}$; in particular the norm of E_λ does not exceed 1. Because $\lim_{\lambda\to\infty} \lambda(\lambda - A)^{-1} g = g$, we have $\lim_{\lambda\to\infty} E_\lambda g = g(0)$. Similarly (comp. Exercise 1.1), $Ag = \lim_{\lambda\to\infty} \lambda(\lambda(\lambda - A)^{-1} g - g)$, $g \in D(A)$ implies $g''(0) = \lim_{\lambda\to\infty} \lambda(E_\lambda g - g(0))$. Introducing for fixed $r > 0$, $h_r(x) = x \wedge r$, we see that this may be written as:

$$
\lim_{\lambda\to\infty} [g''(0) - \lambda F_\lambda g_1 + g(0)\lambda(1 - E_\lambda e)] = 0, \qquad g \in D(A), \tag{3.9}
$$

where $F_\lambda g = E_\lambda(h_r g)$, $g \in C[0,\infty]$ and $g_1(x) = \frac{g(x)-g(0)}{x\wedge r}$, $x > 0$; note that g_1 has a continuous extension to an element of $C[0,\infty]$ with $g_1(0) = g'(0)$. Since the functionals $\|F_\lambda\|^{-1} F_\lambda$ may be identified with probability measures, it follows by the Helly principle that there exists a sequence $(\lambda_n)_{n\geq 1}$ converging to infinity and a probability measure ν such that:

$$
\lim_{n\to\infty} \|F_{\lambda_n}\|^{-1} F_{\lambda_n} g = \int_{[0,\infty]} g\, \mathrm{d}\nu, \qquad g \in C[0,\infty].
$$

Choosing further subsequences if necessary, we find a sequence such that additionally the limits:

$$
\begin{aligned}
\lim_{n\to\infty} \lambda_n \|F_{\lambda_n}\| &=: \alpha \geq 0, \\
\lim_{n\to\infty} \lambda_n (1 - E_{\lambda_n} e) &=: \beta \geq 0,
\end{aligned}
$$

and $\lim_{n\to\infty} \frac{\|F_{\lambda_n}\|}{1 - E_{\lambda_n} e} =: \gamma \geq 0$ exist (but may be infinite). If α and β are finite, taking the limit in (3.9) along the chosen sequence we obtain:

$$
g''(0) - \alpha \int_{[0,\infty]} g_1\, \mathrm{d}\nu + \beta g(0) = 0. \tag{3.10}
$$

In the other case, dividing the expression in brackets in (3.9) by $\lambda\|F_\lambda\| + \lambda(1 - E_\lambda e)$ and taking the limit along the chosen sequence, we obtain:

$$\frac{\gamma}{1+\gamma}\int_{[0,\infty]} g_1 \, d\nu + \frac{1}{1+\gamma} g(0) = 0, \tag{3.11}$$

where $\frac{\gamma}{1+\gamma} = 1$ and $\frac{1}{1+\gamma} = 0$ for $\gamma = \infty$. Hence, in both cases there are non-negative constants a_0, b_0, c_0 such that:

$$a_0 g''(0) - b_0 \int_{[0,\infty]} \frac{g(x) - g(0)}{x \wedge r} \nu(\, dx) + c_0 g(0) = 0, \quad g \in D(A); \tag{3.12}$$

this is the promised, most general form of a boundary condition. Note that $a_0 + b_0$ must be positive, because otherwise we would have $c_0 = 1$, the boundary condition would read $g(0) = 0$, and $D(A)$ couldn't possibly be dense in $C[0, \infty]$.

Now, assume $\int_{(0,\infty]} \frac{1}{x \wedge r} \nu(\, dx) =: d_0$ is finite, and take $a = a_0, b = b_0 \nu(\{0\})$, $c = c_0 + b_0 d_0, d = b_0 d_0$ and $\mu(B) = d_0^{-1} \int_B \frac{1}{x \wedge r} \nu(\, dx)$. Then (3.12) reduces to (3.1) provided $\nu(\{\infty\}) = 0$; by the way, this explains why we assumed $c \geq d$ in (3.1). As earlier, we check that the case $a + b = 0$ is impossible.

In the case in which $d_0 = \infty$, the related process is much trickier to describe in probabilistic terms than it is in the case $d_0 < \infty$, covered earlier – compare, for example, [124, Chapter 4].

Exercise 3.1 A matrix $Q = (q_{ij})_{i,j \in \{1,\dots,n\}}$ is said to be a Kolmogorov matrix or an intensity matrix iff its off-diagonal entries $q_{ij}, i \neq j$ are non-negative and its rows add up to zero. If $\{1, \dots, n\}$ is equipped with the discrete topology, the space $\mathbb{R}^n$ may be identified with the space of continuous functions on $\{1, \dots, n\}$. Check that for any Kolmogorov matrix Q, the operator $\mathbb{R}^n \ni x \mapsto Qx \in \mathbb{R}^n$ satisfies the maximum principle: if $\max_{i \in \{1,\dots,n\}} x_i = x_j$, then the jth coordinate of Qx is nonpositive.

Exercise 3.2 Given a twice continuously differentiable $f \in C[0, \infty]$ with $f'' \in C[0, \infty]$, find a sequence f_n satisfying (3.1), of the form $f_n(x) = f(x) + \alpha_n e^{-\beta_n x}$, where $\beta_n \geq 0$, such that $\lim_{n\to\infty} f_n = f$.

Exercise 3.3 Prove the properties of E_λ stated in the text.

Exercise 3.4 Prove the convergence result of this chapter by proving convergence of the related generators (without calculating resolvents).

Exercise 3.5 A Feller semigroup generated by A in $C[0, \infty]$ is said to be conservative iff it maps $1_{[0,\infty)}$ into itself. Equivalently, $1_{[0,\infty)} \in D(A)$ and $A1_{[0,\infty)} = 0$. Check that the semigroup related to the boundary condition (3.1) is conservative iff $c = d$.

Exercise 3.6 For $a \geq 0$ and a measure μ on $[0, 1)$, let an operator A in $C[0, 1]$ be defined by $Af = f'$ with domain composed of continuously differentiable functions satisfying $f'(1) = a(\int_{[0,1)} f \,\mathrm{d}\mu - f(1))$. Check that A is the generator of a conservative Feller semigroup in $C[0, 1]$. (This is the semigroup governing the Markov process in which a particle in $[0, 1)$ moves to the right with constant speed 1. Upon reaching the boundary $x = 1$ the process stays there for an exponential time with parameter a, and then jumps to one of the points of $[0, 1)$, the distribution of the point after the jump being μ, to continue its motion to the right. For $a = 0$, the process remains in $x = 1$ forever. Turning to the case where μ is the Dirac measure at zero, show that for $t \in [0, 1]$, the semigroup generated by A is given by:

$$\mathrm{e}^{tA} f(x) = \tilde{f}(x + t), \qquad x \in [0, 1] \tag{3.13}$$

where, for $x \in [1, 2]$:

$$\tilde{f}(x) = f(x), x \in [0, 1], f(x) = \mathrm{e}^{-a(x-1)} f(1) + a \int_0^{x-1} \mathrm{e}^{-a(x-1-y)} f(y) \,\mathrm{d}y.$$

☞ Chapter's summary

We have covered the first example of convergence of semigroups based on the theorem of Chapter 2. Feller semigroups generated by a one-dimensional Laplace operator in $C[0, \infty]$ are related to boundary conditions describing behavior of the related process at the boundary where $x = 0$. At the boundary, the process may be reflected, stopped, or killed and removed from the state-space, it may also jump from the boundary to the interior of the right half-line. Even though the boundary conditions influence the domains of the generators of the semigroups, and changing coefficients in the boundary condition changes the domain, the semigroups depend continuously on these coefficients.

4

Semipermeable Membrane

This chapter may be seen as an introduction to the model of fast neurotransmitters of Chapter 33. Its primary aim, however, is to further illustrate Theorem 2.2. We consider particles diffusing on the real line and assume that the physical characteristics of the material and hence the diffusion coefficients on the positive and negative half-axes are different. The half-axes are separated by a semipermeable membrane at $x = 0$, described later.

Because we want to study dynamics of the processes' distributions, a natural concept here is that of a **Markov operator**, which is a linear operator in the $L^1(S) = L^1(S, m)$ space of (equivalence classes) of functions on a measure space (S, m) that are absolutely integrable with respect to the measure m. An operator P is said to be Markov iff it leaves the positive cone invariant (i.e., $Pf \geq 0$ for $f \geq 0$) and preserves the integral there, that is, $\int Pf \,\mathrm{d}m = \int f \,\mathrm{d}m$, for $f \geq 0$. It is easy to see that Markov operators are contractions (Exercise 4.1).

For a densely defined operator A in $L^1(S)$ to generate a semigroup of Markov operators, it is necessary and sufficient for its resolvent to be Markov, which means by definition that all $\lambda(\lambda - A)^{-1}$, $\lambda > 0$ are Markov. This result may be deduced from the Hille–Yosida Theorem (see [234], or Lemma 13.2 in Chapter 13).

In our case, m is the Lebesgue measure on $\mathbb{R}$. For our purposes, however, it will be convenient to identify a member ϕ of $L^1(\mathbb{R})$ with the pair $(\phi_i)_{i\in\mathbb{I}}$ of functions on $\mathbb{R}^+$ defined by $\phi_i(x) = \phi(ix), x \geq 0$. Here $\mathbb{I} = \{-1, 1\}$. Certainly $\phi_i \in L^1(\mathbb{R}^+)$, that is, we identify $L^1(\mathbb{R})$ with $L^1(\mathbb{R}^+) \times L^1(\mathbb{R}^+)$ (see Figure 4.1).

With this identification in mind, and given positive constants k_i, σ_i, $i \in \mathbb{I}$, we follow [57] to define the operator A in $L^1(\mathbb{R}^+)$ by:

$$A\,(\phi_i)_{i\in\mathbb{I}} = \left(\sigma_i^2 \phi_i''\right)_{i\in\mathbb{I}} \tag{4.1}$$

Figure 4.1 $L^1(\mathbb{R}) = L^1(\mathbb{R}^+) \times L^1(\mathbb{R}^+)$ by Marek Bobrowski

with domain composed of $(\phi_i)_{i\in\mathbb{I}} \in W^{2,1}(\mathbb{R}^+) \times W^{2,1}(\mathbb{R}^+)$ satisfying the **transmission conditions**:

$$\sigma_1^2\phi_1'(0) = k_1\sigma_1^2\phi_1(0) - k_{-1}\sigma_{-1}^2\phi_{-1}(0), \qquad \sigma_1^2\phi_1'(0) + \sigma_{-1}^2\phi_{-1}'(0) = 0. \tag{4.2}$$

Here, $W^{2,1}(\mathbb{R}^+)$ is the set of differentiable functions on $\mathbb{R}^+$ whose derivatives are absolutely continuous with second derivatives belonging to $L^1(\mathbb{R}^+)$. Relations (4.2) describe the way the membrane allows the traffic from one half-axis to the other: as Lemma 4.1 shows, the second relation in (4.2) is a balance condition saying that the mass inflow into one half-axis is equal to the mass outflow out of the other one, and guarantees that the resolvent of A preserves the integral (for an alternative explanation, see Exercise 4.6). The other condition in (4.2) describes this flow in more detail. The constants k_i appearing there are permeability coefficients: k_1 describes permeability from the right to the left, and k_{-1} describes permeability from the left to the right (see Chapter 11 and Exercise 11.2 for more details).

We claim that A defined earlier generates a semigroup of Markov operators in $L^1(\mathbb{R})$. We will need the following two lemmas.

Lemma 4.1 *For $\lambda > 0$, let the operator $R_\lambda(\psi_i)_{i\in\mathbb{I}} = (\phi_i)_{i\in\mathbb{I}}$ in $L^1(\mathbb{R})$ be given by:*

$$\phi_i(x) = C_i \mathrm{e}^{\frac{\sqrt{\lambda}}{\sigma_i}x} + D_i \mathrm{e}^{-\frac{\sqrt{\lambda}}{\sigma_i}x} - \frac{1}{\sigma_i\sqrt{\lambda}}\int_0^x \sinh\frac{\sqrt{\lambda}}{\sigma_i}(x-y)\psi_i(y)\,\mathrm{d}y$$
$$= \frac{1}{2\sigma_i\sqrt{\lambda}}\int_0^\infty \mathrm{e}^{-\frac{\sqrt{\lambda}}{\sigma_i}|x-y|}\psi_i(y)\,\mathrm{d}y + D_i\mathrm{e}^{-\frac{\sqrt{\lambda}}{\sigma_i}x}, \qquad x \geq 0. \tag{4.3}$$

Here, $C_i = C_i(\psi_i) = \frac{1}{2\sigma_i\sqrt{\lambda}}\int_0^\infty \mathrm{e}^{-\frac{\sqrt{\lambda}}{\sigma_i}y}\psi_i(y)\,\mathrm{d}y$, and D_i are some functionals on $L^1(\mathbb{R})$. The operators λR_λ preserve the integral iff D_i are chosen so that ϕ_i satisfy the second condition in (4.2).

Proof Integrating (4.3):

$$\int_{\mathbb{R}^+}\phi_i = \frac{\sigma_i}{\sqrt{\lambda}}D_i + \frac{1}{\lambda}\int_{\mathbb{R}^+}\psi_i - \frac{\sigma_i}{\sqrt{\lambda}}C_i.$$

Hence the integral is preserved iff:

$$\sigma_1(D_1 - C_1) + \sigma_{-1}(D_{-1} - C_{-1}) = 0. \tag{4.4}$$

On the other hand, $\phi_i'(0) = \frac{\sqrt{\lambda}}{\sigma_i}(C_i - D_i)$, showing that the second condition in (4.2) is equivalent to (4.4). □

Lemma 4.2 *The R_λ's from the previous lemma leave the positive cone invariant iff:*

$$C_i + D_i \geq 0, \qquad i \in \mathbb{I}. \tag{4.5}$$

Proof This condition means that $C_i(\psi) + D_i(\psi) \geq 0$ provided $\psi \in L^1(\mathbb{R})$ is non-negative. Assuming (4.5) we have:

$$\phi_i(x) \geq \frac{1}{2\sigma_i\sqrt{\lambda}}\int_0^\infty \mathrm{e}^{-\frac{\sqrt{\lambda}}{\sigma_i}|x-y|}\psi_i(y)\,\mathrm{d}y - C_i\mathrm{e}^{-\frac{\sqrt{\lambda}}{\sigma_i}x}$$
$$\geq \frac{1}{2\sigma_i\sqrt{\lambda}}\int_0^\infty \left[\mathrm{e}^{-\frac{\sqrt{\lambda}}{\sigma_i}|x-y|} - \mathrm{e}^{-\frac{\sqrt{\lambda}}{\sigma_i}(x+y)}\right]\psi_i(y)\,\mathrm{d}y$$
$$\geq 0,$$

as long as $\psi_i \geq 0$. Conversely, suppose that $\phi_i \geq 0$ if $\psi_i \geq 0$. Then, $C_i + D_i = \phi_i(0) \geq 0$ since ϕ_i is continuous. □

To show that A generates a semigroup of Markov operators, consider the resolvent equation for A:

$$\lambda\,(\phi_i)_{i\in\mathbb{I}} - A\,(\phi_i)_{i\in\mathbb{I}} = (\psi_i)_{i\in\mathbb{I}}\,, \tag{4.6}$$

where $\psi_i \in L^1(\mathbb{R}^+)$ and $\lambda > 0$ are given. The solution is given by (4.3) where D_i are to be determined so that $(\phi_i)_{i\in\mathbb{I}} \in D(A)$. In particular, by the second condition in (4.2), we must have (4.4). The other condition in (4.2) forces:

$$\sigma_1\sqrt{\lambda}(C_1 - D_1) = k_1\sigma_1^2(C_1 + D_1) - k_{-1}\sigma_{-1}^2(C_{-1} + D_{-1}).$$

These two are satisfied iff:

$$D_i = \frac{\sqrt{\lambda} + k_{-i}\sigma_{-i} - k_i\sigma_i}{\sqrt{\lambda} + k_{-1}\sigma_{-1} + k_1\sigma_1}C_i + \frac{2k_{-i}\sigma_{-i}^3}{\sigma_1\sigma_{-1}\sqrt{\lambda} + k_{-1}\sigma_{-1}^2\sigma_1 + k_1\sigma_1^2\sigma_{-1}}C_{-i}. \tag{4.7}$$

This proves that the resolvent equation has a solution. Moreover, as the coefficient of C_i is no less than -1, we have $D_i + C_i \geq 0$. By Lemmas 4.1 and 4.2, the resolvent of A is Markov. Hence, we are done provided we show that A is densely defined, but this is left as an exercise.

We are interested in the limit as the membrane's permeability increases. Hence, we consider A_n defined by (4.1) and (4.2) with k_i replaced by nk_i. Then, the corresponding sequence of D_i converges to:

$$D_i = \frac{k_{-i}\sigma_{-i} - k_i\sigma_i}{k_{-1}\sigma_{-1} + k_1\sigma_1}C_i + \frac{2k_{-i}\sigma_{-i}^3}{k_{-1}\sigma_{-1}^2\sigma_1 + k_1\sigma_1^2\sigma_{-1}}C_{-i}. \tag{4.8}$$

Therefore, ϕ_i defined by (4.3) and the resolvents of A_n converge also. We check (see Exercise 4.4) that (4.3) with the above D_i is the resolvent of the densely defined operator A_∞ related to the transmission conditions:

$$k_1\sigma_1^2\phi_1(0) = k_{-1}\sigma_{-1}^2\phi_{-1}(0), \qquad \sigma_1^2\phi_1'(0) + \sigma_{-1}^2\phi_{-1}'(0) = 0. \tag{4.9}$$

Because the limit of Markov operators is Markov, the resolvent of A_∞ is Markov, and we conclude that A_∞ generates a semigroup of Markov operators. By Theorem 2.2, the semigroups generated by A_n converge to the semigroup generated by A_∞.

More on the limit semigroup may be found in Exercise 4.5, see also Chapter 11.

Exercise 4.1 Show that Markov operators are contractions.

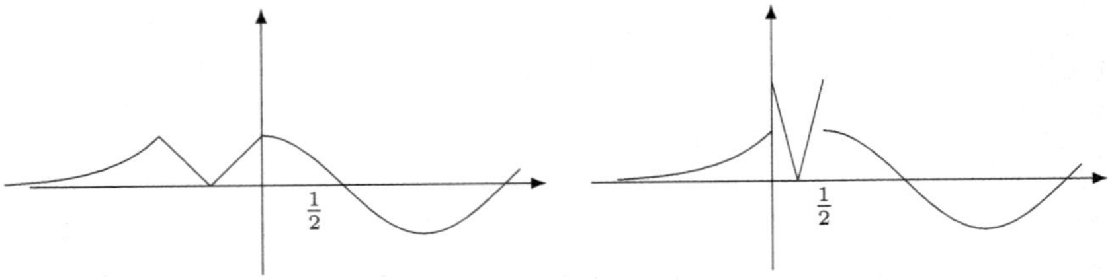

Figure 4.2 The semigroup $\{T(t), t \geq 0\}$ in action: it maps the graph on the left to the graph on the right. Here $\sigma_1 = 1, \sigma_{-1} = 2, t = \frac{1}{2}$. Points on the left half-axis move fast, and need to slow down on the right half-axis: hence, they are congested in the interval $[0, \frac{1}{2}]$. (Redrawn from [57].)

Exercise 4.2 Check that the following relation between transcription rate distribution q and protein number distribution p (taken from [279]):

$$p(x) = \int_0^1 \frac{e^{-xb^{-1}} x^{ah-1}}{b^{ah}\Gamma(ah)} q(h)\,dh,$$

defines a Markov operator from $L^1(0,1)$ to $L^1(0,\infty)$, that is, an operator mapping a non-negative q in the former space to a non-negative p in the latter space, in such a way that $\int_0^\infty p = \int_0^1 q$.

Exercise 4.3 Show that the operator defined by (4.1) and (4.2) is densely defined. *Hint:* argue as in Exercise 3.2, take $\beta_n = n$ for example.

Exercise 4.4 Show that (4.3) with (4.8) is the resolvent of the operator (4.1) with transmission conditions (4.2) replaced by (4.9). Check also that the latter operator is densely defined.

Exercise 4.5

1. Consider a process on $\mathbb{R}$ in which points of $\mathbb{R}^+$ move to the right with speed σ_1 and points of $\mathbb{R}^-$ move to the right with speed σ_{-1}. If ϕ is an initial distribution of such points, then:

$$T(t)\phi(x) = \begin{cases} \phi(x - \sigma_1 t), & x \geq \sigma_1 t, \\ \frac{\sigma_{-1}}{\sigma_1}\phi\left(\frac{\sigma_{-1}}{\sigma_1}x - \sigma_{-1}t\right), & 0 < x < \sigma_1 t, \\ \phi(x - \sigma_{-1}t), & x \leq 0, \end{cases}$$

is their distribution after time $t \geq 0$ (see Figure 4.2). Check that this formula defines a semigroup of Markov operators in $L^1(\mathbb{R})$, and that the generator of this semigroup is $B_{\sigma_{-1},\sigma_1}\phi(x) = \sigma_{\operatorname{sgn} x}\phi'(x)$. Determine the domain of B_{σ_{-1},σ_1}. (Compare Exercise 1.13.)

2. Consider $J \in \mathcal{L}(L^1(\mathbb{R}))$ given by $J\phi(x) = \phi(-x)$. Check that J is a Markov operator with $J^{-1} = J$. Moreover, for $\phi \in D(B_{\sigma_{-1},\sigma_1})$, we have $J\phi \in D(B_{\sigma_1,\sigma_{-1}})$ and $JB_{\sigma_{-1},\sigma_1}J\phi = -B_{\sigma_1,\sigma_{-1}}\phi$. Conclude that $-B_{\sigma_{-1},\sigma_1}$ is similar (or: isomorphic) to $B_{\sigma_1,\sigma_{-1}}$ [49, 128, 129], and generates the semigroup of Markov operators $\{JT(t)J, t \geq 0\}$. Use the generation theorem for groups (see, e.g., [128] p. 79) to see that B_{σ_{-1},σ_1} generates a group of Markov operators.
3. Combine this with Theorems 3.14.15 and 3.14.17 in [9] to see that $B^2_{\sigma_{-1},\sigma_1}$ generates the strongly continuous cosine family:

$$C(t) = \frac{1}{2}\left(\mathrm{e}^{|t|B_{\sigma_{-1},\sigma_1}} + J\mathrm{e}^{|t|B_{\sigma_{-1},\sigma_1}}J\right), \quad t \in \mathbb{R},$$

   and the related semigroup (both composed of Markov operators) defined by the Weierstrass Formula (1.4). The latter semigroup may be thought of as describing diffusion with different coefficients in the two half-axes and no barrier at $x = 0$. Check that the domain of $B^2_{\sigma_{-1},\sigma_1}$ contains functions $\phi \in W^{2,1}(\mathbb{R}^+) \cap W^{2,1}(\mathbb{R}^-)$ with:

$$\sigma_{-1}\phi(0-) = \sigma_1\phi(0+), \qquad \sigma^2_{-1}\phi'(0-) = \sigma^2_1\phi'(0+), \tag{4.10}$$

   and we have $B^2_{\sigma_{-1},\sigma_1}\phi = \sigma^2_{\operatorname{sgn} x}\phi''$.
4. Recall that $L^1(\mathbb{R})$ is isometrically isomorphic to $L^1(\mathbb{R}^+) \times L^1(\mathbb{R}^+)$. Check that the isomorphic image in the latter space of the semigroup generated by $B^2_{\sigma_{-1},\sigma_1}$ is A_∞ provided $k_i\sigma_i = 1$. In other words, A_∞ describes the case of no barrier at $x = 0$ if the influences of diffusion and permeability coefficients cancel out. In general conditions (4.9) do not describe the case of no barrier at $x = 0$, because there is an asymmetry in the way the particles filter in through the membrane from one half-axis to the other.

Exercise 4.6 Consider the functionals F_- and F_+ on $L^1(\mathbb{R}^+)$ given by:

$$F_-\phi = \int_{-\infty}^0 \phi, \qquad F_+\phi = \int_0^\infty \phi.$$

Let $\phi \in D(A)$, where A is given by (4.1) and (4.2), be a density. Check that $\alpha(t) := F_-(\mathrm{e}^{tA}\phi)$, the proportion of probability mass in $\mathbb{R}^-$ at time $t \geq 0$, satisfies $\frac{\mathrm{d}}{\mathrm{d}t}\alpha(t) = \sigma^2_{-1}(\frac{\mathrm{d}}{\mathrm{d}x}\mathrm{e}^{tA}\phi)(0-)$, that is, the quantity $\sigma^2_{-1}(\frac{\mathrm{d}}{\mathrm{d}x}\mathrm{e}^{tA}\phi)(0-)$ describes the intensity of mass inflow into $\mathbb{R}^-$ (or outflow out of $\mathbb{R}^-$) at time t. Derive a similar equation for $\beta(t) = F_+(\mathrm{e}^{tA}\phi)$ and conclude that the second relation in (4.2) is a balance condition saying that the mass inflow into one half-axis is equal to the mass outflow out of the other one.

Exercise 4.7 In the Rotenberg model of **cell proliferation** [299], a cell in a population is characterized by two parameters: $x \in (0, 1)$ denoting her maturity, and $v \in (a, b)$ denoting her speed of maturation, where $0 \le a < b < \infty$ are given minimal and maximal speeds. A cell reaching maximal maturity $x = 1$ divides; her daughter cells start maturing from the state $x = 0$ with random velocity. It is assumed that $K(v, w)$, the probability density of daughter cell's velocity, conditional on v being the velocity of the mother, is given, and we have:

$$\int_a^b K(v, w)\,dw = 1, \qquad v \in (a, b).$$

It is assumed, furthermore, that the average number of surviving daughter cells does not change in time and equals $p \in (0, \infty)$, and that the cellular flux out of $x = 1$ is the same as the flux into $x = 0$. This leads to the boundary condition:

$$v\phi(0, v) = p \int_a^b w\phi(1, w)K(w, v)\,dw, \qquad v \in (a, b), \tag{4.11}$$

where $\phi(x, v)$ is the cells's population density at x and v.

In this exercise the reader will construct a semigroup related to the Rotenberg model in the Banach space $L^1(\Omega)$ of absolutely integrable functions on $\Omega = (0, 1) \times (a, b)$ with the usual norm:

$$\|\phi\|_{L^1(\Omega)} = \int_0^1 \int_a^b v|\phi(x, v)|\,dv\,dx < \infty;$$

see [73] for an alternative approach.

1. For $\phi \in L^1(\Omega)$, let $\tilde{\phi}$ be its extension to $\Omega \cup \Gamma$, where $\Gamma = \{(x, v) \in \mathbb{R}^2; -v < bx < 0, v \in (a, b)\}$ (see Figure 4.3) given by:

$$\tilde{\phi}(x, v) = \frac{p}{v} \int_a^b w\phi(1 + wv^{-1}x, w)K(w, v)\,dw, \qquad (x, v) \in \Gamma.$$

   Define:

$$(T(t)\phi)(x, v) = \tilde{\phi}(x - tv, v), \qquad t \in [0, b^{-1}), \phi \in L^1(\Omega),$$

   and check that $T(t + s) = T(t)T(s)$ for $t + s < b^{-1}$.
2. Prove that the family $\{T(t), t \in [0, b^{-1})\}$ may be (uniquely) extended to a strongly continuous semigroup in $L^1(\Omega)$ and that the semigroup's generator is A given by $(A\phi)(x, v) = -v\frac{\partial \phi}{\partial x}(x, v)$ with domain composed of ϕ such that: (a) for almost all $v \in (a, b)$, $x \mapsto \phi(x, v)$ is absolutely continuous and such that $(x, v) \mapsto -v\frac{\partial \phi}{\partial x}(x, v)$ is a member of $L^1(\Omega)$, and (b) the boundary condition (4.11) is satisfied for almost all $v \in (a, b)$.

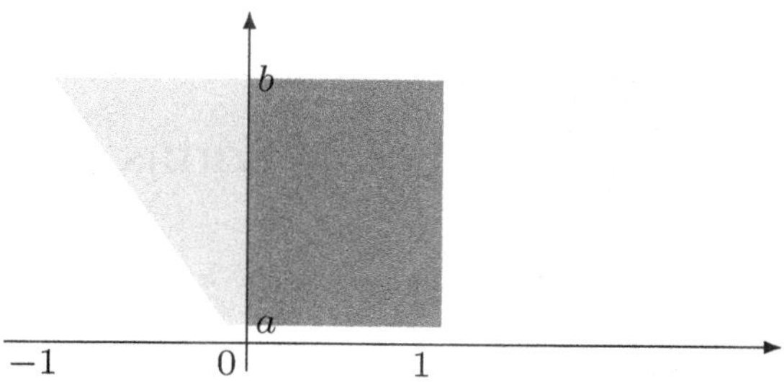

Figure 4.3 Region $\Omega \cup \Gamma$ with Ω in darker shade and Γ in lighter shade

3. Check that for $p = 1$, the semigroup from point 2 is Markov.
4. Rotenberg's model may include the case of cells that degenerate in the sense that they do not divide. With probability $q \le 1$ they survive and are treated as their daughters inheriting the mother cell's velocity. In this case the boundary condition becomes:

$$v\phi(0, v) = q\phi(1, v) + p \int_a^b w\phi(1, w)K(w, v)\,\mathrm{d}w, \qquad v \in (a, b). \tag{4.12}$$

Construct the related semigroup by redefining $\tilde{\phi}$ and following the steps 1–2. This semigroup is Markov provided $p + q = 1$.

☞ Chapter's summary

In our second example of convergence of semigroups, we study dynamics of distributions of Brownian motions on the real line with a semipermeable membrane at $x = 0$, and differing diffusion coefficients in the two half-axes. We prove that as permeability coefficients converge to infinity, the related semigroups converge to a semigroup describing an "almost no membrane case."

5

Convergence of Forms

A familiar situation in which one can deduce convergence of resolvents without calculating them is the one in which the generators are related to sesquilinear forms; folk wisdom is that convergence of forms implies convergence of resolvents. Before diving into details of this implication, let us recall the rudiments of the theory of sectorial forms [8, 99, 107, 201, 215, 282].

Let $\mathbb{X}$ be a *complex* Hilbert space and let $V \subset \mathbb{X}$ be its linear subspace. A map $\mathfrak{a} : V \times V \to \mathbb{C}$ is said to be a sesquilinear form on V if it is linear in the first coordinate and semilinear in the second coordinate:

$$\begin{aligned} \mathfrak{a}(\alpha x + \beta y, z) &= \alpha \mathfrak{a}(x, z) + \beta \mathfrak{a}(y, z), \\ \mathfrak{a}(z, \alpha x + \beta y) &= \overline{\alpha} \mathfrak{a}(z, x) + \overline{\beta} \mathfrak{a}(z, y), \qquad x, y, z \in V, \alpha, \beta \in \mathbb{C}. \end{aligned}$$

The associated quadratic form (denoted by the same letter) $\mathfrak{a} : V \to \mathbb{C}$ given by $\mathfrak{a}[x] = \mathfrak{a}(x, x)$ determines the sesquilinear form by the **polarization principle**:

$$\mathfrak{a}(x, y) = \frac{1}{4}(\mathfrak{a}[x + y] - \mathfrak{a}[x - y] + i\mathfrak{a}[x + iy] - i\mathfrak{a}[x - iy]), \qquad x, y \in V.$$

(Recall [201, p. 49] that if $\mathbb{X}$ is a real Banach space, there is no analogue of this principle, unless the form is symmetric – see further on for the definition of a symmetric form.) The numerical range of $\mathfrak{a}$, denoted $\Theta_\mathfrak{a}$ is the set of all values of the quadratic form for $\|x\| = 1$:

$$\Theta_\mathfrak{a} = \{z \in \mathbb{C}; \exists_{x \in V, \|x\|=1} z = \mathfrak{a}[x]\}.$$

The numerical range is a convex subset of $\mathbb{C}$.

For two forms, say $\mathfrak{a}$ and $\mathfrak{b}$, defined on subspaces $V_\mathfrak{a}$ and $V_\mathfrak{b}$, respectively, their linear combination is defined in a natural way on $V_\mathfrak{a} \cap V_\mathfrak{b}$ by:

$$(\alpha \mathfrak{a} + \beta \mathfrak{b})(x, y) = \alpha \mathfrak{a}(x, y) + \beta \mathfrak{b}(x, y).$$

In particular, we write $\mathfrak{a}+\alpha$ for the form defined (on $V=V_{\mathfrak{a}}$) by:

$$(\mathfrak{a}+\alpha)(x,y)=\mathfrak{a}(x,y)+\alpha(x,y),$$

where (x,y) denotes the scalar product in $\mathbb{X}$. The forms:

$$\mathfrak{a}^{\mathrm{r}}=\frac{1}{2}(\mathfrak{a}+\mathfrak{a}^*) \quad \text{and} \quad \mathfrak{a}^{\mathrm{i}}=\frac{1}{2i}(\mathfrak{a}-\mathfrak{a}^*),$$

where $\mathfrak{a}^*(x,y)=\overline{\mathfrak{a}(y,x)}$, are referred to as the real and imaginary parts of $\mathfrak{a}$. Note, however, that even though $\mathfrak{a}=\mathfrak{a}^{\mathrm{r}}+i\mathfrak{a}^{\mathrm{i}}$, and both forms are symmetric (i.e. $(\mathfrak{a}^{\mathrm{r}})^*=\mathfrak{a}^{\mathrm{r}}$ and $(\mathfrak{a}^{\mathrm{i}})^*=\mathfrak{a}^{\mathrm{i}}$), $\mathfrak{a}^{\mathrm{r}}$ and $\mathfrak{a}^{\mathrm{i}}$ are not real-valued. We merely have $\mathfrak{a}^{\mathrm{r}}[x]=\Re\mathfrak{a}[x]$ and $\mathfrak{a}^{\mathrm{i}}[x]=\Im\mathfrak{a}[x]$, $x\in V$.

A form $\mathfrak{a}$ is said to be *sectorial* if there are constants $\gamma\in\mathbb{R}$ and $\theta\in[0,\frac{\pi}{2})$ such that the numerical range $\Theta_{\mathfrak{a}}$ is contained in the sector:

$$\Sigma_{\theta,\gamma}=\{z\in\mathbb{C};\,|\arg(z-\gamma)|\le\theta\}. \tag{5.1}$$

(In this context, γ is termed the **vertex** of the sector.) In other words, $\mathfrak{a}^{\mathrm{r}}\ge\gamma$ (i.e. $(\mathfrak{a}^{\mathrm{r}}-\gamma)[x]\ge 0, x\in V$) and $|\mathfrak{a}^{\mathrm{i}}[x]|\le\tan\theta(\mathfrak{a}^{\mathrm{r}}-\gamma)[x], x\in V$. The first of these conditions (coupled with symmetry) allows thinking of $\mathfrak{a}^{\mathrm{r}}-\gamma$ as a scalar product. Hence, we have the following version of the Cauchy-Schwarz inequality:

$$|(\mathfrak{a}^{\mathrm{r}}-\gamma)(x,y)|\le\sqrt{(\mathfrak{a}^{\mathrm{r}}-\gamma)[x]}\sqrt{(\mathfrak{a}^{\mathrm{r}}-\gamma)[y]}, \qquad x,y\in V.$$

The second of these conditions allows controlling the imaginary part of the quadratic form by its real part, and yields estimates for the resolvent of the related operator (see later in this chapter).

A form $\mathfrak{a}$ defined on a subspace V is said to be closed if for any sequence $(x_n)_{n\ge1}$ of elements $x_n\in V$ conditions $\lim_{n\to\infty}x_n=x$ and $\lim_{n,m\to\infty}\mathfrak{a}[x_n-x_m]=0$ imply that $x\in V$ and $\lim_{n\to\infty}\mathfrak{a}[x-x_n]=0$. It is clear from this definition that $\mathfrak{a}$ is closed iff $\mathfrak{a}-\gamma$ is closed for some, and hence for all $\gamma\in\mathbb{R}$. Moreover, as the imaginary part of a sectorial form is controlled by its real part, a sectorial form is closed iff its real part is closed.

Example 5.1 Let $\mathbb{X}=L^2[0,1]$ be the space of square integrable complex functions on $[0,1]$ and let $V=H^1[0,1]$ be its subspace of absolutely continuous functions with square integrable distributional derivatives. In other words, $f\in V$ iff there is a $g\in\mathbb{X}$ such that $f(x)=f(0)+\int_0^x g(y)\,\mathrm{d}y$, $x\in[0,1]$; in such a case we write $g=f'$. Given a real number κ we consider the sesquilinear form on V:

$$\mathfrak{a}(f,g)=\kappa(f(1)-f(0))\overline{(g(1)-g(0))}+\int_0^1 f'\overline{g}'. \tag{5.2}$$

The related quadratic form is given by:

$$\mathfrak{a}[f] = \kappa |f(1) - f(0)|^2 + \|f'\|^2_{\mathbb{X}}.$$

It is clear that $\mathfrak{a}$ is symmetric. Moreover, for non-negative κ, the numerical range $\Theta_{\mathfrak{a}}$ is contained in the positive real half-axis, implying that $\mathfrak{a}$ is sectorial (with any θ and any $\gamma \leq 0$). To treat the case in which $\kappa < 0$, we recall (see [201, p. 193]) that for any $\epsilon > 0$ there is a $K = K(\epsilon)$ such that:

$$\sup_{x\in[0,1]} |f(x)| \leq \epsilon \|f'\|_{\mathbb{X}} + K\|f\|_{\mathbb{X}}, \tag{5.3}$$

for $f \in V$. It follows that:

$$\begin{aligned}|f(0) - f(1)|^2 &\leq 4(\epsilon\|f'\|_{\mathbb{X}} + K\|f\|_{\mathbb{X}})^2 \\ &\leq 4(\epsilon^2 + \epsilon K)\|f'\|^2_{\mathbb{X}} + 4(K^2 + \epsilon K)\|f\|^2_{\mathbb{X}}.\end{aligned}$$

In particular, choosing ϵ so that $4(\epsilon^2 + \epsilon K\eta) \leq \frac{-1}{2\kappa}$, we see that there is a $\gamma > 0$ such that:

$$\mathfrak{a}[f] + \gamma\|f\|^2_{\mathbb{X}} \geq \frac{1}{2}\|f'\|^2_{\mathbb{X}} \geq 0, \tag{5.4}$$

proving that $\mathfrak{a}$ is sectorial.

Finally, if for $f_n \in V$ we have:

$$\lim_{n,m\to\infty} \mathfrak{a}[f_n - f_m] = 0 \text{ and } \lim_{n\to\infty} \|f_n - f\|_{\mathbb{X}} = 0,$$

then using (5.4) we see that $(f'_n)_{n\geq 1}$ is a Cauchy sequence in $\mathbb{X}$ and, consequently, $(f_n)_{n\geq 1}$ is a Cauchy sequence in V (equipped with the norm $\|f\|_V = \sqrt{\|f\|^2_{\mathbb{X}} + \|f'\|^2_{\mathbb{X}}}$). This suffices to show that $f \in V$, for V is in fact a Hilbert space (with the norm specified earlier). The relation $\lim_{n\to\infty} \mathfrak{a}[f_n - f] = 0$ is then a direct consequence of (5.3). Hence, $\mathfrak{a}$ is closed. □

Next, we turn to the fundamental notion of the operator associated with a sesquilinear form. Assume that $\mathfrak{a}$ is sectorial and closed, and that its domain $V = V_{\mathfrak{a}}$ is dense in $\mathbb{X}$. Then, there exists a linear, closed operator A with domain $D(A)$ satisfying the following conditions:

1. $D(A) \subset V_{\mathfrak{a}}$ and for $x \in D(A)$, $y \in V_{\mathfrak{a}}$:

$$\mathfrak{a}(x, y) = -(Ax, y),$$

2. if $x \in V_{\mathfrak{a}}$ and there is a $z \in X$ such that $\mathfrak{a}(x, y) = -(z, y)$ for all $y \in V_{\mathfrak{a}}$, then $x \in D(A)$ and $Ax = z$.

Using the Lax–Milgram Lemma, it may be shown that the range of $\lambda - A$ is the whole of $\mathbb{X}$ provided $\Re\lambda > \gamma$ where γ is the vertex of the sector (5.1). Also, A is uniquely determined by the conditions given earlier.

The operator A is then the generator of a semigroup satisfying $\|e^{tA}\| \le e^{-\gamma t}$. (In fact, this semigroup has a holomorphic extension, see [8, 215, 282], comp. Chapter 15.) To see this, take λ with $\Re\lambda > -\gamma$. Then, for $x \in D(A)$ with $\|x\| = 1$:

$$\begin{aligned}\|\lambda x - Ax\| &\ge |(\lambda x - Ax, x)| = |\lambda(x, x) - (Ax, x)| = |\lambda + \mathfrak{a}(x, x)| \\ &\ge \operatorname{dist}(-\lambda, \Theta_{\mathfrak{a}}) \ge \operatorname{dist}(-\lambda, \{z;\, \Re z \ge \gamma\}) = \Re\lambda + \gamma,\end{aligned}$$

showing the claim by the Hille–Yosida Theorem.

For example, the operator associated with the form (5.2) is $Af = f''$ with the domain composed of continuously differentiable functions $f \in \mathbb{X} = L^2[0,1]$ with $f' \in V = H^1[0,1]$ and:

$$f'(0) = f'(1) = \kappa(f(0) - f(1)). \tag{5.5}$$

For, if f satisfies the conditions listed earlier, then $\mathfrak{a}(f, g) = -(f'', g)$ for all $g \in V$ by integration by parts formula. Conversely, if for some $f \in V$ and $h \in L^2[0,1]$ we have $-\mathfrak{a}(f, g) = (h, g)$ for all $g \in H^1[0,1]$, then integrating by parts we obtain:

$$\int_0^1 f'\overline{g}' + \kappa(f(1) - f(0))\overline{(g(1) - g(0))} = \int_0^1 \int_0^x h(y)\,\mathrm{d}y\,\overline{g}'(x)\,\mathrm{d}x - \overline{g(1)}\int_0^1 h. \tag{5.6}$$

We note that the functions $g \in H^1[0,1]$ with $g(1) = g(0) = 0$ are precisely the g of the form $g(x) = \int_0^x k(y)\,\mathrm{d}y$ where $k \in L^2[0,1]$ and $\int_0^1 k = 0$. For such g this formula reduces to:

$$\int_0^1 f'\overline{g}' = \int_0^1 \int_0^x h(y)\,\mathrm{d}y\,\overline{g}'(x)\,\mathrm{d}x,$$

or, which is the same,

$$\int_0^1 f'\overline{k} = \int_0^1 \int_0^x h(y)\,\mathrm{d}y\,\overline{k(x)}\,\mathrm{d}x,$$

for all $k \in L^2[0,1]$ satisfying $\int_0^1 \overline{k} = 0$. The subspace of functions k with this property is the orthogonal complement of the space of constant functions. It follows that $f'(x) = f'(0) + \int_0^x h(y)\,\mathrm{d}y$, $x \in [0,1]$, that is, that $f' \in H^1[0,1]$. Plugging this relation into (5.6) and choosing g so that $g(1) = 0$ and $g(0) = 1$ we obtain $f'(0) = \kappa(f(0) - f(1))$. Similarly, taking $g(1) = g(0) = 1$, we obtain $f'(1) = f'(0)$, completing the proof of the claim.

Remark 5.2 Interestingly, for $\kappa = 1$, operator A is the "minimal," that is, Krein–von Neumann extension [222] of the Laplace operator defined on $C_c^\infty(0, 1)$ – see [4, Example 5.1], see also [266, Example 2.4].

The following main convergence theorem of this chapter is a simple version of Theorem 3.6, p. 455 in [201].

Theorem 5.3 *Let $\mathfrak{a}_n, n \ge 1$ and $\mathfrak{a}$ be sesquilinear, closed sectorial forms defined on the same, dense subspace V of a Hilbert space $\mathbb{X}$, and let A_n, $n \ge 1$ and A be the associated operators. Assume that $\mathfrak{b}_n := \mathfrak{a}_n - \mathfrak{a}$ are uniformly sectorial in the sense that all numerical ranges $\Theta_{\mathfrak{b}_n}$ lie in a single sector* (5.1)*. If, for $x \in V$, $\lim_{n\to\infty} \mathfrak{a}_n[x] = \mathfrak{a}[x]$, then:*

$$\lim_{n\to\infty} \mathrm{e}^{tA_n}x = \mathrm{e}^{tA}x, \qquad x \in \mathbb{X},$$

almost uniformly in $t \ge 0$.

Proof Replacing $\mathfrak{a}_n$ by $\mathfrak{a}_n - \gamma$ results in shifting A_n to $A_n - \gamma$ (and the same concerns $\mathfrak{a}$ and A). Hence, without loss of generality, we assume γ in (5.1) to be zero. In particular, $\mathfrak{a}_n^{\mathrm{r}} \ge \mathfrak{a}^{\mathrm{r}} \ge 0$. By Theorem 2.2, it suffices to show strong convergence of the resolvents $R_{\lambda,n} := (\lambda - A_n)^{-1}$, $n \ge 1$ to $R_\lambda := (\lambda - A)^{-1}$ for $\Re\lambda > 0$ (In fact, we could restrict ourselves to real $\lambda > 0$.) Because $\|R_{\lambda,n}\| \le \frac{1}{\Re\lambda}$ we may restrict our attention to x in the dense subspace V of $\mathbb{X}$.

For $x \in V$ and λ with $\Re\lambda > 0$, $R_{\lambda,n}x$ and R_λ belong to V and, expanding $(\mathfrak{a} + \lambda)[R_{\lambda,n}x - R_\lambda x]$, we have:

$$\begin{aligned} \mathfrak{a}[R_{\lambda,n}x - R_\lambda x] &+ \lambda\|R_{\lambda,n}x - R_\lambda x\|^2 + \mathfrak{b}_n[R_{\lambda,n}x] \\ &= (\mathfrak{a}_n + \lambda)[R_{\lambda,n}x] - (\mathfrak{a} + \lambda)(R_{\lambda,n}x, R_\lambda x) \\ &\quad - (\mathfrak{a} + \lambda)(R_\lambda x, R_{\lambda,n}x) + (\mathfrak{a} + \lambda)[R_\lambda x]. \end{aligned} \tag{5.7}$$

Now, the first and the third terms cancel out (both being equal to $-(x, R_{\lambda,n}x)$). Hence, the expression reduces to:

$$(\mathfrak{a} + \lambda)[R_\lambda x] - (\mathfrak{a}_n + \lambda)(R_{\lambda,n}x, R_\lambda x) + \mathfrak{b}_n(R_{\lambda,n}x, R_\lambda x) = \mathfrak{b}_n(R_{\lambda,n}x, R_\lambda x),$$

since the first and the second terms cancel out in a similar way.

The real part of the left-hand side in (5.7) is composed of three non-negative terms. Hence, if the right-hand side converges to zero, all three of them converge to zero. Forms $\mathfrak{b}_n$ being uniformly sectorial:

$$\begin{aligned} |\mathfrak{b}_n(R_{\lambda,n}x, R_\lambda x)| &\le [1 + \tan\theta]|\mathfrak{b}^{\mathrm{r}}(R_{\lambda,n}x, R_\lambda x)| \\ &\le [1 + \tan\theta]\sqrt{\mathfrak{b}_n^{\mathrm{r}}[R_{\lambda,n}x]}\sqrt{\mathfrak{b}_n^{\mathrm{r}}[R_\lambda x]}. \end{aligned}$$

Because, by assumption, $\lim_{n\to\infty} \mathfrak{b}_n^{\rm r}[R_\lambda x] = 0$, we are left with showing that $(\mathfrak{b}_n^{\rm r}[R_{\lambda,n}x])_{n\ge 1}$ is bounded. To this end, we estimate as follows:

$$0 \le \mathfrak{b}_n^{\rm r}[R_{\lambda,n}x] \le \mathfrak{a}_n^{\rm r}[R_{\lambda,n}x] < (\mathfrak{a}_n^{\rm r} + \Re\,\lambda)[R_{\lambda,n}x]$$
$$= \Re\,(x, R_{\lambda,n}x) \le \frac{1}{\Re\lambda}\|x\|^2.$$

This completes the proof. □

Remark 5.4 The proof presented here is slightly simpler than that of Theorem 3.6, p. 455 in [201], but this comes at the price of assuming convergence of forms on the whole of V and not merely on a core of the limit form. Although in the example considered in this chapter this assumption is satisfied, in general convergence on a core is all we can hope for: As Example 3.10, p. 459 in [201] makes it clear, resolvents may converge despite the fact that the related forms *do not* converge to the limit form on the whole of V. □

For example, let $(\kappa_n)_{n\ge 1}$ be a sequence of positive numbers converging to zero, and let $\mathfrak{a}_n$ be the forms in $H^1[0,1] \subset L^2[0,1]$ given by (5.2) with κ replaced by κ_n, and let $\mathfrak{a}$ correspond to $\kappa = 0$. Then, the forms $\mathfrak{b}_n(f,g) = \mathfrak{a}_n(f,g) - \mathfrak{a}(f,g) = \kappa_n(f(1) - f(0))\overline{(g(1) - g(0))}$ have their numerical ranges in the positive real half-line, and therefore are uniformly sectorial. Because $\lim_{n\to\infty} \mathfrak{a}_n = \mathfrak{a}$, it follows that:

$$\lim_{n\to\infty} \mathrm{e}^{tA_n} f = \mathrm{e}^{tA} f, \qquad f \in L^2[0,1], \tag{5.8}$$

where A_n are Laplace operators in $L^2[0,1]$ with boundary conditions (5.5) and A is the **Neumann Laplace operator**. Looking back to Chapter 3, we see that the semigroups involved here describe Brownian motions on $[0,1]$ that, after being reflected "many times" at $x = 1$, jump to $x = 0$, and vice versa. Formula (5.8) simply says that as the "frequency" of such jumps decreases to zero, the corresponding Brownian motions converge to the Brownian motion with two reflecting barriers.

For $\kappa < 0$ the boundary condition (5.5) is not associated with a stochastic process. (For $\kappa = -1$ this boundary condition and the related cosine operator function was introduced in [71].) Nevertheless, it is interesting whether (5.8) may be extended to the case where $\kappa_n < 0$. As the following theorem, taken from [201] as well, shows, the answer is in the affirmative.

Theorem 5.5 *Suppose that $\mathfrak{a}$ and $\mathfrak{a}_n, n \ge 1$ are closed, symmetric forms on a common domain V, and that $\mathfrak{a}_n \uparrow \mathfrak{a}$, that is:*

$$\mathfrak{a}_{n+1}^{\rm r}[x] \ge \mathfrak{a}_n^{\rm r}[x] \quad \textit{and} \quad \lim_{n\to\infty} \mathfrak{a}_n[x] = \mathfrak{a}[x], x \in V.$$

Assume also that these forms are bounded from below (see (5.10)*). Then, for the related semigroups we have:*

$$\lim_{n\to\infty} \mathrm{e}^{tA_n}x = \mathrm{e}^{tA}x, \qquad x \in \mathbb{X}. \tag{5.9}$$

Remark 5.6 Because the theorem is not of primary importance in what follows, we merely sketch its proof, taking for granted some additional material on forms. The reader will also need to forgive us for using some information on pseudoresolvents to be covered in Chapter 7; we could move this example after Chapter 7, but this is not really the place it belongs to. A complete proof of a more general result (Theorem 58.2) will be given in Chapter 58.

Proof By assumption, there is γ such that:

$$\gamma \|x\|^2 \le \mathfrak{a}_n^{\mathrm{r}}[x] \le \mathfrak{a}^{\mathrm{r}}[x], \qquad x \in V. \tag{5.10}$$

In particular, as $\mathfrak{a}_n^{\mathrm{i}} = \mathfrak{a}^{\mathrm{i}} = 0$, the numerical ranges $\Theta_{\mathfrak{a}}$ and $\Theta_{\mathfrak{a}_n}$ are contained in the half-line $[\gamma, \infty)$. Hence, for $\lambda > -\gamma$, $(\lambda - A)^{-1}$ and $(\lambda - A_n)^{-1}$ exist and:

$$\| (\lambda - A)^{-1} \| \le \frac{1}{\lambda + \gamma}, \qquad \| (\lambda - A_n)^{-1} \| \le \frac{1}{\lambda + \gamma}.$$

Fixing such λ and an $x \in V$, we let $x_n := (\lambda - A_n)^{-1}x$. Then:

$$\big((\lambda - A_n)^{-1} x, x\big) = (x_n, \lambda x_n - A_n x_n) = (\mathfrak{a}_n + \lambda)[x_n] \ge 0; \tag{5.11}$$

in particular $(\lambda - A_n)^{-1}$ is self-adjoint (see [294, p. 195]), that is:

$$((\lambda - A_n)^{-1} x, y) = (x, (\lambda - A_n)^{-1} y),$$

for $x, y \in \mathbb{X}$. Moreover, for $m \ge n$:

$$\begin{aligned} ((\lambda - A_m)^{-1} x, x) &= (x_m, \lambda x_n - A_n x_n) = (\mathfrak{a}_n + \lambda)(x_n, x_m) \\ &\le \sqrt{(\mathfrak{a}_n + \lambda)[x_n](\mathfrak{a}_n + \lambda)[x_m]} \\ &\le \sqrt{(\mathfrak{a}_n + \lambda)[x_n](\mathfrak{a}_m + \lambda)[x_m]} \\ &= \sqrt{\big((\lambda - A_n)^{-1} x, x\big)\big((\lambda - A_m)^{-1} x, x\big)}, \end{aligned} \tag{5.12}$$

implying $((\lambda - A_m)^{-1}x, x) \le ((\lambda - A_n)^{-1}x, x)$. This means that the operator-valued sequence $((\lambda - A_n)^{-1})_{n\ge 1}$ bounded by 0 from below and nonincreasing. The same argument shows that the sequence is in fact bounded from below by $(\lambda - A)^{-1}$. It follows (see, e.g., [49, p. 108], [201, p. 254], or [264, Chapter 4]) that the strong limit:

$$R_\lambda = \lim_{n\to\infty} (\lambda - A_n)^{-1}$$

exists and is a bounded, self-adjoint operator. The kernel of R_λ is trivial, because if $x \in Ker\, R_\lambda$ then:

$$0 = (R_\lambda x, x) \geq \left((\lambda - A)^{-1} x, x\right) \geq 0,$$

and taking the self-adjoint square root ([49, p. 109], [107, p. 173], or [294, p. 196]), say B_λ, of $(\lambda - A)^{-1}$ we see that $\|B_\lambda x\|^2 = \left((\lambda - A)^{-1} x, x\right) = 0$; this implies $x \in Ker\, B_\lambda \subset Ker\,(\lambda - A)^{-1} = \{0\}$. Hence (see Exercise 7.7) there is a closed operator, say A_0, such that $R_\lambda = (\lambda - A_0)^{-1}$. By Theorem 2.21, p. 330 in [201], for the related form $\mathfrak{a}_0$ we have $\mathfrak{a}_n[x] \leq \mathfrak{a}_0[x] \leq \mathfrak{a}[x], x \in V$ (in particular, the domain of definition of $\mathfrak{a}_0$ is V). Combining this with the assumption, we obtain $\mathfrak{a}_0 = \mathfrak{a}$, implying $A = A_0$ and completing the proof. □

Exercise 5.1 Show that a sesquilinear form is symmetric iff the corresponding quadratic form has real values.

☞ Chapter's summary

"Convergence of forms implies convergence of resolvents," but beware: "the form of the limit form" on some vectors may be misleading. If caution is exercised, however, convergence of forms is an easy way to prove convergence of semigroups. For more recent results in this direction, see, for example, [267] and the references cited there.

6

Uniform Approximation of Semigroups

This chapter is a detour from the main subject. To explain, we recall that our main question is that of convergence: given a sequence of semigroups, we are interested in finding its limit. In this chapter, we consider the related question of approximation: given a semigroup we want to find a sequence of semigroups converging to this semigroup and satisfying some additional properties.

For simplicity, we restrict ourselves to bounded semigroups to consider the problem of uniform approximation in operator norm. Recall from Theorem 2.2 that semigroups by nature converge *almost* uniformly in *strong topology*: the question is whether we may claim *uniform* convergence in *operator norm*. In other words, we want to know if given a bounded, strongly continuous semigroup (e^{tA}) in a Banach space $\mathbb{X}$, we may find bounded, strongly continuous semigroups (e^{tA_n}), $n \geq 1$ such that:

$$\lim_{n\to\infty} \sup_{t\geq 0} \|\mathrm{e}^{tA} - \mathrm{e}^{tA_n}\| = 0,$$

where $\|\cdot\|_{\mathcal{L}(\mathbb{X})}$ denotes the operator in $\mathcal{L}(\mathbb{X})$. To avoid trivial solutions we require that $A_n \neq A$ for all $n \geq 1$. In this chapter, we will characterize semigroups that can be approximated in this way.

If we let $Semi_{b,sc}(\mathbb{X})$ denote the set of all bounded strongly continuous semigroups on $\mathbb{X}$ made into a metric space by means of the metric:

$$d(S, \tilde{S}) = \sup_{t\geq 0} \|S(t) - \tilde{S}(t)\|_{\mathcal{L}(\mathbb{X})},$$

then our question turns into that of characterizing isolated points of the space $Semi_{b,sc}(\mathbb{X})$. It transpires that the characterization involves **scalar semigroups**; by definition, a semigroup $S = \{S(t), t \geq 0\}$ is said to be scalar if each $S(t)$ is a scalar multiple of $I_{\mathbb{X}}$. If $\mathbb{X}$ is complex, denoting by $S_{[z]} = \{S_{[z]}(t)\}_{t\geq 0}$ the semigroup on $\mathbb{X}$ given by:

$$S_{[z]}(t) = e^{zt} I_{\mathbb{X}} \quad (t \geq 0, z \in \mathbb{C}),$$

we see that every scalar semigroup in $Semi_{b,sc}(\mathbb{X})$ is of the form $S_{[z]}$ for some z with $\Re z \leq 0$. If $\mathbb{X}$ is real, then all scalar members of $Semi_{b,sc}(\mathbb{X})$ are of the form $S_{[z]}$ where z is real and nonpositive.

Our main theorem in this chapter, taken from [63], says that isolated points of $Semi_{b,sc}(\mathbb{X})$ constitute a fraction of the set of all scalar semigroups in this space. (This is not so in the case of cosine families: all bounded, scalar, strongly continuous cosine families are isolated points of the metric space of strongly continuous, bounded cosine families equipped with the metric of uniform convergence in operator norm; see [63].)

Theorem 6.1 *If $\mathbb{X}$ is a complex Banach space, then the isolated points of $Semi_{b,sc}(\mathbb{X})$ are precisely the scalar semigroups:*

$$S_{[ia]}(t) = e^{iat} I_{\mathbb{X}} \quad (t \geq 0, \ a \in \mathbb{R}). \tag{6.1}$$

If $\mathbb{X}$ is a real Banach space, then the only isolated point of $Semi_{b,sc}(\mathbb{X})$ is the identity semigroup defined by $S(t) = I_{\mathbb{X}}$ for each $t \geq 0$.

For the proof of this theorem we need two lemmas. We start with an auxiliary result that is part of the folklore of operator theory: For a Banach space $\mathbb{X}$, let $\mathbb{X}'$ be the dual space of $\mathbb{X}$. Given $x \in \mathbb{X}$ and $x' \in \mathbb{X}'$, we denote by $\langle x, x' \rangle$ the value of the functional x' at x. Let $Z(\mathcal{L}(\mathbb{X}))$ be the center of the algebra of $\mathcal{L}(\mathbb{X})$, that is:

$$Z(\mathcal{L}(\mathbb{X})) = \{A \in \mathcal{L}(\mathbb{X}) \mid AB = BA \quad \text{for each } B \in \mathcal{L}(\mathbb{X})\}.$$

Lemma 6.2 *If $\mathbb{X}$ is a Banach space, then $Z(\mathcal{L}(\mathbb{X}))$ consists precisely of all scalar multiples of $I_{\mathbb{X}}$.*

Proof Without loss of generality, we may assume that $\mathbb{X}$ is nonzero. It is clear that any scalar multiple of $I_{\mathbb{X}}$ is in $Z(\mathcal{L}(\mathbb{X}))$. To prove the converse statement, suppose that $A \in Z(\mathcal{L}(\mathbb{X}))$. For any $x' \in \mathbb{X}'$ and any $y \in \mathbb{X}$, let $T_{x',y}$ be the operator in $\mathcal{L}(\mathbb{X})$ given by:

$$T_{x',y}x = \langle x, x' \rangle y \quad (x \in \mathbb{X}).$$

Then $AT_{x',y} = T_{x',y}A$ for all $x' \in \mathbb{X}'$ and all $y \in \mathbb{X}$, or equivalently:

$$\langle x, x' \rangle Ay = \langle Ax, x' \rangle y, \tag{6.2}$$

for all $x' \in \mathbb{X}'$ and all $x, y \in \mathbb{X}$. Fix $x_0 \in \mathbb{X} \setminus \{0\}$ arbitrarily and next, employing the Hahn–Banach Theorem, select $x_0' \in \mathbb{X}'$ so that $\langle x_0, x_0' \rangle = 1$. If we now let

$\lambda = \langle Ax_0, x_0' \rangle$, then (6.2) yields $Ay = \lambda y$ for all $y \in \mathbb{X}$, or equivalently, $A = \lambda I_{\mathbb{X}}$. The lemma follows. □

In our next lemma we show directly that the semigroup (6.1) with $a = 0$ is an isolated point of $Semi_{b,sc}(\mathbb{X})$. We remark that condition (6.3) is optimal: for any $\lambda > 0$, we have $\sup_{t\geq 0} \|I_{\mathbb{X}} - \mathrm{e}^{-\lambda t} I_{\mathbb{X}}\| = 1$, while $\mathrm{e}^{-\lambda t} \neq 1$ whenever $t > 0$.

Lemma 6.3 *For any semigroup $T = \{T(t)\}_{t\geq 0}$ in $Semi_{b,sc}(\mathbb{X})$, if:*

$$\sup_{t\geq 0} \|T(t) - I_{\mathbb{X}}\| < 1, \tag{6.3}$$

then $T(t) = I_{\mathbb{X}}$ for every $t \geq 0$.

Proof There are many proofs of this result; two of them are outlined in Exercises 6.2 and 6.3, another one may be found in [63]. Arguably the simplest reasoning is based on the following identity (see [269, p. 54] or [357]):

$$2(T(t) - I_{\mathbb{X}}) = T(2t) - I_{\mathbb{X}} - (T(t) - I_{\mathbb{X}})^2$$

(of course, it is much easier to prove this formula than to devise it as a tool for proving the lemma). The identity shows that for $\alpha := \sup_{t\geq 0} \|T(t) - I_{\mathbb{X}}\|$ we have $2\alpha \leq \alpha + \alpha^2$, that is, $\alpha \leq \alpha^2$. Combining this with assumption, we obtain $\alpha = 0$. □

A generalization of Lemma 6.3 will be given in Exercise 14.3 in Chapter 14, where the notion of an integrated semigroup is introduced.

Proof of Theorem 6.1

Necessity We first show that an isolated point of $Semi_{b,sc}(\mathbb{X})$ is necessarily a scalar semigroup. It suffices to prove that if a semigroup $S = \{S(t)\}_{t\geq 0}$ is not scalar, then S is not an isolated point of $Semi_{b,sc}(\mathbb{X})$. Let $s > 0$ be such that $S(s)$ is not a scalar multiple of $I_{\mathbb{X}}$. Then, by Lemma 6.2, there exists $B \in \mathcal{L}(\mathbb{X})$ such that $BS(s) \neq S(s)B$. Let $(\epsilon_n)_{n\in\mathbb{N}}$ be a sequence in $(0, \|B\|^{-1})$ converging to 0, and let $I_n = I_{\mathbb{X}} + \epsilon_n B$ for every $n \in \mathbb{N}$. Then each operator I_n has a bounded inverse and:

$$S_n(t) = I_n^{-1} S(t) I_n \quad (t \geq 0,\ n \in \mathbb{N})$$

defines a sequence of bounded strongly continuous semigroups on $\mathbb{X}$. Moreover, $\lim_{n\to\infty} d(S_n, S) = 0$. As, for each $n \in \mathbb{N}$, $S_n(s) = S(s)$ holds if and only if $S(s)B = BS(s)$, we see that $S_n(s) \neq S(s)$ for all $n \in \mathbb{N}$. Thus S is not an isolated point of $Semi_{b,sc}(\mathbb{X})$.

We next show that if a semigroup $S_{[z]}$ is an isolated point of the space $Semi_{b,sc}(\mathbb{X})$, then necessarily $\Re z = 0$. This is the same as claiming that $S_{[-\lambda+ia]}$ is not an isolated point provided $\lambda > 0$. If $0 < \mu < \lambda$, then, as is easily checked, the function $t \mapsto e^{-\mu t} - e^{-\lambda t}$ on $[0, \infty)$ is non-negative and attains its maximum at $t_{\lambda,\mu} = (\ln \mu - \ln \lambda)/(\mu - \lambda)$. Because $\lim_{\mu \to \lambda} t_{\lambda,\mu} = \lambda^{-1}$, we see that, for any $a \in \mathbb{R}$, the expression:

$$\sup_{t \geq 0} \|S_{[-\mu+ia]}(t) - S_{[-\lambda+ia]}(t)\| = e^{-\mu t_{\lambda,\mu}} - e^{-\lambda t_{\lambda,\mu}}$$

converges to 0 as $\mu \to \lambda$, proving our claim.

Sufficiency. Finally, we show that, for each $a \in \mathbb{R}$, $S_{[ia]}$ is an isolated point of $Semi_{b,sc}(\mathbb{X})$. To this end, we note that if S is a semigroup in $Semi_{b,sc}(\mathbb{X})$ such that $\sup_{t \geq 0} \|S(t) - e^{iat} I_{\mathbb{X}}\| < 1$, then:

$$\sup_{t \geq 0} \|e^{-iat} S(t) - I_{\mathbb{X}}\| < 1$$

and this, by Lemma 6.3, implies that $e^{-iat} S(t) = I_{\mathbb{X}}$ for each $t \geq 0$ so that $S = S_{[ia]}$, with the immediate consequence that $S_{[ia]}$ is an isolated point of $Semi_{b,sc}(\mathbb{X})$.

Finally, we note that if the space $\mathbb{X}$ is real, then every scalar semigroup in $Semi_{b,sc}(\mathbb{X})$ is of the form $S_{[-\lambda]}$ for some $\lambda \geq 0$. This observation along with straightforward modifications of the proof given thus far establishes the result in the real case. □

As a by-product of the proof, we obtain the following corollary.

Corollary 6.4 *For any $a \in \mathbb{R}$, the open ball in $Semi_{b,sc}(\mathbb{X})$ with radius 1 and center at $S_{[ia]}$ defined by (6.1) is composed solely of its center.*

Exercise 6.1 Using the Weierstrass Formula, show that for a strongly continuous cosine family $\{C(t), t \in \mathbb{R}\}$, condition $\sup_{t \in \mathbb{R}} \|C(t) - I_{\mathbb{X}}\| < 1$ implies $C(t) = I_{\mathbb{X}}, t \in \mathbb{R}$. (See [63, 64, 89, 131, 308, 309] for more general results for cosines.)

Exercise 6.2

A) Let $\mathbb{X}$ be a normed algebra with unity e. Check that for any $x \in \mathbb{X}$:

$$x - e = \frac{x^n - e}{n} + \frac{x - e}{n} \sum_{i=1}^{n-1} (e - x^i).$$

B) Conclude that if $\alpha_n = \|e - x^n\|$ satisfies:

$$\lim_{n \to \infty} \frac{\alpha_n}{n} = 0 \qquad \text{and} \qquad \liminf_{n \to \infty} \frac{1}{n} \sum_{i=1}^{n-1} \alpha_i < 1,$$

then $x = e$. In particular, $x = e$ provided $\sup_{n\geq 1} \|x^n - e\| < 1$. (This result is due to L. J. Wallen [336], who generalized a result of Cox [96] concerning the case of square matrices. Cox's Theorem was extended to bounded operators on Hilbert space by Nakamura and Yoshida [271] and to an arbitrary normed algebra by Hirschfeld [181], and – as already mentioned – by Wallen [336]. Wils [348], Chernoff [82], Nagisa and Wada [270], and Kalton et al. [194] provided further generalizations.)

C) Deduce Lemma 6.3 from the result of Wallen.

Exercise 6.3 (H. Zwart, personal communication, see also [308].) Prove Lemma 6.3 by proceeding as follows.

A) Recall that if $\mathbb{X}$ is a Banach algebra with unit e, then for $a \in \mathbb{X}$, $\|a - e\| < 1$ implies that a is invertible and $\|a^{-1} - e\| \leq \frac{\|a-e\|}{1-\|a-e\|}$. (More generally, if a is invertible and $\|a - b\| < \frac{1}{\|a^{-1}\|}$, then so is b and $\|a^{-1} - b^{-1}\| \leq \frac{\|a^{-1}\|^2\|a-b\|}{1-\|a^{-1}\|\|a-b\|}$.)

B) Integrate to see that the assumption forces:

$$\sup_{\lambda>0} \|\lambda(\lambda - A)^{-1} - I_{\mathbb{X}}\| =: q < 1.$$

C) Combine A) and B) to conclude that A is bounded and:

$$\left\|\frac{A}{\lambda}\right\| = \left\|\frac{\lambda - A}{\lambda} - I_{\mathbb{X}}\right\| < \frac{q}{1-q}.$$

Since $\lambda > 0$ is arbitrary, this is possible only if $A = 0$.

☞ Chapter's summary

On a little detour from our main subject, we characterize bounded strongly continuous semigroups that can be approximated uniformly in the operator topology. It transpires that these are all bounded, strongly continuous semigroups save the "purely imaginary" scalar semigroups given by (6.1). In particular, if the underlying Banach space is real, it is only the identity semigroup $S(t) = I_{\mathbb{X}}, t \geq 0$ that cannot be approximated in the way described earlier. Interestingly, as W. Chojnacki has shown (see [63]) in separable Banach spaces, all bounded, strongly continuous semigroups can be approximated uniformly in the *strong* topology. For a unified, novel approach to convergence rates in approximation theory for operator semigroups, see [164].

7

Convergence of Resolvents

The situation in which, as in Theorem 2.2, from the very beginning we know the form of the limit semigroup $(e^{tA})_{t\geq 0}$ is not too common. It happens rather often that we are interested in convergence of a sequence of semigroups $(e^{tA_n})_{t\geq 0}$ satisfying (2.1), but have no clue as to what the limit semigroup might be. In particular, even if the limit semigroup exists, there is no reason to expect a priori that it is defined on the whole of the original Banach space.

In view of Theorem 2.2, a natural thing to do first in such cases is to check convergence of $(\lambda - A_n)^{-1}$. Alternatively, if the resolvents are not available in a manageable form, following the idea of M. Sova [320] and T. G. Kurtz [229] one should consider the "operator":

$$A_{ex}x = \lim A_n x_n, \tag{7.1}$$

where $x_n \in D(A_n)$ are such that $\lim_{n\to\infty} x_n = x$ and $\lim_{n\to\infty} A_n x_n$ exists; the domain of A_{ex} is composed of x for which at least one such sequence exists. In general, A_{ex}, termed the **extended limit** of A_n, $n \geq 0$, is not an operator in the usual sense, as the defining limit may depend on the choice of x_n. However, it is customary and convenient to think of A_{ex} as a "multivalued operator." The reader will check that A_{ex} is linear and closed (Exercise 7.2). In what follows, by $(x, y) \in A_{ex}$ or $A_{ex}x = y$ we mean that $x \in D(A_{ex})$ and y is one of possible values of $A_{ex}x$. Also, by the range of $\lambda - A_{ex}$, where $\lambda > 0$, we mean the set of $z \in \mathbb{X}$ such that there exists an $(x, y) \in A_{ex}$ with $z = \lambda x - y$.

The approaches via extended limit and via convergence of resolvents are equivalent in the sense of Proposition 7.2 (later on) but the latter has, of course, the advantage of allowing us to work without actually calculating resolvents. First, however, we need the following fundamental lemma. The reader will find that its proof is a generalization of the argument used in Theorem 2.2.

Lemma 7.1 *Fix $\lambda > 0$.*

a) Suppose $(x, y) \in A_{ex}$. Then $\lim_{n\to\infty}(\lambda - A_n)^{-1}(\lambda x - y)$ exists and equals x.
b) Conversely, if the limit $w = \lim_{n\to\infty}(\lambda - A_n)^{-1}z$ exists for some $z \in \mathbb{X}$, then $(w, \lambda w - z) \in A_{ex}$.

Proof $(x, y) \in A_{ex}$ iff there exist $x_n \in D(A_n)$ such that:

$$\lim_{n\to\infty}(x_n, A_n x_n) = (x, y),$$

or, equivalently, $\lim_{n\to\infty}(x_n, \lambda x_n - A_n x_n) = (x, \lambda x - y)$. The limits:

$$\lim_{n\to\infty} (\lambda - A_n)^{-1} (\lambda x - y)$$
$$\lim_{n\to\infty} (\lambda - A_n)^{-1} (\lambda x_n - y_n) = \lim_{n\to\infty} x_n$$

exist simultaneously, and are equal, for we have $\|(\lambda - A_n)^{-1}\| \le M\lambda^{-1}$. As the latter limit is x, a) is proved.

Conversely, if $w = \lim_{n\to\infty}(\lambda - A_n)^{-1}z$, then $w_n := (\lambda - A_n)^{-1}z \in D(A_n)$ and $\lim_{n\to\infty} A_n w_n = \lim_{n\to\infty}(\lambda(\lambda - A_n)^{-1}w_n - z) = \lambda w - z$, by the resolvent equation. □

Proposition 7.2 *Let $\lambda > 0$ be given. $(\lambda - A_n)^{-1}$ converge strongly iff the range of $\lambda - A_{ex}$ is dense in $\mathbb{X}$.*

Proof Sufficiency follows from Lemma 7.1 a) and density argument, because $\|(\lambda - A_n)^{-1}\| \le M\lambda^{-1}$. For necessity, we take:

$$w = \lim_{n\to\infty} (\lambda - A_n)^{-1} z, z \in \mathbb{X},$$

to see, by Lemma 7.1 b), that $w \in D(A_{ex})$ and $\lambda w - A_{ex}w = z$. □

Our next result tells us that it suffices to check convergence of the operators $(\lambda - A_n)^{-1}$ for a single $\lambda > 0$.

Proposition 7.3 *If the limit $\lim_{n\to\infty}(\lambda - A_n)^{-1}$ exists for a $\lambda > 0$, then it does for all $\lambda > 0$.*

Proof Suppose that the limit exists for a particular $\lambda > 0$. Since $(\lambda - A_n)(\lambda - A_n)^{-1}x = x, x \in \mathbb{X}$, we have:

$$(\mu - A_n)^{-1} = \sum_{i=1}^{\infty}(\lambda - A_n)^{-i}(\lambda - \mu)^{i-1}, \qquad |\mu - \lambda| < \lambda, \tag{7.2}$$

with the series converging because of the bound $\|(\lambda - A_n)^{-i}\| \le M\lambda^{-i}$. Because of that bound, we also obtain that $\lim_{n\to\infty}(\lambda - A_n)^{-i}$ exists for all $i \ge 1$. Hence, by the Lebesgue Dominated Convergence Theorem,

$\lim_{n\to\infty}(\mu - A_n)^{-1}$ exists for $\mu \in (0, 2\lambda)$. An easy induction argument shows now that the same is true for $\mu \in (0, 2^k\lambda)$, $k \geq 1$ and we are done. □

Assume now that the strong limit:

$$R_\lambda = \lim_{n\to\infty} (\lambda - A_n)^{-1} \tag{7.3}$$

exists or, equivalently, that:

$$\lambda - A_{\mathrm{ex}}$$

has a dense range for a single, hence for all $\lambda > 0$. (This condition is known as the **range condition**.) Then R_λ, $\lambda > 0$ satisfies the Hilbert Equation (1.2), since $(\lambda - A_n)^{-1}$, $\lambda > 0$, $n \geq 1$ do. We will say that R_λ, $\lambda > 0$ is a **pseudoresolvent**. It is easy to check that the range:

$$\mathcal{R} = Range\, R_\lambda$$

and the kernel:

$$\mathcal{K} = Ker R_\lambda$$

do not depend on $\lambda > 0$. The notation just introduced allows us to formulate the following corollary to Lemma 7.1.

Corollary 7.4 *We have:*

$$D(A_{\mathrm{ex}}) = \mathcal{R}.$$

Also, all possible values of A_{ex} on $x \in D(A_{\mathrm{ex}})$ are of the form $\lambda R_\lambda y - y$ where $x = R_\lambda y$. Equivalently, all possible values of A_{ex} on $R_\lambda y$ are of the form $\lambda R_\lambda y - y + z$ where $z \in \mathcal{K}$.

Proof The first part is immediate. For the rest we note that, if (x, z_1) and (x, z_2) belong to A_{ex} then so does $(0, z_1 - z_2)$ and, by Lemma 7.1 a), $z_1 - z_2 \in \mathcal{K}$. Because, by Lemma 7.1 b), $(R_\lambda y, \lambda R_\lambda y - y) \in A_{\mathrm{ex}}$, it follows that all possible values of $x = R_\lambda y$ are of the form $\lambda R_\lambda y - y + z$ where $z \in \mathcal{K}$. Then $x = R_\lambda(y - z)$ and $\lambda R_\lambda y - y + z = \lambda R_\lambda(y - z) - (y - z)$. □

The vital question is: does existence of the limit in (7.3) imply convergence of the related semigroups? As the following two examples show, the answer is in the negative.

Example 7.5 Let $(\mathrm{e}^{tA})_{t\geq 0}$, $\|\mathrm{e}^{tA}\| \leq M$, be a semigroup acting in a complex Banach space $\mathbb{X}_0$. Define $\mathbb{X}$ as the Cartesian product $\mathbb{X}_0 \times \mathbb{C}$ where $\mathbb{C}$ is the field of complex numbers and let, for $x \in \mathbb{X}_0$, $z \in \mathbb{C}$, $t \geq 0$, $n \geq 1$:

$$T_n(t)\binom{x}{z} = \binom{T(t)x}{e^{itn}z}.$$

For $\lambda > 0$, we have:

$$\lim_{n\to\infty} \int_0^\infty e^{-\lambda t} T_n(t)\binom{x}{z}\, dt = \lim_{n\to\infty} \begin{pmatrix} (\lambda - A)^{-1}x \\ \frac{1}{\lambda - ni} z \end{pmatrix} = \begin{pmatrix} (\lambda - A)^{-1}x \\ 0 \end{pmatrix},$$

even in the operator topology. However, $T_n(t)\binom{x}{z}$ does not converge, as $n \to \infty$, either in the strong or in the weak topology as long as $z \neq 0$. □

Example 7.6 Let $\mathbb{X} = C_0(0, \infty)$ be the space of continuous functions x with $\lim_{\tau\to 0} x(\tau) = \lim_{\tau\to\infty} x(\tau) = 0$. For $n \geq 1$, let:

$$T_n(t)x(\tau) = 1_{\mathbb{R}^+}(\tau - nt)x(\tau - nt), \qquad t \geq 0,$$

and let A_n be the generators of these semigroups. The set $\mathbb{Y} = \{x \in \mathbb{X} | \exists K(x) > 0 \text{ such that } \tau > K(x) \Rightarrow x(\tau) = 0\}$ is dense in $\mathbb{X}$. For $x \in \mathbb{Y}$,

$$\begin{aligned}(\lambda - A_n)^{-1}x(\tau) &= \int_0^\infty e^{-\lambda t} T_n(t)x(\tau)\, dt = \int_0^{\frac{\tau}{n}} e^{-\lambda t} x(\tau - nt)\, dt \\ &= \frac{1}{n}\int_0^\tau e^{-\frac{\lambda}{n}(\tau - \sigma)} x(\sigma)\, d\sigma \leq \frac{1}{n} K(x)\|x\|,\end{aligned}$$

which tends to 0 as $n \to \infty$. By the estimate $\|(\lambda - A_n)^{-1}\| \leq \frac{1}{\lambda}$, we have $\lim_{n\to\infty}(\lambda - A_n)^{-1}x = 0$, for all $x \in \mathbb{X}$. Thus, if the strong limit of $T_n(t)$ exists it is equal to 0, too. But, for all $n \geq 1$, $t > 0$, and $x \in \mathbb{X}$, $\|T_n(t)x\| = \|x\|$, a contradiction. □

Exercise 7.1 Show that $(x, 0) \in A_{ex}$ iff $\lim_{n\to\infty}(\lambda - A_n)^{-1}x$ exists and equals $\lambda^{-1}x$.

Exercise 7.2 Show that A_{ex} is linear and closed.

Exercise 7.3 Find A_{ex} in Examples 7.5 and 7.6.

Exercise 7.4 Let $c(\mathbb{X})$ denote the space of all convergent $\mathbb{X}$-valued sequences, with supremum norm. Define the operator $\mathcal{A}$ in $c(\mathbb{X})$ by $\mathcal{A}(x_n)_{n\geq 1} = (A_n x_n)_{n\geq 1}$ with domain:

$$D(\mathcal{A}) = \{(x_n)_{n\geq 1} \,|x_n \in D(A_n),\, (A_n x_n)_{n\geq 1} \in c(\mathbb{X})\}.$$

Also let $L : c(\mathbb{X}) \to \mathbb{X}$ be given by $L(x_n)_{n\geq 1} = \lim_{n\to\infty} x_n$. Show that x belongs to the domain of the extended limit of A_n, $n \geq 1$, iff there exists an

$(x_n)_{n\geq 1} \in \mathcal{D}(\mathcal{A})$ such that $L(x_n)_{n\geq 1} = x$. What are the possible values of A_{ex} on such an x?

Exercise 7.5 Let A be closed and B be bounded, and let $(b_n)_{n\geq 1}$ be a sequence of reals with $\lim_{n\to\infty} b_n = \infty$. Show that the domain of the extended limit of $A_n = A + b_n B$ is contained in the kernel of B.

Exercise 7.6 Show that the kernel and the range of a pseudoresolvent do not depend on λ.

Exercise 7.7 Assume that the kernel of a pseudoresolvent is trivial: $\mathcal{K} = \{0\}$. Introducing $A_\lambda = \lambda - R_\lambda^{-1}$ show that it is a closed operator and its definition does not depend on λ. Conclude that the pseudoresolvent turns out to be the resolvent of this operator.

Exercise 7.8 Assume R_λ, $\lambda > 0$ is a pseudoresolvent satisfying $\|R_\lambda^n\| \leq M\lambda^{-n}$, $n \geq 1$.

a) Using the Hilbert Equation show that $(0, \infty) \ni \lambda \to R_\lambda$ is a continuous function, and then that it is inifinitely many times differentiable with $\frac{\mathrm{d}^n R_\lambda}{\mathrm{d}\lambda^n} = (-1)^n n! R_\lambda^{n+1}$, $n \geq 0$.
b) Prove by induction the Taylor Formula with integral reminder:

$$f(t) = f(s) + f'(s)(t-s) + f''(s)\frac{(t-s)^2}{2!} + \cdots + \int_s^t \frac{(t-u)^{n-1}}{(n-1)!} f^{(n)}(u)\,\mathrm{d}u \tag{7.4}$$

where f is an n times continuously differentiable vector-valued function on, say, $(0, \infty)$. Using this show that, in analogy to (7.2):

$$R_\mu = \sum_{i=1}^{\infty} R_\lambda^i (\lambda - \mu)^{i-1}, \qquad |\mu - \lambda| < \lambda. \tag{7.5}$$

Exercise 7.9 Let $\mathcal{L}$ be an algebra over a commutative ring $\mathbb{K}$. A map $\lambda \mapsto R_\lambda \in \mathcal{L}$ of a nonempty subset $D(R)$ of $\mathbb{K}$ is called a pseudoresolvent if the Hilbert equality: $R_\lambda - R_\mu = (\mu - \lambda)R_\lambda R_\mu$ is satisfied for $\lambda, \mu \in D(R)$. A pseudoresolvent is said to be maximal if it does not possess a proper extension to a pseudoresolvent. Let $\to$ be a relation in $\mathbb{K} \times \mathcal{L}$ defined by:

$$(\lambda, A) \to (\mu, B) \text{ iff } A - B = (\mu - \lambda)AB.$$

1. Show that $\to$ is reflexive and transitive.
2. Show that $\leftrightarrow$ defined by $(\lambda, A) \leftrightarrow (\mu, B)$ if $(\lambda, A) \to (\mu, B)$ and $(\mu, B) \to (\lambda, A)$ is an equivalence relation.

3. Show that $(\lambda, A) \leftrightarrow (\lambda, B)$ implies $A = B$ and conclude that nonempty subsets of equivalence classes of $\leftrightarrow$ are (graphs of) pseudoresolvents.
4. Check to see that graphs of $\mathcal{L}$-valued pseudoresolvents are contained in equivalence classes of $\leftrightarrow$.
5. Check to see that a subset of $\mathbb{K} \times \mathcal{L}$ is an equivalence class of $\leftrightarrow$ iff it is a graph of a maximal $\mathcal{L}$-valued pseudoresolvent.
6. Conclude that every pseudoresolvent has a unique maximal extension. This result is due to J. Kisyński [217].

☞ Chapter's summary

We commence the study of the case where the limit semigroup is not a priori given. As a first step we introduce criteria of convergence of the resolvents in terms of the extended limit operator. The resolvents converge iff for some $\lambda > 0$ (and then for all $\lambda > 0$) the range of $\lambda - A_{ex}$ is dense in $\mathbb{X}$. We also note the first failure: the semigroups may diverge even though the resolvents converge.

8

(Regular) Convergence of Semigroups

Although the sole convergence of the resolvents (7.3) does not imply convergence of the semigroups, our case is not completely hopeless. For there is always a subspace $\mathbb{X}_0$ of $\mathbb{X}$ where the semigroups converge; in the second example of the previous chapter $\mathbb{X}_0 = \{0\}$. It turns out that using R_λ, $\lambda > 0$, or A_{ex} we may *characterize* $\mathbb{X}_0$.

To this end, let $\mathbb{X}_0$ be composed of those x for which:

$$T(t)x = \lim_{n\to\infty} \mathrm{e}^{tA_n}x \tag{8.1}$$

exists strongly and almost uniformly in $t \in \mathbb{R}^+$. $\mathbb{X}_0$ is termed the **regularity space** (of the limit pseudoresolvent, or of A_{ex}, or of the semigroups $(\mathrm{e}^{tA_n})_{t\geq 0}$), or the **Hille–Yosida space**. We check that $\mathbb{X}_0$ is a closed linear subspace and that the operators $T(t), t \geq 0$ leave $\mathbb{X}_0$ invariant – it suffices to check that $\lim_{n\to\infty} \mathrm{e}^{tA_n}T(s)x = T(t+s)x$, $x \in \mathbb{X}_0$ almost uniformly in $s \in \mathbb{R}^+$. $(T(t))_{t\geq 0}$ is a strongly continuous semigroup in $\mathbb{X}_0$, being almost uniform limit of semigroups with the same property.

Theorem 8.1 *Assume that the limit (7.3) exists. Then:*

$$\mathbb{X}_0 = cl\, D(A_{ex}) = cl\, \mathcal{R} \tag{8.2}$$

where 'cl' stands for the closure.

Proof In the proof of b) and d) $\Rightarrow$ a) in Theorem 2.2, we have seen in fact that $x \in D(A_{ex})$ implies $x \in \mathbb{X}_0$. This forces $cl\, D(A_{ex}) \subset \mathbb{X}_0$. To prove the converse, it suffices to show $D(A_0) \subset D(A_{ex})$, where A_0 is the generator of $(T(t))_{t\geq 0}$. Taking $x \in D(A_0)$ and $\lambda > 0$, however, we may find a $y \in \mathbb{X}_0$ such that $x = (\lambda - A_0)^{-1}y$. Then, $(\lambda - A_0)^{-1}y = \int_0^\infty \mathrm{e}^{-\lambda t}T(t)y\,\mathrm{d}t = \lim_{n\to\infty}\int_0^\infty \mathrm{e}^{-\lambda t}\mathrm{e}^{tA_n}y\,\mathrm{d}t = \lim_{n\to\infty}(\lambda - A_n)^{-1}y = R_\lambda y \in \mathcal{R} = D(A_{ex})$. □

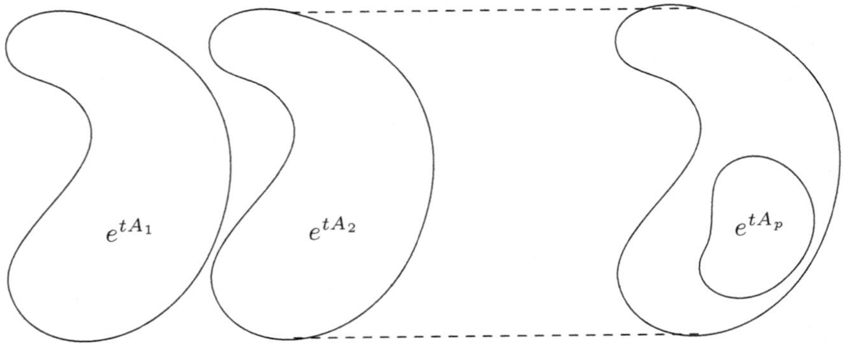

Figure 8.1 The limit semigroup

Corollary 8.2 *The generator of the limit semigroup equals A_p, the part of A_{ex} in $\mathbb{X}_0$, defined as $A_p = (A_{ex})_{|D(A_p)}$ where $D(A_p)$ is composed of $x \in D(A_{ex})$ such that one of the values of $A_{ex}x$ belongs to $\mathbb{X}_0$. In particular, A_p is single-valued.*

Proof By Corollary 7.4, for $x \in D(A_p)$ and $\lambda > 0$, there exists a y such that $x = R_\lambda y$ and $A_p x = \lambda R_\lambda y - y$ belongs to $\mathbb{X}_0$. The latter condition holds iff $y \in \mathbb{X}_0$. However, as we have seen in the proof of the theorem, for $y \in \mathbb{X}_0$, $(\lambda - A_0)^{-1}y = R_\lambda y$. Thus, $x \in D(A_0)$ and $A_p x = \lambda R_\lambda y - y = \lambda(\lambda - A_0)^{-1}y - y = A_0 x$. Hence, A_p is a restriction of A_0 (in particular, A_p is single-valued).

Conversely, if $x \in D(A_0)$ then for some $\lambda > 0$ and $y \in \mathbb{X}_0$, $x = (\lambda - A_0)^{-1}y = R_\lambda y \in D(A_{ex})$ and one of possible values of A_{ex} on x is $\lambda R_\lambda y - y \in \mathbb{X}_0$, completing the proof. □

Corollary 8.3 *If A is an operator in $\mathbb{X}_0$ such that $(\lambda - A)^{-1}$ exists for $\lambda > 0$ and coincides with R_λ on $\mathbb{X}_0$, then A is the generator of the limit semigroup.*

Proof By Corollary 8.2, $(\lambda - A)^{-1}x = R_\lambda x = (\lambda - A_p)^{-1}x$, $x \in \mathbb{X}_0$. Hence, the domains of A and A_p coincide and we have $A(\lambda - A)^{-1}x = \lambda(\lambda - A)^{-1}x - x = \lambda(\lambda - A_p)^{-1}x - x = A_p(\lambda - A_p)^{-1}x = A_p(\lambda - A)^{-1}x$. □

In the terminology introduced in Chapter 2, Theorem 8.1 together with Corollaries 7.4 and 8.3 constitutes the second Sova–Kurtz and Trotter–Kato Theorems, respectively.

We turn to an important example. Suppose that an A satisfies all the assumptions of the Hille–Yosida Theorem except for the one concerning density of its domain. (Such operators are often called **Hille–Yosida operators**; apparently, some of the first to study such operators were Da Prato and Sinestrari [291].)

In particular, $\lambda - A$ is invertible with bounded left and right inverse $(\lambda - A)^{-1}$ satisfying:

$$\|(\lambda - A)^{-n}\| \le M\lambda^{-n}. \tag{8.3}$$

For $\mu > 0$ we define $A_\mu = \mu^2(\mu - A)^{-1} - \mu$. Operators A_μ or their exponents are known as the **Yosida approximation**. We note that:

$$\|e^{tA_\mu}\| = \left\| e^{-t\mu} \sum_{n=0}^{\infty} \frac{\mu^{2n} t^n (\mu - A)^{-n}}{n!} \right\| \le e^{-\mu t} \sum_{n=0}^{\infty} M \frac{t^n \mu^n}{n!} = M,$$

so that $(e^{tA_\mu})_{t\ge 0}$ are equibounded. Also, the resolvent equation implies:

$$(\lambda - A_\mu)^{-1} = \frac{1}{\lambda + \mu} + \left(\frac{\mu}{\lambda+\mu}\right)^2 \left(\frac{\lambda\mu}{\lambda+\mu} - A\right)^{-1}, \mu, \lambda > 0 \tag{8.4}$$

so that $\lim_{\mu\to\infty}(\lambda - A_\mu)^{-1} = (\lambda - A)^{-1}$, $\lambda > 0$. Clearly, the regularity space here is the closure of the domain of $D(A)$ and by Corollary 7.4, $A = A_{ex}$. Hence, by Corollary 8.2, A_{p} generates a strongly continuous semigroup in $\mathbb{X}_0$, being the limit of the Yosida approximation.

This generalizes the Hille–Yosida Theorem and shows that Theorem 8.1 combined with Corollary 8.2 may be seen also as a generation theorem. The same remark applies to the following result.

Corollary 8.4 *Suppose that A is a densely defined operator such that some $\lambda > 0$ the range of $\lambda - A$ is dense in $\mathbb{X}$. If $A \subset A_{\mathrm{ex}}$ for the extended limit of generators of $(e^{tA_n})_{t\ge 0}$ satisfying (2.1), then A is closable and its closure $\overline{A}$ generates the semigroup:*

$$T(t)x = \lim_{n\to\infty} e^{tA_n} x, \qquad x \in \mathbb{X}, t \ge 0. \tag{8.5}$$

Proof By assumption, Proposition 7.2 and Theorem 8.1, the limit $R_\lambda = \lim_{n\to\infty}(\lambda - A_n)^{-1}$ exists, and the regularity space equals $\mathbb{X}$. Let B $(= A_{\mathrm{ex}})$ be the generator of the limit semigroup (8.5). Clearly B extends A, and since B is closed, A is closable. By Lemma 7.1, $R_\lambda(\lambda x - Ax) = x, x \in D(A)$, and since $D(A)$ is a core for $\overline{A}$ (i.e., for $x \in D(\overline{A})$ there are $x_n \in D(A)$ such that $\lim_{n\to\infty} x_n = x$ and $\lim_{n\to\infty} Ax_n = \overline{A}x$), we obtain $R_\lambda(\lambda x - \overline{A}x) = x, x \in D(\overline{A})$.

The proof will be complete once we show that the range of $\lambda - \overline{A}$ is $\mathbb{X}$. To see that, given $x \in D(B)$ and $\lambda > 0$ we find $y \in \mathbb{X}$ such that $x = R_\lambda y$. If the claim made above is true, we may find $x' \in D(\overline{A})$ such that $\lambda x' - \overline{A}x' = y$. Then $x = R_\lambda(\lambda x' - \overline{A}x') = x'$, showing that B is not a proper extension of $\overline{A}$.

Because the range of $\lambda - A$ is dense in $\mathbb{X}$, we are left with showing that the range of $\lambda - \overline{A}$ is closed. To this end, we argue as in Lemma 2.2 in [132]. Consider $x_n \in D(A_n)$ converging to $x \in D(A)$ in such a way that $\lim_{n\to\infty} A_n x_n = Ax$.

By (2.1), $M\|\lambda x_n - A_n x_n\| \ge \lambda\|x_n\|$. Thus $M\|\lambda x - Ax\| \ge \lambda\|x\|, x \in D(A)$ and, by a similar argument, $M\|\lambda x - \overline{A}x\| \ge \lambda\|x\|, x \in D(\overline{A})$. Therefore, if $z_n = \lambda w_n - \overline{A}w_n, n \ge 1$, where $w_n \in D(\overline{A})$, converges, then so does $w_n, n \ge 1$. This in turn implies existence of $\lim_{n\to\infty} \overline{A}w_n$, and then $z := \lim_{n\to\infty} z_n = w - \overline{A}w$ where $w = \lim_{n\to\infty} w_n$, $\overline{A}$ being closed. This establishes the claim. □

The final corollary of this chapter may be explained as follows. Because the convergence we obtain at the end concerns merely a subspace of $\mathbb{X}$, we really do not need convergence of the resolvents on the whole of $\mathbb{X}$, do we? Indeed, we don't, but then we need to strengthen the assumptions we make on the extended limit, for example (see [17] and Chapter 25 for more on this subject).

The following result is apparently in the spirit of the first Sova–Kurtz Theorem (see Chapter 2), because we assume the existence of the limit semigroup. However, its proof would be quite impossible without the notion of extended limit, and therefore the corollary fits in this chapter quite well. Note also that here we are not able to characterize $\mathbb{X}_0$: what we do is guess, or sometimes even *choose*, the subspace where convergence takes place (see Chapter 18). Compare [232], Theorem 1.10.

Corollary 8.5 *Suppose A is closable, and its closure $\overline{A}$ is the generator of a strongly continuous semigroup $(\mathrm{e}^{t\overline{A}})_{t\ge 0}$ in a subspace $\mathbb{X}_0$ of a Banach space $\mathbb{X}$, and A_n are the generators of equibounded semigroups $(\mathrm{e}^{tA_n})_{t\ge 0}$ in $\mathbb{X}$. Assume $A \subset A_{\mathrm{ex}}$. Then:*

$$\lim_{n\to\infty} \mathrm{e}^{tA_n}x_n = \mathrm{e}^{t\overline{A}}x, \tag{8.6}$$

provided $\lim_{n\to\infty} x_n = x \in \mathbb{X}_0$.

Proof Because A_{ex} is closed, $\overline{A} \subset A$. First we show that if $\lim_{n\to\infty} x_n = x \in \mathbb{X}_0$, then $\lim_{n\to\infty}(\lambda - A_n)^{-1}x_n = (\lambda - \overline{A})^{-1}x$. Let $y \in D(\overline{A})$ be such that $\lambda y - \overline{A}y = x$, and let $y_n \in D(A_n)$ be chosen so that $\lim_{n\to\infty} y_n = y$ and $\lim_{n\to\infty} A_n y_n = \overline{A}y$. Then $x_n' := \lambda y_n - A_n y_n, n \ge 1$ converge to x and $(\lambda - A_n)^{-1}x_n' = y_n, n \ge 1$ converge to y. Because the operators $(\lambda - A_n)^{-1}$, $n \in \mathbb{N}$ are equibounded, $\lim_{n\to\infty}(\lambda - A_n)^{-1}x_n = \lim_{n\to\infty}(\lambda - A_n)^{-1}x_n' = y = (\lambda - \overline{A})^{-1}x$.

Note that the proof presented here is an adaptation of the argument leading from d) to b) in Theorem 2.2. Similarly, arguing as in the proof of 'b) and d) implies a)' in Theorem 2.2, we obtain (8.6) for $x \in D(\overline{A})$, and then extend this result to all $x \in \mathbb{X}_0$ by density argument. □

This chapter makes it clear that the limit semigroup may in a natural way be defined on a subspace $\mathbb{X}_0$ of the space $\mathbb{X}$ where the approximating semigroups are given. See [251] for a natural biological example in which the limit semigroup acts in a space that is *larger* than $\mathbb{X}$.

Exercise 8.1 Show that the regularity space is closed.

Exercise 8.2 Check that if $\lim_{n\to\infty} x_n = x \in \mathbb{X}_0$, then $\lim_{n\to\infty} e^{tA_n} x_n = T(t)x$.

Exercise 8.3 Find the regularity space in Examples 7.5 and 7.6.

Exercise 8.4 Use the Hilbert Equation to show:

$$\mathbb{X}_0 = \{x \in \mathbb{X};\, x = \lim_{\lambda\to\infty} \lambda R_\lambda x\}.$$

Exercise 8.5 Show (8.4). *Hint:* compare (31.8).

Exercise 8.6 Fill out the details in the second part of the proof of Corollary 8.5.

Exercise 8.7 Let B be the generator of a C_0 semigroup $(S(t))_{t\ge 0}$ of equi-bounded operators in a Banach space $\mathbb{X}$, and assume that the limit:

$$\lim_{\lambda\to 0} \lambda \int_0^\infty e^{-\lambda t} S(t)x \,dt = Px, \tag{8.7}$$

exists for all $x \in \mathbb{X}$. Conclude that P is bounded and idempotent ($P^2 = P$), and *Range* P = *Ker* B (in particular: $BPx = 0$ for $x \in \mathbb{X}$). Moreover, $PBx = 0$ for $x \in D(B)$, $S(t)P = PS(t)$ for $t \ge 0$ and $\overline{\textit{Range}\, B} = \textit{Ker}\, P$.

Hint: Assumption (8.7) implies that the resolvents of the operators nB converge to the pseudoresolvent $R_\lambda = \frac{1}{\lambda}P$, $\lambda > 0$; this implies $P^2 = P$. The rest follows directly by (8.7) if we use the resolvent equation; check for instance that $\lim_{\lambda\to 0} B(\lambda - B)^{-1}x = x$, for $x \in$ *Ker* P.

☞ Chapter's summary

The failure of the previous chapter turns into a characterization theorem. Assuming convergence of the resolvents we are able to characterize (a) the subspace $\mathbb{X}_0$, termed the regularity space, where the semigroups converge almost uniformly in $t \in [0, \infty)$, and (b) the generator of the limit semigroup.

9

A Queue in Heavy Traffic

As an example of the application of Theorem 8.1, consider a **queue** in which service times are independent and exponential with parameter a, and times between customers' arrivals are independent and independent of service times, and exponentially distributed with parameter b (the so-called M/M/1 queue). If $b > a$ the queue grows, and if $b < a$ the queue decreases in size (this statement, though appealing, is quite incorrect: see for example Problem 10.25, p. 190 in [343] for a precise statement). The most interesting case is therefore the one where $a = b$.

Let us make some preparatory remarks for the appropriate analysis. Consider first the times where the size of the queue may change (at least potentially), that is, the times where a new customer arrives or (a potential) service is completed. These form a sequence of independent, exponentially distributed random variables with parameter $a + b$. This is because the minimum of two independent, exponentially distributed random variables is exponentially distributed with the parameter being the sum of the two parameters involved.

Also, using the joint distribution of two independent exponentially distributed random variables we see that the probability that a customer arrives before a potential service is completed is $\frac{b}{a+b}$. Hence, if the queue's length is x, then after service/arrival event the expected value of the weighted length is $\frac{a}{a+b}f(x-1) + \frac{b}{a+b}f(x+1)$, provided $x \geq 1$, where f is a weight function. If $x = 0$, the service is merely potential for there were no customers in the queue.

It will be convenient to take the weight functions from $C[0, \infty]$ and think also of noninteger, but non-negative lengths x. To avoid considering case $x < 1$

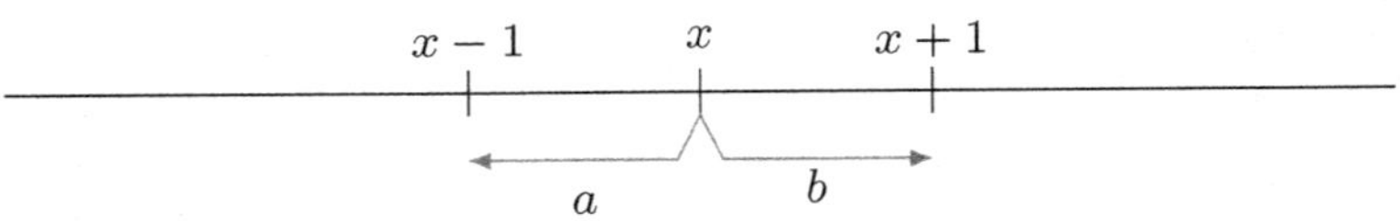

Figure 9.1 Queue's length

Figure 9.2 A queue in heavy traffic by Marek Bobrowski

separately, we extend f to an even function on $\mathbb{R}$ (denoted by the same letter) and introduce $B \in \mathcal{L}(C[0,\infty])$ by $Bf(x) = \frac{a}{a+b}f(x-1) + \frac{b}{a+b}f(x+1)$.

The consecutive powers of the contraction B describe the random process of the queue's length at (discrete) times of service/arrivals. The full process then is described by the contraction semigroup:

$$T(t) = \sum_{n=0}^{\infty} \mathrm{e}^{-(a+b)t} \frac{(a+b)^n t^n}{n!} B^n, \qquad t \geq 0. \tag{9.1}$$

Here, $\mathrm{e}^{-(a+b)t}\frac{(a+b)^n t^n}{n!}$ is the probability that n service/arrival events occurred up to time t, these events forming a Poisson process. In other words, $T(t) = \mathrm{e}^{tA_0}$ where $A_0 f(x) = (a+b)Bf(x) - (a+b)f(x) = a[f(x-1) - f(x)] + b[f(x+1) - f(x)]$.

To have some insight into what happens when $a = b$, we consider a sequence of scaled processes evolving in sped up time, describing **heavy traffic** [238, 295]. More specifically, we assume that:

- service's intensity equals a_n and arrival time's intensity is $b_n < a_n$,
- the limits $\lim_{n\to\infty} a_n = a > 0$ and $\lim_{n\to\infty} \sqrt{n}(a_n - b_n) = c > 0$ exist,
- service/arrival events change the queue's length by $\pm\frac{1}{\sqrt{n}}$, and
- time is sped up n times.

This gives a sequence of contraction semigroups $(e^{tA_n})_{t\geq 0}$ where

$$A_n f(x) = na_n \left[f\left(x - \frac{1}{\sqrt{n}}\right) - f(x) \right] + nb_n \left[f\left(x + \frac{1}{\sqrt{n}}\right) - f(x) \right], \quad n \geq 1.$$

Now, let A be the operator:

$$Af = -cf' + af, \tag{9.2}$$

defined on $D(A)$ composed of twice differentiable $f \in C[0,\infty]$ with $f'' \in C[0,\infty]$ and such that $f'(0) = 0$. The even extension of an $f \in D(A)$ is twice differentiable with the second derivative in $C[-\infty,\infty]$, and using the Taylor Formula (together with uniform continuity of members of the latter space) we check that:

$$\lim_{n\to\infty} A_n f = Af, \qquad f \in D(A). \tag{9.3}$$

Moreover, for $\lambda > 0$, the range of $\lambda - A$ is $C[0,\infty]$ (Exercise 9.2). Therefore, since A_{ex} extends A, the range of $(\lambda - A_{\mathrm{ex}})$ is $C[0,\infty]$ also, so that we are in the set-up of Theorem 8.1. The regularity space, containing the closure of $D(A)$, is $C[0,\infty]$. Also, by Corollary 8.4 and since A is closed, A generates the semigroup in $C[0,\infty]$ being the limit of $(e^{tA_n})_{t\geq 0}$.

This semigroup is related to reflected Brownian motion with drift coefficient $-c$, the information on reflection being hidden in the condition $f'(0) = 0$ imposed on the domain of the generator A. In other words, in heavy traffic, the scaled queue's length seen in the sped-up time performs a Brownian motion on $[0,\infty)$ with drift toward zero (a tendency to shorten the queue), being reflected immediately upon touching the barrier $x = 0$.

We note again that we have made much progress from Chapter 2, where we had to assume existence of the limit semigroup. Here, its existence and the form of its generator are obtained as by-products of approximation procedure.

Exercise 9.1 Check (9.3).

Exercise 9.2 For A introduced in (9.2), find the explicit, unique solution of the equation $\lambda f - Af = g$, $f \in D(A)$ where $g \in C[0,\infty]$ is given. Also, check that A is densely defined.

Exercise 9.3

(a) Let $A \in \mathcal{L}(C[0,\infty])$ be given by:

$$Af(x) = af(x+1) - af(x), \qquad x \geq 0,$$

where $a > 0$ is given. Check that:

$$e^{tA} f(x) = E\, f(x + N(t)), \qquad t \ge 0,$$

where $N(t)$ is a Poisson-distributed random variable: $\Pr(N(t) = n) = e^{-at} \frac{a^n t^n}{n!}$.

(b) The space $C[0, \infty]$ is isometrically isomorphic to $C[0, 1]$ with isomorphism $I : C[0, 1] \to C[0, \infty]$ given by $If(x) = f(e^{-x})$. Use (a) and Exercise 1.7 to check that:

$$e^{tB} f(x) = E\, f(xe^{-N(t)}), \qquad t \ge 0,$$

for $B \in \mathcal{L}(C[0, 1])$ given by $Bf(x) = af(e^{-1}x) - af(x)$.

☞ Chapter's summary

As a first application of Theorem 8.1, we study semigroups related to Markov chains describing the length of a simple Markovian queue with one server, and find their limit in heavy traffic. In this example, the regularity space is still the entire space, but the main difference between the analysis carried out in this chapter and those of Chapters 3 and 4 is that here the existence of the limit semigroup is obtained as a bonus, and didn't need to be assumed a priori.

10

Elastic Brownian Motions

We continue with illustrations of Theorem 8.1. Here, $\mathbb{X} = C[0, \infty]$ is the space of continuous functions $f : \mathbb{R}^+ \to \mathbb{R}$ with finite limits at infinity, equipped with the supremum norm. Given $\epsilon > 0$, we define the domain $D(A_\epsilon)$ of an operator A_ϵ as the set of twice continuously differentiable functions f with $f'' \in \mathbb{X}$ satisfying the **Robin boundary condition**: $f(0) = \epsilon f'(0)$, and set $A_\epsilon f = \frac{1}{2} f''$. From Chapter 3 we know that $2A_\epsilon$ is the generator of a Feller semigroup in $\mathbb{X}$. Therefore, so is A_ϵ and using (3.4)–(3.6) with $a = d = 0, b = \epsilon, c = 1$ we obtain:

$$(\lambda - A_\epsilon)^{-1} f(x) = \frac{1}{\sqrt{2\lambda}} \int_0^\infty \mathrm{e}^{-\sqrt{2\lambda}|x-y|} f(y)\,\mathrm{d}y + \frac{\epsilon\sqrt{2\lambda} - 1}{\epsilon\sqrt{2\lambda} + 1} \frac{1}{\sqrt{2\lambda}} \int_0^\infty \mathrm{e}^{-\sqrt{2\lambda}(y+x)} f(y)\,\mathrm{d}y. \tag{10.1}$$

The semigroups $(\mathrm{e}^{tA_\epsilon})_{t \ge 0}, \epsilon > 0$ are related to elastic Brownian motions and, as explained in Chapter 3, the parameter ϵ measures the degree in which the barrier $x = 0$ is "sticky." The smaller ϵ is, the less time the process spends at $x = 0$. This suggests that as $\epsilon \to 0$, the semigroups should converge to the one related to minimal Brownian motion, where the Brownian traveler disappears from the space immediately upon touching the boundary. The related semigroup may be constructed as follows; see [44].

Let A be the generator of the Brownian motion semigroup (2.2). Since the Gaussian distribution is symmetric, the semigroup commutes with the symmetry P given by $Pf(x) = f(-x)$. In particular, the semigroup leaves the subspace of odd functions invariant. Hence, as restricted to this subspace, it is a strongly continuous semigroup there, with the generator being the part of A in this subspace. On the other hand, the subspace of odd functions is isometrically isomorphic to $C_0(0, \infty] \subset \mathbb{X}$ where $f(0) = 0$. Hence, there exists the isometrically isomorphic image of the latter semigroup in $C_0(0, \infty]$, and its generator is $A_0 f = \frac{1}{2}f''$ on $D(A_0) = \{f \in C_0(0,\infty];\ f'' \in C_0(0,\infty]\}$. By (2.3) we also have:

$$\begin{aligned}
(\lambda - A_0)^{-1} f(x) &= \int_0^\infty e^{-\lambda t} e^{tA} \tilde{f}(x)\,dt = (\lambda - A)^{-1} \tilde{f}(x) \\
&= \frac{1}{\sqrt{2\lambda}} \int_0^\infty e^{-\sqrt{2\lambda}|x-y|} f(y)\,dy \qquad (10.2)\\
&\quad - \frac{1}{\sqrt{2\lambda}} \int_0^\infty e^{-\sqrt{2\lambda}(x+y)} f(y)\,dy,
\end{aligned}$$

where $\tilde{f}$ is the odd extension of f, and $\lambda > 0, x \geq 0$.

Turning to the convergence hypothesis we note that R_λ given formally by the same formula as $(\lambda - A_0)^{-1}$:

$$R_\lambda f(x) = \frac{1}{\sqrt{2\lambda}} \int_0^\infty e^{-\sqrt{2\lambda}|x-y|} f(y)\,dy - \frac{1}{\sqrt{2\lambda}} \int_0^\infty e^{-\sqrt{2\lambda}(x+y)} f(y)\,dy, \tag{10.3}$$

where $\lambda > 0, x \geq 0$, maps $C[0,\infty]$ into itself. Also, comparing (10.1) and (10.3):

$$(\lambda - A_\epsilon)^{-1} f(x) - R_\lambda f(x) = \frac{2\epsilon}{\epsilon\sqrt{2\lambda} + 1} \int_0^\infty e^{-\sqrt{2\lambda} y} f(y)\,dy\, e^{-\sqrt{2\lambda} x}.$$

Therefore, for every $f \in \mathbb{X}$ and $\lambda > 0$:

$$\begin{aligned}
\left\|(\lambda - A_\epsilon)^{-1} f - R_\lambda f\right\| &= \left|\frac{2\epsilon}{\epsilon\sqrt{2\lambda} + 1}\right| \sup_{x\geq 0} \left| \int_0^\infty e^{-\sqrt{2\lambda} y} f(y)\,dy\, e^{-\sqrt{2\lambda} x}\right| \\
&\leq \left|\frac{2\epsilon}{\epsilon\sqrt{2\lambda} + 1}\right| \frac{1}{\sqrt{2\lambda}} \|f\| \underset{\epsilon\to 0}{\longrightarrow} 0.
\end{aligned}$$

Hence, if $\lim_{n\to\infty} \epsilon_n = 0$, then $\lim_{n\to\infty} (\lambda - A_{\epsilon_n})^{-1} = R_\lambda$ and we are in the setup of Theorem 8.1. Because $R_\lambda f(0) = 0$, $\mathbb{X}_0 \subset C_0(0,\infty]$. By contrast, because $(\lambda - A_0)^{-1} = (R_\lambda)_{|C_0(0,\infty]}$ is the resolvent of the generator of a strongly continuous semigroup, the range of R_λ is dense in $C_0(0,\infty]$, showing that $\mathbb{X}_0 = C_0(0,\infty]$. By the same token, $(\lambda - A_0)^{-1}$ coincides with the resolvent of the limit semigroup in $\mathbb{X}_0$. Hence, by Corollary 8.3, A_0 is the generator of the limit semigroup.

Exercise 10.1 Let $f \in BUC(\mathbb{R})$ (the space of bounded, uniformly continuous functions on $\mathbb{R}$). Show that:

$$\lim_{\mu\to\infty} \mu \int_{-\infty}^\infty e^{-\mu|x-y|} f(y)\,dy = f(x)$$

uniformly in $x \in \mathbb{R}$.

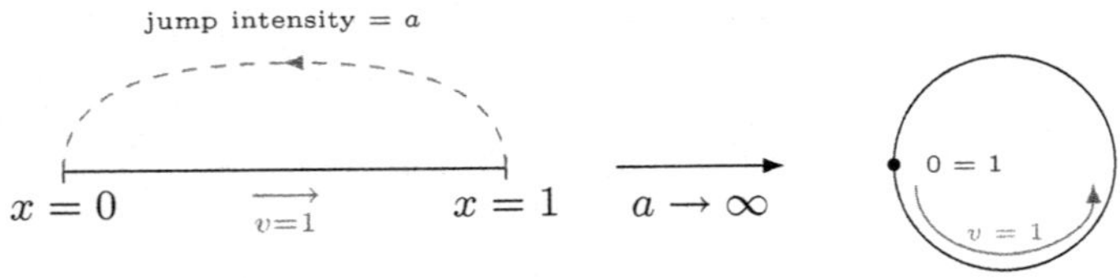

Figure 10.1 The approximating process and the limit process for Exercises 10.4–10.5

Exercise 10.2 (Compare [201, Example 3.10, p. 459].) For A_ϵ defined in this chapter, let $B_\epsilon = \epsilon A_\epsilon$. Show that:

$$\lim_{\epsilon\to 0} (\lambda - B_\epsilon)^{-1} f = \lambda^{-1} f, \qquad f \in C_0(0,\infty], \tag{10.4}$$

to conclude, by Corollary 8.5, that $\lim_{\epsilon\to 0} \mathrm{e}^{tB_\epsilon} f = f$, $f \in C_0(0,\infty]$. To this end, check that the resolvent of B_ϵ is given by the right-hand side of (10.1) with f replaced by $\epsilon^{-1} f$ and λ replaced by $\epsilon^{-1}\lambda$, and then check that (10.4) will follow once we show that:

$$\lim_{\mu\to\infty} \mu \int_0^\infty \mathrm{e}^{-\mu|x-y|} f(y)\,\mathrm{d}y - \mu \int_0^\infty \mathrm{e}^{-\mu(x+y)} f(y)\,\mathrm{d}y = f(x)$$

uniformly in $x \ge 0$; this can be deduced from Exercise 10.1.

Exercise 10.3 Show that in the previous exercise the limit of resolvents does not exist for $f \in C[0,\infty] \setminus C_0(0,\infty)$. (*Hint:* show that the limit does not exist for $f_0 \equiv 1$, and then write any $f \in C[0,\infty]$ as $f = f(0)f_0 + (f - f(0)f_0)$).

Exercise 10.4 For a sequence $(a_n)_{n\ge 1}$ of positive numbers converging to infinity, let A_n, $n \ge 1$ be the generators of semigroups of Exercise 3.6 with a replaced by a_n, and μ being the Dirac delta at $x = 0$, so that members of $D(A_n)$ satisfy $f'(1) = a_n(f(0) - f(1))$. Show that the semigroups $(\mathrm{e}^{tA_n})_{t\ge 0}$ converge to the semigroup describing the motion with constant velocity 1 on a circle of radius $\frac{1}{2\pi}$ (see Figure 10.1). To this end:

1. Show that $(\lambda - A_n)^{-1} g(x) = C_n \mathrm{e}^{\lambda x} + \int_x^1 \mathrm{e}^{\lambda(x-y)} g(y)\,\mathrm{d}y$, $x \in [0,1]$, $g \in C[0,1]$, where $C_n = \frac{g(1) + a_n \int_0^1 \mathrm{e}^{-\lambda y} g(y)\,\mathrm{d}y}{\lambda \mathrm{e}^{\lambda} + a_n(\mathrm{e}^{\lambda} - 1)}$, and find the limit pseudoresolvent.
2. Show that $\mathbb{X}_0 \subset \{f \in C[0,1] \mid f(0) = f(1)\}$, and that for $g \in \mathbb{X}_0$, the limit pseudoresolvent coincides with the resolvent of the semigroup:

$$T(t)f(x) = f(x \oplus_1 t), \qquad t \ge 0, x \in [0,1],$$

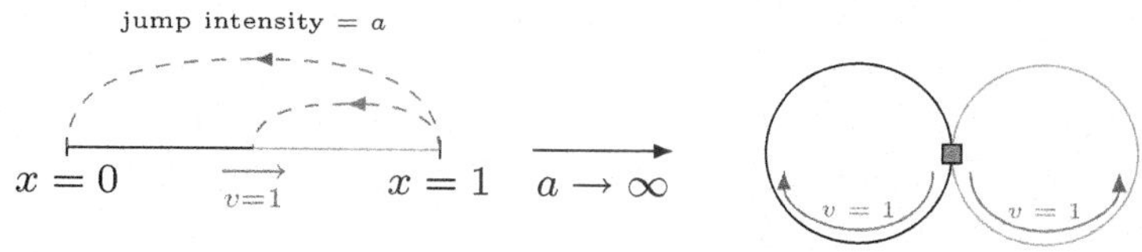

Figure 10.2 In the approximating processes considered in Exercise 10.6, a jump from $x = 1$ to $x = 0$ or $x = \frac{1}{2}$ occurs after an exponential time with probabilities p and $1 - p$, respectively. In the limit process $x = 1$ "connects immediately" to $x = 0$ or $x = \frac{1}{2}$ with the same probabilities. As a result the limit process may be described as follows: A toy train speeds on a railway composed of two adjacent circles connected at a junction marked with a small square (corresponding to amalgamated points $x = 0$, $x = \frac{1}{2}$, and $x = 1$; strictly speaking, however, $x = 0$ and $x = \frac{1}{2}$ are separate points, and $x = 1$ is not in the state-space; otherwise, the limit process is not Markov). When on the left-hand circle, the train moves clockwise and after completing the full circle continues its motion counterclockwise on the right-hand one. When it approaches the amalgamated points junction from the right, a random switch directs it either to the left-hand circle where it continues clockwise or to the right-hand one, where it continues counterclockwise.

of operators in $\{f \in C[0, 1] \mid f(0) = f(1)\}$, where $\oplus_1$ denotes addition modulo 1. To this end write the resolvent in the form:

$$\int_0^\infty e^{-\lambda t} g(x \oplus_1 t)\, dt = \int_x^1 e^{\lambda(x-y)} g(y)\, dy + \sum_{n=1}^{\infty} e^{-\lambda n} \int_0^1 e^{\lambda(x-y)} g(y)\, dy.$$

3. Conclude that $\mathbb{X}_0 = \{f \in C[0, 1] \mid f(0) = f(1)\}$ and that:

$$\lim_{n\to\infty} e^{tA_n} f = T(t)f, \qquad f \in \mathbb{X}_0.$$

Exercise 10.5 Redo Exercise 10.4 by characterizing the extended limit operator. (*Hint:* show first that its domain is contained in $C^1[0, 1] \cap \{f \in C[0, 1] \mid f(0) = f(1)\}$.) Deduce the same result directly from (3.13): to this end prove convergence for $t \in [0, 1]$ first, and then proceed by induction.

Exercise 10.6 Consider the semigroups of Exercise 3.6 with μ being a convex combination of Dirac measures at 0 and $\frac{1}{2}$; say $\mu = p\delta_0 + (1 - p)\delta_{1/2}$, where $p \in (0, 1)$. Show that as $a \to \infty$, the related semigroups converge to the semigroup describing the process in which the traveler moving with velocity 1 in the unit interval, after reaching $x = 1$ jumps swiftly to 0 or $\frac{1}{2}$ with probabilities p and $1 - p$, respectively. The limit semigroup is defined in the subspace of $f \in C[0, 1]$ satisfying $f(1) = pf(0) + (1 - p)f(1/2)$. Somewhat informally (see Figure 10.2), the limit process may be identified with the motion with unit velocity along the curve composed of two adjacent circles, of total length 1, obtained from the unit interval by joining points $x = 0$, $x = 1/2$, and $x = 1$.

Discuss an analogous result for the case where μ is a convex combination of Dirac measures at three points, and generalize immediately to $k \geq 1$ points. The state-space of the limit process should be a flower with k petals.

☞ Chapter's summary

Our second application of Theorem 8.1 involves convergence of elastic Brownian motion semigroups to the minimal Brownian motion semigroup. It is worth stressing that here, in contrast to the previous example, the regularity space $\mathbb{X}_0 \subsetneq \mathbb{X}$ is a proper subspace of $\mathbb{X}$.

11

Back to the Membrane

We come back to the semipermeable membrane of Chapter 4, but now instead of dealing with the dynamics of distributions we will study the dynamics of expected values of related stochastic processes (see [57]). Turning from $L^1(\mathbb{R}^+) \times L^1(\mathbb{R}^+)$ to its predual $C[0,\infty] \times C[0,\infty]$ we will be able to explain the role of transmission conditions and establish another convergence result.

Note that by considering $\mathbb{X} = C[0,\infty] \times C[0,\infty]$ we become more specific concerning the phase space of the stochastic process involved: instead of thinking of $\mathbb{R}$ we think of two copies of $\mathbb{R}^+$, with boundary $x = 0$ split into two points, say $0-$ and $0+$, representing positions to the immediate left and to the immediate right from the membrane. Alternatively, the state-space of the process is $S = [-\infty, 0-] \cup [0+, \infty]$ and $\mathbb{X}$ may be naturally identified with $C(S)$.

Given four positive numbers $\sigma_i, k_i, i \in \mathbb{I} = \{-1, 1\}$ we define A by:

$$A\,(f_i)_{i\in\mathbb{I}} = \left(\sigma_i^2 f_i''\right)_{i\in\mathbb{I}}, \tag{11.1}$$

Figure 11.1 Semipermeable membrane by Marek Bobrowski

with domain composed of $(f_i)_{i\in\mathbb{I}} \in C^2[0,\infty] \times C^2[0,\infty]$ satisfying:

$$f_i'(0) = k_i f_i(0) - k_i f_{-i}(0), \qquad i \in \mathbb{I}; \tag{11.2}$$

the $C^2[0,\infty]$ denoting twice continuously differentiable functions on $\mathbb{R}^+$ with second derivative in $C[0,\infty]$. As in Chapter 4, the numbers σ_i are diffusion coefficients in the left and right half-axis, and the k_i are membrane's permeability coefficients as we explain in more detail now. (See Exercise 11.2 for the relationship between A introduced here and the generator discussed in Chapter 4.)

Viewing a member of $\mathbb{X}$ as a continuous function on S, and writing condition (11.2) corresponding to $i = 1$ as:

$$-f'(0+) + k_1 f(0+) - k_1 \int_S f \, \mathrm{d}\delta_{0-} = 0,$$

where δ_{0-} is the Dirac measure at $0-$, we see that they have the form similar to (3.1), except that now the distribution μ of the point right after the jump from the boundary is a measure on S. Hence, according to the interpretation provided in Chapter 3, we may describe the process related to A as follows. (Stochastic analysis of the related *snapping out Brownian motion* may be found in [239]. As it transpires, analogues of transition conditions (11.2) are used in modelling flows between cells – see [5, 130, 148] cited after [239].) Conditional on starting in $[0+,\infty)$, the particles initially perform a Brownian motion in this half-axis with diffusion coefficient σ_1. Upon touching the boundary at $0+$ they are reflected but after a random time spent "at the boundary," they jump to $0-$ and start to perform a Brownian motion in $(-\infty, 0-]$, where the diffusion coefficient is σ_{-1}. The k_1 is the permeability coefficient in the sense that the larger is k_1, the quicker particles filter through the membrane from $0+$ to $0-$. (As in Chapter 3, the probability that the particle will still be performing the Brownian motion in $[0+,\infty)$ at time $t > 0$ is $\mathrm{e}^{-k_1 t^+(t)}$ where t^+ is the Lévy local time the particle spends at $0+$. Of course, in so saying we do not count the particles that jumped to $0-$ and later on returned.) A particle starting in $(-\infty, 0-]$ behaves similarly, except that the diffusion and permeability coefficients are different here. (See Figure 11.1.)

We have not yet proved that A generates a semigroup. Because density of its domain and the maximum principle may be shown as in Chapter 3 (see Exercise 11.1), we restrict ourselves to checking the range condition. Given $(g_i)_{i\in\mathbb{I}} \in \mathbb{X}$ and $\lambda > 0$, we want to find $(f_i)_{i\in\mathbb{I}} \in D(A)$ such that $\lambda(f_i)_{i\in\mathbb{I}} - A(f_i)_{i\in\mathbb{I}} = (g_i)_{i\in\mathbb{I}}$, that is:

$$\lambda f_i - \sigma_i^2 f_i'' = g_i, \qquad i \in \mathbb{I}.$$

As in Chapter 4, we look for $(f_i)_{i\in\mathbb{I}}$ of the form:

$$\begin{aligned} f_i(x) &= C_i e^{\frac{\sqrt{\lambda}}{\sigma_i}x} + D_i e^{-\frac{\sqrt{\lambda}}{\sigma_i}x} - \frac{1}{\sigma_i\sqrt{\lambda}}\int_0^x \sinh\frac{\sqrt{\lambda}}{\sigma_i}(x-y) g_i(y)\,\mathrm{d}y \\ &= \frac{1}{2\sigma_i\sqrt{\lambda}}\int_0^\infty e^{-\frac{\sqrt{\lambda}}{\sigma_i}|x-y|} g_i(y)\,\mathrm{d}y + D_i e^{-\frac{\sqrt{\lambda}}{\sigma_i}x}, \qquad x\geq 0, \end{aligned} \tag{11.3}$$

where $C_i := \frac{1}{2\sigma_i\sqrt{\lambda}}\int_0^\infty e^{-\frac{\sqrt{\lambda}}{\sigma_i}x} g_i(y)\,\mathrm{d}y$ and D_i are to be determined. Conditions (11.2) now impose $(\sigma_i\sqrt{\lambda}+k_i)D_i - k_iD_{-i} = (\sigma_i\sqrt{\lambda}-k_i)C_i + k_iC_{-i}$, $i\in\mathbb{I}$. This is satisfied iff:

$$D_i = \frac{\sqrt{\lambda}+k_{-i}\sigma_{-i}-k_i\sigma_i}{\sqrt{\lambda}+k_{-1}\sigma_{-1}+k_1\sigma_1}C_i + \frac{2k_i\sigma_i}{\sqrt{\lambda}+k_{-1}\sigma_{-1}+k_1\sigma_1}C_{-i},$$

completing the task.

Now, as in Chapter 4, we would like to see what happens when permeability of the membrane becomes infinite. Replacing k_i with nk_i in (11.2) and defining the related operators A_n, we check that the corresponding sequence $D_i = D_i(n)$ converges to:

$$D_i = \frac{k_{-i}\sigma_{-i}-k_i\sigma_i}{k_{-1}\sigma_{-1}+k_1\sigma_1}C_i + \frac{2k_i\sigma_i}{k_{-1}\sigma_{-1}+k_1\sigma_1}C_{-i}. \tag{11.4}$$

Therefore, the resolvents of A_n converge to the operator defined by (11.3) with D_i introduced above. However, we cannot apply Theorem 2.2 here. The reason is that f_i defined by (11.3) and D_i given here satisfy:

$$\begin{aligned} f_1(0) = C_1 + D_1 &= \frac{2k_{-1}\sigma_{-1}}{k_{-1}\sigma_{-1}+k_1\sigma_1}C_1 + \frac{2k_1\sigma_1}{k_{-1}\sigma_{-1}+k_1\sigma_1}C_{-1} \\ &= C_{-1}+D_{-1} = f_{-1}(0). \end{aligned}$$

It follows that the range of the limit R_λ of $(\lambda - A_n)^{-1}$ is not dense in $\mathbb{X}$ and cannot be the resolvent of a densely defined operator. It may be shown (Exercise 11.3) though, that R_λ is the resolvent of the operator A_∞ defined by (11.1) with (11.2) replaced by:

$$k_{-1}f_1'(0) + k_1 f_{-1}'(0) = 0, \quad f_1(0) = f_{-1}(0). \tag{11.5}$$

Notably, A_∞ is not densely defined.

We will show that the closure of the range of R_λ is the subspace $\mathbb{X}_0$ of $(f_i)_{i\in\mathbb{I}}\in\mathbb{X}$ satisfying $f_1(0) = f_{-1}(0)$, and then characterize the limit semigroup in this subspace. To this end first note that, by (11.3) and (11.4), the coordinates of

$R_\lambda (g_i)_{i\in\mathbb{I}}$ are:

$$f_i(x) = \frac{1}{2\sigma_i\sqrt{\lambda}} \int_{-\infty}^{\infty} \mathrm{e}^{-\frac{\sqrt{\lambda}}{\sigma_i}|x-y|} g_i^*(y)\,\mathrm{d}y, \qquad x \ge 0, \tag{11.6}$$

where

$$g_i^*(x) = \frac{k_{-i}\sigma_{-i} - k_i\sigma_i}{k_{-1}\sigma_{-1} + k_1\sigma_1} g_i(-x) + \frac{2k_i\sigma_i}{k_{-1}\sigma_{-1} + k_1\sigma_1} g_{-i}\left(-\frac{\sigma_{-i}}{\sigma_i}x\right), \qquad x < 0.$$

For $(g_i)_{i\in\mathbb{I}} \in \mathbb{X}_0$, g_i^* is a continuous function, in fact a member of $C[-\infty, \infty]$, the space of continuous functions on $\mathbb{R}$ with limits at plus and minus infinity.

The operators $B_i = \sigma_i^2 \frac{\mathrm{d}^2}{\mathrm{d}x^2}$ with domain $C^2[-\infty, \infty]$ generate strongly continuous semigroups in $C[-\infty, \infty]$, and their resolvents are (see (2.3)):

$$(\lambda - B_i)^{-1}\, g(x) = \frac{1}{2\sigma_i\sqrt{\lambda}} \int_{-\infty}^{\infty} \mathrm{e}^{-\frac{\sqrt{\lambda}}{\sigma_i}|x-y|} g(y)\,\mathrm{d}y, \qquad \lambda > 0, x \in \mathbb{R}. \tag{11.7}$$

In particular, $\lim_{\lambda\to\infty} \lambda(\lambda - B_i)^{-1} g = g$ in $C[-\infty, \infty]$. This implies that for $(g_i)_{i\in\mathbb{I}} \in \mathbb{X}_0$, λf_i, where f_i is defined by (11.6), converges in $C[0, \infty]$ to g_i and shows that the range of R_λ is dense in $\mathbb{X}_0$, as claimed.

Hence, in $\mathbb{X}_0$ there is a semigroup being the limit of the semigroups generated by A_n. Its generator is the part A_p of A_∞ in $\mathbb{X}_0$.

Exercise 11.1 Show that A given by (11.1) and (11.2) is densely defined and satisfies the maximum principle.

Exercise 11.2 Let (only in this exercise) the operator in $L^1(\mathbb{R})$ defined by (4.1) and (4.2) be denoted by A^*. Also, identify a member $(f_i)_{i\in\mathbb{I}}$ of $C[0, \infty] \times C[0, \infty]$ with the function on $\mathbb{R}$ defined by $f(x) = f_1(x), x \ge 0$ and $f(-x) = f_{-1}(x), x > 0$; note that f is right-continuous at 0, and has the left limit. With this identification, check that:

$$\int_{\mathbb{R}} f A^* \phi = \int_{\mathbb{R}} \phi A f$$

for all $\phi \in D(A^*)$ and $f \in D(A)$, where A is defined by (11.1) and (11.2).

Exercise 11.3 Show directly that the closure of the domain of A_∞ is $\mathbb{X}_0$, and that R_λ is the resolvent of A_∞.

Exercise 11.4 Check that $\mathbb{X}_0$ is isometrically isomorphic to $C[-\infty, \infty]$. Show directly that (the isomorphic image of) A_p is the generator of a conservative Feller semigroup in $C[-\infty, \infty]$.

☞ Chapter's summary

We study convergence of Feller semigroups that are in a sense 'dual' to those of Chapter 4. Interestingly, although the original Markov semigroups of Chapter 4 converge on the whole of $L^1(\mathbb{R}^+) \times L^1(\mathbb{R}^+)$ (where the approximating semigroups are defined), for the Feller semigroups the regularity space is a proper subspace of $C[0, \infty] \times C[0, \infty]$.

12

Telegraph with Small Parameter

Our next example is taken from [44]; for a different approach to the same singular perturbation, see [139, Chapter VI] and references given at the end of this chapter. Let A be the infinitesimal generator of a cosine operator function $\{C(t), t \in \mathbb{R}\}$ in a Banach space $\mathbb{X}$ such that for some $M > 0$:

$$\|C(t)\| \le M, \qquad t \in \mathbb{R}. \tag{12.1}$$

Given $\epsilon \in (0, 1)$, let us consider the abstract **telegraph equation with small parameter** in $\mathbb{X}$:

$$\epsilon^2 x''(t) + x'(t) = Ax(t), \qquad x(0) = x_0, x'(0) = x_1. \tag{12.2}$$

(In this equation, we take ϵ^2 instead the usual ϵ merely for the later notational convenience.) It is natural to expect (see [17, 44, 139] for selected bibliography on the subject) that as $\epsilon \to 0$, the solutions to this equation tend to those of:

$$x'(t) = Ax(t), \qquad x(0) = x_0. \tag{12.3}$$

This hypothesis may be stated as a result concerning convergence of semigroups. To this end, introducing $y(t) = \frac{\mathrm{d}x(t)}{\mathrm{d}t}$, the second order Cauchy problem (12.2), is transformed to that of order one in $\mathbb{B} = \mathbb{X}_{\mathrm{Kis}} \times \mathbb{X}$:

$$\frac{\mathrm{d}}{\mathrm{d}t}\begin{pmatrix} x(t) \\ y(t) \end{pmatrix} = \begin{pmatrix} y(t) \\ \epsilon^{-2}Ax(t) - \epsilon^{-2}y(t) \end{pmatrix} =: B_\epsilon \begin{pmatrix} x(t) \\ y(t) \end{pmatrix}, \quad \begin{pmatrix} x(0) \\ y(0) \end{pmatrix} = \begin{pmatrix} x_0 \\ x_1 \end{pmatrix}, \tag{12.4}$$

where $\mathbb{X}_{\mathrm{Kis}}$ is the related Kisyński space (see Chapter 1), x and y are $\mathbb{X}_{\mathrm{Kis}}$-and $\mathbb{X}$-valued functions, respectively, and the right-most equation defines an operator B_ϵ in $\mathbb{B}$ with domain $D(B_\epsilon) = D(A) \times \mathbb{X}_{\mathrm{Kis}}$. Because $\epsilon^{-2}A$ generates the cosine family $\{C(\epsilon^{-1}t), t \in \mathbb{R}\}$ in $\mathbb{X}$, by the Kisyński Theorem, the operator $\tilde{B}_\epsilon$ given by $\tilde{B}_\epsilon\binom{x}{y} = \binom{y}{\epsilon^{-2}Ax}$ on the domain $D(B_\epsilon)$ generates a group of operators in $\mathbb{B}$ and, by the Phillips Perturbation Theorem, the same is true of B_ϵ so that (12.4) is well-posed. However, the semigroups $(e^{tB_\epsilon})_{t\geq 0}$ are *not* equibounded. Hence, we consider the semigroups generated by A_ϵ with $D(A_\epsilon) = D(B_\epsilon)$ where:

$$e^{tA_\epsilon} = \begin{pmatrix} 1 & 0 \\ 0 & \epsilon \end{pmatrix} S_\epsilon(t) \begin{pmatrix} 1 & 0 \\ 0 & \epsilon^{-1} \end{pmatrix}, \quad A_\epsilon \begin{pmatrix} x \\ y \end{pmatrix} = \begin{pmatrix} \epsilon^{-1} y \\ \epsilon^{-1} Ax - \epsilon^{-2} y \end{pmatrix}, \qquad (12.5)$$

which turn out to be equibounded. (It should be noted that the proof of the fact that $\{T_\epsilon(t), t \geq 0\}$ are equibounded while $\{S_\epsilon(t), t \geq 0\}$ are not is nontrivial and requires either knowing probabilistic Kac-type formulae for solutions of (12.4), or intelligent positivity arguments – see [44, 49, 192, 212, 214].)

Let $(\epsilon_n)_{n\geq 1}$ be a sequence converging to zero. To settle the question of convergence of $(e^{tA_{\epsilon_n}})_{t\geq 0}$ we consider the part A_p of A in $\mathbb{X}_{\text{Kis}}$:

$$A_p x = Ax; \qquad D(A_p) = \{x \in D(A) | Ax \in \mathbb{X}_{\text{Kis}}\}.$$

Lemma 12.1 *Regardless of the choice of $(\epsilon_n)_{n\geq 1}$:*

$$D(A_{\text{ex}}) = D(A_p) \times \{0\}$$

and a vector in $\mathbb{B}$ is a value of A_{ex} on $\binom{x}{0}$ where $x \in D(A_p)$ iff it is of the form $\binom{Ax}{z}$, $z \in \mathbb{X}$.

Proof For $\binom{x}{y} \in D(A_{\text{ex}})$ there exist $\binom{x_n}{y_n} \in D(A) \times \mathbb{X}_{\text{Kis}}, n \geq 1$ such that $\lim_{n\to\infty} \binom{x_n}{y_n} = \binom{x}{y}$ and $\lim_{n\to\infty} \binom{\epsilon_n^{-1} y_n}{\epsilon_n^{-1} Ax_n - \epsilon_n^{-2} y_n} = \binom{w}{z}$ for some $w \in \mathbb{X}_{\text{Kis}}$ and $z \in \mathbb{X}$. Then $y = \lim_{n\to\infty} y_n = 0$ and $\lim_{n\to\infty} Ax_n = \lim_{n\to\infty} \epsilon_n^{-1} y_n = w$. By closedness of A, we have $x \in D(A)$ and $Ax = w \in \mathbb{X}_{\text{Kis}}$, i.e., $x \in D(A_p)$. Therefore, $D(A_{\text{ex}}) \subset D(A_p) \times \{0\}$ and the only possible values of A_{ex} on $\binom{x}{0}$ where $x \in D(A_p)$ are $\binom{Ax}{z}$, $z \in \mathbb{X}$.

Conversely, taking $x \in D(A_p)$ and $z \in \mathbb{X}_{\text{Kis}}$ we define:

$$\begin{pmatrix} x_n \\ y_n \end{pmatrix} := \begin{pmatrix} x \\ \epsilon_n Ax - \epsilon_n^2 z \end{pmatrix}$$

to see that $\lim_{n\to\infty} \binom{x_n}{y_n} = \binom{x}{0}$ while:

$$\lim_{n\to\infty} A_{\epsilon_n} \begin{pmatrix} x_n \\ y_n \end{pmatrix} = \lim_{n\to\infty} \begin{pmatrix} Ax - \epsilon_n z \\ z \end{pmatrix} = \begin{pmatrix} Ax \\ z \end{pmatrix}.$$

This shows that $\binom{x}{0} \in D(A_{\text{ex}})$ and $A_{\text{ex}}\binom{x}{0} = \binom{Ax}{z}$. Because $\mathbb{X}_{\text{Kis}}$ is dense in $\mathbb{X}$ and A_{ex} is closed, the result extends to any $z \in \mathbb{X}$. □

The lemma combined with properties of A_p characterizes the limit behavior of the semigroups involved. Namely, it is straightforward to check with the help of the Weierstrass Formula that $(e^{tA})_{t\ge0}$ leaves $\mathbb{X}_{\rm Kis}$ invariant and, as restricted to $\mathbb{X}_{\rm Kis}$ is a strongly continuous semigroup in the $\mathbb{X}_{\rm Kis}$ topology. Moreover, the generator of this restricted semigroup is A_p. In particular, the range of $\lambda - A_p$ is $\mathbb{X}_{\rm Kis}$ and so the range of $\lambda - A_{\rm ex}$ is $\mathbb{B}$ for all $\lambda > 0$. Hence, we are in the setup of Chapter 8. Moreover, $D(A_p)$ is dense in $\mathbb{X}_{\rm Kis}$ so that, by Lemma 12.1, the regularity space $\mathbb{B}_0$ equals $\mathbb{X}_{\rm Kis} \times \{0\}$. Finally, it is clear that the part of $A_{\rm ex}$ in the regularity space is the map $\binom{x}{0} \mapsto \binom{Ax}{0}$, $x \in D(A_p)$, the generator of the semigroup $T(t)\binom{x}{0} = \binom{e^{tA_p}}{0}$, $t \ge 0$ in $\mathbb{X}_{\rm Kis} \times \{0\}$. Hence, we obtain:

$$\lim_{\epsilon\to 0} e^{tA_\epsilon}\begin{pmatrix} x \\ 0 \end{pmatrix} = \begin{pmatrix} e^{tA_p}x \\ 0 \end{pmatrix}, \qquad x \in \mathbb{X}_{\rm Kis}. \tag{12.6}$$

It is instructive to prove that same result using the Trotter–Kato Theorem. To this end we solve the resolvent equation for A_ϵ to see that:

$$(\lambda - A_\epsilon)^{-1} = \begin{pmatrix} (\epsilon^2\lambda + 1)(\lambda^2\epsilon^2 + \lambda - A)^{-1} & \epsilon(\lambda^2\epsilon^2 + \lambda - A)^{-1} \\ A\epsilon(\lambda^2\epsilon^2 + \lambda - A)^{-1} & \epsilon^2(\lambda^2\epsilon^2 + \lambda - A)^{-1} \end{pmatrix},$$

for $\lambda > 0$. This implies:

$$\lim_{\epsilon\to 0} (\lambda - A_\epsilon)^{-1}\begin{pmatrix} x \\ y \end{pmatrix} = \begin{pmatrix} (\lambda - A)^{-1}x \\ 0 \end{pmatrix}, \qquad \lambda > 0, \begin{pmatrix} x \\ y \end{pmatrix} \in \mathbb{B}, \tag{12.7}$$

so that (12.6) follows, but a thorough analysis is more time consuming than that needed in the Sova–Kurtz approach.

Exercise 12.1 Fill out the details needed to show (12.7).

Exercise 12.2 Use (12.6) to show that the solution to (12.2) converges almost uniformly with the first derivative to that of (12.3) iff $x_0 \in D(A_p)$ and $x_1 = Ax_0$.

☞ Chapter's summary

Convergence of semigroups describing solutions of the abstract telegraph equation to those of the abstract diffusion equation is established. Here, again, the regularity space is a proper subspace of the underlying Banach space. This is a reflection of the fact that while the approximating semigroups are related to hyperbolic differential equations requiring two initial conditions, the limit semigroup is related to a parabolic equation requiring one initial condition. Hence, the convergence is uniform around $t = 0$ if the initial conditions are chosen appropriately (see Exercise 12.2). It should be noted here that convergence of solutions of the abstract telegraph equation to those of the abstract diffusion equation has of course been studied by many authors and in various contexts, not necessarily involving semigroups of operators or the Trotter–Kato Theorem. See, for example, [14, 17, 27, 85, 137, 138, 139, 140, 192, 210, 212, 243, 244, 258, 272, 281, 317, 321, 355], [49] pp. 358–361, [132] pp. 468–471, [288] pp. 1–6, and references given therein.

13

Minimal Markov Chains

In this chapter we show another application of Theorem 8.1. Let $\mathbb{I} \subset \mathbb{N}$ be a set of indexes and let l^1 be the space of absolutely summable sequences $(\xi_i)_{i\in\mathbb{I}}$ with the norm $\|(\xi_i)_{i\in\mathbb{I}}\| = \sum_{i\in\mathbb{I}} |\xi_i|$. Let F_1 be the (positive) functional on l^1 given by $F_1 x = \sum_{i\in\mathbb{I}} \xi_i$, $x = (\xi_i)_{i\in\mathbb{I}} \in l^1$. Non-negative sequences such that $F_1 x = 1$ are termed **densities** or **distributions**; the set of distributions is denoted by $\mathcal{D}$. As in Chapter 4, a bounded linear operator in l^1 is said to be a **Markov operator** if it leaves $\mathcal{D}$ invariant.

A matrix $Q = (q_{ij})_{i,j\in\mathbb{I}}$ is said to be a **Kolmogorov matrix** or an **intensity matrix** or a **Q-matrix** iff its off-diagonal entries q_{ij}, $i \neq j$ are non-negative and its rows add up to zero. If $\mathbb{I}$ is finite, Q may be identified with the bounded operator $l^1 \ni x \mapsto x \cdot Q = (\sum_{j\in\mathbb{I}} x_j q_{ji})_{i\in\mathbb{I}}$, where the dot denotes matrix multiplication. It may be shown that $(\mathrm{e}^{tQ})_{t\geq 0}$ is then a semigroup of Markov operators in l^1 (i.e., $x \cdot \mathrm{e}^{tQ}$ is a density provided x is); the converse statement is also true (see Exercise 13.2).

We would like to have a similar result in the case where $\mathbb{I}$ is infinite. (Without loss of generality we assume $\mathbb{I} = \mathbb{N}$ in what follows.) However, the situation here is not so simple; the argument used in the proof for finite $\mathbb{I}$ carries over only if $\sup_{n\geq 1}(-q_{n,n}) < \infty$. In fact, in general the operator $x = (\xi_n)_{n\geq 1} \mapsto xQ = (\sum_{j\geq 1} \xi_i q_{in})_{n\geq 1}$ defined on the maximal domain $D_{\max} = \{x \in l^1; xQ \in l^1\}$ is not a generator of a semigroup. $D_{\max}$ is simply too large and $\lambda - Q$ may not be one-to-one. Probabilistically, the situation is as follows. When $\mathbb{I}$ is finite, the semigroup $(\mathrm{e}^{tQ})_{t\geq 0}$ maps densities into densities, because Q determines a Markov chain with state-space $\mathbb{I}$. The chain, starting at $i \in \mathbb{I}$ waits there for an exponential time τ_1 with parameter $-q_{i,i}$ to jump to a state j with probability $\frac{q_{i,j}}{-q_{i,i}}$; if $q_{i,i} = 0$ the process stays at i forever. At j, the process starts afresh: waits for an (independent!) exponential time τ_2 to jump to some other state, and so on. Because the number of states is finite, the sum of all τ_n is infinite and the process is well defined for all $t > 0$ (see Exercise 13.4). For

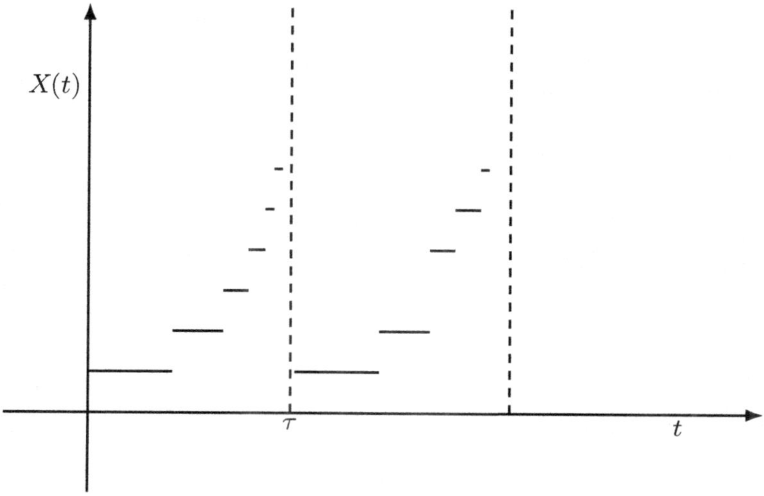

Figure 13.1 Explosion

infinite $\mathbb{I}$, generally Q determines the process only up to a certain (random) time.

For example, if $-q_{n,n} = q_{n,n+1} = a_n$ for some sequence a_n, $n \ge 1$, and the remaining entries of Q are zero (which is the intensity matrix of a **pure birth process**), then the process increases at exponential times τ_n with $E\tau_n = a_n^{-1}$. If a_n are chosen so that $\sum_{n=1}^{\infty} a_n^{-1} < \infty$, $\tau = \sum_{n\ge 1} \tau_n$ is finite a.s. Because after τ the process is left undefined by Q, it must be determined by additional rules such as, for example, starting afresh at 1 at time τ. This phenomenon is often called **explosion**. Such rules have their reflections in the domain of the generator; without restrictions on its domain, Q instead of being related to a single Markov chain, is related to a whole class of chains.

A theorem due to Kato [180, 197] to which this chapter is devoted describes the **minimal semigroup** related to the minimal Markov chain, that is, the chain where the process is undefined after τ so that the probability of observing this process at $t > 0$ is less than 1 (this is to say that the semigroup is **sub-Markov**). Following Banasiak and Lachowicz [24], we present an abstract form of this theorem, valid in Kantorovič–Banach spaces. The idea is to see the problem as a perturbation result: we introduce the operators A and B corresponding to the diagonal and off-diagonal parts of Q:

$$A\,(\xi_n)_{n\ge 1} = \left(q_{n,n}\xi_n\right)_{n\ge 1}, \qquad B\,(\xi_n)_{n\ge 1} = \left(\sum_{m\ne n} q_{m,n}\xi_m\right)_{n\ge 1} \tag{13.1}$$

with common domain $D = \{(\xi_n)_{n\ge1} \in l^1; \sum_{n\ge1} |q_{n,n}\xi_n| < \infty\}$. We note that $D \subset D_{\max}$ and that A generates a contraction semigroup $e^{tA}(\xi_n)_{n\ge1} = (e^{tq_{n,n}}\xi_n)_{n\ge1}$. A short calculation shows that:

$$\|Bx\| \le \|Ax\|, \quad x \in D. \tag{13.2}$$

If we had a stronger relation, say $\|Bx\| \le r\|Ax\|$ with $r \in (0, 1)$, we would be able to show that $A + B$ generates a contraction semigroup (see the proof of Theorem 13.3). Relation (13.2) allows for a more delicate result: there is an extension Q_0 of $A + B$ that generates a contraction semigroup, and later we will see that $(e^{tQ_0})_{t\ge0}$ is in a sense minimal.

The argument uses the fact that l^1 is a lattice, and that $-A$ and B are positive operators. To recall, a vector space is said to be **ordered** iff the following conditions are satisfied:

(a) $x \le x$,
(b) $x \le y$ and $y \le x$ imply $x = y$,
(c) $x \le y$ and $y \le z$ imply $x \le z$,
(d) $x \le y$ implies $x + z \le x + z$,
(e) $x \ge 0$ and $\alpha \ge 0$ imply $\alpha x \ge 0$,

where x, y, and z are arbitrary members of the space, and α is a real number.

An ordered vector space is said to be a **vector lattice** iff there exist minima and maxima of all two-element sets. The minimum and maximum of the subset $\{x, y\}$ of a vector lattice is denoted $x \wedge y$ and $x \vee y$, respectively. This allows defining, for all x in the lattice, $x^+ := x \vee 0$ and $x^- := (-x) \vee 0$, $|x| := x \vee (-x)$ and checking that (see Exercise 13.6):

$$x = x^+ - x^- \quad \text{and} \quad |x| = x^+ + x^-. \tag{13.3}$$

A vector lattice is said to be a **Banach lattice** if it is also a Banach space with a norm such that:

$$|x| \le |y| \Longrightarrow \|x\| \le \|y\|. \tag{13.4}$$

Finally, a Banach lattice is said to be a **KB-space** or a **Kantorovič–Banach space** if any sequence x_n, $n \ge 1$ of its elements with $0 \le x_n \le x_{n+1}$, $n \ge 1$ and $\|x_n\| \le M$, $n \ge 1$ for some $M > 0$, converges. l^1 is an example of a KB-space (Exercise 13.8).

An operator A in an ordered vector space is said to be positive if $Ax \ge 0$ for $x \ge 0$; we then write $A \ge 0$. For two operators A and B, $A \ge B$ means $A - B \ge 0$. It turns out that positive linear operators defined on the whole of a Banach lattice (with values in an ordered normed vector space) are necessarily bounded, [16] p. 54. We will, however, merely need the following two lemmas; before continuing, the reader should solve Exercises 13.5–13.7.

Lemma 13.1 *For a (bounded) positive operator A in a Banach lattice, $\|A\| = \sup_{x\geq 0, \|x\|=1} \|Ax\|$.*

Proof We need to show that $\sup_{x\geq 0, \|x\|=1} \|Ax\| \geq \sup_{\|x\|=1} \|Ax\|$. Let us consider an x with $\|x\| = 1$. Because $|x| \geq 0$ and $\| |x| \| = \|x\| = 1$, it suffices to show that $\|Ax\| \leq \|A|x|\,\|$. However, $-(x^-) \leq x \leq x^+$ implies $-A(x^-) \leq Ax \leq A(x^+)$. Hence $|Ax| = (Ax) \vee 0 + (-Ax) \vee 0 \leq (A(x^+)) \vee 0 + (A(x^-)) \vee 0 = A(x^+) + A(x^-) = A|x|$ and the result follows. □

Lemma 13.2 *A strongly continuous semigroup $(T(t))_{t\geq 0}$ in a Banach lattice $\mathbb{X}$ is positive (i.e., $T(t) \geq 0$ for all $t \geq 0$) iff its resolvent is positive for sufficiently large λ ($R_\lambda \geq 0$ for such λ).*

Proof We show first that the **positive cone** where $x \geq 0$ (or, equivalently, where $x^+ = x$) is closed in $\mathbb{X}$. To this end, it suffices to show that $|x^+ - y^+| \leq |x - y|$ because this implies that $\|x^+ - y^+\| \leq \|x - y\|$, for $x, y \in \mathbb{X}$. We have $0 \leq y^+ + |x - y|$ and, since $y \leq y^+$, $x \leq y^+ + x - y \leq y^+ + |x - y|$. Therefore, $x^+ \leq y^+ + |x - y|$. By symmetry, the claim follows.

By this, positivity of the semigroup implies positivity of the resolvent, the resolvent being the (Laplace transform) integral of the semigroup. The converse is true as the semigroup is the limit of appropriate Yosida approximation semigroups built from the resolvent. □

Here is our main theorem of this chapter.

Theorem 13.3 *Let $\mathbb{X}$ be a KB-space, and let A and B be linear operators in $\mathbb{X}$ with common domain D such that:*

(1) A generates a semigroup of positive contractions,
(2) $-A$ and B are positive,
(3) $\|Bx\| \leq \|Ax\|$ for $x \in D, x \geq 0$,
(4) for each nonzero $x \in D, x \geq 0$ there exists a bounded, positive functional F such that $\|F\| = 1$, $F(x) = \|x\|$ and $F(Ax + Bx) \leq 0$.

Then there exists an extension Q_0 of $A + B$ that generates a positive contraction semigroup in $\mathbb{X}$.

Proof Let us consider $C_r = A + rB$, $r \in [0, 1)$ with domain D. First, we show that each C_r generates a positive contraction semigroup.

For $x \geq 0$, as $-A$ and $(\lambda - A)^{-1}$ are positive, $0 \leq -A(\lambda - A)^{-1}x = x - \lambda(\lambda - A)^{-1}x \leq x$. Hence, $|-A(\lambda - A)^{-1}x| \leq x = |x|$ implying the estimate $\|A(\lambda - A)^{-1}x\| \leq \|x\|$. Hence, for $x \in D, x \geq 0$, $\|B(\lambda - A)^{-1}x\| \leq \|A(\lambda - A)^{-1}x\| \leq \|x\|$. By Lemma 13.1, the positive operator $B(\lambda - A)^{-1}$ is a contraction. Hence:

$$R(r, \lambda) = (\lambda - A)^{-1} \sum_{n=0}^{\infty} r^n \left[B(\lambda - A)^{-1}\right]^n, \quad r \in [0, 1), \lambda > 0 \qquad (13.5)$$

is well defined and a standard calculation shows $R(r, \lambda) = (\lambda - C_r)^{-1}$. Clearly, $R(r, \lambda) \geq 0$ but by the sole definition it is hard to see that $\|\lambda R(r, \lambda)\| \leq 1$.

The latter estimate may be obtained from condition (4). Indeed, taking non-negative $x \in D$ we have $\|\lambda x - C_r x\| \geq F(\lambda x - C_r x) = \lambda F(x) - F(Ax + Bx) + (1 - r)F(Bx) \geq \lambda F(x) = \lambda\|x\|$ where F is as in (4), because B and F are positive. Hence, for a positive $y \in \mathbb{X}$, taking $x = (\lambda - C_r)^{-1}y$ we obtain $\|y\| \geq \lambda\|(\lambda - C_r)^{-1}y\|$ and, by Lemma 13.1, $\|(\lambda - C_r)^{-1}\| \leq \frac{1}{\lambda}, \lambda > 0$. By the Hille–Yosida Theorem, C_r's generate contraction semigroups. Also, e^{tC_r} are positive operators, as so are $(\lambda - C_r)^{-1}$.

Let $r_n, n \geq 1$ be an increasing sequence with $\lim_{n\to\infty} r_n = 1$ and let $A_n = C_{r_n}$. Then:

$$(\lambda - A_n)^{-1} x \leq (\lambda - A_{n+1})^{-1} x \quad \text{and} \quad \|(\lambda - A_n)^{-1} x\| \leq \lambda^{-1}\|x\|$$

for all $x \geq 0$. Because $\mathbb{X}$ is a KB-space, $\lim_{n\to\infty}(\lambda - A_n)^{-1}x$ exists. Using decomposition (13.3) we check that the same is true for all $x \in \mathbb{X}$. Hence, we are in the setup of Theorem 8.1.

Moreover, for $x \in D$ we may define $x_n = x \in D(A_n), n \geq 1$ to see that $\lim_{n\to\infty} A_n x_n = \lim_{n\to\infty}(Ax + r_n Bx) = Ax + Bx$. Hence, the extended limit A_{ex} of A_n is an extension of $A + B$. In particular, as D is dense in $\mathbb{X}$, $\mathbb{X}_0 = \mathbb{X}$. Thus, A_{ex} is single valued and generates the limit semigroup. As we have seen, A_{ex} is an extension of $A + B$, completing the proof. □

Proposition 13.4 *If Q_1 is another extension of $A + B$ generating a positive contraction semigroup, then $\mathrm{e}^{tQ_0} \leq \mathrm{e}^{tQ_1}, t \geq 0$.*

Proof By assumption, $(\lambda - Q_1)^{-1}$ exists for sufficiently large λ and is positive. For $y \in D$, we may write:

$$(1 - r)By = (\lambda - A - rB)y - (\lambda - Q_1)y.$$

Taking $y = (\lambda - A - rB)^{-1}x, x \in \mathbb{X}$, and applying $(\lambda - Q_1)^{-1}$ to both sides of this equality:

$$(1 - r)(\lambda - Q_1)^{-1}B(\lambda - A - rB)^{-1}x = (\lambda - Q_1)^{-1}x - (\lambda - A - rB)^{-1}x.$$

Because all operators on the left-hand side are non-negative, for large $\lambda > 0$ we have $(\lambda - A - rB)^{-1} \le (\lambda - Q_1)^{-1}$. Because $(\lambda - A - rB)^{-1}$ converges to $(\lambda - Q_0)^{-1}$ as $r \to 1$, we have $(\lambda - Q_0)^{-1} \le (\lambda - Q_1)^{-1}$, and this implies the claim via Yosida approximation, as in Lemma 13.2. □

The assumptions of the abstract setup of Theorem 13.3 and Proposition 13.4 are clearly satisfied in the original problem related to intensity matrices. In fact, merely condition (4) needs to be checked (see also Exercise 13.10): in this case, the functional F is universal for all $x \ge 0$ and equals F_1 introduced at the beginning of this chapter.

Exercise 13.1 A matrix $(p_{ij})_{i,j\in\mathbb{I}}$ may be identified with an operator P in l^1 defined by:

$$P(\xi_i)_{i\in\mathbb{I}} = \left(\sum_{j\in\mathbb{I}} p_{ij}\xi_j\right)_{i\in\mathbb{I}}.$$

Show P is a Markov operator iff the coordinates in the matrix are non-negative, and in each column they add up to 1.

Exercise 13.2 Show that if $\mathbb{I}$ is finite and $Q = (q_{ij})_{i,j\in\mathbb{I}}$ is a matrix then $(\mathrm{e}^{tQ})_{t\ge 0}$ is composed of Markov operators in l^1 iff Q is an intensity matrix.

Exercise 13.3 If we exclude the trivial zero matrix, all 2×2 Kolmogorov matrices are of the form $Q = \begin{pmatrix} -a & a \\ b & -b \end{pmatrix}$ where $a, b \ge 0$, $a + b > 0$. Show that:

$$\mathrm{e}^{tQ} = \frac{1}{a+b}\begin{pmatrix} b + a\mathrm{e}^{-(a+b)t} & a - a\mathrm{e}^{-(a+b)t} \\ b - b\mathrm{e}^{-(a+b)t} & a + b\mathrm{e}^{-(a+b)t} \end{pmatrix}.$$

Exercise 13.4 Let τ_n be independent, exponential random variables with parameters a_n, $n \ge 1$. Show that $\sum_{n=1}^{\infty} \tau_n < \infty$ (a.s.) iff $\sum_{n=1}^{\infty} \frac{1}{a_n} < \infty$.

Exercise 13.5 Show that in a vector lattice $x \le y$ implies $-y \le -x$, $x^+ \le y^+$ and $y^- \le x^-$.

Exercise 13.6 Show (13.3). To this end, prove first:

$$x \vee y + z = (x + z) \vee (y + z), \quad x, y, z \in \mathbb{X}.$$

Exercise 13.7 Show that in a Banach lattice, $\| |x| \| = \|x\|$.

Exercise 13.8 Show that l^1 is a KB-space.

Exercise 13.9 A Banach lattice is said to have **order continuous norm** if any sequence $(x_n)_{n\ge 1}$ of its elements satisfying $0 \le x_{n+1} \le x_n$, $n \ge 1$, converges.

Show that any KB-space has order continuous norm and that c_0 has order continuous norm even though it is not a KB-space.

Exercise 13.10 Show that A defined in (13.1) generates the positive contraction semigroup $e^{tA}(\xi_n)_{n\geq 1} = (e^{tq_{n,n}}\xi_n)_{n\geq 1}$.

For a nicely written introduction to ordered vector spaces, see [16] and references given there. For more on positive semigroups, see [269], and for positive semigroups with applications in L^1 setting, see [234], see also [127]. Other applications may be found in the already cited [16].

☞ Chapter's summary

Approximating procedures are common tools for proving generation theorems: For example, the Hille–Yosida Theorem is often proved with the help of the Yosida approximation. Using an approximation procedure introduced by T. Kato in his proof of existence of a minimal Markov chain semigroup related to a Kolmogorov matrix, we follow J. Banasiak and M. Lachowicz to extend Kato's result to an abstract generation theorem valid in Kantorovič–Banach spaces.

14

Outside of the Regularity Space: A Bird's-Eye View

And as they were going to buy, the bridegroom came; and those who were ready went with him to the wedding feast. And the door was shut. And later the rest of the virgins came also, saying, Lord, lord, open to us! But he answered and said, Truly I say to you, I do not know you.

(Matt. 25:10–12. Holy Bible. Recovery Version.)

So far, we have proved that the necessary condition for convergence of (equi-bounded) semigroups is convergence of the related resolvents. Also, in terms of the limit pseudoresolvent and/or the extended limit operator, we were able to characterize the regularity subspace where the semigroups have a limit, and the generator of the limit semigroup. The natural question arises: what can we say about convergence outside of the regularity space?

Theorem 14.1 *Let* $(\mathrm{e}^{tA_n})_{t\geq 0}$ *be a sequence of semigroups satisfying* (2.1), *and suppose that* $(\lambda - A_n)^{-1}$ *converge. Then, for* $x \in \mathbb{X}$, *there exist the following limits*:

a) $\lim_{n\to\infty} \int_0^t \mathrm{e}^{sA_n} x \,\mathrm{d}s = U(t)x, t \geq 0,$

b) $\lim_{n\to\infty} H_n(\phi)x = H(\phi)x, \phi \in L^1(\mathbb{R}^+),$

where $H_n(\phi)x = \int_0^\infty \phi(s)\mathrm{e}^{tA_n} x \,\mathrm{d}s$ *is a Bochner integral, and* $L^1(\mathbb{R}^+)$ *is the space of (classes of) absolutely (Lebesgue) integrable functions.*

Proof Clearly, a) is a special case of b) for $\phi = 1_{[0,t]}$. Let $e_\lambda(t) = \mathrm{e}^{-\lambda t}, t \geq 0, \lambda > 0$. By assumption, the limits:

$$\lim_{n\to\infty} H_n(e_\lambda) = \lim_{n\to\infty} (\lambda - A_n)^{-1}, \lambda > 0$$

exist. Because $\|H_n(\phi)\| \leq M\|\phi\|_{L^1(\mathbb{R}^+)}$, the map $\phi \mapsto H_n(\phi)$ is linear, and the set of $e_\lambda, \lambda > 0$ is linearly dense in $L^1(\mathbb{R}^+)$, a three-epsilon argument shows that $H_n(\phi)x$ is a Cauchy sequence. □

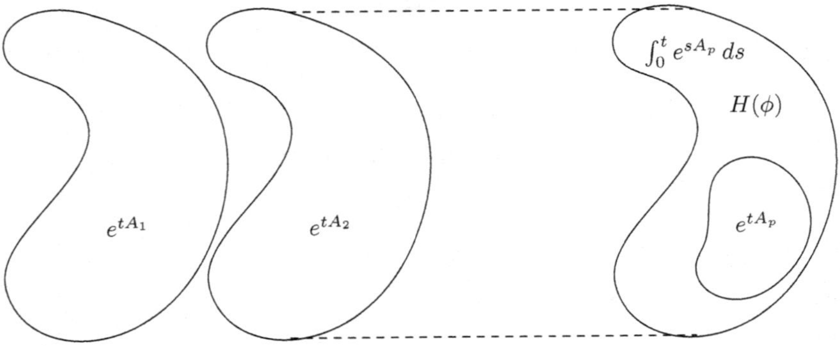

Figure 14.1 Convergence outside of $\mathbb{X}_0$

Condition a) is due to T. G. Kurtz [230]; condition b) is due to J. Kisyński. From the proof it is clear that *b*) is in fact equivalent to convergence of resolvents, and so is a) as the set of $u(t) = 1_{[0,t]}$ is linearly dense in $L^1(\mathbb{R}^+)$. The operator $L^1(\mathbb{R}^+) \ni \phi \mapsto H_n(\phi)$ is termed a **homomorphism** of the algebra $L^1(\mathbb{R}^+)$ related to the semigroup $(\mathrm{e}^{tA_n})_{t\geq 0}$. We recall that $L^1(\mathbb{R}^+)$ is a convolution algebra (see Appendix A); it is easy to check that $H_n(\phi * \psi) = H_n(\phi)H_n(\psi)$, where $*$ denotes convolution. Hence, $H(\phi * \psi) = H(\phi)H(\psi)$; H is a homomorphism related to the limit pseudoresolvent.

Example 14.2 Let us consider, as in Chapter 8, an operator A satisfying all the assumptions of the Hille–Yosida Theorem except for the one involving density of its domain. Then, for the Yosida approximation A_μ of A, the limit $\lim_{\mu\to\infty} \int_0^t \mathrm{e}^{tA_\mu s}\,\mathrm{d}s = U(t), t \geq 0$ exists. It is easy to see that $\lambda \int_0^\infty \mathrm{e}^{-\lambda t} U(t)\,\mathrm{d}t = (\lambda - A)^{-1}, \lambda > 0$. The family $(U(t))_{t\geq 0}$ is termed the **once integrated semigroup** related to A, and its existence was proved first by W. Arendt [6]. This family is Lipschitz continuous: $\|U(t) - U(s)\| \leq \limsup_{\mu\to\infty} \int_s^t \|\mathrm{e}^{uA_\mu}\|\,\mathrm{d}u \leq M(t-s), 0 \leq s \leq t$. Since for $x \in clD(A)$, $\lim_{\mu\to\infty} \mathrm{e}^{tA_\mu}x = T(t)x$, where $(T(t))_{t\geq 0}$ is the semigroup generated by A_p, we have $U(t) = \int_0^t T(s)x\,\mathrm{d}s, x \in \mathbb{X}_0, t \geq 0$. The limit $H(\phi) = \lim_{\mu\to\infty} \int_0^\infty \phi(t)\mathrm{e}^{tA_\mu}\,\mathrm{d}t$ is the homomorphism related to A.

To summarize, if the resolvents of a sequence of equibounded semigroups in a Banach space converge, then on the whole of this space the limit b) in Theorem 14.1 exists; additionally, for x from the regularity space, the semigroups converge, too. Extending the idea of Kisyński [211] (see also [230]), all these results may be seen from the perspective of a single generation theorem, as explained below [45].

Let $c(\mathbb{X})$ be the space of all convergent $\mathbb{X}$-valued sequences with supremum norm. Given a sequence of semigroups $(e^{tA_n})_{t\ge0}$ satisfying (2.1), we define the operator $\mathcal{A}$ in $c(\mathbb{X})$ by:

$$\mathcal{A}\,(x_n)_{n\ge1} = (A_n x_n)_{n\ge1}\,,$$
$$D(\mathcal{A}) = \{(x_n)_{n\ge1}\,|x_n \in D(A_n),\,(A_n x_n)_{n\ge1} \in c(\mathbb{X})\}. \tag{14.1}$$

The assumption that the limit $\lim_{n\to\infty}(\lambda - A_n)^{-1}$ exists allows defining the bounded linear operators $\mathcal{R}_\lambda, \lambda > 0$ in $c(\mathbb{X})$ by $\mathcal{R}_\lambda (x_n)_{n\ge1} = ((\lambda - A_n)^{-1}x_n)_{n\ge1}$. A straightforward computation shows:

$$\mathcal{R}_\lambda(\lambda - \mathcal{A})(x_n)_{n\ge1} = (x_n)_{n\ge1}, \qquad (x_n)_{n\ge1} \in D(\mathcal{A}) \text{ and}$$
$$(\lambda - \mathcal{A})\mathcal{R}_\lambda(x_n)_{n\ge1} = (x_n)_{n\ge1}, \qquad (x_n)_{n\ge1} \in c(\mathbb{X}),$$

so that $\mathcal{R}_\lambda, \lambda > 0$ is the resolvent of $\mathcal{A}$. Moreover, since $\|(\lambda - A_n)^{-n}\| \le M\lambda^{-n}$, we have $\|\mathcal{R}_\lambda^n\| \le M\lambda^{-n}$. Hence, $\mathcal{A}$ satisfies assumptions of Example 14.2 and its part in $cl\,D(\mathcal{A})$ generates a semigroup $(\mathcal{T}(t))_{t\ge0}$ there. It may be shown that this closure is the space of all sequences with limits in $\mathbb{X}_0$ and that $\mathcal{T}(t)(x_n)_{n\ge1} = (T_n(t)x_n)_{n\ge1}, (x_n)_{n\ge1} \in cl\,D(\mathcal{A})$. This is a counterpart of Theorem 8.1. Moreover, $\mathcal{A}$ generates an integrated semigroup $(\mathcal{U}(t))_{t\ge0}$; we check that $\mathcal{U}(t)(x_n)_{n\ge1} = (\int_0^t e^{sA_n}x_n\,ds)_{n\ge1}$. Finally, the homomorphism $\mathcal{H}$ related to $\mathcal{A}$ is given by $\mathcal{H}(\phi)(x_n)_{n\ge1} = (\int_0^\infty \phi(t)e^{tA_n}x_n\,dt)_{n\ge1}$. These are restatements of a) and b) in Theorem 14.1.

Exercise 14.1 Deduce from Lemma 2.1 that condition a) in Theorem 14.1 is equivalent to convergence of the resolvents. Conclude that a) implies b).

Exercise 14.2 Prove the claims made here concerning $\mathcal{A}$.

Exercise 14.3 A Banach algebra-valued function $[0,\infty) \ni t \mapsto p(t) \in \mathbb{A}$ is said to be an integrated semigroup if it is integrable on any finite interval:

$$p(t)p(s) = \int_t^{s+t} p(r)\,dr - \int_0^s p(r)\,dr, \qquad s,t \ge 0,$$

and $p(0) = 0$.

(a) Check that:

$$p_0(t) = 1_{[0,t)} \in L^1(\mathbb{R}^+), \tag{14.2}$$

is an example of an integrated semigroup.[1]

[1] As a result of the Kisyński Homomorphism Theorem (see for example [216] or one of [46, 47, 49, 88] or Chapter 62), there is a one-to-one correspondence between Lipschitz continuous integrated semigroups (i.e., integrated semigroups such that there is M such that:

$$\|p(t) - p(s)\|_{\mathbb{A}} \le M|t-s|,\, t,s \ge 0$$

(b) Let p be an integrated semigroup in a unital Banach algebra $\mathbb{A}$ with unit u. Show the following Wallen-type formula [336]: for $s > t$:

$$(s^{-1}p(s) - u)(t^{-1}p(t) - u)$$
$$= (st)^{-1}\int_s^{s+t} p(r)\,\mathrm{d}r - (st)^{-1}\int_0^t p(r)\,\mathrm{d}r - s^{-1}p(s) - t^{-1}p(t) + u.$$

(c) Suppose that for an integrated semigroup p the following conditions are satisfied:

1. $\liminf_{s\to\infty} \|s^{-1}p(s) - u\|_{\mathbb{A}} < 1$.
2. For each $t > 0$, $\lim_{s\to\infty} s^{-1}\|t^{-1}\int_s^{s+t} p(r)\,\mathrm{d}r - p(s)\|_{\mathbb{A}} = 0$.

Using (b) show that $p(t) = tu, t \geq 0$.

(d) Deduce Lemma 6.3 from (c).

☞ Chapter's summary

Under assumption of resolvents' convergence, the related semigroups converge strongly and almost uniformly on the regularity space equal to the closure of the range of the limit pseudoresolvent. As we know from the previous chapters, outside of the regularity space, the semigroups in general diverge. As it turns out, a weaker type convergence occurs outside of this space: the semigroups do not converge here but the related homomorphisms do. A nice, panoramic view of this fact may be seen from the high peak of the space of convergent sequences.

and homomorphisms of the convolution algebra $L^1(\mathbb{R}^+)$. In fact, each Lipschitz continuous integrated semigroup is of the form:

$$p = H \circ p_0,$$

where H is the corresponding homomorphism of $L^1(\mathbb{R}^+)$; the norm of H is the smallest Lipschitz constant for p.

15

Hasegawa's Condition

Assume, as we do throughout this book, that $(e^{tA_n})_{t\geq 0}$ are strongly continuous semigroups satisfying (2.1). Assume also that we may show that $D(A_{ex})$ is dense in $\mathbb{X}$. Unfortunately, this condition alone does not imply convergence of the semigroups. In fact, it does not even imply convergence of the resolvents. It is clear from Proposition 7.2 and from the examples of previous chapters that the main difficulty is establishing that the range of $\lambda - A_{ex}$, $\lambda > 0$ is the whole of $\mathbb{X}$. To repeat, this is called the **range condition**. Fortunately, sometimes we may avoid checking the range condition, having been able to check the alternative **condition of Hasegawa** [171]. The latter is:

$$\lim_{n,m\to\infty} \|e^{sA_n}e^{tA_m}x - e^{tA_m}e^{sA_n}x\| = 0, \quad x \in \mathbb{X}, \tag{15.1}$$

almost uniformly in $s, t \geq 0$, and its weaker version in terms of resolvents is (see Exercise 15.1):

$$\lim_{n,m\to\infty} \| (\lambda - A_n)^{-1} (\lambda - A_m)^{-1}x - (\lambda - A_m)^{-1} (\lambda - A_n)^{-1} x\| = 0, \tag{15.2}$$

for $\lambda > 0, x \in \mathbb{X}$.

Theorem 15.1 (Hasegawa) *The semigroups $(e^{tA_n})_{t\geq 0}$ converge on $\mathbb{X}$ iff the set $D(A_{ex})$ is dense in $\mathbb{X}$, and condition* (15.2) *holds.*

Proof Necessity of $cl\, D(A_{ex}) = \mathbb{X}$ is known from Theorem 8.1. Also, assuming that $T(t)x = \lim_{n\to\infty} e^{tA_n}x$ exists for all x and using:

$$\begin{aligned}&\|e^{sA_n}e^{tA_m}x - T(s+t)x\| \\ &\qquad \leq M\|e^{tA_m}x - T(t)x\| + \|e^{sA_n}T(t)x - T(s)T(t)x\|\end{aligned}$$

we obtain $\lim_{n,m\to\infty} e^{sA_n}\, e^{tA_m}x = T(s+t)x$. This limit is almost uniform in $s, t \in [0, \infty)$ because, for any $t_0 > 0$, the range of $[0, t_0] \ni t \mapsto T(t)x \in \mathbb{X}$ is compact. By symmetry, this implies necessity of (15.1).

Hence, it is sufficiency that is of interest here: by Theorem 8.1 we need to show that $\lim_{n\to\infty}(\lambda - A_n)^{-1}$ exists. We follow the argument of Kurtz [230]. Let $l^\infty(\mathbb{X})$ be the space of bounded $\mathbb{X}$-valued sequences equipped with supremum norm, and let the operators $\mathcal{R}_\lambda, \lambda > 0$ in $l^\infty(\mathbb{X})$ be defined by $\mathcal{R}_\lambda(x_n)_{n\geq 1} = ((\lambda - A_n)^{-1}x_n)_{n\geq 1}$; we need to show that $\mathcal{R}_\lambda, \lambda > 0$ map $c(\mathbb{X})$ into itself.

Assumption on $D(A_{\mathrm{ex}})$ combined with density of all $D(A_n)$ implies that $D(\mathcal{A})$ defined in (14.1) is dense in $c(\mathbb{X})$. Because $\|\mathcal{R}_\lambda\| \leq M\lambda^{-1}, \lambda > 0$, our task reduces to showing that $\mathcal{R}_\lambda$ maps $D(\mathcal{A})$ into $c(\mathbb{X})$.

If $(x_n)_{n\geq 1} \in c(\mathbb{X})$, then for any permutation $\pi_n, n \geq 1$ of positive integers $\lim_{n\to\infty}(x_n - x_{\pi_n}) = 0$. Conversely, the last condition implies that $(x_n)_{n\geq 1} \in l^\infty(\mathbb{X})$ is Cauchy, hence convergent. Such a permutation $\pi_n, n \geq 1$ induces a bounded linear operator $\mathcal{P}$ on $l^\infty(\mathbb{X})$ given by $\mathcal{P}(x_n)_{n\geq 1} = (x_{\pi_n})_{n\geq 1}$ with bounded inverse $\mathcal{P}^{-1}$ related to the inverse permutation. Hence, introducing $c_0(\mathbb{X})$ as the subspace of $\xi = (x_n)_{n\geq 1} \in c(\mathbb{X})$ with $\lim_{n\to\infty} x_n = 0$, we have that $\xi \in c(\mathbb{X})$ iff $\mathcal{P}\xi - \xi \in c_0(\mathbb{X})$ for all operators $\mathcal{P}$ induced by permutations, and our task reduces to showing:

$$\mathcal{P}\mathcal{R}_\lambda - \mathcal{R}_\lambda : D(\mathcal{A}) \to c_0(\mathbb{X}),$$

for all such operators. We note that in this context (15.2) can be stated as

$$\mathcal{R}_\lambda\mathcal{P}^{-1}\mathcal{R}_\lambda\mathcal{P}\xi - \mathcal{P}^{-1}\mathcal{R}_\lambda\mathcal{P}\mathcal{R}_\lambda\xi \in c_0(\mathbb{X}) \quad \text{for all} \quad \xi \in c(\mathbb{X}). \tag{15.3}$$

For $\xi \in D(\mathcal{A})$, $\lambda > 0$ and a permutation induced $\mathcal{P}$:

$$\begin{aligned}\mathcal{P}\mathcal{R}_\lambda\xi - \mathcal{R}_\lambda\xi &= \mathcal{P}\mathcal{R}_\lambda(I - \mathcal{P}^{-1})\xi + \mathcal{R}_\lambda(\mathcal{P} - I)\xi \\ &\quad + \mathcal{P}\mathcal{R}_\lambda\mathcal{P}^{-1}\mathcal{R}_\lambda(\lambda - \mathcal{A})\xi - \mathcal{R}_\lambda\mathcal{P}\mathcal{R}_\lambda\mathcal{P}^{-1}(\lambda - \mathcal{A})\xi \\ &\quad - \mathcal{R}_\lambda\mathcal{P}\mathcal{R}_\lambda(I - \mathcal{P}^{-1})(\lambda - \mathcal{A})\xi.\end{aligned}$$

Here, as $\xi \in c(\mathbb{X})$, $(\mathcal{P} - I)\xi$ belongs to $c_0(\mathbb{X})$ and so does $\mathcal{R}_\lambda(\mathcal{P} - I)\xi$ because $\mathcal{R}_\lambda$ maps $c_0(\mathbb{X})$ into itself. Because $\mathcal{P}$ has the same property, both terms in the first line above belong to $c_0(\mathbb{X})$. Similarly, the third line belongs to $c_0(\mathbb{X})$ because $\lambda\xi - \mathcal{A}\xi$ is a member of $c(\mathbb{X})$. Because the second line belongs to $c_0(x)$ by assumption (15.3), we are done. □

As an application, let us consider the question of existence of an infinite product of commuting contraction semigroups. This question was first discussed in Arendt et al. [10], a motivating example being the heat semigroup in infinite dimensions introduced by Cannarsa and Da Prato [76]. Let $B_n, n \geq 1$ be generators of *commuting* contraction semigroups:

$$\mathrm{e}^{tB_n}\mathrm{e}^{sB_k} = \mathrm{e}^{sB_k}\mathrm{e}^{tB_n}, \qquad s, t \geq 0, k, n \geq 1.$$

Then $T_n(t) = \prod_{k=1}^n e^{tB_k}$ is a strongly continuous contraction semigroup with the generator $A_n = \overline{\sum_{k=1}^n B_k}$, the closure of $\sum_{k=1}^n B_k$ defined on $\bigcap_{k=1}^n D(B_k)$ (see, e.g., [269] A-1.3.8). We say that the infinite product $T(t) = \prod_{k=1}^\infty e^{tB_k}$ exists iff the semigroups $(T_n(t))_{t\geq 0}$ converge.

The following theorem from [10] was proved there in a different way, the authors being apparently unaware of Hasegawa's result.

Theorem 15.2 *Define an operator A by $Ax = \sum_{k=1}^\infty B_k x$ with domain:*

$$D(A) = \left\{x \in \bigcap_{k=1}^{\infty} D(B_k);\ \sum_{k=1}^{\infty} \|B_k x\| < \infty\right\}$$

and suppose $D(A)$ is dense in $\mathbb{X}$. Then, the infinite product of $(e^{tB_n})_{t\geq 0}$ exists and its generator is the closure of A.

Proof Clearly, A is a restriction of A_{ex} and the assumption implies that $D(A_{ex})$ is dense in $\mathbb{X}$. Because $(e^{tA_n})_{t\geq 0}$ commute, Hasegawa's condition is satisfied and the infinite product exists by Theorem 15.1. Also, the generator of the product, equaling A_{ex}, is an extension of A. Hence, we are left with proving that $D(A)$ is a core for A_{ex}. To this end, by Exercise 15.2, it suffices to show that $D(A)$ is left invariant by $T(t) = \prod_{k=1}^\infty e^{tB_k}$.

Fix $k \in \mathbb{N}$ and $x \in D(B_k)$. For any $n \in \mathbb{N}$ and $t \geq 0$, $T_n(t)x$ belongs to $D(B_k)$ and $B_k T_n(t)x = T_n(t)B_k x$ as $(e^{tB_k})_{t\geq 0}$ commutes with $(T_n(t))_{t\geq 0}$. Closedness of B_k implies now $T(t)x \in D(B_k)$ and $B_k T(t)x = T(t)B_k x$. Thus, for $x \in D(A)$, $T(t)x$ belongs to $\bigcap_{k=1}^\infty D(B_k)$ and $\sum_{k=1}^\infty \|B_k T(t)x\| \leq \sum_{k=1}^\infty \|B_k x\| < \infty$, proving that $T(t)x \in D(A)$. This completes the proof. □

As our second application we establish existence of the **boundary value of a holomorphic semigroup**. To recall, a closed operator A in a *complex* Banach space is said to be **sectorial** of angle $\delta_0 \in (0, \frac{\pi}{2}]$ if the sector:

$$\Sigma_{\frac{\pi}{2}+\delta_0} = \left\{\lambda \in \mathbb{C} \setminus \{0\};\ |\arg \lambda| < \frac{\pi}{2} + \delta_0\right\}$$

is contained in the resolvent set $\rho(A)$ of A, and for all $\delta \in (0, \delta_0)$ there exists $M_\delta > 0$ such that:

$$\|(\lambda - A)^{-1}\| \leq \frac{M_\delta}{|\lambda|}, \qquad \lambda \in \Sigma_{\frac{\pi}{2}+\delta}. \tag{15.4}$$

If A is additionally densely defined (see [9] pp. 157–158, [128] p. 96, [201] p. 488, or [353] p. 257), it generates a strongly continuous semigroup $(e^{tA})_{t\geq 0}$ that may be extended to the complex t in Σ_{δ_0}. For $t \in \Sigma_\delta$, where $\delta < \delta_0$, the

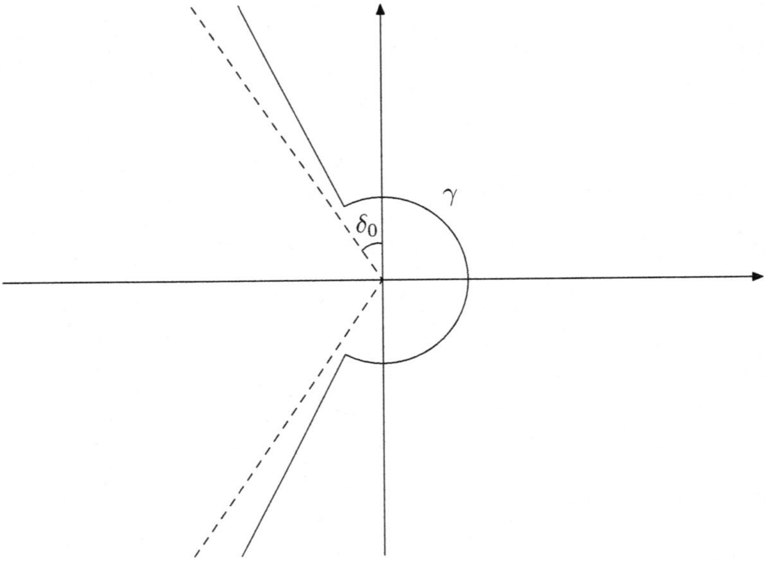

Figure 15.1 A typical γ

semigroup may be expressed as the Dunford integral:

$$e^{tA} = \frac{1}{2\pi i}\int_\gamma e^{t\lambda}(\lambda - A)^{-1}\,d\lambda \tag{15.5}$$

where γ is a piecewise smooth curve running, within the closure of $\Sigma_{\frac{\pi}{2}+\delta}$ (excluding $\lambda = 0$), that is, within:

$$\left\{\lambda \in \mathbb{C} \setminus \{0\};\ |\arg\lambda| \le \frac{\pi}{2} + \delta\right\}$$

from infinity with $\arg\lambda = -(\frac{\pi}{2} + \delta)$ to infinity with $\arg\lambda = \frac{\pi}{2} + \delta$ (see Figure 15.1). By (15.4), the integral here is absolutely convergent and by Cauchy's Theorem, it does not depend on the choice of the curve. Moreover, $\lim_{t\to 0, t\in\Sigma_\delta} e^{tA}x = x$ for all $x \in \mathbb{X}$ and $\delta \in (0, \delta_0)$, and the semigroup is bounded in each sector $\Sigma_{\frac{\pi}{2}+\delta}$, $\delta \in (0, \delta_0)$. Hence, A is said to generate a **bounded holomorphic semigroup**.

Let A be sectorial of angle δ_0. Then for $z \in \mathbb{C}$ with $|z| = 1$ and $|\arg z| < \delta_0$, zA generates a strongly continuous semigroup $(e^{tzA})_{t\ge 0}$. This, however, does not need to be so if $\arg z = \delta_0$. The question of existence of the semigroup generated by z_0A where $z_0 = e^{i\delta_0}$ is known as that of existence of the boundary value of a holomorphic semigroup. It turns out that the boundary value semigroup exists iff:

$$K := \sup_{t \in \Sigma_{\delta_0}, \Im t \geq 0, |t| \leq 1} \|e^{tA}\| \tag{15.6}$$

is finite. A simple proof of necessity of this condition may be found in [9] p. 174, and we will restrict ourselves to proving the sufficiency.

Let z_n satisfy $|z_n| = 1$, $|\arg z_n| < \delta_0$ and $\lim_{n\to\infty} z_n = z_0$. By assumption, $\|e^{tz_nA}\| \leq K$ for $t \in [0, 1]$ and $n \geq 1$. This implies that there exist $M \geq 1$ and $\omega \in \mathbb{R}$ such that $\|e^{tz_nA}\| \leq Me^{\omega t}, t \geq 0, n \geq 1$, and the semigroups generated by $A_n = z_nA - \omega I$ are equibounded. The semigroups generated by z_nA, being "cut out of" the single semigroup generated by A, commute. Hence, A_n's satisfy the Hasegawa condition. Also $\lim_{n\to\infty} A_nx = z_0Ax - \omega x$ for $x \in D(A)$, and $D(A)$ is dense by assumption. Therefore, the strong limit $\lim_{n\to\infty} e^{tA_n}$ exists by Theorem 15.1, the generator of the limit semigroup is the extended limit A_{ex} of A_n, and $z_0A - \omega I \subset A_{ex}$. Because all $z_n \neq 0$, closedness of A implies $A_{ex} \subset z_0A - \omega I$. Thus $z_0A = \omega I + A_{ex}$ generates a semigroup as well.

Exercise 15.1 Show that (15.1) implies (15.2).

Exercise 15.2 Let A be the generator of a semigroup $(e^{tA})_{t\geq 0}$ in a Banach space $\mathbb{X}$, and $D \subset D(A)$ be dense in $\mathbb{X}$. Show that if $e^{tA}D \subset D, t \geq 0$, then D is a core for A, that is, for any $x \in D(A)$ there are $x_n \in D$ such that $\lim_{n\to\infty} x_n = x$ and $\lim_{n\to\infty} Ax_n = Ax$. (See e.g. [49] p. 261, [128] p. 53, or Proposition 3.3 in [132].)

Exercise 15.3 For $a < b$, let A in $C[a, b]$ be given by $Af = f''$ on the domain composed of twice continuously differentiable functions satisfying $f'(a) = f'(b) = 0$. Show that A is sectorial of angle $\frac{\pi}{2}$. Also, A generates a Feller semigroup.

☞ Chapter's summary

The condition of Hasegawa gives a way to circumvent checking the range condition. As proved by two examples given in this chapter, knowing that Hasegawa's condition is satisfied substantially simplifies analysis of convergence.

16

Blackwell's Example

The explosion described in Chapter 13 does not exhaust the list of curious phenomena accompanying general Markov chains. In this chapter, we would like to discuss Blackwell's Example of an honest Markov chain with countably many states, all of which are instantaneous. (See the original paper [37] or *e.g.* [150, p. 297] or [240, p. 65]. Predecessors of this example may be found in [114] and [147].) The example involves infinite product of commuting semigroups as in Chapter 15; intriguingly, it is easier to prove existence of this product directly by finding an explicit formula than it is to do it by using Theorem 15.2. In fact, in notations of the latter theorem, it is quite difficult to find a single, nonzero vector $x \in D(A)$. It's even worse: it is difficult to find a nonzero x for which $\sum_{n=1}^{\infty} B_n x$ converges, despite the fact that all B_n's are bounded.

To begin the presentation of the example, let S be the set of functions $s : \mathbb{N} \to \{0, 1\}$ admitting value 1 finitely many times. Since S is countable, functions $x : S \to \mathbb{R}$ will be called sequences. By l^1 we denote the space of absolutely summable sequences on S: x is a member of l^1 if $\|x\| := \sum_{s \in S} |x(s)| < \infty$. We note that the members δ_s, $s \in S$ of l^1 defined by:

$$\delta_s(s') = \begin{cases} 1, & s' = s, \\ 0, & s' \neq s, \end{cases}$$

form a Schauder basis for this space (see [78] for the definition); in particular, for each $x \in l^1$:

$$x = \sum_{s \in S} x(s)\delta_s.$$

Given positive numbers $\alpha_n, \beta_n, n \geq 1$ we construct a sequence of Markov semigroups in l^1. To this end, we need some notation. For $n \geq 1$ let F_n be the map $F_n : S \to S$ changing the n-th coordinate of an s from 0 to 1 and vice versa. Also, for a finite set $E = \{i_1, \ldots, i_k\} \subset \mathbb{N}$, let:

$$F_E := F_{i_1} \circ \cdots \circ F_{i_k},$$

be the map that changes an $s \in S$ at its coordinates $i_1, \ldots, i_k$. Finally, let $G_n : S \to \{0, 1\}$ assign to an s its n-th coordinate.

For $n \geq 1$, let B_n be the bounded linear operator determined by its values on the Schauder basis by:

$$B_n\delta_s = \begin{cases} -\beta_n\delta_s + \beta_n\delta_{F_n(s)}, & \text{if } G_n(s) = 0, \\ -\alpha_n\delta_s + \alpha_n\delta_{F_n(s)}, & \text{if } G_n(s) = 1. \end{cases} \tag{16.1}$$

B_n is the generator of a Markov chain on S in which the n-th coordinate of an $s \in S$ jumps between 0 and 1, the intensity of the "forward" jump being β_n and that of the "backward" jump being α_n. In other words, B_n is the generator of the Markov semigroup in l^1 determined by (compare Exercise 13.3):

$$\mathrm{e}^{tB_n}\delta_s = \begin{cases} p_n(t)\delta_s + (1 - p_n(t))\delta_{F_n(s)}, & \text{if } G_n(s) = 0, \\ q_n(t)\delta_s + (1 - q_n(t))\delta_{F_n(s)}, & \text{if } G_n(s) = 1, \end{cases}$$

where:

$$p_n(t) = \frac{\alpha_n}{\alpha_n + \beta_n} + \frac{\beta_n}{\alpha_n + \beta_n}\mathrm{e}^{-(\alpha_n+\beta_n)t},$$
$$q_n(t) = \frac{\beta_n}{\alpha_n + \beta_n} + \frac{\alpha_n}{\alpha_n + \beta_n}\mathrm{e}^{-(\alpha_n+\beta_n)t}.$$

It is clear that B_n's commute, so that we are in the setup of the first example of the previous chapter, that is, in the setup of [10]. Moreover, since B_n's are bounded, the generator of the strongly continuous semigroup $T_n(t) = \prod_{k=1}^n \mathrm{e}^{tB_k}$ is $A_n = \sum_{k=1}^n B_k$. This semigroup describes n combined independent Markov chains, each changing one of the first n coordinates of s as described earlier, and is determined by:

$$T_n(t)\delta_s = \sum_{E\subset\{1,\ldots,n\}} p_{E,n}(t, s)\delta_{F_E(s)}, \tag{16.2}$$

with summation over 2^n subsets E of $\{1, \ldots, n\}$ and:

$$p_{E,n}(t, s) = \prod_{i=1}^n r_i(t, s, E),$$

where:

$$r_i(t,s,E)=\begin{cases} p_i(t), & \text{if } G_i(s)=0 \text{ and } i\notin E,\\ 1-p_i(t), & \text{if } G_i(s)=0 \text{ and } i\in E,\\ q_i(t), & \text{if } G_i(s)=1 \text{ and } i\notin E,\\ 1-q_i(t), & \text{if } G_i(s)=1 \text{ and } i\in E. \end{cases} \tag{16.3}$$

This formula simply says that, if the initial state of the Markov chain related to A_n is given, its state at time $t \ge 0$ may be characterized by listing the coordinates, which are different from the original ones – this is the role of the set E. Because each coordinate evolves independently from the other ones, the probability of such a change is the product of probabilities of change or no change on the first, second, third, and remaining coordinates.

Assume now that:

$$\sum_{n=1}^{\infty} \frac{\beta_n}{\alpha_n+\beta_n} < \infty, \tag{16.4}$$

and fix s and a *finite* subset E of $\mathbb{N}$. Then, for n large enough, $E \subset \{1,\dots,n\}$. Also, as s has finitely many coordinates different from 0, for sufficiently large n and $i > n$, $r_i(t,s,E)$ describes the probability of no change from initial 0 in the i-th coordinate of s (after perhaps many changes back and forth in the meantime), that is, $r_i(t,s,E) = p_i(t)$. Recalling that (see, e.g., [312, Theorem 118]) the product $\prod_{n=1}^{\infty} a_n$, where $a_n \in (0,1]$, converges if the series $\sum_{n=1}^{\infty}(1-a_n)$ also does, we see that the limit:

$$p_E(t,s) := \lim_{n\to\infty} p_{E,n}(t,s) \tag{16.5}$$

exists, because the finite number of terms in the product does not influence convergence, whereas:

$$\sum_{n=1}^{\infty}(1-p_n(t)) = \sum_{n=1}^{\infty} \frac{\beta_n}{\alpha_n+\beta_n}(1-\mathrm{e}^{-(\alpha_n+\beta_n)t}) \le \sum_{n=1}^{\infty} \frac{\beta_n}{\alpha_n+\beta_n} < \infty,$$

by assumption.

Lemma 16.1 *As $n \to \infty$, the right-hand side of* (16.2) *converges to:*

$$T(t)\delta_s := \sum_E p_E(t,s)\delta_{F_E(s)},$$

where the sum is over all finite subsets E of $\mathbb{N}$.

Proof Clearly, $\sum_{E\subset\{1,\dots,n\}} p_{E,n}(t,s) = 1$ for $s \in S, t \ge 0$ and $n \ge 1$, and the summands are non-negative. Therefore, by (16.5) and Scheffé's Lemma (see, e.g., [347, p. 55] or [49, p. 199]), all we need to show is that:

$$\sum_E p_E(t, s) = 1, \qquad \text{for} \qquad t \geq 0, s \in S. \tag{16.6}$$

The proof of this relation will be more transparent if we assume $s = (0, 0, \ldots)$; this we do without loss of generality. Also, for simplicity of notation, we suppress dependence on t and s and write p_n and p_E instead of $p_n(t)$ and $p_E(t, s)$, respectively. Thus:

$$p_E = \prod_{n \in E}(1 - p_n) \prod_{n \notin E} p_n.$$

For a (finite) $E \subset \mathbb{N}$, let:

$$E_1 = \begin{cases} E \setminus \{1\}, & \text{if } 1 \in E, \\ E \cup \{1\}, & \text{if } 1 \notin E. \end{cases}$$

Then:

$$p_E + p_{E_1} = \prod_{n \in E, n \neq 1} (1 - p_n) \prod_{n \notin E, n \neq 1} p_n.$$

Therefore, letting $L = \sum_{\text{finite } E \subset \{1,2,\ldots\}} p_E$, we have:

$$L = \sum_{\text{finite } E \subset \{2,3,\ldots\}} \prod_{n \in E}(1 - p_n) \prod_{n \notin E, n \neq 1} p_n.$$

Repeating this argument k times, we see that:

$$L = \sum_E \prod_{n \in E}(1 - p_n) \prod_{n \in (\mathbb{N} \setminus \{1,\ldots,k\}) \setminus E} p_n,$$

where $E \subset \{k+1, k+2, \ldots\}$ are finite sets. Because the sum includes the component corresponding to $E = \emptyset$, $L \geq \prod_{n=k+1}^{\infty} p_n$. On the other hand, the last product may be chosen as close to 1 as we wish by taking k large enough. This completes the proof. □

Theorem 16.2 *The semigroups $(T_n(t))_{t \geq 0}$ converge to the semigroup $(T(t))_{t \geq 0}$ given by*:

$$T(t)x = \sum_{s \in S} x(s) T(t) \delta_s$$

where $T(t)\delta_s$ is defined in Lemma 16.1. The limit semigroup is strongly continuous and is composed of Markov operators.

Proof The statement concerning convergence is clear by Lemma 16.1. Also, $(T(t))_{t \geq 0}$ is a semigroup, being a limit of semigroups, and its Markovian nature is a direct consequence of (16.6). Hence, we are left with showing that $(T(t))_{t\geq 0}$ is strongly continuous, and for this it suffices to show that $\lim_{t\to 0} T(t)\delta_s = \delta_s$, $s \in S$. Let $s \in S$, and let k be so large that all coordinates of s with indexes larger than k are zeros. Then:

$$p_\emptyset(t, s) = \prod_{i=1}^{n} r_i(t, s, E) \prod_{i=n+1}^{\infty} p_i(t) \geq \prod_{i=1}^{n} r_i(t, s, E) \prod_{i=n+1}^{\infty} \frac{\alpha_i}{\alpha_i + \beta_i}$$

for $n \geq k$, where (see (16.3)):

$$r_i(t, s, \emptyset) = \begin{cases} p_i(t), & \text{if } G_i(s) = 0, \\ q_i(t), & \text{if } G_i(s) = 1. \end{cases}$$

Since $\lim_{t\to 0} p_i(t) = \lim_{t\to 0} q_i(t) = 1$ for all $i \geq 1$, we have:

$$\liminf_{t\to 0} p_\emptyset(t) \geq \prod_{i=n+1}^{\infty} \frac{\alpha_i}{\alpha_i + \beta_i},$$

and the latter product may be chosen as close to 1 as desired. This shows that $\lim_{t\to 0} p_\emptyset(t, s, E) = 1$. Therefore, since $\|\delta_{s'}\| = 1$, for all $s' \in S$, (use (16.6)):

$$\|T(t)\delta_s - \delta_s\| \leq \|(p_\emptyset(t, s, \emptyset) - 1)\delta_s\| + \sum_{E\neq\emptyset} p_E(t) = 2(1 - p_\emptyset(t, s, \emptyset))$$

converges to 0, as $t \to 0$. □

Finally, assume:

$$\sum_{n=1}^{\infty} \beta_n = \infty,$$

(note that this assumption does not contradict (16.4)), and consider $s \in S$. Let k be so large that all coordinates of s with indexes $n \geq k$ are zero. For such n, $B_n\delta_s$ is given by the first formula in (16.1). Since all $F_n(s)$'s are pairwise different and different from s, it follows that for $n \geq k$:

$$\|A_n\delta_s\| = \left\| \sum_{i=1}^{n} B_i x \right\| \geq \sum_{i=1}^{k-1} \gamma_i + \sum_{i=k}^{n} \beta_i,$$

where γ_i is either β_i or α_i, depending on whether $G_i(s) = 0$ or $G_i(s) = 1$. In any case, $\lim_{n\to\infty} \|A_n\delta_s\| = \infty$. In other words, no δ_s belongs to the extended limit of the generators A_n. This explains why it is difficult to use the Sova–Kurtz Theorem here: all members of $D(A_{\text{ex}})$ are 'skew' to the members of the natural basis. As a result, it is also quite hard to characterize the generator of the limit semigroup.

☞ Chapter's summary

Following [58], Blackwell's Example is presented in the context of the infinite product of semigroups of Chapter 15. In this example, it is hard to prove existence of the limit semigroup using Theorem 15.2 but it is quite easy to prove it directly. All states s of the Markov chain involved are instantaneous, which in terms of the semigroup means that none of δ_s belongs to the domain of the generator. As a result, a manageable characterization of the generator of the semigroup is missing. It should be perhaps noted here that examples of semigroups given by explicit formulae, where it is hard to characterize the generator in a full way, are known and include for instance the Cauchy semigroup (known also as a Poisson semigroup) in the space of continuous functions [343, Chapter 10] (but not in the L^p spaces, see [256, p. 99]). However, in these examples one is at least able to find a core of the generator; moreover, the form of the generator on the core is known (see, e.g., Hunt's Theorem on generators of convolution semigroups: [178] p. 262 or [183]). A similar, but simpler, semigroup with this property is presented in [101] pp. 349–352. In Blackwell's Example, it is difficult to find an explicit form of a single nonzero member of the domain, not to mention the form of the generator.

17

Wright's Diffusion

In spite of mutations that – given the state of a population – occur independently in each individual, members of (especially: small) populations exhibit striking similarities. This is due to **genetic drift**, one of the most important forces of population genetics. Simply put, the reason for this phenomenon is that in a population, on the one hand, new variants are introduced randomly by (neutral) mutations, and, on the other, many variants are also randomly lost as not all members of the current generation pass their genetic material to the next one.

This situation is clearly presented in the following model of Wright and Fisher [49, 134, 141]. We suppose the population in question to be composed of $2N$ individuals; in doing so we identify individuals with chromosomes (that come in pairs), or even with corresponding loci on these chromosomes. We assume there are only two possible **alleles** (variants) at this locus: A and a. The size of the population is kept constant all the time, and we consider its evolution in discrete non-overlapping generations formed as follows: an individual in the daughter generation is the same as its parent (reproduction is asexual) and the parent is assumed to be chosen from the parent generation randomly, with all parents being equally probable. In other words, the daughter generation is formed by $2N$ independent draws with replacement from the parent generation. In each draw all parents are equally likely to be chosen and daughters have the same allele as their parents. It should be noted here that such sampling procedure models the genetic drift by allowing some parents not to be selected for reproduction, and hence not contributing to the genetic pool.

Then, the state of the population at time $n \geq 0$ is conveniently described by a single random variable X_n with values in $\{0, \dots, 2N\}$ being equal to the number of individuals of type A. The sequence $X_n, n \geq 0$ is a time-homogeneous

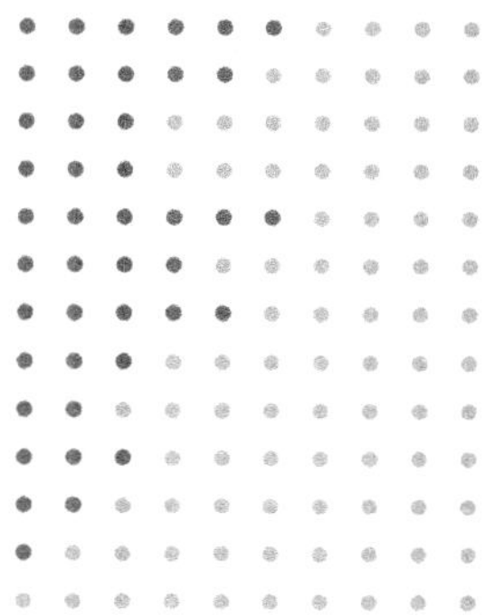

Figure 17.1 Wright–Fisher model: each row depicts a generation sampled with replacement from the one lying above it

Markov chain with transition probabilities:

$$p_{kl} := \Pr\{X_{n+1} = l | X_n = k\} = \binom{2N}{l} p_k^l (1 - p_k)^{2N-l}, \text{ where } p_k = \frac{k}{2N}. \tag{17.1}$$

In other words, if $X_n = k$, then X_{n+1} is a binomial random variable with parameter $\frac{k}{2N}$. A typical realization of the Wright–Fisher chain is depicted in Figure 17.1, where $N = 5$ and dark and light balls represent A and a alleles, respectively. The figure illustrates also the fact that the states 0 and $2N$ are absorbing. This is to say that if for some n, $X_n = 0$ we must have $X_m = 0$ for all $m \geq n$, and a similar statement is true if $X_n = 2N$. Genetically, this expresses the fact that in finite populations, in the absence of other genetic forces, genetic drift (i.e., random change of allele frequencies) reduces variability of population by fixing one of the existing alleles.

Now, imagine that the number $2N$ of individuals is quite large and the individuals are placed on the unit interval $[0, 1]$ with distances between neighboring individuals equal to $\frac{1}{2N}$, so that the kth individual is placed at $\frac{k}{2N}$. Imagine also that the time that elapses from one generation to the other is $\frac{1}{2N}$, so that there are $2N$ generations in a unit interval. Then the process we observe bears more and more resemblance to a continuous-time, continuous-path diffusion process on $[0, 1]$, as depicted in Figure 17.2.

Since, conditional on $X_n = k$, the expected single-step displacement $E\,\frac{\Delta X_n}{2N}$ of the approximating process equals $\frac{1}{2N} 2N p_k - \frac{k}{2N} = 0$ with variance $Var\,\frac{\Delta X_n}{2N} = \frac{1}{2N} p_k(1 - p_k)$, in the limit we expect the process starting at $x \in [0, 1]$ to have infinitesimal variance $x(1 - x)$ and infinitesimal displacement 0. Indeed, for each N, a point $x \in [0, 1]$ may be identified with $\frac{k}{2N}$ where

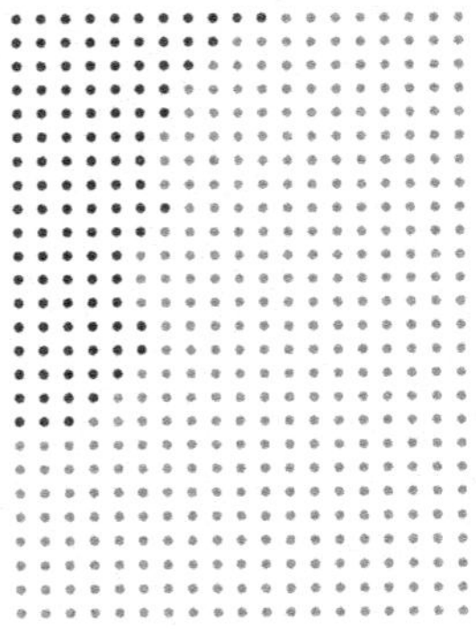

Figure 17.2 Diffusion approximation: the border line between the light and dark balls resembles a path of a diffusion process

$k = [2Nx]$, and letting $N \to \infty$ in $\frac{[2Nx]}{2N}(1 - \frac{[2Nx]}{2N})$, which is the infinitesimal variance in a unit time-interval, we obtain $x(1-x)$. Moreover, the limiting process should inherit its boundary behavior from the approximating Markov chains. Therefore, we conjecture that in the limit we will obtain the process related to the semigroup in $C[0, 1]$, generated by:

$$Af(x) = x(1-x)f''(x) \tag{17.2}$$

with domain $D(A)$ composed of twice continuously differentiable functions on (0,1) such that:

$$\lim_{x\to 0+} x(1-x)f''(x) = \lim_{x\to 1-} x(1-x)f''(x) = 0.$$

(The requirement that the process is absorbed at $x = 0$ and $x = 1$ is expressed in the fact that $Af(0) = Af(1) = 0$.) This conjecture is proved, e.g. in Section 8.4.20 of [49] and in [128] pp. 224–226; see also [240] p. 120. The limiting process is often referred to as **Wright's diffusion** or the **Wright–Fisher diffusion** (without mutations).

A noteworthy variant of this limit procedure arises when mutations are allowed. Suppose, namely, that in passing from one generation to the other an individual with allele A may change its state to allele a with probability $\frac{\alpha}{2N}$, and an individual with allele a may change its state to allele A with probability $\frac{\beta}{2N}$, where α and β are non-negative numbers. Then, the number of A alleles is still a Markov chain with transition probabilities of the form (17.1), but p_k is now changed to:

$$p_k = \frac{k}{2N}\left(1 - \frac{\alpha}{2N}\right) + \left(1 - \frac{k}{2N}\right)\frac{\beta}{2N}.$$

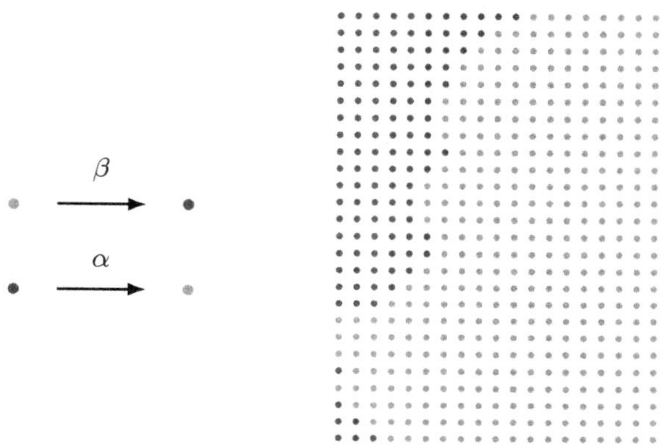

Figure 17.3 Diffusion approximation to Wright–Fisher model with mutations

Of course, this influences the expected single-step displacement and its variance (conditional on $X_n = k$):

$$E\frac{\Delta X_n}{2N} = p_k - \frac{k}{2N} = \frac{1}{2N}(\beta - (\alpha+\beta)\frac{k}{2N}),$$
$$Var\left(\frac{\Delta X_n}{2N}\right) = \frac{1}{2N}p_k(1-p_k). \tag{17.3}$$

This suggests that the limit process starting at x will have an infinitesimal unit-time displacement of $\beta - (\alpha+\beta)x$ and the corresponding variance $x(1-x)$. We will show this conjecture in Chapter 20.

An even more interesting phenomenon occurs at the boundaries $x = 0$ and $x = 1$. For the scaled Wright–Fisher process with mutation, after reaching $\frac{0}{2N}$, will not stay there forever but for a geometric time τ: $\Pr(\tau \geq n) = (1 - \frac{\beta}{2N})^{2Nn}$ and then resume his motion (see Figure 17.3). Hence, we should expect a similar behavior of the limit process: paths should be able to return to the interior of the unit interval after a random time spent at the boundary.

We will come back to Wright's diffusion in Chapter 20, after discussing more general issues in Chapter 18.

☞ Chapter's summary

We discuss briefly the classical, discrete-time Markov–chain-based, Wright–Fisher model of population genetics and provide intuitions of its limit behavior for large number of individuals and scaled space and time. We argue that in the limit we should obtain a continuous-time Markov process (a diffusion process) on a unit interval; the details of this approximation are postponed for a time.

18

Discrete-Time Approximation

In the previous chapter, a discrete-time process is approximated by a continuous-time process. To establish a general scheme for such an approximation, consider non-negative numbers h_n such that $\lim_{n\to\infty} h_n = 0$ and equipowerbounded operators T_n, $n \geq 1$ in a Banach space $\mathbb{X}$:

$$\|T_n^k\| \leq M, \qquad n,k \geq 1. \tag{18.1}$$

The related families of operators $V_n(t) := T_n^{[t/h_n]}$ and e^{tA_n}, $t \geq 0, n \geq 1$, where $A_n = \frac{T_n - I}{h_n}$, are then also composed of equibounded operators: $\|V_n(t)\| \leq M$, and:

$$\|\mathrm{e}^{tA_n}\| = \mathrm{e}^{-t/h_n} \|\sum_{k=0}^{\infty} \frac{(T_n t)^k}{h_n^k k!}\| \leq M.$$

The key observation concerning the limit behavior of these families is that (see Exercise 18.1, or [49] p. 265):

$$\int_0^\infty \mathrm{e}^{-\lambda t} V_n(t)x \, \mathrm{d}t = \frac{\lambda_n}{\lambda}(\lambda_n - A_n)^{-1}x, \qquad \lambda > 0, n \geq 1, x \in \mathbb{X}, \tag{18.2}$$

where $\lambda_n = \frac{\mathrm{e}^{\lambda h_n} - 1}{h_n}$. Since $\lim_{n\to\infty} \lambda_n = \lambda$, this relation makes it clear that Laplace transforms of V_n and $\mathrm{e}^{\cdot A_n}$ converge simultaneously, and their limits coincide.

Theorem 18.1 *Suppose the limit $R_\lambda := \lim_{n\to\infty}(\lambda - A_n)^{-1}$ exists. (Equivalently: the range of $\lambda - A_{\mathrm{ex}}$ is dense in $\mathbb{X}$.) Then, the limit semigroup T on the regularity space $\mathbb{X}_0$ and the related integrated semigroup U (on $\mathbb{X}$) may be approximated as follows:*

$$U(t)x = \lim_{n\to\infty} \int_0^t V_n(s)x \, \mathrm{d}s = \lim_{n\to\infty} \int_0^t T_n^{[u/h_n]} x \, \mathrm{d}u, \qquad x \in \mathbb{X}, t \geq 0, \tag{18.3}$$

$$T(t)x = \lim_{n\to\infty} V_n(t)x = \lim_{n\to\infty} T_n^{[t/h_n]} x, \qquad x \in \mathbb{X}_0, t \geq 0, \tag{18.4}$$

with the limit almost uniform in $t \in \mathbb{R}^+$.

Proof Since $t \mapsto \int_0^t V_n(s)\,ds$ is Lipschitz continuous, the assumption coupled with (18.2) implies that the limit on the right-hand side of (18.3) exists and is almost uniform in $t \in \mathbb{R}^+$. Equality must hold, because both sides are continuous and their Laplace transforms coincide.

Analogously, to show (18.4) we need to show merely that the limit on the right-hand side exists. For $x \in \mathbb{X}_0$, $U(t)x = \int_0^t T(s)x\,ds$. Hence, by the strong continuity of $\{T(t), t \geq 0\}$, $\lim_{t\to 0} \frac{1}{t}U(t)x = x$. Therefore, it suffices to show (18.4) for x of the form $x = U(s)y$, $y \in \mathbb{X}$, $s > 0$. In view of (18.3) and $\|T_n^{[t/h_n]}\| \leq M$, this will be done once we show that $\lim_{n\to\infty} T_n^{[t/h_n]} \int_0^s T_n^{[u/h_n]} y\,du$ exists. However, we have:

$$
\begin{aligned}
T_n^{\left[\frac{t}{h_n}\right]} \int_0^s T_n^{\left[\frac{u}{h_n}\right]} y\,du &= \int_0^s T_n^{\left[\frac{u+h_n\left[\frac{t}{h_n}\right]}{h_n}\right]} y\,du = \int_{h_n\left[\frac{t}{h_n}\right]}^{s+h_n\left[\frac{t}{h_n}\right]} T_n^{\left[\frac{u}{h_n}\right]} y\,du \\
&= \int_0^{s+h_n\left[\frac{t}{h_n}\right]} T_n^{\left[\frac{u}{h_n}\right]} y\,du - \int_0^{h_n\left[\frac{t}{h_n}\right]} T_n^{\left[\frac{u}{h_n}\right]} y\,du \\
&\underset{n\to\infty}{\longrightarrow} U(t+s)y - U(t)y,
\end{aligned}
$$

and the limit is almost uniform in $t \in \mathbb{R}^+$, by the first part. By density argument the same is true of convergence for arbitrary $x \in \mathbb{X}_0$. □

Perhaps the most famous corollary of this theorem is the following **Chernoff's Product Formula**.

Corollary 18.2 *Suppose that $V_n(t), t \geq 0, n \geq 1$ is a family of equi-powerbounded operators:*

$$\|(V_n(t))^k\| \leq M, \qquad t \geq 0, n \geq 1, k \geq 1,$$

and assume that $V_n(0) = I_{\mathbb{X}}, n \geq 1$. Also, let A be a densely defined linear operator satisfying the range condition: for some $\lambda > 0$ the range of $\lambda - A$ is dense in $\mathbb{X}$. Assume that for any sequence $(h_n)_{n\geq 1}$ of positive numbers converging to zero:

$$\lim_{n\to\infty} h_n^{-1}(V_n(h_n)x - x) = Ax, \qquad x \in D(A). \tag{18.5}$$

Then, A is closable and its closure generates the strongly continuous semigroup:

$$T(t)x = \lim_{n\to\infty} (V_n(t/n))^n x, \qquad t \geq 0, x \in \mathbb{X}, \tag{18.6}$$

where the limit is almost uniform in $t \in \mathbb{R}^+$.

Proof The operators $A_n = n(V_n(1/n) - I_{\mathbb{X}})$ are generators of semigroups satisfying (2.1). Hence, by Corollary 8.4, A is closable and its closure generates a semigroup $\{T(t), t \geq 0\}$. Suppose (18.6) is not true for this semigroup. Then, there exists an $x \in \mathbb{X}$, an $\epsilon > 0$ and a bounded sequence $\{t_n\}$ such that:

$$\|T(t_n)x - (V_n(t_n/n))^n x\| \geq \epsilon. \tag{18.7}$$

Without loss of generality, we may assume that the limit $\lim_{n\to\infty} t_n =: t$ exists. Then $A_n' = \frac{V_n(h_n)-I}{h_n}$, where $h_n = t_n/n$, generate equibounded semigroups and by assumption and Corollary 8.4,

$$T(t)x = \lim_{n\to\infty} \mathrm{e}^{tA_n'}x, \qquad x \in \mathbb{X}.$$

By Theorem 18.1,

$$T(s)x = \lim_{n\to\infty} (V_n(h_n))^{[s/h_n]}x, \qquad s \geq 0$$

almost uniformly in $s \in \mathbb{R}^+$. This contradicts (18.7) if we take $s = t_n$. □

In most applications, the family V does not depend on n, and condition (18.5) simplifies to $\lim_{t\to 0+} t^{-1}(V(t)x - x) = Ax$, $x \in D(A)$. Here, however, is an example where dependence on n allows more flexibility. This is taken from the monograph [215], where the original idea is attributed to S. Kowalski.[1]

Let $\{T(t), t \geq 0\}$ be a semigroup of contractions generated by an operator A. Define $V_n(t) = (1 - nt)I_{\mathbb{X}} + ntT(\frac{1}{n})$, for $t \leq \frac{1}{n}$ and $V_n(t) = T(t)$, for $t > \frac{1}{n}$; clearly, all $V_n(t)$'s are contractions, as well. We note that if $h_n \leq \frac{1}{n}$ then $h_n^{-1}(V_n(h_n)x - x) = n(T(\frac{1}{n})x - x)$, and in the other case $h_n^{-1}(V_n(h_n)x - x) = h_n^{-1}(T(h_n)x - x)$. It follows that (18.5) holds, and Corollary 18.2 applies. In particular,

$$\begin{aligned} T(t)x &= \lim_{n\to\infty} \left((1-t)I_{\mathbb{X}} + tT\left(\frac{1}{n}\right)\right)^n x \\ &= \lim_{n\to\infty} \sum_{k=0}^{n} \binom{n}{k} t^k (1-t)^{n-k} T\left(\frac{k}{n}\right) x \end{aligned} \tag{18.8}$$

for all $x \in \mathbb{X}$, uniformly in $t \in [0, 1]$. The approximation on the right-hand side here is an analogue of Bernstein's polynomial approximation of continuous functions. In particular, taking $\{T(t), t \geq 0\}$ to be the semigroup of translations to the left in $C[0, \infty]$, and noting that any member of $C[0, 1]$ may be extended

[1] Then a freshman in mathematics at University of Warsaw, Poland, today Kowalski is a quite well-known sociologist.

to a member of $C[0, \infty]$, we obtain

$$x(t) = T(t)x(0) = \lim_{n\to\infty} \sum_{k=0}^{n} \binom{n}{k} t^k (1-t)^{n-k} x\left(\frac{k}{n}\right),$$

uniformly in $t \in [0, 1]$, that is, we recover the original **Bernstein's Approximation Theorem**. Moreover, by considering the same semigroup in spaces like $L^p(\mathbb{R}^+)$, we obtain other, nonstandard versions of the theorem. However, our result applies to all contraction semigroups, not merely to translation semigroups.

Chernoff's Product Formula implies the following **Trotter's Product Formula**, known also as the **Lie–Trotter Formula**.

Corollary 18.3 *Suppose that A and B and C are generators of C_0 semigroups $\{S(t), t \ge 0\}$, $\{T(t), t \ge 0\}$, and $\{U(t), t \ge 0\}$, respectively, in a Banach space $\mathbb{X}$. Suppose also that $\mathcal{D}$ is a core for C, $\mathcal{D} \subset \mathcal{D}(A) \cap \mathcal{D}(B)$ and $Cx = Ax + Bx$ for $x \in \mathcal{D}$. Finally, assume that:*

$$\|(S(t)T(t))^n\| \le M, \qquad t \ge 0, n \ge 1. \tag{18.9}$$

Then,

$$U(t)x = \lim_{n\to\infty} \left[S\left(\frac{t}{n}\right) T\left(\frac{t}{n}\right)\right]^n x, \qquad t \ge 0, x \in \mathbb{X}$$

strongly and almost uniformly in $t \in \mathbb{R}^+$.

Proof Take $V(t) = S(t)T(t)$. Then $t^{-1}(V(t)x - x) = t^{-1}S(t)(T(t)x - x) + t^{-1}(S(t)x - x) \to Bx + Ax$, as $t \to 0$ for $x \in \mathcal{D}$. Hence, the result follows by Corollary 18.2. □

Condition (18.9) is often referred to as **stability condition**. Without a form of stability condition, Trotter Product Formula does not hold [77, 226]. Interestingly, there seems to be no Trotter product analogue for cosine operator functions.

Exercise 18.1 Show (18.2) by direct computation.

Exercise 18.2 Let A be the generator of a semigroup such that $\|e^{tA}\| \le M, t \ge 0$. Show that $e^{tA} = \lim_{n\to\infty}(1 - \frac{t}{n}A)^{-n}, t \ge 0$. (This is the **Hille approximation**. An analogue for cosine families may be found in [340].) *Hint:* Take $V_n(t) = V(t) = (1 - tA)^{-1}$ in Corollary 18.2 and use Exercise 1.1.

Exercise 18.3 Prove the following result. Assume (18.1) and let A be the generator of a strongly continuous semigroup in a subspace $\mathbb{X}_0$ of $\mathbb{X}$. If $A \subset A_{ex}$, then:

$$T(t)x = \lim_{n\to\infty} T_n^{[t/h_n]}x, \qquad x \in \mathbb{X}_0, t \geq 0, \tag{18.10}$$

with the limit almost uniform in $t \in \mathbb{R}^+$. *Hint:* from (18.2) and Corollary 8.5 we obtain convergence of the Laplace transforms of $T_n^{[\cdot/h_n]}x_n$ where $\lim_{n\to\infty} x_n = x, x \in \mathbb{X}_0$. As in Theorem 18.1, this implies convergence of $\int_0^t T_n^{[s/h_n]}x\,\mathrm{d}s, x \in \mathbb{X}_0$ and then convergence of $T_n^{[t/h_n]}x, x \in \mathbb{X}$.

Exercise 18.4 (Chernoff's Formula of kth order, J. Goldstein [162][2]) Let $k \geq 1$ be an integer and let A be a densely defined linear operator in a Banach space $\mathbb{X}$, satisfying the range condition: for some $\lambda > 0$ the range of $\lambda - A$ is dense in $\mathbb{X}$. Suppose also that $\{V(t), t \geq 0\}$ is a family of contractions and that for $x \in D(A)$, $t \mapsto V(t)x$ is k times continuously differentiable in a neighborhood of 0 and that $V^{(i)}(0) = 0, i = 0, \ldots, k-1$, while $V^{(k)}(0)x = k!Ax$. Conclude that:

$$\lim_{n\to\infty} V\left(\frac{t}{n}\right)^{n^k} x = \mathrm{e}^{t^k A}x, \qquad x \in \mathbb{X}.$$

Hint: $W(t) = V(t^k)$ satisfies the assumptions of Chernoff's Formula.

☞ Chapter's summary

The main theorem of this chapter establishes, under assumption of convergence of resolvents, convergence of discrete-parameter semigroups and their averages, to a semigroup and the related integrated semigroup, respectively. This result is then used to prove Chernoff's Product Formula and the Lie–Trotter Formula. A noteworthy abstract version of the Bernstein Approximation Theorem is presented as a first application.

[2] To quote [162]: "The case $k = 2$ is due, independently, to Goldstein [159] and Nelson [274]. Chernoff [83] later published a variant of the $k = 2$ case. The above theorem is new for $k \geq 3$."

19

Discrete-Time Approximation: Examples

Before returning to the motivating case of Wright's diffusion, let us examine some more elementary applications of Theorem 18.1.

Example 19.1 To begin with, let l^1 be the space of absolutely summable sequences $(\xi_k)_{k\geq 0}$. Given $p_n \in [0, 1]$ such that $\lim_{n\to\infty} np_n = \lambda > 0$, consider $T_n = p_n R + (1 - p_n)I$, where R is translation to the right (i.e., $R(\xi_k)_{k\geq 0} = (\xi_{k-1})_{k\geq 0}$; by convention $\xi_{-1} = 0$). Then, $\lim_{n\to\infty} n[T_n - I] = \lambda(R - I)$, and Theorem 18.1 yields:

$$\lim_{n\to\infty} T_n^{[nt]} = \mathrm{e}^{\lambda t(R-I)}. \tag{19.1}$$

If $(\xi_k)_{k\geq 0}$ is interpreted as a distribution of a random variable Y, then $T_n(\xi_k)_{k\geq 0}$ is the distribution of $Y + X_n$ where X_n is an independent Bernoulli variable with parameter p_n (i.e., $P(X_n = 1) = p_n$, $P(X_n = 0) = 1 - p_n$). Therefore, $T_n^n(\xi_k)_{k\geq 0}$ is the distribution of $Y + Z_n$ where Z_n is an independent binomial variable: $P(Z_n = i) = \binom{n}{i} p_n^i (1 - p_n)^{n-1}$, $i = 0, \ldots, n$. Similarly, $\mathrm{e}^{\lambda(R-I)}(\xi_k)_{k\geq 0}$ is the distribution of $Y + Z$ where Z is Poisson with parameter λ. In other words, (19.1) with $t = 1$ establishes the Poisson approximation to binomial theorem.

Example 19.2 Next, as in Chapter 2, let h_n and Δ_n satisfy the condition $\lim_{n\to\infty} \Delta_n = 0$ with $\lim_{n\to\infty} \frac{\Delta_n^2}{h_n} = 1$, and let $A = \frac{1}{2}\frac{\mathrm{d}^2}{\mathrm{d}x^2}$ be the generator of the Brownian motion semigroup. We have (see Exercise 2.2):

$$\lim_{n\to\infty} h_n^{-1}(T_n f - f) = \frac{1}{2} f'', \qquad C^2[-\infty, \infty] \tag{19.2}$$

where:

$$T_n f(x) = \frac{1}{2}(f(x + \Delta_n) + f(x - \Delta_n)), \qquad x \in \mathbb{R}. \tag{19.3}$$

Therefore, Theorem 18.1 yields:

$$\lim_{n\to\infty} T_n^{[t/h_n]} = \mathrm{e}^{tA}, \quad t \geq 0.$$

This formula says that Brownian motion may be approximated by properly scaled symmetric random walks.

Example 19.3 A similar approximation may be given for the reflected Brownian motion semigroup: this is the semigroup in $C[0, \infty]$ generated by the operator $A_\mathrm{r} f = \frac{1}{2} f''$ with domain composed of $f \in C^2[0, \infty]$ satisfying $f'(0) = 0$ (see Chapter 3). To this end consider a sequence of random walks where a particle at $x \geq \Delta_n$ moves with equal probability either to $x + \Delta_n$ or to $x - \Delta_n$ while the particle at $x \in [0, \Delta_n)$ moves either to $x + \Delta_n$ or tries to move to the left and is reflected at the boundary $x = 0$ to land at $\Delta_n - x$. In other words:

$$T_{n,\mathrm{r}} f(x) = \begin{cases} \frac{1}{2} f(x + \Delta_n) + \frac{1}{2} f(x - \Delta_n), & x \geq \Delta_n, \\ \frac{1}{2} f(x + \Delta_n) + \frac{1}{2} f(\Delta_n - x), & 0 \leq x < \Delta_n. \end{cases} \tag{19.4}$$

A simple trick allows deducing:

$$\lim_{n\to\infty} T_{n,\mathrm{r}}^{[t/h_n]} = \mathrm{e}^{tA_\mathrm{r}}, \quad t \geq 0, \tag{19.5}$$

from (19.2). Namely, note that $T_{n,\mathrm{r}} f = T_n f_\mathrm{e}$, where f_e is the even extension of f. Because $f \in D(A_\mathrm{r})$, f_e belongs to $C^2[-\infty, \infty]$, (19.2) implies (19.5).

Example 19.4 In our next example, we would like to find a similar sequence of Markov chains approximating the elastic Brownian motion semigroup of Chapter 10, to shed some more light on the meaning of the Robin boundary condition. To this end, we assume that in the approximating Markov chain, a particle starting at $x \geq \Delta_n$ moves with the same probability either to $x - \Delta_n$ or to $x + \Delta_n$. If the chains starts at $0 \leq x < \Delta_n$, it jumps with probability $\frac{1}{2}$ to $x + \Delta_n$ or is reflected from the boundary $x = 0$ to land at $\Delta_n - x$; however, at the boundary it may be 'killed' with non-negative probability p_n, and hence it lands at $\Delta_n - x$ with probability $\frac{1}{2}(1 - p_n)$. As suggested by Mandl [247] assuming $\lim_{n\to\infty} \frac{2\Delta_n}{p_n} = \epsilon > 0$, we should obtain that these chains approximate the elastic Brownian motion with parameter ϵ (see Chapter 10).

The trouble with this approximation is that, even though the natural choice for the Banach space here is $C[0, \infty]$ where the elastic Brownian motion semigroup acts, the operators $T_{n,\mathrm{el}}$ related to the approximating Markov chains lead out of $C[0, \infty]$. We have:

$$T_{n,\mathrm{el}} f(x) = \begin{cases} \frac{1}{2} f(x + \Delta_n) + \frac{1}{2} f(x - \Delta_n), & x \geq \Delta_n, \\ \frac{1}{2} f(x + \Delta_n) + \frac{1}{2}(1 - p_n) f(\Delta_n - x), & 0 \leq x < \Delta_n, \end{cases} \tag{19.6}$$

and the resulting function is discontinuous at $x = \Delta_n$. We could use Exercise 18.3 with $\mathbb{X}$ equal to the space $BM(\mathbb{R}^+)$ of bounded measurable functions on $\mathbb{R}^+$, and $\mathbb{X}_0 = C[0, \infty]$. However, this approximation will simply not work in the supremum norm. For, using Taylor's Theorem it is easy to check that:

$$\lim_{n\to\infty} \sup_{x\geq\Delta_n} \left| \frac{T_{n,\mathrm{el}} f(x) - f(x)}{h_n} - \frac{1}{2} f''(x) \right| = 0,$$

and in particular, taking $x = \Delta_n$,

$$\lim_{n\to\infty} \left| \frac{\frac{1}{2} f(2\Delta_n) + \frac{1}{2} f(0) - f(\Delta_n)}{h_n} - \frac{1}{2} f''(\Delta_n) \right| .$$

By contrast, the analogous supremum over $x \in [0, \Delta_n)$ is no less than $|\frac{\frac{1}{2} f(2\Delta_n)+\frac{1}{2}(1-p_n)f(0)-f(\Delta_n)}{h_n} - \frac{1}{2} f''(\Delta_n)|$. If this were to converge to 0, we would need to have $\lim_{n\to\infty} \frac{p_n}{h_n} = 0$, because in general $f(0) \neq 0$. However, by assumption $\lim_{n\to\infty} \frac{p_n}{h_n} = \infty$.

Turning to the dual[1] space $L^1(\mathbb{R})$ we check first (Exercise 19.1) that the dual T_n^* to the operator T_n in (19.3) is given formally by the same formula and we have

$$\lim_{n\to\infty} \frac{1}{h_n} (T_n \phi - \phi) = A^* \phi, \tag{19.7}$$

implying $\lim_{n\to\infty} T_n^{[t/h_n]} = \mathrm{e}^{tA^*}$, where A^* is defined as $\frac{1}{2} \frac{\mathrm{d}^2}{\mathrm{d}x^2}$ for $\phi \in W^{2,1}(\mathbb{R})$. A^* generates a strongly continuous Markov semigroup given formally by the same formula as its predual (see (2.2)).

Analogously, A_ϵ^* defined as $\frac{1}{2} \frac{\mathrm{d}^2}{\mathrm{d}x^2}$ for $\phi \in W^{2,1}(\mathbb{R}^+)$ satisfying $\phi'(0) = \epsilon\phi(0)$ generates a sub-Markov semigroup in $L^1(\mathbb{R}^+)$ and the dual $T_{n,\mathrm{el}}^*$ is formally given by the same formula as $T_{n,\mathrm{el}}$. The relation:

$$\lim_{n\to\infty} \frac{1}{h_n} (T_{n,\mathrm{el}}^* \phi - \phi) = A_\epsilon^* \phi,\ \phi \in D(A_\epsilon^*), \tag{19.8}$$

proving that $\lim_{n\to\infty} T_{n,\mathrm{el}}^{[t/h_n]} = \mathrm{e}^{tA_\epsilon^*}$, may be deduced from (19.7) as follows. A $\phi \in D(A_\epsilon^*)$ may be extended to a $\phi \in W^{2,1}(\mathbb{R})$ by:

$$\phi(-x) = \phi(x) - 2\epsilon^{-1} \int_0^x \mathrm{e}^{-\epsilon^{-1}(x-y)} \phi(y)\, \mathrm{d}y,\ x > 0$$

(comp. Lemma 60.1). Since:

$$T_{n,\mathrm{el}}^* \phi(x) = T_n^* \phi(x) + \frac{1}{2} [(1 - p_n)\phi(\Delta_n - x) - \phi(x - \Delta_n)] 1_{x<\Delta_n}(x), \quad x \geq 0,$$

[1] More specifically, $L^1(\mathbb{R})$ is a *subspace* of the dual to $C[-\infty, \infty]$.

(19.7) reduces our task to showing that:

$$\lim_{n\to\infty} h_n^{-1} \int_0^{\Delta_n} \left| 2\epsilon^{-1} \int_0^x \mathrm{e}^{-\epsilon^{-1}(x-y)} \phi(y)\,\mathrm{d}y - \frac{1}{2} p_n \phi(x) \right| \mathrm{d}x = 0. \tag{19.9}$$

This, however, is an exercise in calculus (Exercise 19.2). For a different approach to this approximation, see Chapter 24.

Example 19.5 In this example we follow the idea of Nittka [277] to express the elastic Brownian motion semigroup in terms of the unrestricted Brownian motion semigroup in $C[-\infty, \infty]$ (generated by $A = \frac{1}{2}\frac{\mathrm{d}^2}{\mathrm{d}x^2}$ with maximal domain $C^2[-\infty, \infty]$). To this end, given $\epsilon > 0$ and $f \in C[0, \infty]$, we define $\tilde{f}_\epsilon \in C[-\infty, \infty]$ by:

$$\tilde{f}_\epsilon(x) = \begin{cases} f(x), & x \ge 0 \\ \mathrm{e}^{2x\epsilon^{-1}} f(-x), & x < 0. \end{cases}$$

It is clear that $\|\tilde{f}_\epsilon\|_{C[-\infty,\infty]} = \|f\|_{C[0,\infty]}$ and that $\tilde{f}_\epsilon$ is twice continuously differentiable in $x \in (-\infty, 0)$, provided that so is f in $x \in (0, \infty)$, and we have:

$$(\tilde{f}_\epsilon)'(x) = \frac{2}{\epsilon}\mathrm{e}^{2x\epsilon^{-1}} f(-x) - \mathrm{e}^{2x\epsilon^{-1}} f'(-x), \text{ and}$$

$$(\tilde{f}_\epsilon)''(x) = \frac{4}{\epsilon^2}\mathrm{e}^{2x\epsilon^{-1}} f(-x) - \frac{4}{\epsilon}\mathrm{e}^{2x\epsilon^{-1}} f'(-x) + \mathrm{e}^{2x\epsilon^{-1}} f''(-x), \qquad x < 0.$$

It follows that if $f \in D(A_\epsilon)$ (in particular: $f(0) = \epsilon f'(0)$), then $\tilde{f}_\epsilon \in D(A)$.

Thus, $V(t) = R\mathrm{e}^{tA}E_\epsilon$, $t \ge 0$ is a strongly continuous family of contractions in $C[0, \infty]$, where $E_\epsilon : C[0, \infty] \to C[-\infty, \infty]$ assigns $\tilde{f}_\epsilon$ to an f, and R maps a $g \in C[-\infty, \infty]$ into its restriction to $[0, \infty]$. Clearly, $V(0) = I_{C[0,\infty]}$ and for $f \in D(A_\epsilon)$ we have:

$$\begin{aligned} \lim_{t\to 0} t^{-1}(V(t)f - f) &= R \lim_{t\to 0} t^{-1}(\mathrm{e}^{tA}E_\epsilon f - E_\epsilon f) = RAE_\epsilon f = R\frac{1}{2}(\tilde{f}_\epsilon)'' \\ &= A_\epsilon f. \end{aligned}$$

Hence, by Chernoff's Product Formula,

$$\mathrm{e}^{tA_\epsilon} f = \lim_{n\to\infty} (R\mathrm{e}^{\frac{t}{n}A}E_\epsilon)^n f, \qquad f \in C[0, \infty], \tag{19.10}$$

establishing Nittka's approximation.

It is clear that Nittka's argument hinges on finding a contraction $E_\epsilon : C[0, \infty] \to C[-\infty, \infty]$, mapping $D(A_\epsilon)$ into $D(A)$, and Nittka shows how to find an analogue of this operator in the case where $(0, \infty)$ is replaced by an arbitrary bounded, class C^∞, open subset of $\mathbb{R}^n$ (and $C[-\infty, \infty]$ is replaced by $C_0(\mathbb{R}^n)$). In Lemma 60.1 in Chapter 60 we find a more carefully chosen E_ϵ

(which, however, is not of norm 1) to express a cosine family generated by A_ϵ in terms of the cosine family generated by A; the Weierstrass Formula allows then an alternative way of connecting the semigroup generated by A_ϵ with the semigroup generated by A. In fact, we have used this E_ϵ in our Markov chain approximation of the elastic Brownian motion.

For our next example, we need the following lemma (see [63, Lemma 6], see also [47, Example 6.2]).

Lemma 19.6 *Let $(\epsilon_k)_{k\in\mathbb{N}}$ be a sequence of mutually independent Rademacher random variables satisfying $\mathrm{P}(\epsilon_k = -1) = \mathrm{P}(\epsilon_k = 1) = 1/2$ for each $k \in \mathbb{N}$. Let $\{C(t)\}_{t\in\mathbb{R}}$ be a cosine operator family in a Banach space $\mathbb{X}$. Then:*

$$C(t)^n = \mathrm{E}[C(\epsilon_1 t + \cdots + \epsilon_n t)]$$

for each $t \in \mathbb{R}$ and each $n \in \mathbb{N}$.

Proof We proceed by induction on n, with $t \in \mathbb{R}$ being fixed arbitrarily. The statement is true for $n = 1$ because:

$$\mathrm{E}[C(\epsilon_1 t)] = \frac{1}{2}(C(t) + C(-t)) = C(t).$$

Assume that the statement holds for n. Then:

$$\begin{aligned} C(t)^{n+1} &= C(t)^n C(t) = \mathrm{E}[C(\epsilon_1 t + \cdots + \epsilon_n t)C(t)] \\ &= \mathrm{E}\left[\frac{1}{2}\big(C(\epsilon_1 t + \cdots + \epsilon_n t + t) + C(\epsilon_1 t + \cdots + \epsilon_n t - t)\big)\right]. \end{aligned}$$

Because the ϵ_k's are mutually independent, we have:

$$\begin{aligned} &\frac{1}{2}\big(C(\epsilon_1 t + \cdots + \epsilon_n t + t) + C(\epsilon_1 t + \cdots + \epsilon_n t - t)\big) \\ &\qquad = \mathrm{E}[C(\epsilon_1 t + \cdots + \epsilon_{n+1} t) \mid \epsilon_1, \ldots, \epsilon_n], \end{aligned}$$

where $\mathrm{E}[C(\epsilon_1 t + \cdots + \epsilon_{n+1} t) \mid \epsilon_1, \ldots, \epsilon_n]$ denotes the conditional expectation of $C(\epsilon_1 t + \cdots + \epsilon_{n+1} t)$ given $\epsilon_1, \ldots, \epsilon_n$. Now the statement for $n + 1$ follows from the law of total expectation (which is a particular case of the tower property of conditional expectation):

$$\mathrm{E}[C(\epsilon_1 t + \cdots + \epsilon_{n+1} t)] = \mathrm{E}\big[\mathrm{E}[C(\epsilon_1 t + \cdots + \epsilon_{n+1} t) \mid \epsilon_1, \ldots, \epsilon_n]\big].$$

The induction is complete and so is the proof. □

Example 19.7 Suppose that $\{C(t), t \in \mathbb{R}\}$ is a cosine operator family of contractions. For x in the domain of its generator, say A, $t \mapsto C(\sqrt{2t})x$ is twice continuously differentiable with first derivative at 0 being equal 0, and the second derivative equaling $2Ax$. The second order Chernoff's Formula (Example 18.4) implies that for all $t > 0$,

$$\lim_{n\to\infty}\left[C\left(\frac{\sqrt{2}t}{n}\right)\right]^{n^2} = \mathrm{e}^{t^2A} = \frac{1}{2\sqrt{\pi t^2}}\int_{-\infty}^{\infty}\mathrm{e}^{-\frac{s^2}{4t^2}}C(s)\,\mathrm{d}s,$$

where the last equality is the Weierstrass Formula (1.4). Combined with Lemma 19.6, this says that if $(\epsilon_k)_{k\in\mathbb{N}}$ is an infinite sequence of mutually independent Rademacher's random variables, then:

$$\lim_{n\to\infty}\mathrm{E}\,C\left(\frac{\sqrt{2}t}{n}\sum_{k=1}^{n^2}\epsilon_k\right) = \frac{1}{2\sqrt{\pi t^2}}\int_{-\infty}^{\infty}\mathrm{e}^{-\frac{s^2}{4t^2}}C(s)\,\mathrm{d}s.$$

This in turn is another disguise of the Central Limit Theorem. Hence, Lemma 19.6 is a discrete counterpart of the Weierstrass Formula. (For further connections between limit theorems of probability theory and approximation theorems of semigroup theory see, for example, [91, 161, 164].)

Exercise 19.1 Prove (19.7). *Hint:* write:

$$T_n\phi(x) - \phi(x) = \int_0^{\Delta_n}\int_{-y}^{y}\phi''(x+z)\,\mathrm{d}z\,\mathrm{d}y.$$

Exercise 19.2 Prove (19.9). *Hint:* consider the cases $\phi(0) > 0$, $\phi(0) < 0$, and $\phi(0) = 0$ separately. Show that for a continuous function ϕ,

$$\lim_{n\to\infty}\frac{2}{h_n}\int_0^{\Delta_n}\int_0^{x}\mathrm{e}^{-\epsilon^{-1}(x-y)}\phi(y)\,\mathrm{d}y\,\mathrm{d}x = \phi(0),$$

$$\lim_{n\to\infty}\frac{p_n}{2h_n}\int_0^{\Delta_n}\phi(x)\,\mathrm{d}x = \epsilon^{-1}\phi(0).$$

Exercise 19.3 Consider $\mathbb{X}$, μ, b, M, and B of Exercise 1.5. Assuming $b(0) = 0$ we have, by Exercise 1.3, $\mathrm{e}^{tB}f = f + \frac{f(0)}{b(0)}(\mathrm{e}^{b(0)t} - 1)b$, $f \in \mathbb{X}, t \geq 0$.

(a) Combine this with (1.7), assuming additionally $\mu(0) = b(0) =: \beta > 0$, to see that:

$$\mathrm{e}^{tB}\mathrm{e}^{tM}f = \mathrm{e}^{tM}f + \beta^{-1}f(0)(1 - \mathrm{e}^{-\beta t})b, \qquad t \geq 0, f \in \mathbb{X}.$$

(b) Using $\mathrm{e}^{tB}\mathrm{e}^{tM}f(0) = f(0)$, conclude that:

$$\left(\mathrm{e}^{tB}\mathrm{e}^{tM}\right)^2 f = \mathrm{e}^{2tM}f + \beta^{-1}f(0)(1 - \mathrm{e}^{-\beta t})\mathrm{e}^{tM}b + \beta^{-1}f(0)(1 - \mathrm{e}^{-\beta t})b,$$

and, by induction,

$$\left(e^{tB}e^{tM}\right)^n f = e^{ntM} f + \beta^{-1} f(0)(1 - e^{-\beta t}) \sum_{k=0}^{n-1} e^{ktM} b,$$

for $t \geq 0, n \geq 0, f \in \mathbb{X}$.

(c) By the Trotter Product Formula, (b) suggests:

$$\begin{aligned} e^{t(B+M)} f(x) &= \lim_{n\to\infty} \left(e^{\frac{t}{n}B} e^{\frac{t}{n}M}\right)^n f(x) \\ &= e^{-\mu(x)t} f(x) + \beta^{-1} f(0)(1 - e^{-\mu(x)t}) \lim_{n\to\infty} \frac{1 - e^{-\frac{\beta}{n}t}}{1 - e^{-\frac{\mu(x)}{n}t}} \\ &= e^{-\mu(x)t} f(x) + f(0)(1 - e^{-\mu(x)t}) b(x), \end{aligned}$$

see (1.8), where we have encountered equality between the first and the last expression here. However, unless we check the stability condition, we cannot claim the first equality here. If, rather than in convergence, we are interested in the sole formula (1.8), checking the latter directly is an easier task.

☞ Chapter's summary

Various applications of the results of the previous chapter are given, including approximating a few types of Brownian motions by Markov chains.

20

Back to Wright's Diffusion

Let $C[0, 1]$ be the space of continuous functions on the closed interval $[0, 1]$ and let A in $C[0, 1]$ be given by:

$$Af(x) = \frac{1}{2}x(1-x)f''(x) + (\beta - (\alpha+\beta)x)f'(x)$$

on the domain composed of all polynomials. If f is a polynomial $f(x) = \sum_{i=0}^{n} a_i x^i$, then Af is the polynomial:

$$Af(x) \tag{20.1}$$

$$= \beta a_1 + \sum_{i=1}^{n-1} \left\{ \left[\beta(i+1) + \frac{i}{2}(i+1) \right] a_{i+1} - \left[(\alpha+\beta)i + \frac{i}{2}(i-1) \right] a_i \right\} x^i$$

$$- \left[(\alpha+\beta)n + \frac{n}{2}(n-1) \right] a_n x^n. \tag{20.2}$$

In particular, A with such domain cannot possibly satisfy the range condition, and thus does not generate a semigroup. However, a well-known criterion (see, e.g., [49, 132, 193]) says that a densely defined operator A satisfying the maximum principle and such that for all $\lambda > 0$, the range of $\lambda - A$ is dense in $C[0, 1]$, is closable, and its closure generates a Feller semigroup in this space. Because our operator is clearly densely defined and satisfies the maximum principle (see Exercise 20.1), we want to show that the range of $\lambda - A$ contains all polynomials. By (20.2), for a polynomial $g(x) = \sum_{i=0}^{n} b_i x^i$, the resolvent equation $\lambda f - Af = g$, where $\lambda > 0$, is satisfied iff the coefficients satisfy the system:

$$\left[\lambda + (\alpha+\beta)n + \frac{n}{2}(n-1) \right] a_n = b_n,$$

$$\left[\lambda + (\alpha+\beta)i + \frac{i}{2}(i-1) \right] a_i = b_i + \left[\beta(i+1) + \frac{i}{2}(i+1) \right] a_{i+1},$$

$$i = 1, \dots, n-1,$$

$$\lambda a_0 = b_0 + \beta a_1.$$

The system clearly having a (unique) solution, our claim is proved. The semigroup generated by the closure of A is related to Wright's diffusion with mutation.

We will show, as anticipated in Chapter 17, that Wright's diffusion is a limit of the Markov chains of the Wright–Fisher model. To this end, we consider contraction operators in $C[0, 1]$:

$$T_N f(x) = \sum_{k=0}^{2N} f(k/2N)\, p_{x,N}^k (1 - p_{x,N})^{2N-k},$$

where $p_{x,N} = x(1 - \frac{\alpha}{2N}) + (1 - x)\frac{\beta}{2N} \in [0, 1]$, so that $p_{x,N}$ tends to x uniformly in $[0, 1]$. For $A_N := 2N(T_N - I)$ where I is the identity in $C[0, 1]$ we want to show that:

$$\lim_{N\to\infty} A_N f = Af, \qquad f \in D(A).$$

Then, Theorem 18.1 (together with Corollary 8.4) will force:

$$e^{tA} f = \lim_{N\to\infty} T_N^{[2Nt]} f,\ f \in C[0, 1], t \geq 0,$$

as claimed.

By the Taylor Theorem,

$$\begin{aligned} A_N f(x) &= 2NE\left[f(Y_{N,x}/2N) - f(x)\right] \\ &= 2Nf'(x)E\,Z_{x,N} + NE\,f''(x + \Theta Z_{x,N})Z_{x,N}^2 \\ &= f'(x)[\beta - (\alpha + \beta)x] + NE\,f''(x + \Theta Z_{x,N})Z_{x,N}^2 \end{aligned}$$

where $Y_{N,x}$ is a binomial variable with parameters $2N$ and $p_{x,N}$, $Z_{x,N} := Y_{x,N}/2N - x$, and $\Theta = \Theta(x, N)$ is a random variable with values in $[0, 1]$. Since $NEZ_{x,N}^2 = \frac{1}{2}p_{x,N}(1 - p_{x,N}) + \frac{[\beta-(\alpha+\beta)x]^2}{4N}$ converges uniformly to $\frac{1}{2}x(1 - x)$, we need to show that $NE\,[f''(x + \Theta Z_{x,N}) - f''(x)]Z_{x,N}^2$ converges uniformly to zero.

To this end, given $\epsilon > 0$ we find a $\delta > 0$ such that $|f''(x) - f''(y)| < \epsilon$, provided $x, y \in [0, 1]$ and $|x - y| < \delta$, and writing $Z_{x,N} = Y_{x,N}/2N - p_{x,N} + p_{x,N} - x$, we choose N so large that $\sup_{x\in[0,1]} |p_{x,N} - x| < \delta/2$. Then calculating the relevant expected value over the event where $|Y_{x,N}/2N - p_{x,N}| \geq \delta/2$ and its complement we obtain that the expression involved does not exceed:

$$\begin{aligned} &N\epsilon EZ_{x,N}^2 + 2N\|f''\|P(|Y_{x,N}/2N - p_{x,N}| \geq \delta/2) \\ &\leq const.\,\epsilon + 32\delta^{-4}N\|f''\|E\,(Y_{x,N}/2N - p_{x,N})^4 \end{aligned}$$

where in the first line we have used the fact that $|Z_{x,N}| \leq 1$. Since for a binomial variable with parameters n and p, $E(X - np)^4 = 3n^2p^2q^2 + npq(1 - 6pq)$, where $q = 1 - p$ (see, e.g., [190] p. 110) the second summand above does not exceed $const. N^{-2}$, and the claim follows.

For a much more general diffusion approximation theorem see [132], p. 415. A thorough discussion of the behavior of the Wright's diffusion at the boundaries, and of the possibility of the diffusions' return from the boundary to the interior of the interval, is contained in [196].

Exercise 20.1 Show that A satisfies the positive maximum principle.

Exercise 20.2 Let A be the operator in $C[-1, 1]$ defined by:

$$Af(x) = (\alpha_1 - \alpha_2 x)f'(x) + \beta(1 - x^2)f''(x)$$

for all polynomials f, where α_1, α_2 and $\beta > 0$ are constants satisfying $\alpha_2 \geq |\alpha_1|$. Show that A is closable, and its closure generates a conservative Feller semigroup in $C[-1, 1]$.

Exercise 20.3 Let $A : D(A) \to \mathbb{X}$ be a linear operator in a Banach space $\mathbb{X}$. An operator B in the same space is said to be **relatively bounded** with respect to A, or simply A-bounded, if $D(A) \subset D(B)$ and there are non-negative constants a, b such that:

$$\|Bx\| \leq a\|Ax\| + b\|x\|, \qquad x \in D(A). \tag{20.3}$$

A well-known lemma (see, e.g., [128, p. 171]) says that if this condition holds with $a < 1$, and A is closed then so is $A + B$ (with domain $D(A)$). Use this to show that in the space $C[1, 2]$ of continuous functions on the interval $[1, 2]$, the operator $f \mapsto h^2 f'' + hf'$ where $h(x) = x$, $x \in [1, 2]$, is closed on the domain composed of twice continuously differentiable functions such that $f'(1) = f'(2) = 0$.

Hint: The operator $f \mapsto f''$ is closed (see Exercise 15.3), and so is $f \mapsto h^2 f''$, since $f \mapsto h^2 f$ is continuous with bounded inverse. Also, for the Bielecki-type norm [34, 125] in $C[1, 2]$ given by:

$$\|f\|_\lambda = \sup_{x \in [1,2]} |e^{-\lambda x} f(x)|, \qquad \lambda > 0$$

we have:

$$\begin{aligned} \|hf'\|_\lambda &\leq \sup_{x\in[1,2]} e^{-\lambda x} \int_1^2 e^{\lambda y} e^{-\lambda y} y^2 |f''(y)| \, dy \\ &\leq 2 \sup_{x\in[1,2]} \int_0^{x-1} e^{-\lambda z} \, dz \, \|h^2 f''\|_\lambda \leq \frac{2}{\lambda} \|h^2 f''\|_\lambda, \end{aligned}$$

and the norm is equivalent to the original supremum norm.

Exercise 20.4 Show that the operator of the previous exercise generates a conservative Feller semigroup in $C[1, 2]$. To this end check that:

(a) A is densely defined and satisfies the positive maximum principle.
(b) For fixed noninteger $\lambda > 0$, and for any integer $n \geq 0$, there are constants C_1 and C_2 such that:

$$f(x) = C_1 x^{\sqrt{\lambda}} + C_2 x^{-\sqrt{\lambda}} + \frac{1}{\lambda - n^2} x^n, \qquad x \in [1, 2],$$

defines a member of $D(A)$ such that $\lambda f(x) - Af(x) = x^n, x \in [1, 2]$.
(c) Functions $k_n(x) = x^n, x \in [1, 2], n \geq 0$ form a linearly dense set in $C[1, 2]$ (because so do $k_n^\sharp(x) = (x + 1)^n$ in $C[0, 1]$).

☞ Chapter's summary

Using the theory developed in Chapter 18, we are able to make intuitions of Chapter 17 more clear, and to show convergence of scaled Markov chains involved in the Wright–Fisher model to Wright–Fisher's diffusion on [0, 1].

21

Kingman's n-Coalescent

Individuals in the Wright–Fisher model are not independent: tracing back their genealogical lines, we discover that some of them descend from a single common ancestor, from whom they inherit most of their genetical makeup. This is to say that the structure of dependence is hidden in (random) genealogical trees. This structure is described by a beautiful mathematical object named **Kingman's n-coalescent**. As a first step toward its construction to which this chapter is devoted, we define the related pure death process.

We consider the Wright–Fisher population of size $M = 2N$, and observe n individuals sampled from generation 0. We are interested in the number $X_M(k), k \geq 1$, of ancestors of this sample k generations back; we assume that the process is well defined for all $k \geq 0$, that is, that the population has evolved according to the Wright–Fisher rules for an infinitely long time. $X_M(k), k \geq 0$, is a discrete-time Markov chain with values in $\{1, \ldots, n\}$ and transition probabilities $p_{i,j} = p_{i,j}(M) = M^{-i}\left\{{i \atop j}\right\}\binom{M}{j} j!$, where $\left\{{i \atop j}\right\}$ is the Stirling number of the second kind – see e.g. [49, 165]. Indeed, M^i is the number of all possible ways i members may choose their parents, and the number of ways exactly j parents may be chosen is the product of three numbers. The first of them is the number of ways the set of i elements may be partitioned into j subsets, that is, the Stirling number of the second kind. The second is the number of ways j parents may be chosen from the population of M individuals – the binomial coefficient $\binom{M}{j}$, and the third is the number of possible assignments of j parents to j subsets.

The process $X_M(k), k \geq 0$, is a **pure death process** in that its paths are non-increasing sequences. We will show that $X_M([tM]), t \geq 0$, converges to a continuous-time (pure death) process with intensity matrix $Q = (q_{i,j})$, where:

$$q_{i,i} = -\binom{i}{2}, \, i = 1, \ldots, n, \; q_{i,i-1} = \binom{i}{2}, \, i = 2, \ldots, n \qquad (21.1)$$

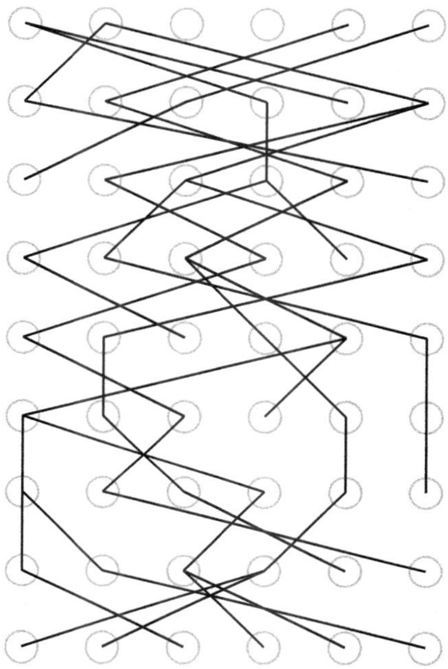

Figure 21.1 Random genealogies in the Wright–Fisher model

and $q_{i,j} = 0$ otherwise. To this end, we note first that to prove that

$$M\left[\left(p_{i,j}\right)_{1\leq i,j\leq n} - I\right]$$

converges to Q it suffices to show that the corresponding entries of these matrices converge. Moreover,

$$p_{i,i} = \prod_{k=1}^{i-1}\left(1 - \frac{k}{M}\right) = 1 - \sum_{k=1}^{i-2}\frac{k}{M} + h_1$$

where $|h_1| \leq 2^{i-1}\sum_{l=2}^{i-1}\frac{(i-1)^l}{M^l}$, so that $\lim_{M\to\infty} Mh_1 = 0$. Similarly,

$$p_{i,i-1} = \binom{i}{2}\frac{1}{M}\prod_{k=1}^{i-2}\left(1 - \frac{k}{M}\right) = \frac{1}{M}\binom{i}{2} + h_2 \tag{21.2}$$

where $|h_2| \leq 2^{i-1}\binom{i}{2}\frac{1}{M}\sum_{l=1}^{i-2}\frac{(i-2)^l}{M^l}$ so that $\lim_{M\to\infty} Mh_2 = 0$. This shows that $\lim_{M\to\infty} M[p_{i,i} - 1] = -\binom{i}{2} = -\lim_{M\to\infty} Mp_{i,i-1}$. Moreover, since:

$$\sum_{j=1}^{n} p_{i,j} = \sum_{j=1}^{i} p_{i,j} = 1, \tag{21.3}$$

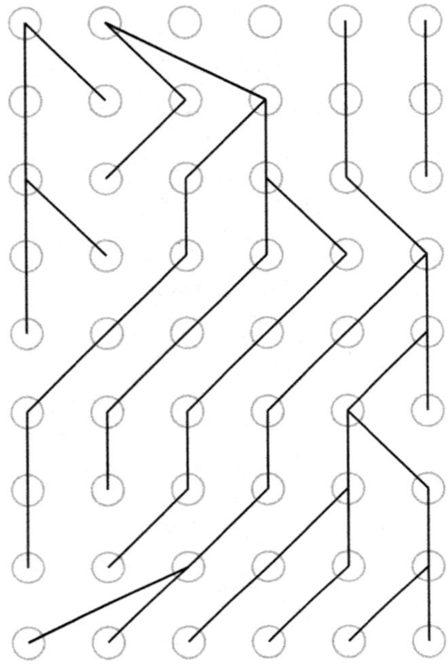

Figure 21.2 Untangled genealogies (redrawn from [66])

for $j \neq i, i-1$, $Mp_{i,j} \leq M \sum_{j \neq i, i-1} p_{i,j} = M(1 - p_{i,i} - p_{i,i-1}) \leq M|h_1| + M|h_2| \longrightarrow 0$ as $M \to \infty$, as desired.

A more careful analysis allows tracing of the whole genealogy of a sample. To this end, for a sample of n individuals we consider the Markov chain $\mathcal{R}_M(k), k \geq 0$, of equivalence relations in $\{1, \ldots, n\}$; the pair (i, j) belongs to the equivalence relation $\mathcal{R}_M(k)$ iff the individuals i and j have a common ancestor k generations ago. Each equivalence class corresponds to a member of a population that lived k generations ago, yet the opposite statement is not true because some members of this generation may not have had descendants. $\mathcal{R}_M(0)$ is the main diagonal in the square $\{(i, j)|1 \leq i, j \leq n\}$ and by this analysis, $\mathcal{R}_M(k)$ eventually reaches the full equivalence relation, that is, the whole square (see Figure 21.2 and Table 21.1).

We follow Kingman [207], [209] (compare [325]) to show that the process $\mathcal{R}_M([Mt]), t \geq 0$, converges, as $M \to \infty$, to the continuous-time Markov chain with intensity matrix Q given by:

$$q_{\mathcal{E},\mathcal{E}'} = \begin{cases} -\binom{|\mathcal{E}|}{2}, & \text{if } \mathcal{E} = \mathcal{E}', \\ 1, & \text{if } \mathcal{E} \prec \mathcal{E}', \\ 0, & \text{otherwise}, \end{cases} \tag{21.4}$$

Table 21.1 *Equivalence relations in Figure 21.2 (redrawn from [66])*

equivalence relation	equivalence classes
$\mathcal{R}_6(0)$	$\{1\}\{2\}\{3\}\{4\}\{5\}\{6\}$
$\mathcal{R}_6(1)$	$\{1,2\}\{3\}\{4\}\{5,6\}$
$\mathcal{R}_6(2)$	$\{1,2\}\{3,4\}\{5,6\}$
$\mathcal{R}_6(3)$ and $\mathcal{R}_6(4)$	$\{1,2\}\{3,4,5,6\}$
$\mathcal{R}_6(5)$ and consecutive ones	$\{1,2,3,4,5,6\}$

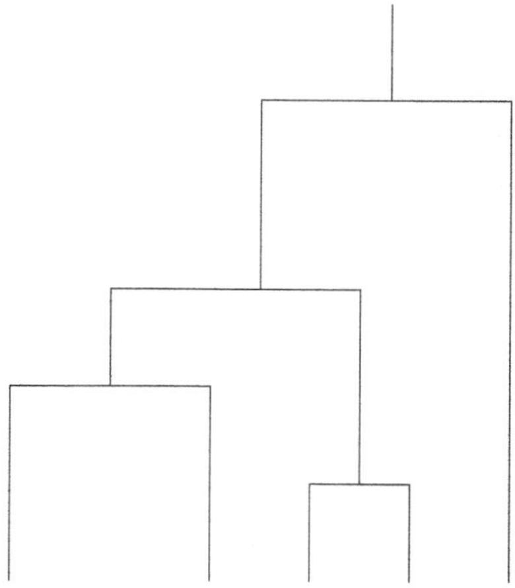

Figure 21.3 Converging genealogical lines

where $|\mathcal{E}|$ denotes the number of equivalence classes in an equivalence relation $\mathcal{E}$ and we write $\mathcal{E} \prec \mathcal{E}'$ iff $\mathcal{E} \subset \mathcal{E}'$ and $\mathcal{E}'$ is formed by amalgamating (exactly) two equivalence classes of $\mathcal{E}$. The Markov chain with intensity matrix (21.4) is called the **n-coalescent of Kingman**.

To this end we note that $p_{\mathcal{E},\mathcal{E}'}$, the transition probability of the chain $\mathcal{R}_M$, is zero if $\mathcal{E} \not\subset \mathcal{E}'$. Also if $\mathcal{E} \subset \mathcal{E}'$ yet $\mathcal{E} \not\prec \mathcal{E}'$, then $|\mathcal{E}| - |\mathcal{E}'| \geq 2$, and $p_{\mathcal{E},\mathcal{E}'} \leq p_{|\mathcal{E}|,|\mathcal{E}'|}$ where $p_{i,j}$ is the transition probability of the related pure death chain. Hence, $\lim_{M\to\infty} M p_{\mathcal{E},\mathcal{E}'} = 0$. Moreover, $p_{\mathcal{E},\mathcal{E}} = p_{|\mathcal{E}|,|\mathcal{E}|}$, so that $\lim_{M\to\infty} M(p_{\mathcal{E},\mathcal{E}} - 1) = -\binom{|\mathcal{E}|}{2}$. Finally, if $\mathcal{E} \prec \mathcal{E}'$, $p_{\mathcal{E},\mathcal{E}'} = \binom{|\mathcal{E}|}{2}^{-1} p_{|\mathcal{E}|,|\mathcal{E}|-1}$ (we do know which two equivalence classes are to be amalgamated), so that $\lim_{M\to\infty} M p_{\mathcal{E},\mathcal{E}'} = 1$, as desired.

This result may be used to derive many classic formulae for sampling distributions in population genetics (see [134]) and in particular the famous Ewens Sampling Formula (see [208]).

☞ Chapter's summary

As another application of the theory developed in Chapter 18, we establish existence of the n-coalescent of J. F. C. Kingman, a continuous-time Markov process, being the limit of a sequence of discrete-time processes. This is a process whose state-space is the set of equivalence relations on a set of n elements. After arriving at an equivalence relation with k equivalence classes, the process waits for an exponential time with parameter $\binom{k}{2}$, to amalgamate two random equivalence classes into one, and then continues in the same manner until it arrives at the absorbing state where all elements of the set are in relation with each other.

22

The Feynman–Kac Formula

One of the most interesting results that can be deduced from the Lie–Trotter Approximation Theorem of Chapter 18 is the famous **Feynman–Kac Formula**. The latter fact is nowadays usually formulated in the context of real (or vector-valued) Markov processes, and there are well-developed and intuitive tools of stochastic analysis for its proof [240, 298]. For our purposes, however, it will be more appropriate to think of Lévy processes with values in a locally compact group.

To explain the original Feynman–Kac Formula, we note that the half of the one-dimensional Laplacian $A = \frac{1}{2}\frac{\mathrm{d}^2}{\mathrm{d}x^2}$ of Chapter 2 may also be considered in the space $C_0(-\infty, \infty)$ of functions vanishing at infinity: when considered with maximal domain $C_0^2(-\infty, \infty)$, A generates the Brownian motion semigroup:

$$\mathrm{e}^{tA} f(x) = Ef(x + w_t) = \frac{1}{\sqrt{2\pi t}} \int_{-\infty}^{\infty} \mathrm{e}^{-\frac{y^2}{2t}} f(x + y)\,\mathrm{d}t, \tag{22.1}$$

where $w_t, t \geq 0$ is a standard Brownian motion. For $s \geq 0$, the solution to the related nonautonomous Cauchy problem:

$$\frac{\partial u(t,x)}{\partial t} = \frac{\partial^2 u(t,x)}{\partial x^2} + V(t,x)u(t,x), \quad u(s,x) = f(x), x \in \mathbb{R}, t \geq s, \tag{22.2}$$

Figure 22.1 Feynman–Kac Formula by Radek Bobrowski

is given by the Feynman–Kac Formula:

$$u(t,x) = E \exp\left[\int_s^t V(r, x + w_{t-r})\,\mathrm{d}r\right] f(w_{t-s}), \tag{22.3}$$

provided V is sufficiently regular.

As indicated earlier, in this chapter it is our aim to show that a similar formula holds for a much broader class of semigroups, namely, those related to certain Lévy processes in locally compact groups. Our main tool is the Lie–Trotter-type formula for evolution families of operators given later, and we follow [43].

Let A be the generator of a strongly continuous semigroup $(\mathrm{e}^{tA})_{t\ge 0}$ in a Banach space $\mathbb{X}$ and suppose that, for some $\lambda \in \mathbb{R}$,

$$\|\mathrm{e}^{tA}\| \le \mathrm{e}^{\lambda t}, \quad t \ge 0. \tag{22.4}$$

Furthermore, let $\Theta > 0$ be given and $V(t) \in L(\mathbb{X}), t \in [0, \Theta]$ be a strongly continuous family of operators such that, for every $x \in \mathcal{D}(A)$, the function $t \to V(t)x$ is strongly continuously differentiable. By the classic result of Kato (see, e.g., [223] ch. 2, Theorems 3.6 and 3.7, or the corollary at p. 102 in [326]), for $x \in D(A)$ and $s \in [0, \Theta)$ the nonautonomous Cauchy problem

$$\frac{\mathrm{d}u(t)}{\mathrm{d}t} = Au(t) + V(t)u(t), \quad u(s) = x, t \in [s, \Theta] \tag{22.5}$$

has a unique (classical) solution. Here, we restrict ourselves to proving uniqueness of the solutions: suppose that $u : [s, \Theta] \to \mathbb{X}$ is strongly continuously differentiable, $u(r) \in D(A)$ for $r \in [s, \Theta]$, and (22.5) is satisfied. Replacing t in (22.5) by r, applying $\mathrm{e}^{(t-r)A}$ to both sides, and integrating from s to t, we see that u satisfies the following **Volterra Equation**:

$$u(t) = \mathrm{e}^{(t-s)A}x + \int_s^t \mathrm{e}^{(t-r)A}V(r)u(r)\,\mathrm{d}r, \qquad t \in [s, \Theta]. \tag{22.6}$$

We will show that *for each* $x \in \mathbb{X}$, there exists a unique continuous u satisfying (22.6). (In general, such a u needs not satisfy (22.5)).

To this end, let $\mathcal{X}$ be the space of continuous $\mathbb{X}$-valued functions on $[s, \Theta]$ equipped with Bielecki-type norm [34, 125]:

$$\|u\|_\mu = \sup_{r\in[s,\Theta]} \mathrm{e}^{-\mu r}\|u(r)\|,$$

where $\mu > \lambda$, and let K be the operator in $\mathcal{X}$ mapping u into the function defined by the right-hand side of (22.6). Then,

$$\begin{aligned}\|Ku_1 - Ku_2\|_\mu &\leq M \sup_{t\in[s,\Theta]} e^{-\mu t} \int_s^t e^{\lambda(t-r)} \|u_1(r) - u_2(r)\| \,dr \\ &\leq M \sup_{t\in[s,\Theta]} \int_s^t e^{-(\mu-\lambda)(t-r)} \,dr \, \|u_1 - u_2\|_\mu \\ &\leq \frac{M}{\mu - \lambda} \|u_1 - u_2\|_\mu, \qquad u_1, u_2 \in \mathcal{X},\end{aligned}$$

where $M = \sup_{r\in[0,\Theta]} \|V(r)\|$. It follows that for large μ, K is a contraction, and Banach's Fixed Point Theorem implies that there is a unique u satisfying (22.6).[1]

Therefore the map $x \mapsto u(t) \in \mathbb{X}$, denoted $S(t, s)$, is linear and continuous. The operators $S(t, s)$, $\Theta \geq t \geq s \geq 0$ form the **evolution family** generated by $A + V(t), t \in [0, \Theta]$. Uniqueness of solutions of (22.5) implies:

$$S(t, s) = S(t, r)S(r, s), \qquad \Theta \geq t \geq r \geq s.$$

Theorem 22.1 (Lie–Trotter-type formula) *We have*

$$S(t, s)x = \lim_{n\to\infty} S_n(t, s)x, \qquad x \in \mathbb{X},$$

where

$$S_n(t, s) = e^{\Delta_n A} e^{\Delta_n V(t-\Delta_n)} e^{\Delta_n A} e^{\Delta_n V(t-2\Delta_n)} \cdots e^{\Delta_n A} e^{\Delta_n V(s)},$$

and $\Delta_n = \frac{t-s}{n}$.

Remark 22.2 In the "true" Lie–Trotter Formula, S_n would need to be defined by:

$$S_n(t, s) = e^{\Delta_n A} R(t, t - \Delta_n) e^{\Delta_n A} R(t - \Delta_n, t - 2\Delta_n) \cdots e^{\Delta_n A} R(s + \Delta_n, s),$$

where R is the evolution family generated by $V(t), t \in [0, \Theta]$ (compare [28], Theorems 2.2 and 3.2). However, with the definition we use, the proofs of Theorem 22.1 and of the Feynman–Kac Formula are slightly shorter (comp. [43]).

Proof (Compare [128] pp. 227–229, see also [275].) Let $\mathcal{X}$ be the space of $\mathbb{X}$-valued continuous functions on $[s, \Theta]$. Identifying a member f of $\mathcal{X}$ with its extension to $(-\infty, \Theta]$ given by $f(r) = f(s), r < s$, we define: $[\mathcal{S}(t)f](r) = e^{tA} f(r - t), t \geq 0$ and $[\mathcal{S}_\mu(t)f](r) = e^{tA_\mu} f(r - t), r \in [s, \Theta]$, where $A_\mu, \mu > 0$ is the Yosida approximation of A. Then $(\mathcal{S}(t))_{t\geq 0}$ and $(\mathcal{S}_\mu(t))_{t\geq 0}$ are strongly continuous semigroups in $\mathcal{X}$, and we have $\lim_{\mu\to\infty} \mathcal{S}_\mu(t) = \mathcal{S}(t)$ strongly in

[1] Local uniqueness of solutions follows by Gronwall's Lemma.

$\mathcal{X}$. Let $\mathcal{A}$ and $\mathcal{A}_\mu$ be their generators and let $\mathcal{B}$ be the bounded operator in $\mathcal{X}$ given by:

$$(\mathcal{B}f)(s) = V(s)f(s), s \in [0, \Theta].$$

The operators $\mathcal{A}_\mu$ have the common domain composed of continuously differentiable functions, and for such functions we have $[\mathcal{A}_\mu f](s) = A_\mu f(s) - f'(s)$.

Therefore u_μ, $\mu > 0$ defined by $u_\mu(t) = [\mathrm{e}^{(t-s)(\mathcal{A}_\mu + \mathcal{B})} f](t)$, $t \in [s, \Theta]$ where $f(r) = x$, $r \in [s, \Theta]$, are differentiable and satisfy:

$$\begin{aligned}\frac{\mathrm{d}u_\mu(t)}{\mathrm{d}t} &= [(\mathcal{A}_\mu + \mathcal{B})\mathrm{e}^{(t-s)(\mathcal{A}_\mu+\mathcal{B})} f](t) + [\mathrm{e}^{(t-s)(\mathcal{A}_\mu+\mathcal{B})} f]'(t) \\ &= A_\mu u_\mu(t) + V(t)u_\mu(t), \qquad u_\mu(s) = x.\end{aligned}$$

It follows that $u_\mu(t) = \mathrm{e}^{(t-s)A_\mu}x + \int_s^t \mathrm{e}^{(t-r)A_\mu} V(r)u_\mu(r)\,\mathrm{d}r$. Introducing $u(t) := [\mathrm{e}^{(t-s)(\mathcal{A}+\mathcal{B})} f](t)$, we see that $\lim_{\mu\to\infty} u_\mu(t) = u(t)$. Hence, u satisfies (22.6), implying $u(t) = S(t, s)x$.

On the other hand, since $\|S(t)\| \le \mathrm{e}^{\lambda t}$ and $\|\mathrm{e}^{t\mathcal{B}}\| \le \mathrm{e}^{Kt}$ where $K = \sup_{r\in[s,\Theta]} \|V(r)\|$, the semigroups generated $\mathcal{A} - \lambda$ and $\mathcal{B} - K$ satisfy the stability condition (18.9). Therefore, by the Trotter Product Formula for semigroups, $u(t) = \lim_{n\to\infty}[(\mathrm{e}^{\Delta_n\mathcal{A}}\mathrm{e}^{\Delta_n\mathcal{B}})^n f](t)$. Since:

$$\left[\left(\mathrm{e}^{\Delta_n\mathcal{A}}\mathrm{e}^{\Delta_n\mathcal{B}}\right)^n f\right](t) = \mathrm{e}^{\Delta_n A}\mathrm{e}^{\Delta_n V(t-\Delta_n)}\mathrm{e}^{\Delta_n A}\mathrm{e}^{\Delta_n V(t-2\Delta_n)}\cdots\mathrm{e}^{\Delta_n A}\mathrm{e}^{\Delta_n V(s)}x,$$

we are done. □

Coming back to our main subject, let $\mathcal{G}$ be a locally compact group, and let g_t, $t \ge 0$ be a Lévy process in $\mathcal{G}$. The latter means that:

(a) the random variables $g_{t+h}g_t^{-1}$ and g_t, $t, h \ge 0$ are independent, and
(b) the random variables $g_{t+h}g_t^{-1}$ and g_h, $t, h \ge 0$ have the same distribution,
(c) sample paths $t \to g_t(\omega)$ have left limits and are right continuous for almost all ω in the probability space Ω where the process is defined.

It follows that the Borel measures on $\mathcal{G}$ defined by $\mu_t(B) = \Pr(g_t \in B)$, form a convolution semigroup: $\mu_t * \mu_s = \mu_{t+s}$.

Let $\mathbb{X}$ be a commutative Banach algebra, and let U be a strongly continuous representation of $\mathcal{G}$ by linear continuous automorphisms of $\mathbb{X}$:

$$U(g_1)U(g_2) = U(g_1g_2), \qquad g_1, g_2 \in \mathcal{G}.$$

Then the formula:

$$T(t)x = E\,U(g_t)x = \int_{\mathcal{G}} U(g)x\,\mu_t(\mathrm{d}g), \qquad t \ge 0, x \in \mathbb{X} \tag{22.7}$$

defines a semigroup of operators in $\mathbb{X}$; using (c) in the definition of a Lévy process we check that the semigroup is strongly continuous. We let A denote its generator and assume that (22.4) holds.

Theorem 22.3 (The Feynman–Kac Formula) *Suppose that W is a strongly continuous representation of the group $\mathcal{G}$ by linear continuous automorphisms of $\mathbb{X}$ such that for some $M > 0$:*

$$\|W(g)\| \leq M, \quad g \in \mathcal{G}, \tag{22.8}$$

and, for $x, y \in \mathbb{X}$ and $g \in \mathcal{G}$,

$$U(g)(xy) = [W(g)x][U(g)y], \qquad W(g)(xy) = [W(g)x][W(g)y]. \tag{22.9}$$

If $[0, \infty) \ni t \mapsto V(t) \in \mathbb{X}$ is strongly continuously differentiable, then the evolution family related to $A + V(t), t \geq 0$ is given by (we identify $V(t) \in \mathbb{X}$ with the related multiplication operator):

$$S(t, s)x = E \exp\left[\int_s^t W(g_{t-r})V(r)\,\mathrm{d}r\right] U(g_{t-s})x. \tag{22.10}$$

Proof (Comp. [163] pp. 54–55.) By (22.9),

$$W(g)(\mathrm{e}^y) = W(g)\sum_{n=0}^{\infty}\frac{y^n}{n!} = \sum_{n=0}^{\infty}\frac{[W(g)y]^n}{n!} = \mathrm{e}^{W(g)y}, \quad g \in \mathcal{G}, y \in \mathbb{X}. \tag{22.11}$$

The same formula implies by induction argument that, given $g_i \in \mathcal{G}$ and $x, y_i \in \mathbb{X}$, $i = 1, 2, \ldots, n$, we have:

$$\begin{aligned} &U(g_n)y_nU(g_{n-1})y_{n-1}\ldots U(g_1)y_1x \\ &\quad = [W(g_n)y_n]\,[W(g_ng_{n-1})y_{n-1}] \\ &\qquad \ldots [W(g_ng_{n-1}\ldots g_1)y_1][U(g_ng_{n-1}\ldots g_1)x]. \end{aligned} \tag{22.12}$$

Here, to avoid an excessive number of brackets, we agree to calculate "from the right to the left." This means that for operators A and B, and vectors x, y, and z, the expression $BzAyx$ is calculated by multiplying vectors x and y, taking the value of A on xy, multiplying it by z and calculating B on $zAyx$. The value of $Bz(Ay)x$ is calculated by multiplying the vectors x, Ay, and z, and calculating B on this product.

To use the Lie–Trotter-type formula, fix $x \in \mathbb{X}$, $t > s \geq 0$ and $n \in \mathbb{N}$. Let $V_i = \Delta V(s_i)$, $s_i = s + (i-1)\Delta$, $i = 1, 2, \ldots, n$, $\Delta = \Delta_n = \frac{t-s}{n}$. By (22.7),

$$\begin{aligned} S_n(t, s)x &= T(\Delta)\mathrm{e}^{V_n}T(\Delta)\mathrm{e}^{V_{n-1}}\cdots T(\Delta)\mathrm{e}^{V_1}x \\ &= \int_\Omega \cdots \int_\Omega U(g_\Delta(\omega_n))\mathrm{e}^{V_n}\cdots U(g_\Delta(\omega_1))\mathrm{e}^{V_1}x\,\mathrm{d}P(\omega_1)\ldots\,\mathrm{d}P(\omega_n). \end{aligned}$$

By (22.12), the integrand is the product of:

$$
\begin{gathered}
W(g_\Delta(\omega_n))e^{V_n} \\
W(g_\Delta(\omega_n)g_\Delta(\omega_{n-1}))e^{V_{n-1}} \\
\vdots \\
W(g_\Delta(\omega_n)g_\Delta(\omega_{n-1})\dots g_\Delta(\omega_1))e^{V_1} \\
U(g_\Delta(\omega_n)g_\Delta(\omega_{n-1})\dots g_\Delta(\omega_1))x.
\end{gathered}
$$

By (22.11), this is the product of $U(g_\Delta(\omega_n)g_\Delta(\omega_{n-1})\dots g_\Delta(\omega_1))x$ and of the exponent of:

$$
W(g_\Delta(\omega_n))V_n + W(g_\Delta(\omega_n)g_\Delta(\omega_{n-1}))V_{n-1} \\
+ \cdots + W(g_\Delta(\omega_n)g_\Delta(\omega_{n-1})\cdots g_\Delta(\omega_1))V_1,
$$

because $\mathbb{X}$ is commutative. Since $g_t, t \geq 0$ has independent, identically distributed increments, $S_n(t, s)$ is the expected value of the random vector

$$
F_n(t, s) = U(g_{t-s})x \ \exp\left[\Delta_n \sum_{i=1}^{n} W(g_{t-s_i})V(s_i)\right] \in \mathbb{X}.
$$

By (c) in the definition of the Lévy process, $r \mapsto W(g_{t-r})V(r)$, $r \in [s, t]$ is left continuous and has right limits, and therefore it is Riemann integrable; the expression in the exponent is its integral sum. Hence $F_n(t, s)$ a.s. tends to:

$$
F(t, s) = \exp\left[\int_s^t W(g_{t-r})V(r)\, dr\right] U(g_{t-s})x.
$$

Since $\|F_n(t, s)x\| \leq e^{ML(t-s)}\|U(g_{t-s})x\|$, where $L := \sup_{s \leq r \leq t} \|V(r)\|$, the Lebesque Dominated Convergence Theorem completes the proof. □

Returning to the Brownian motion semigroup, we note that (22.1) may be written as:

$$
T(t)f = E\ U(w_t)f,
$$

where U is the strongly continuous representation of $\mathcal{G} = \mathbb{R}$ given by $U(t)f(x) = f(x + t)$, $x, t \in \mathbb{R}$, and the Brownian motion is a standard example of a Lévy process. Since:

$$
U(t)(f_1 f_2) = [U(t)f_1][U(t)f_2], \qquad f_1, f_2 \in C_0(-\infty, \infty),
$$

condition (22.9) is satisfied with $W = U$. Since $U(t), t \in \mathbb{R}$ are contractions, (22.4) is satisfied also, and (22.3) turns out to be a particular case of(22.10).

In the next chapter, we present an example where $W \neq U$.

Exercise 22.1 Show that the domains of $\mathcal{A}_\mu$ considered in the proof of Theorem 22.1 are composed of continuously differentiable members of $\mathcal{X}$.

Exercise 22.2 Show (22.12).

☞ Chapter's summary

The Chernoff Product Formula of Chapter 18 is an elegant means for proving the celebrated Feynman–Kac Formula [163, pp. 54–55]. Here, we establish the latter in a quite general situation of a Lévy process on a locally compact group.

23

The Two-Dimensional Dirac Equation

Our main goal in this chapter is the proof of the Feynman–Kac-type formula for the two-dimensional Dirac Equation. Introducing multiplication in $\mathcal{G} := \mathbb{R} \times \mathbb{Z}$ by:

$$(a, m)(b, n) = (a(-1)^n + b, n + m)$$

we see that $\mathcal{G}$ is a noncommutative locally compact group. Let $N(t), t \geq 0$ be a homogenous Poisson process with the mean value $EN(t) = t$, and let $g_t, t \geq 0$ be the process in $\mathcal{G}$ defined by:

$$g_t = \Big(\xi(t), N(t)\Big),$$

where $\xi(t) = \int_0^t (-1)^{N(s)} \, ds$ is sometimes termed the telegraph process. A short calculation shows that $g_t, t \geq 0$ has independent, identically distributed increments in $\mathcal{G}$, that is, conditions (a) and (b) of the definition of a Lévy process are satisfied (see the previous chapter). Because the points of discontinuity of the process $g_t, t \geq 0$ are exactly the points of jumps of the Poisson process, it follows that $g_t, t \geq 0$ is a Lévy process.

Let A be the generator of a strongly continuous group $G(t), t \in \mathbb{R}$ of bounded linear automorphisms of a Banach space $\mathbb{X}$, B be a linear, bounded automorphism of this space, and U be the mapping from $\mathcal{G}$ into $L(\mathbb{X})$ defined by:

$$U(a, m) = B^m G(a).$$

Theorem 23.1 *Suppose*

$$G(t)B = BG(-t), \qquad t \in \mathbb{R}. \tag{23.1}$$

Then U is a strongly continuous representation of the group $\mathcal{G}$ by linear automorphisms of $\mathbb{X}$. Furthermore, for $t \geq 0$, the one-parameter group of operators *generated by the operator $A + B$ has the form:*

$$T(t)x = \int_{\mathcal{G}} \mathrm{e}^t U(g)x \, \mu_t(\mathrm{d}g) = E \, \mathrm{e}^t U(g_t)x, \qquad x \in \mathbb{X}, \tag{23.2}$$

where μ_t is the distribution of g_t.

Proof We have:

$$\begin{aligned} U(a, m)U(b, n) &= B^m G(a) B^n G(b) = B^{m+n} G\big((-1)^n a + b\big) \\ &= U\big((a, m)(b, n)\big), \end{aligned}$$

proving that U is a representation. Hence, (23.2) defines a continuous semigroup of operators. Since $\lim_{t \to 0} g_t = (0, 0)$ (almost surely), μ_t converges to the Dirac measure at $(0, 0)$. This implies that the semigroup is strongly continuous, and we are left with showing that $A + B$ is its generator.

To this end we define the Borel measures on $\mathbb{R}$ by:

$$\mu_t^n(\mathcal{B}) = \mu_t\big(\mathcal{B} \times \{n\}\big), \qquad n \geq 0, t \geq 0.$$

Then:

$$T(t)x = \sum_{n=0}^{\infty} \int_{\mathbb{R}} \mathrm{e}^t B^n G(s)x \, \mu_t^n(\mathrm{d}s) \tag{23.3}$$

and:

$$\begin{aligned} &\mu_t^0(\mathcal{B}) = \mathrm{e}^{-t} \delta(\mathcal{B} - t), \\ &\operatorname{supp} \mu_t^n \subset [-t, t], \quad t \in \mathbb{R}, n \in \mathbb{N}, \\ &\mu_t^n(\mathbb{R}) = P\big(N(t) = n\big) = \frac{t^n}{n!} \mathrm{e}^{-t}. \end{aligned}$$

Thus:

$$\int_{\mathbb{R}} \mathrm{e}^t G(s)x \, \mu_t^0(\mathrm{d}s) = G(t)x \text{ and } \lim_{t \to 0} \left\| \frac{1}{t} \sum_{n=2}^{\infty} \int_{\mathbb{R}} \mathrm{e}^t B^n G(s)x \, \mu_t^n(\mathrm{d}s) \right\| = 0. \tag{23.4}$$

Given $\varepsilon > 0$, there exists a $\delta(\varepsilon) > 0$ such that the inequality $\|G(s)x - x\| < \frac{\varepsilon}{\|B\|}$ holds for $|s| < \delta(\varepsilon)$. Therefore, for t small enough,

$$\begin{aligned} \left\| \int_{\mathbb{R}} \frac{\mathrm{e}^t BG(s)x \, \mu_t^1(\mathrm{d}s)}{t} - Bx \right\| &= \left\| \int_{\mathbb{R}} \frac{BG(s)x - Bx}{t} \, \mathrm{e}^t \, \mu_t^1(\mathrm{d}s) \right\| \\ &\leq \int_{\mathbb{R}} t^{-1} \|B\| \, \|G(s)x - x\| \, \mathrm{e}^t \mu_t^1(\mathrm{d}s) \leq \varepsilon. \end{aligned} \tag{23.5}$$

Combining (23.3), (23.4) and (23.5),

$$\lim_{t\to 0}\left(\frac{T(t)x-x}{t}-\frac{G(t)x-x}{t}\right)=Bx, \quad x\in\mathbb{X}.$$

It follows that the domain of the generator of the semigroup T is the domain of A, and that the generator of T is the operator $A+B$. □

For example, let $C_0(\mathbb{R}\times\{-1,1\})$ be the space of continuous complex functions on $\mathbb{R}\times\{-1,1\}$ vanishing at infinity. Also, let $G(t)f(x,k)=f(x+kt,k)$ and $Bf(x,k)=zf(x,-k)$, $t,x\in\mathbb{R}$, $k=-1,1$ where z is a given complex constant. Then, (23.1) holds and the representation U has the form:

$$U(t,n)f(x,k)=z^k f(x+kt,(-1)^n k).$$

Thus:

$$T(t)f(x,k)=E\,\mathrm{e}^t z^{N(t)}f(x+k\xi(t),(-1)^{N(t)}k).$$

Turning to the related Feynman–Kac Formula, we note that since $N(t)$ is Poisson distributed with parameter t, (22.4) holds with $\lambda=|z|$. Moreover, $\mathbb{X}$ is a commutative Banach algebra with multiplication $f_1f_2(x,k)=f_1(x,k)f_2(x,k)$. Introducing:

$$W(t,n)f(x,k)=f(x+kt,(-1)^n k),$$

we see that W is a representation of $\mathcal{G}$ and (22.8) holds with $M=1$. Also,

$$\begin{aligned}[U(t,n)(f_1f_2)](x,k)&=z^k f_1(x+kt,(-1)^n k)f_2(x+kt,(-1)^n k)\\&=[W(t,n)f_1](x,k)\cdot[U(t,n)f_2](x,k),\end{aligned}$$

proving the first condition in (22.9). The second condition there is immediate.

Therefore, if $\mathbb{R}^+\ni t\mapsto V(t)\in\mathbb{X}$ is strongly continuously differentiable, the solution of the two-dimensional Dirac Equation:

$$\frac{\partial u}{\partial t}(t,x,k)=k\frac{\partial u}{\partial x}(t,x,k)+zu(t,x,-k)+V(t,x,k)u(t,x,k),$$

where $x\in\mathbb{R}$, $t\geq s\geq 0$, with initial condition $u(s,x,k)=f(x,k)$ is:

$$\begin{aligned}u(t,x,k)=E\,&\mathrm{e}^{t-s}z^{N(t-s)}f(x+k\xi(t-s),k(-1)^{N(t-s)})\\&\times\exp\left[\int_s^t V(r,x+\xi(t-r),(-1)^{N(t-r)}k)\,\mathrm{d}r\right]\end{aligned}\tag{23.6}$$

This formula was originally obtained in [38, 39, 40, 41]; see also [43, 354].

Exercise 23.1 Show that for $a \in \mathbb{R}$ the probabilistic formula for the equation:

$$\frac{\partial u(t,x)}{\partial t} = a\frac{\partial u(t,x)}{\partial x} + bu(t,-x) + V(t,x)u(t,x), \qquad u(s,x) = f(x),$$

is:

$$u(t,x) = E\, \mathrm{e}^{t-s} b^{N(t-s)} f\big[(-1)^{N(t-s)}(x - a\xi(t-s))\big] \\ \times \exp\left[\int_s^t V\left(r, (-1)^{N(t-r)}(x - a\xi(t-r)\right) \mathrm{d}r\right],$$

where $x \in \mathbb{R}$, $t \geq s \geq 0$.

Exercise 23.2 Show that (23.1) is equivalent to each of the following two conditions: (a) B leaves $D(A)$ invariant and $ABx = -BAx$ for $x \in D(A)$, and (b) $B(\lambda + A)^{-1} = (\lambda - A)^{-1}B$, $\lambda > 0$.

Exercise 23.3 Let $\mathcal{H} := \mathbb{R} \times \{-1, 1\}$ be the Kisyński group with multiplication defined by:

$$(\xi, k)(\eta, l) = (\xi l + \eta, kl) \quad \xi, \eta \in \mathbb{R},\ k, l = \pm 1,$$

and let $\mathcal{H} \times \mathbb{Z}$ be the direct product of $\mathcal{H}$ and the group of integers $\mathbb{Z}$. Check that:

$$\mathcal{G}_0 = \left\{(\xi, k, z) : k = (-1)^z\right\}$$

is a subgroup of $\mathcal{H} \times \mathbb{Z}$, and that this subgroup is isomorphic to the group $\mathcal{G}$ defined at the beginning of this chapter.

☞ Chapter's summary

We prove the Feynman–Kac Formula for a telegraph process related to the two-dimensional Dirac Equation.

24

Approximating Spaces

In many applications, especially in the context of discrete approximation, the approximating operators are naturally defined in different spaces than the limit semigroup. Until now, we have chosen to ignore this fact, because our aim was to stress that the limit semigroup may act on a subspace of the original space; the latter phenomenon may also be described in the more general setup of approximating spaces but the issuing formulation is not so intuitively appealing. On the other hand, one may not ignore facts for too long without paying a price. For example, we could have avoided the trouble we had with Markov chains approximating elastic Brownian motion (see Chapter 18) had we chosen different spaces for the related operators.

Here are the details. Given $\Delta_n > 0$ and $p_n > 0$, consider a random walk on the mesh $M_n = \{0, \Delta_n, 2\Delta_n, \dots\}$, moving to the left or right with equal probability when started at $k\Delta_n$, $k \geq 1$; conditional on starting at 0 the process is either killed with probability $\frac{1}{2}p_n$ or moves to Δ_n. The related operators in

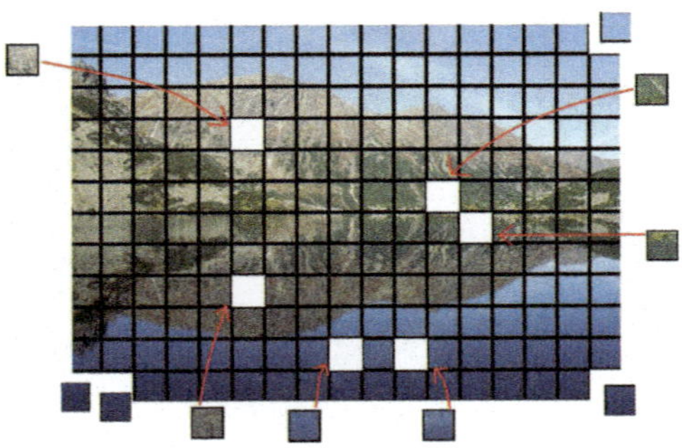

Figure 24.1 Approximating spaces by Marek Bobrowski

the space C_n of continuous functions on M_n (i.e., in the space of convergent sequences) are:

$$T_{n,\mathrm{el}}f(k\Delta_n) = \begin{cases} \frac{1}{2}f(k\Delta_n + \Delta_n) + \frac{1}{2}f(k\Delta_n - \Delta_n), & k \ge 1, \\ (1 - \frac{1}{2}p_n)f(\Delta_n), & k = 0. \end{cases} \tag{24.1}$$

Consider also the operators $P_n : C[0, \infty] \to C_n$ given by $P_n f(k\Delta_n) = f(k\Delta_n)$. We have $\|P_n\| \le 1$ and it is natural to think of a sequence $(f_n)_{n\ge1}$ such that $f_n \in C_n$ as converging to $f \in C[0, \infty]$ iff:

$$\lim_{n\to\infty} \|P_n f - f_n\|_n = 0,$$

where $\|\cdot\|_n$ denotes the (supremum) norm in C_n. It is clear from Chapter 18 that for $A_{n,\mathrm{el}} = h_n^{-1}(T_{n,\mathrm{el}} - I)$ we have:

$$\lim_{n\to\infty} \sup_{k\ge1} |A_{n,\mathrm{el}}f(k\Delta_n) - \frac{1}{2}f''(k\Delta_n)| = 0,$$

for $f \in D(A_\epsilon)$, provided $\lim_{n\to\infty} \Delta_n = 0$ and $\lim_{n\to\infty} \frac{\Delta_n^2}{h_n} = 1$. Moreover, if $\Delta_n = \frac{\epsilon p_n}{2-p_n}$, then $|A_{n,\mathrm{el}}f(0) - \frac{1}{2}f''(0)| = \frac{1}{2}|\frac{\Delta_n^2}{h_n}f''(\Theta_n\Delta_n) - f''(0)| \to 0$, where $\Theta_n \in [0, 1]$. Hence, under such conditions,

$$\lim_{n\to\infty} \|A_{n,\mathrm{el}}P_n f - P_n A_\epsilon f\|_n = 0, \qquad f \in D(A_\epsilon), \tag{24.2}$$

and no supremum-norm-difficulties emerge.

In this context it is vital to ask whether (24.2) implies $\lim_{n\to\infty} T_n^{[t/h_n]} f = \mathrm{e}^{tA_\epsilon} f$, that is, $\lim_{n\to\infty} \|T_n^{[t/h_n]} P_n f - P_n \mathrm{e}^{tA_\epsilon} f\|_n = 0$, and happily the answer is in the positive.

In fact, from the very beginning of this book we could have assumed that the approximating semigroups $(\mathrm{e}^{tA_n})_{t\ge0}$ are defined in Banach spaces $\mathbb{X}_n$ with norms $\|\cdot\|_n$, respectively, and that there exist equi-bounded operators P_n mapping a limit Banach space $\mathbb{X}$ into $\mathbb{X}_n$. Had we agreed to say that $(x_n)_{n\ge1}$, where $x_n \in \mathbb{X}_n$, converges to an $x \in \mathbb{X}$ iff $\lim_{n\to\infty} \|x_n - P_n x\|_n = 0$, all the results presented in the book would remain true with obvious modifications (this is what S. N. Ethier and T. G. Kurtz do in their monograph [132]).[1] Here is another example of such an approximation.

Let us consider a financial market **herding model** [187], where N agents are active on a financial market. Each of them is either optimistic or pessimistic about the market future, but may change his attitude on his own or by being

[1] However, it seems to be an open problem whether there is a trick allowing to deduce results pertaining to the latter setup from the one involving a single Banach space.

influenced by other agents, with probability for the other change being proportional to the number of agents of opposite views. In other words, an optimistic agent changes his mind with intensity $\alpha_P + \beta N_P$, where N_P is the number of pessimistic colleagues, and α_P and β are certain constants. Similarly, a pessimistic agents' intensity is $\alpha_O + \beta N_O$, where $N_O = N - N_P$ is the number of optimistic agents; β is a herding constant.

We assume that agents' decisions to change their minds are independent of the decisions of the other agents, and consider the random process $X_N(t), t \ge 0$ describing the difference between the numbers of optimistic and pessimistic agents. Hence, conditional on $X_N = k \in \mathcal{N} = \{-N, -N+2, \dots, N-2, N\}$, the time to the first change is exponential with parameter:

$$\lambda_{N,k} = p_{N,k} + q_{N,k},$$

where:

$$p_{N,k} = N_O(\alpha_P + \beta N_P) = \frac{N+k}{2}\left(\alpha_P + \beta\frac{N-k}{2}\right), \quad \text{and}$$
$$q_{N,k} = N_P(\alpha_O + \beta N_O) = \frac{N-k}{2}\left(\alpha_O + \beta\frac{N+k}{2}\right).$$

At the time of change, X_N either decreases or increases by two with probabilities $p_{N,k}/\lambda_{N,k}$ and $q_{N,k}/\lambda_{N,k}$, respectively. The generator of this process in the space of continuous functions f on $\mathcal{N}$ is given by $A_N f(k) = p_{N,k}[f(k-2) - f(k)] + q_{N,k}[f(k+2) - f(k)]$ with obvious changes when $|k| = N$.

We will show that as $N \to \infty$, $N^{-1}X_N$ converges to the diffusion process on $[-1, 1]$ with generator:

$$Af(x) = [\alpha_O - \alpha_P - (\alpha_O + \alpha_P)x]f'(x) + \beta(1 - x^2)f''(x),$$

introduced in Exercise 20.2. To this end, let $P_N : C[-1, 1] \to C(N^{-1}\mathcal{N})$ be given by $P_N f(k/N) = f(k/N)$. Identifying A_N's with their isomorphic images in $C(N^{-1}\mathcal{N})$, by Taylor's Formula we obtain:

$$A_N f(k/N) = \frac{2}{N}(q_{N,k} - p_{N,k})f'(k/N)$$
$$+ \frac{2p_{N,k}}{N^2}f''(k/N - \theta_1) + \frac{2q_{N,k}}{N^2}f''(k/N + \theta_2),$$

where $\theta_i \in (0, \frac{2}{N})$. Therefore,

$$A_N f(k/N) - P_N A f(k/N) = \frac{2p_{N,k}}{N^2}f''(k/N - \theta_1) + \frac{2q_{N,k}}{N^2}f''(k/N + \theta_2)$$
$$- \beta(1 - \frac{k^2}{N^2})f''(k/N).$$

Since f'' is uniformly continuous on $[-1, 1]$ and:

$$\left|\frac{2}{N^2}(p_{N,k} + q_{N,k}) - \beta(1 - \frac{k^2}{N^2})\right| \le \frac{2(\alpha_O \vee \alpha_P)}{N},$$

it follows that $\lim_{N\to\infty} \|A_N f - P_N A f\|_{C(N^{-1}\mathcal{N})} = 0$ for all polynomials f. Thus $\lim_{N\to\infty} \|e^{tA_N} f - P_N e^{tA} f\|_{C(N^{-1}\mathcal{N})}$ for all $f \in C[-1, 1]$, as claimed.

Similarly, as in the case of Wright's diffusion, it is clear from the form of the approximating Markov chains that in the limit diffusion process there should be a possibility for a path to reemerge in $(-1, 1)$ after reaching the boundary at -1 or $+1$. See the discussion in [187].

Exercise 24.1 In notations of Chapter 12, let $\mathbb{B}_\epsilon$ be the Cartesian product of $\mathbb{X}_{\mathrm{Kis}}$ and $\mathbb{X}$ equipped with the norm $\|\binom{x}{y}\|_\epsilon = \|x\|_{\mathbb{X}_{\mathrm{Kis}}} + \epsilon\|y\|_{\mathbb{X}}$. Also, let P_ϵ: $\mathbb{X}_{\mathrm{Kis}} \to \mathbb{B}_\epsilon$ be given by $P_\epsilon x = \binom{x}{0}$. Show that:

(a) The spaces $\mathbb{B}_\epsilon$ approximate $\mathbb{X}_{\mathrm{Kis}}$ in the sense specified in this chapter.
(b) The semigroups generated by B_ϵ are equi-bounded (when considered in the spaces $\mathbb{B}_\epsilon$). *Hint:* use the fact that the semigroups defined in (12.5) are equi-bounded.
(c) The semigroup $(e^{tA})_{t\ge 0}$ leaves $\mathbb{X}_{\mathrm{Kis}}$ invariant, and is a strongly continuous semigroup in $\mathbb{X}_{\mathrm{Kis}}$. The generator of the restricted semigroup is the part of A in $\mathbb{X}_{\mathrm{Kis}}$.
(d) For $x \in \mathbb{X}_{\mathrm{Kis}}$, $\lim_{\epsilon\to 0} \|e^{tB_\epsilon} P_\epsilon x - P_\epsilon e^{tA} x\|_\epsilon = 0$.

☞ Chapter's summary

Results concerning regular convergence of semigroups of operators may be naturally generalized to the case where the approximating semigroups are defined in Banach spaces approximating in a sense the Banach space where the limit semigroup is defined. Such situations occur naturally in many contexts, especially when one deals with approximation of diffusion processes via Markov chains. (However, Exercise 24.1 is of different nature.) As an example, we present a financial market herding model, where a sequence of Markov chains approximates a diffusion on the unit interval.

25

Boundedness, Stabilization

Throughout the foregoing chapters, we have always assumed that the semigroups under consideration were equi-bounded. Hence, it is a high time for us to explain that this is in a sense a *sine qua non* condition for strong and almost uniform convergence of semigroups. To this end, assume that $(\mathrm{e}^{tA_n})_{t\geq 0}$, $n \geq 1$ are strongly continuous semigroups in a Banach space $\mathbb{X}$ such that the strong limit:

$$\lim_{n\to\infty} \mathrm{e}^{tA_n}x$$

exists for all $x \in \mathbb{X}$ and is almost uniform in $t \geq 0$. Let $C([0, 1], \mathbb{X})$ be the space of continuous functions $f : [0, 1] \to \mathbb{X}$, with the usual supremum norm. Also, let $\mathcal{X}$ be the space of $C([0, 1], \mathbb{X})$-valued convergent sequences $(f_k)_{k\geq 1}$ (with the norm $\|(f_k)_{k\geq 1}\| = \sup_{k\geq 1} \|f_k\|$). Consider the linear operator $\mathcal{A} : \mathbb{X} \to \mathcal{X}$ given by:

$$\mathcal{A}x = (f_k)_{k\geq 1} \in \mathcal{X},$$

where $f_k(t) = \mathrm{e}^{tA_k}x$, $t \in [0, 1]$. Since convergence in $\mathcal{X}$ implies coordinate convergence, and convergence in $C([0, 1], \mathbb{X})$ implies pointwise convergence, $\mathcal{A}$ is closed. By the Closed Graph Theorem, $\mathcal{A}$ being defined on the whole of $\mathbb{X}$, $\mathcal{A}$ is bounded. It follows that there is a constant $M \geq 1$ such that:

$$\|\mathrm{e}^{tA_n}\| \leq M, \qquad t \in [0, 1],$$

(note that $M \geq \|I_X\| = 1$), and so for any $t \geq 0$, taking l to be the largest integer $\leq t$,

$$\|\mathrm{e}^{tA_n}\| \leq \|\underbrace{\mathrm{e}^{tA} \cdots \mathrm{e}^{tA}}_{l \text{ times}} \mathrm{e}^{(t-l)A_n}\| \leq M^{l+1} = M\mathrm{e}^{l\ln M} \leq M\mathrm{e}^{\omega t},$$

where $\omega = \ln M$.

To recapitulate, almost uniform convergence of semigroups forces existence of $M \geq 1$ and $\omega \in \mathbb{R}$ such that:

$$\|e^{tA_n}\| \leq Me^{\omega t}, \qquad t \geq 0, n \geq 1. \tag{25.1}$$

Such a **stability condition** is thus necessary for almost uniform convergence. We note that by shifting A_n's by ω (i.e., taking $A_n' = A - \omega I_X$) we obtain generators of equi-bounded semigroups, thus reducing the general problem of convergence to that of convergence of equi-bounded semigroups.

The argument presented above is valid, certainly, if we assume that the semigroups converge on the whole of $\mathbb{X}$ (and the limit is almost uniform in $t \geq 0$). As we have seen, though, convergence often takes place merely on a subspace $\mathbb{X}_0$ of $\mathbb{X}$. It is clear that in such a case, equi-boundedness is not necessary. For example, given a converging sequence $(a_n)_{n\geq 1}$ of non-negative numbers, we see that the semigroups:

$$e^{tA_n}(x, y) = (e^{-a_n t}x, e^{nt}y)$$

defined in $\mathbb{X} = \mathbb{R}^2$ with, say, Euclidean norm, satisfy (25.1) for no M and ω, and yet they converge on the subspace $\mathbb{X}_0$ of (x, y) with $y = 0$. In cases like that (25.1) should be replaced by:

$$\|e^{tA_n}x\| \leq Me^{\omega t}\|x\|, \qquad t \geq 0, n \geq 1, x \in \mathbb{X}_0. \tag{25.2}$$

Having said this, we recall from Chapters 7 and 8 that under assumption of equi-boundedness the sole range condition for A_{ex} implies existence of the limit semigroup on a regularity space. In particular, if a densely defined operator $A \subset A_{\text{ex}}$ satisfies the range condition, it automatically is a restriction of the generator of a semigroup (i.e., of the limit semigroup). Interestingly, sometimes asymptotic analysis (see Chapter 55) may be successfully applied in the case where the semigroups involved are not a priori equi-bounded but we know that the limit operator generates a semigroup [17]. This suggests that the assumptions of equi-boundedness and of the existence of the limit generator are to some extent complementary – compare Pazy's comment in [284, p. 86]; see also [128, 163]. Indeed, if existence of the limit semigroup may be assumed, proving convergence of approximating semigroups is much easier, as then they may be compared with the limit object [229, 284]; in such a case, the assumption of some kind of equi-boundedness plays a secondary, but still essential, role. We illustrate this idea in the following Theorem 25.1 where we prove convergence under a blending of assumptions involving existence of the limit semigroup and equi-boundedness. For an example of application see [17], where this theorem comes from.

Theorem 25.1 *Let* $(e^{tA_n})_{t\geq 0}$ *be (not necessarily equibounded) semigroups in a Banach space* $\mathbb{X}$ *and let* $(e^{tA})_{t\geq 0}$ *be a semigroup in a (closed) subspace* $\mathbb{X}_0$ *of* $\mathbb{X}$ *such that for some* $M > 0$,

$$\|e^{tA_n}x\| \leq M\|x\|, \qquad \|e^{tA}\| \leq M\|x\|, \qquad x \in \mathbb{X}_0,\ t \geq 0. \tag{25.3}$$

Further, suppose that:

a) $(\lambda - A_n)^{-1}, n \geq 1$ *exist and are equibounded for each* $\lambda > 0$;
b) for any x *in a core* D *of* A *there exist* $x_n \in D(A_n)$ *such that*

$$\lim_{n\to\infty} A_n x_n = Ax, \tag{25.4}$$

$$\lim_{n\to\infty} e^{tA_n}(x_n - x) = 0, \qquad t \geq 0, \tag{25.5}$$

$$\exists_{M_x}\forall_{n\geq 1, t\geq 0}\ \|e^{tA_n}A_n x_n\| \leq M_x. \tag{25.6}$$

Then $\lim_{n\to\infty} e^{tA_n}x = e^{tA}x, x \in \mathbb{X}_0$.

Proof For $x \in D$, the sequence $(x_n)_{n\geq 1}$ described in assumption b) satisfies $\lim_{n\to\infty} x_n = x$. Since $(\lambda - A_n)^{-1}$ are equibounded, we may argue as in Lemma 7.1 a) to see that $\lim_{n\to\infty}(\lambda - A_n)^{-1}(\lambda x - Ax) = x, x \in \mathbb{X}_0$. The set of y of the form $y = \lambda x - Ax, x \in D$ is dense in $\mathbb{X}_0$, since D is a core for A. Using equiboundedness of $(\lambda - A_n)^{-1}$ coupled with density of D, we obtain existence of the limit $\lim_{n\to\infty}(\lambda - A_n)^{-1}$ on the whole of $\mathbb{X}_0$. Moreover, we have $\lim_{n\to\infty}(\lambda - A_n)^{-1}x_n = (\lambda - A)^{-1}x$ as well as:

$$\lim_{n\to\infty} (\lambda - A_n)^{-1}A_n x_n = (\lambda - A)^{-1}Ax, \qquad x \in D. \tag{25.7}$$

We observe that,

$$e^{tA_n}x_n - e^{tA}x = \int_0^t (e^{sA_n}A_n x_n - e^{tA}Ax)\,ds + x_n - x \tag{25.8}$$

where, by (25.6), the integrands $s \mapsto e^{sA_n}A_n x_n - e^{tA}Ax$ on the right-hand side are equibounded. Hence, the right-hand sides are Lipschitz continuous with the same Lipschitz constant. By Lemma 2.1 and (25.7) it follows that $\lim_{n\to\infty} \int_0^t (e^{sA_n}A_n x_n - e^{sA}Ax)\,ds = 0$, and (25.8) implies $\lim_{n\to\infty} e^{tA_n}x_n = e^{tA}x$. Using again (25.5) we get $\lim_{n\to\infty} e^{tA_n}x = e^{tA}x$ for any $x \in D$ and the claim follows by density and (25.3). □

We conclude this chapter by discussing the situation where the limit semigroup $(e^{tA})_{t\geq 0}$ is defined on the whole of the space, and yet the approximating semigroups $(e^{tA_n})_{t\geq 0}, n \geq 1$ fail to satisfy (25.1). Of course, in such a case we cannot expect to have strong and almost uniform convergence. What we can hope for, however, is that there are **stabilizing operators** W_n, $n \geq 1$, that is, bounded linear operators such that $\lim_{n\to\infty} W_n = I_{\mathbb{X}}$ and:

$$\lim_{n\to\infty} W_n \mathrm{e}^{tA_n} = \mathrm{e}^{tA}.$$

This idea, coming from [255], is certainly very much related to the theory of regularized semigroups [111].

The following example of such a situation, originally presented in [255], has been slightly modified to make the phenomenon yet more transparent. Consider the space $\mathbb{X} = c_0$ of complex sequences converging to zero, equipped with the usual supremum norm. The vectors:

$$e_n = \left(\delta_{n,k}\right)_{k\geq 1}$$

(where $\delta_{n,k}$ is the Kronecker symbol) form a **Schauder basis** in c_0 (see, e.g., [78]). In particular, for any $(\xi_k)_{k\geq 1} \in c_0$,

$$(\xi_k)_{k\geq 1} = \sum_{k\geq 1} \xi_k e_k.$$

Consider the strongly continuous semigroup $\{T(t), t \geq 0\}$ in c_0 given by

$$T(t)(\xi_k)_{k\geq 1} = \left(\mathrm{e}^{ikt}\xi_k\right)_{k\geq 1} = \sum_{k\geq 1} \mathrm{e}^{ikt}\xi_k e_k.$$

It is an easy task to show that the generator of this semigroup is:

$$A(\xi_k)_{k\geq 1} = (ik\xi_k)_{k\geq 1}$$

with domain composed of $(\xi_k)_{k\geq 1} \in c_0$ such that $(k\xi_k)_{k\geq 1} \in c_0$. Now, for $n \geq 1$ let:

$$B_n(\xi_k)_{k\geq 1} = n\xi_n e_n$$
$$W_n(\xi_k)_{k\geq 1} = \sum_{k\neq n} \xi_k e_k.$$

Then, $A_n := A + B_n$ (with domain $D(A)$) is the generator of a semigroup, and since A and B_n commute, $\mathrm{e}^{tA_n} = \mathrm{e}^{tA}\mathrm{e}^{tB_n}$ is given by:

$$\mathrm{e}^{tA_n}(\xi_k)_{k\geq 1} = \sum_{k\neq n} \mathrm{e}^{ikt}\xi_k e_k + \mathrm{e}^{(in+n)t}\xi_n e_n.$$

In particular, $\|\mathrm{e}^{tA_n}\| \geq \|\mathrm{e}^{tA_n}e_n\| = \mathrm{e}^{nt}$. It follows that the semigroups $(\mathrm{e}^{tA_n})_{t\geq 0}$, $n \geq 1$ cannot converge, despite the fact that for $(\xi_k)_{k\geq 1} \in D(A)$,

$$\|A_n(\xi_k)_{k\geq 1} - A(\xi_k)_{k\geq 1}\| = \|B_n(\xi_k)_{k\geq 1}\| = |n\xi_n| \underset{n\to\infty}{\longrightarrow} 0.$$

On the other hand, $W_n, n \geq 1$ is a stabilizing sequence of operators because:

$$\|W_n(\xi_k)_{k\geq 1} - (\xi_k)_{k\geq 1}\| = |\xi_n| \underset{n\to\infty}{\longrightarrow} 0$$

and:

$$\|W_n e^{tA_n}(\xi_k)_{k\geq 1} - e^{tA}(\xi_k)_{k\geq 1}\| = |e^{int}\xi_n| \underset{n\to\infty}{\longrightarrow} 0,$$

for all $(\xi_k)_{k\geq 1} \in c_0$, proving that $\lim_{n\to\infty} W_n = I_{\mathbb{X}}$ and $\lim_{n\to\infty} W_n e^{tA_n} = e^{tA}$ (strongly).

Here is a general theorem covering situations analogous to the one previously presented. It is a simple generalization of Theorem 2.4 in [255].

Theorem 25.2 *Suppose* $(e^{tA_n})_{t\geq 0}, n \geq 1$ *are strongly continuous semigroups in a Banach space, and* $W_n, n \geq 1$ *are bounded linear operators such that* $\lim_{n\to\infty} W_n = I_{\mathbb{X}}$ *(strongly). Assume, furthermore, that there is an* $M > 0$ *such that:*

$$\|W_n e^{tA_n}\| \leq M, \qquad t \geq 0. \tag{25.9}$$

Finally, let $A \subset A_{\text{ex}}$ *be a densely defined operator such that* $\lambda - A$ *is dense in* $\mathbb{X}$ *for all* $\lambda > 0$. *Then* A *is closable,* $\overline{A}$ *generates a strongly continuous semigroup, and* $\lim_{n\to\infty} W_n e^{tA_n} = e^{t\overline{A}}$ *strongly and almost uniformly in* $t \geq 0$.

Proof We argue as in Theorem 2.2. The Laplace transform of $t \mapsto W_n e^{tA_n}x$ defines a family of bounded linear operators:

$$R_{n,\lambda}x = \int_0^\infty e^{-\lambda t} W_n e^{tA_n} x \,dt, \qquad \lambda > 0, n \geq 1, x \in \mathbb{X},$$

with $\|R_{n,\lambda}\| \leq \frac{M}{\lambda}$. We claim that the strong limit $R_\lambda := \lim_{n\to\infty} R_{n,\lambda}$ exists for all $\lambda > 0$. To prove this claim we take $\lambda > 0$ and note that it suffices to check that $\lim_{n\to\infty} R_{n,\lambda}y$ exists for y of the form $y = \lambda x - Ax$, $x \in D(A)$, such y forming a dense subspace by assumption. Taking $x_n \in D(A_n)$ such that $\lim_{n\to\infty} x_n = x$ and $\lim_{n\to\infty} A_n x_n$ we see that for $y_n = \lambda x_n - A_n x_n$ we have $\lim_{n\to\infty} y_n = y$. By the estimate for the norm of $R_{n,\lambda}$, the sequences $(R_{n,\lambda}y)_{n\geq 1}$ and $(R_{n,\lambda}y_n)_{n\geq 1}$ converge or diverge simultaneously. On the other hand,

$$\begin{aligned} R_{n,\lambda}y_n &= \int_0^\infty W_n e^{-(\lambda - A_n)t}(\lambda x_n - A_n x_n)\,dt \\ &= \int_0^\infty \frac{d}{dt}(W_n e^{-(\lambda - A_n)t} x_n)\,dt = W_n x_n \underset{n\to\infty}{\longrightarrow} x, \end{aligned}$$

proving the claim. As a by-product of the proof we obtain:

$$R_\lambda(\lambda x - Ax) = x, \qquad x \in D(A). \tag{25.10}$$

Next, for $x \in D(A)$ we take $x_n \in D(A_n)$ as earlier, and consider:

$$f_n(t) = W_n e^{tA_n} x_n = W_n x_n + \int_0^t W_n e^{sA_n} A_n x_n \, ds, \qquad n \geq 1, t \geq 0. \quad (25.11)$$

These functions are Lipschitz continuous with common constant:

$$L = M \sup_{n \geq 1} \|A_n x_n\|.$$

Moreover, their Laplace transforms converge, by the first part of the proof. Hence, by Lemma 2.1, the limit $\lim_{n\to\infty} W_n e^{tA_n} x_n$ exists and is almost uniform in $t \geq 0$. By (25.9), it follows that the same is true if x_n is replaced by x. By the same token, since $D(A)$ is dense in $\mathbb{X}$, for any $x \in \mathbb{X}$, the limit:

$$T(t)x = \lim_{n\to\infty} W_n e^{tA_n} x$$

exists and is almost uniform in $t \geq 0$. In particular, $\{T(t), t \geq 0\}$ is a strongly continuous family of bounded linear operators with $\|T(t)\| \leq M$, and:

$$\lim_{t\to 0} T(t)x = x, \quad R_\lambda x = \int_0^\infty e^{-\lambda t} T(t)x \, dt, \qquad x \in \mathbb{X}, \lambda > 0. \quad (25.12)$$

Using (25.10), for $x \in D(A)$ and λ, $\mu > 0$:

$$(\lambda - \mu) R_\mu R_\lambda (\lambda x - Ax) = (\lambda - \mu) R_\mu x, \qquad \text{and}$$

$$(R_\mu - R_\lambda)(\lambda x - Ax) = R_\mu(\mu x - Ax) + R_\mu(\lambda - \mu)x - x$$
$$= (\lambda - \mu) R_\mu x.$$

It follows that the bounded linear operators $(\lambda - \mu) R_\mu R_\lambda$ and $R_\mu - R_\lambda$ coincide on a dense set, and therefore are equal. In other words, the Hilbert Equation (1.2) is satisfied. We know from Chapter 7 that this implies that R_λ, $\lambda > 0$ have common range and kernel. On the other hand, by (25.12),

$$\lim_{\lambda\to\infty} \lambda R_\lambda x = x, \qquad x \in \mathbb{X}. \quad (25.13)$$

In particular, the kernel of R_λ, $\lambda > 0$ is trivial. Defining $B = \lambda - (R_\lambda)^{-1}$ on the range of R_λ, we obtain a closed linear operator, and, by the Hilbert Equation, the definition of B does not depend on $\lambda > 0$. Relation (25.13) shows that B is densely defined. Since $R_\lambda = (\lambda - B)^{-1}$, the second part of (25.12) may be written as:

$$(\lambda - B)^{-1} x = \int_0^\infty e^{-\lambda t} T(t)x \, dt, \qquad x \in \mathbb{X}, \lambda > 0.$$

It follows (see, e.g., [9, Thm 3.1.7]) that $\{T(t), t \geq 0\}$ is a strongly continuous semigroup, and B is its generator.

Finally, taking $x \in D(A)$ and letting $n \to \infty$ in (25.11), we obtain $T(t)x = x + \int_0^t T(s)Ax\,\mathrm{d}s$, $t > 0$. Dividing by t and letting $t \to 0$, yields then $x \in D(B)$ and $Bx = Ax$. Hence, B is a closed extension of A, and so A is closable. Clearly, $\overline{A} \subset B$. To check that B cannot be a proper extension of $\overline{A}$ we argue as in Corollary 8.4. □

☞ Chapter's summary

Boundedness condition may be relaxed to some extent, if additional information on the limit operator and/or semigroup is available. In fact, assumptions of boundedness and existence of the limit semigroup are complementary. In general, convergence requires some kind of boundedness and some information on the extended limit. Moreover, if there is no hope for any kind of boundedness we may still resort to stabilizing operators.

PART II

Irregular convergence

26

First Examples

Can a sequence of semigroups converge outside of the regularity space? A casual reader may quickly answer in the negative: didn't we define the regularity space as the space where the semigroups converge? No, we did not. The regularity space is the space where the semigroups converge *almost uniformly in* $t \in [0, \infty)$. Hence, the semigroups may and in fact quite often do converge outside of the regularity space, though they do not converge almost uniformly in $t \in [0, \infty)$. Such convergence will be referred to as **irregular**. Unfortunately, the methods developed in the first part of the book are in general useless in proving such convergence: we have reached the limits of the theory of strongly continuous semigroups.

Let us start by presenting a simple and yet surprising example showing that semigroups may converge even if the regularity space is trivial, [52].

Example 26.1 Let $\{T_0(t), t \geq 0\}$ be a semigroup of equibounded operators in a Banach space $\mathbb{X}$, generated by an operator A_0, and let $A_n = A_0 - nI_{\mathbb{X}}$. Then, $\mathrm{e}^{A_n t} = \mathrm{e}^{-nt}T_0(t)$ and the semigroups $(\mathrm{e}^{tA_n})_{t\geq 0}$, $n \geq 0$ are equibounded. Moreover, $R_\lambda = \lim_{n\to\infty}(\lambda - A_n)^{-1} = \lim_{n\to\infty}(\lambda + n - A_0)^{-1} = 0$, (even in the operator norm) since $\|(\lambda + n - A_0)^{-1}\| \leq \frac{M}{\lambda+n}$ for appropriate constant M. Hence $\mathbb{X}_0 = \{0\}$ and all we can infer from the Trotter–Kato Theorem is that $\lim_{n\to\infty} \mathrm{e}^{tA_n}0 = 0$, while in fact we have $\lim_{n\to\infty} \mathrm{e}^{tA_n} = 0, t > 0$ in the operator norm. □

Example 26.2 For given $a, b \geq 0$, let:

$$A_n = \begin{pmatrix} -a & a \\ b & -b-n \end{pmatrix}, n \geq 0, \quad I = \begin{pmatrix} 1 & 0 \\ 0 & 1 \end{pmatrix}, \text{ and } J = \begin{pmatrix} 1 & 0 \\ 0 & 0 \end{pmatrix}.$$

From Chapter 13 we know that $\|\exp(tA_0)\| = 1$, where A_0 is treated as an operator in $\mathbb{R}^2$ equipped with the norm $\|\binom{x}{y}\| = |x| + |y|$. (Recall that the value of A_0 on $\binom{x}{y} \in \mathbb{R}^2$ is obtained by multiplying A_0 by (x, y) from the left; in agreement with notations used throughout the book we write, however, $A_0\binom{x}{y}$ for this value.) Moreover, $\|J\| = 1$ and so, by the Phillips Perturbation Theorem, $\|\exp[t(A_0 + nJ)]\| \le \mathrm{e}^{nt}$. Hence, as $A_n = A_0 + nJ - nI$, $\|\exp(tA_n)\| = \mathrm{e}^{-nt}\|\exp[t(A_0 + nJ)]\| \le 1$.

Let us consider a sequence $\binom{x_n}{y_n}, n \ge 1$ of points of $\mathbb{R}^2$. The sequences $\binom{x_n}{y_n}, n \ge 1$ and $A_n\binom{x_n}{y_n}, n \ge 1$ converge simultaneously iff there exists the limit $\lim_{n\to\infty} ny_n =: l$; in particular we have then $\lim_{n\to\infty} y_n = 0$. In this case, $\lim_{n\to\infty} A_n\binom{x_n}{y_n} = \binom{-ax}{ax-l}$, where $x = \lim_{n\to\infty} x_n$. This shows that $\mathcal{D}(A_{\mathrm{ex}}) = \mathbb{X}_0$ is composed of $\binom{x}{y}$ with $y = 0$, and A_{ex} restricted to $\mathbb{X}_0$ is given by $A_{\mathrm{ex}}\binom{x}{0} = \binom{-ax}{0}$. In other words, we obtain:

$$\lim_{n\to\infty} \exp(tA_n)\begin{pmatrix} x \\ 0 \end{pmatrix} = \begin{pmatrix} \mathrm{e}^{-at}x \\ 0 \end{pmatrix}, \qquad x \in \mathbb{R}, t \ge 0. \tag{26.1}$$

On the other hand, recalling a formula from page 7 of [269],

$$\exp(tA_n) = \mathrm{e}^{r_n t}\left\{\frac{\sinh(tp_n)}{p_n}A_n + \left[\cosh(tp_n) - \frac{r_n}{p_n}\sinh(tp_n)\right]I\right\}, \tag{26.2}$$

where $r_n = \frac{\mathrm{trace}(A_n)}{2} = -\frac{a+b+n}{2}$, $p_n = \sqrt{r_n^2 - d_n}$ and $d_n = \det(A_n) = an$. Since $\lim_{n\to\infty} \frac{r_n}{p_n} = -1$, $\lim_{n\to\infty} \frac{n}{p_n} = 2$, and $\lim_{n\to\infty}(r_n + p_n) = -a$, we obtain:

$$\lim_{n\to\infty} \exp(tA_n)\begin{pmatrix} x \\ y \end{pmatrix} = \begin{pmatrix} \mathrm{e}^{-at}x \\ 0 \end{pmatrix}, \qquad x, y \in \mathbb{R}, t > 0, \tag{26.3}$$

that is, that the semigroups converge on the whole of $\mathbb{X}$.

To explain the probabilistic intuition behind this example, we note that A_0 is an intensity matrix of a (most general) two-state honest Markov chain. For $n \ge 1$, A_n is an intensity matrix of a sub-Markov chain behaving as follows. Starting in state 1, the chain remains there for an exponential time with parameter a, and then jumps to the state 2. At 2, the chain waits for an exponential time with parameter $b + n$; at the moment of jump it goes to state 1 with probability $\frac{b}{b+n}$ or disappears with probability $\frac{n}{b+n}$. It is clear that as $n \to \infty$, such processes approach the process which, at the state 1 waits for an exponential time with parameter a to jump to state 2, and at state 2 it immediately disappears from the state-space. □

Example 26.3 Let $\mathbb{X} = l^\infty$ be the space of bounded sequences $x = (\xi_n)_{n\ge1}$ equipped with the supremum norm, and A be the operator given by $Ax = (-n\xi_n)_{n\ge1}$ defined on the domain $\mathcal{D}(A) \subset l^\infty$ composed of all x such that $(-n\xi_n)_{n\ge1}$ belongs to l^∞. We have $\lambda(\lambda - A)^{-1}(\eta_n)_{n\ge1} = (\frac{\lambda}{\lambda+n}\eta_n)_{n\ge1}$, and so

$\|\lambda(\lambda - A)^{-1}\| \le 1, \lambda > 0$, proving that the estimate (8.3) is satisfied with $M = 1$. Hence, A is a Hille–Yosida operator.

From Chapter 8 we know that for the Yosida approximation of A, A_{ex} is A and the regularity space $\mathbb{X}_0$ is the closure of $\mathcal{D}(A)$. Here, any $x \in \mathcal{D}(A)$ is also a member of c_0 (the space of sequences converging to zero) because $|\xi_n| \le \frac{1}{n}\|(n\xi_n)_{n\ge 1}\|$, and so the closure of $D(A)$ is contained in c_0. Considering sequences $(\xi_n)_{n\ge 1}$ that are eventually zero, we prove that $\mathbb{X}_0 = cl\, D(A) = c_0 \subsetneq l^1$.

Despite this fact the Yosida approximation of A converges for all $x \in l^\infty$. This can be seen by noting first that:

$$\mathrm{e}^{A_\lambda t}x = \left(\mathrm{e}^{-\lambda t}\mathrm{e}^{\frac{\lambda^2}{\lambda+n}t}\xi_n\right)_{n\ge 1} = \left(\mathrm{e}^{-\frac{\lambda n}{\lambda+n}t}\xi_n\right)_{n\ge 1}$$

and that $\lim_{\lambda\to\infty} \mathrm{e}^{-\frac{\lambda n}{\lambda+n}t} = \mathrm{e}^{-nt}, t \ge 0$. Defining:

$$T(t)x = \left(\mathrm{e}^{-nt}\xi_n\right)_{n\ge 1}, x \in l^\infty, \tag{26.4}$$

we have, for any $k \in \mathbb{N}$,

$$\|T(t)x - \mathrm{e}^{A_\lambda t}x\| \le \sup_{1\le n<k} |\mathrm{e}^{-\frac{\lambda n}{\lambda+n}t} - \mathrm{e}^{-nt}|\, \|x\| + \mathrm{e}^{-\frac{\lambda k}{\lambda+k}t}\|x\|. \tag{26.5}$$

This is because for $n \ge k$,

$$\left|\mathrm{e}^{-\frac{\lambda n}{\lambda+n}t} - \mathrm{e}^{-nt}\right| = \mathrm{e}^{-\frac{\lambda n}{\lambda+n}t} - \mathrm{e}^{-nt} < \mathrm{e}^{-\frac{\lambda n}{\lambda+n}t} < \mathrm{e}^{-\frac{\lambda k}{\lambda+k}t}$$

since $\frac{\lambda n}{\lambda+n} \le n$ and the sequence $\mathrm{e}^{-\frac{\lambda n}{\lambda+n}t}, n \ge 1$, decreases. Now, for arbitrary $t > 0$ and $\epsilon > 0$ we choose a k so that $\mathrm{e}^{-kt}\|x\| < \frac{\epsilon}{2}$; in fact one such k may be chosen for all t larger than a given t_0. With this k fixed we may choose a λ large enough so that the supremum in (26.5) is less than $\frac{\epsilon}{2}$ and so is $\mathrm{e}^{-\frac{k\lambda}{\lambda+k}t}\|x\|$, uniformly in any interval $[t_0, t_1]$ where $0 < t_0 < t_1$.

Thus, the Yosida approximation converges for $x \in l^\infty$ uniformly in any interval $[t_0, t_1]$. If x belongs to c_0, the term $\mathrm{e}^{-\frac{\lambda k}{\lambda+k}t}\|x\|$ in (26.5) may be replaced by $\sup_{n\ge k}|\xi_n|$. Since for such x, $\lim_{k\to\infty}\sup_{n\ge k}|\xi_n| = 0$, in this case the convergence is almost uniform in $t \ge 0$. As we shall see in Chapter 28, this situation is typical: convergence outside of $\mathbb{X}_0$ is always almost uniform in $t \in (0, \infty)$.

Example 26.4 Let $\{T(t), t \ge 0\}$ be an asymptotically stable semigroup in a Banach space $\mathbb{X}$: that is, assume that $\lim_{t\to\infty} T(t)x = 0$ strongly for $x \in \mathbb{X}$. By the Banach–Steinhauss Theorem, this implies that the operators involved are equibounded and so are the semigroups $\{T_n(t), t \ge 0\}$ defined by $T_n(t) = T(nt), t \ge 0, n \ge 1$. Moreover, we have $\lim_{n\to\infty} T_n(t)x = 0, t > 0$ and $\lim_{n\to\infty} T_n(0)x = x$. The limit here is not almost uniform in $[0, \infty)$, except in the case where $x = 0$ – it is almost uniform in $(0, \infty)$, even uniform in any interval of the form $[\delta, \infty)$ where $\delta > 0$. In other words, $\mathbb{X}_0 = \{0\}$, but the semigroups converge on the whole of $\mathbb{X}$.

Example 26.5 Let $\mathbb{X}$ be a Banach lattice with order continuous norm and let A_0 be the generator of a positive strongly continuous contraction semigroup in $\mathbb{X}$. Suppose that B_n is a sequence of positive increasing sequence of operators, that is, that $\mathcal{D}(B_n) \subset \mathcal{D}(B_{n+1})$ and $0 \le B_n x \le B_{n+1}x$, for non-negative $x \in \mathcal{D}(B_n)$, $n \ge 1$. Assume finally that the operators $A_n = A_0 - B_n, n \ge 1$ with domains $\mathcal{D}(A_0) \cap \mathcal{D}(B_n)$, respectively, generate positive strongly continuous contraction semigroups in $\mathbb{X}$. (This scheme is discussed in [24] in the context of difficulties with constructing semigroups related to fragmentation models with spatial dependence and external fields.) These assumptions imply $0 \le (\lambda - A_{n+1})^{-1} \le (\lambda - A_n)^{-1}, n \ge 1, \lambda > 0$. To see this we note that for $x \in \mathbb{X}$, the $x_n = (\lambda - A_n)^{-1}x$ satisfy $\lambda x_n - A_0 x_n + B_n x_n = x$. Hence, if $x \ge 0$, we have $x_n \ge 0$ and:

$$\begin{aligned} x_{n+1} &= (\lambda - A_{n+1})^{-1} (\lambda x_n - A_0 x_n + B_{n+1}x_n + B_n x_n - B_{n+1}x_n) && (26.6)\\ &\le (\lambda - A_{n+1})^{-1} (\lambda x_n - A_0 x_n + B_{n+1}x_n) = x_n && (26.7) \end{aligned}$$

since $B_n x_n \le B_{n+1}x_n$. Therefore, the limit $\lim_{n\to\infty}(\lambda - A_n)^{-1}x$ exists for non-negative, hence all elements $x \in \mathbb{X}$. On the other hand, the fact that $(\lambda - A_n)^{-1}, n \ge 1$ is decreasing implies $\mathrm{e}^{tA_{n+1}} \le \mathrm{e}^{tA_n}, t \ge 0, n \ge 1$ and establishes strong convergence of $\mathrm{e}^{tA_n}, n \ge 1$ for each $t \ge 0$.

By the Sova–Kurtz Theorem, these semigroups converge almost uniformly in $[0, \infty)$ for $x \in \mathbb{X}_0$, which is the closure of $D(A_{\mathrm{ex}})$ – and the authors of [24] make an assumption to the effect that $\mathbb{X}_0 = \mathbb{X}$. However, as we have already noted, the semigroups do converge on the whole of $\mathbb{X}$ regardless of whether we make such an assumption or not.

Example 26.6 The semigroups considered here are building blocks of the model of stochastic gene expression of Chapter 46 and constitute a special example of the Tikhonov Theorem of Chapter 44, but are worthwhile presenting independently. Let $\mathbb{X} = C([0, 1]^2)$ and for given $r > 1$, $i \in \{0, 1\}$, let the semigroup $\{S_{i,r}(t), t \ge 0\}$ be given by $S_{i,r}(t)f(x, y) = f(x_i(t), y_{i,r}(t))$ where $x_i(t) = i + (x - i)\mathrm{e}^{-t}$ and $y_{i,r}(t) = y\mathrm{e}^{-rt} + \frac{r}{r-1}(x - i)(\mathrm{e}^{-t} - \mathrm{e}^{-rt}) + i(1 - \mathrm{e}^{-rt}), t \ge 0$. We note that $(x_i(t), y_{i,r}(t))$ is the solution of the Cauchy problem:

$$\begin{aligned} \frac{\mathrm{d}x_i}{\mathrm{d}t} &= i - x_i,\\ \frac{\mathrm{d}y_{i,r}}{\mathrm{d}t} &= r(x_i - y_{i,r}), \qquad x_i(0) = x, y_{i,r}(0) = y \in [0, 1]. && (26.8) \end{aligned}$$

The second equation here implies that $y_{i,r}$ chases x_i with efficiency r. Hence, as r becomes infinite, $y_{i,r}$ should become equal to x_i, and in the limit we should obtain the semigroup related to the Cauchy problem $\frac{\mathrm{d}x_i}{\mathrm{d}t} = i - x_i, x_i(0) = x$.

And this is so: we check that the domain of the extended limit, as $r \to \infty$, of generators of these semigroups is contained in the subspace $\mathbb{X}_0$ of functions $f \in \mathbb{X}$ that do not depend on y: $f(x, y) = f(x, x), x, y \in [0, 1]$. Moreover, the domain of the part of the extended generator A in $\mathbb{X}_0$, contains functions such that $[0, 1] \ni x \mapsto f(x, x)$ is continuously differentiable (with left-hand and right-hand derivatives at 1 and 0, respectively), and for such f is given by $Af(x, x) = (i - x)\frac{\mathrm{d}f(x,x)}{\mathrm{d}x}$. It follows that the regularity space here is the space $\mathbb{X}_0$ of functions that do not depend on y. The proof of these statements is straightforward but quite lengthy, especially for what we obtain as a result: for $f \in \mathbb{X}_0$, $\lim_{r\to\infty} S_{i,r}(t)f(x, y) = f(x_i(t), x_i(t))$. For, by direct inspection we have, for $f \in \mathbb{X}_0$, $S_{i,r}(t)f(x, y) = f(x_i(t), y_{i,r}(t)) = f(x_i(t), x_i(t))$.

On the other hand, the distance between:

$$(x_i(t), y_{i,r}(t)) \quad \text{and} \quad (x_i(t), x_i(t))$$

is:

$$\left|(y - i)\mathrm{e}^{-rt} + \frac{1}{r-1}(x - i)\mathrm{e}^{-t} - \frac{r}{r-1}(x - i)\mathrm{e}^{-rt}\right| \le (1 + \frac{r}{r-1})\mathrm{e}^{-rt} + \frac{1}{r-1}.$$

Hence, by uniform continuity of $f \in \mathbb{X}$, we obtain that for *any* $f \in \mathbb{X}$, $\lim_{r\to\infty} S_{i,r}(t)f(x, y) = f(x_i(t), x_i(t))$, with the limit uniform in any interval of the form $t \in [\delta, \infty)$, $\delta > 0$ but in general not almost uniform in $[0, \infty)$.

In other words, despite the fact that $\mathbb{X}_0 \subsetneq \mathbb{X}$ the semigroups converge on the whole of $\mathbb{X}$. □

Example 26.7 Irregular convergence may also take place in the context of Chernoff's Product Formula, or the Trotter Product Formula; see [198, 200, 277]. To consider just one simple example, let us come back to Nittka's approximation (19.10), and let P_ϵ be the contraction in $C[-\infty, \infty]$ given by $P_\epsilon = E_\epsilon R$. The family $S_\epsilon(t) = E_\epsilon \mathrm{e}^{tA_\epsilon} R, t \ge 0$ is then a semigroup in $C[-\infty, \infty]$, which is degenerate in the sense that $\lim_{t\to 0} S_\epsilon(t)f = f$ merely for $f \in C[-\infty, \infty]$ satisfying $P_\epsilon f = f$. Since convergence in (19.10) is almost uniform, instead of t we may take $t_n = \frac{nt}{n+1}$, to conclude that $\lim_{n\to\infty} \mathrm{e}^{tA_\epsilon} f = \lim_{n\to\infty}(R\mathrm{e}^{\frac{t_n}{n}A}E_\epsilon)^n f, f \in C[0, \infty]$. Replacing f by $Rf, f \in C[-\infty, \infty]$ and applying E_ϵ to both sides of this relation, we see that:

$$\begin{aligned} S_\epsilon(t)f &= \lim_{n\to\infty} E_\epsilon(R\mathrm{e}^{\frac{t}{n+1}A}E_\epsilon)^n Rf = \lim_{n\to\infty}(P_\epsilon \mathrm{e}^{\frac{t}{n+1}A})^n P_\epsilon f \\ &= \lim_{n\to\infty}(P_\epsilon \mathrm{e}^{\frac{t}{n+1}A})^{n+1} f = \lim_{n\to\infty}(P_\epsilon \mathrm{e}^{\frac{t}{n}A})^n f,\ f \in C[-\infty, \infty], t > 0, \end{aligned}$$

and the same argument applies to Nittka's general setup (see [277] p. 1439). Clearly, the limit cannot be uniform around $t = 0$ for $f \neq P_\epsilon f$, and we are dealing with the Trotter's Product Formula involving two semigroups, one of which $(T(t) := P_\epsilon, t \geq 0)$ is degenerate. (In particular, here $\mathbb{X}_0$ is the space where $f = P_\epsilon f$.)

Exercise 26.1 Using the fact that A_n in Example 26.2 satisfy $A_n^2 + (a + b + n)A_n + anI = 0$ (which is a particular case of the Cayley–Hamilton Theorem [236]) show (26.2). *Hint:* show $(\lambda_{1,n} - \lambda_{2,n})A_n^k = (\lambda_{1,n}^k - \lambda_{2,n}^k)A_n + (\lambda_{1,n}\lambda_{2,n}^k - \lambda_{2,n}\lambda_{1,n}^k)I$, $n, k \geq 1$, where $\lambda_{i,n} = r_n - (-1)^i p_n$, $i = 1, 2$.

Exercise 26.2 Prove (26.1) by checking that:

$$(\lambda - A_n)^{-1} = \frac{1}{\lambda^2 + \lambda(a+b) + n(\lambda + a)} \begin{pmatrix} \lambda + b + n & a \\ b & \lambda + a \end{pmatrix}.$$

Exercise 26.3 Consider the system of equations:

$$\begin{aligned} \frac{\mathrm{d}x}{\mathrm{d}t} &= \alpha - \beta x, \\ \frac{\mathrm{d}y}{\mathrm{d}t} &= \gamma x - \delta y, \qquad x(0) = x_0, y(0) = y_0 \in \mathbb{R}, \end{aligned} \tag{26.9}$$

describing dynamics of mRNA (x) and protein (y) levels, provided the gene responsible for the protein's production is turned on (comp. Exercise 2.2, p. 23 in [3] and the model of gene expression in Chapter 46, equation (46.3)). Here, α, β, γ, and δ are positive constants describing production rate and natural decay of mRNA and the protein. Use the variation of constants formula to find the solution of this equation. Consider also the case where production and degradation of mRNA takes place on the much faster scale than that of the protein, that is, replace α and β by $n\alpha$ and $n\beta$, respectively, and check that the corresponding solutions converge to those of:

$$\begin{aligned} x &= \frac{\alpha}{\beta}, \\ \frac{\mathrm{d}y}{\mathrm{d}t} &= \frac{\gamma\alpha}{\beta} - \delta y, \qquad y(0) = y_0 \in \mathbb{R}. \end{aligned} \tag{26.10}$$

Is this convergence uniform in t? In which intervals?

☞ Chapter's summary

A number of examples are given showing that semigroups may and in fact quite often do converge outside of the regularity space. Of course, outside the regularity space convergence fails to be uniform and is merely point-wise. As we shall see later, it is around $t = 0$ that the limit is not uniform.

27

Extremely Strong Genetic Drift

In this chapter, we discuss another example of a family of semigroups converging outside of the regularity space. This family is related to the notion of genetic drift and comes from [66, 67].

As formula (21.1) shows, if the population size $M = 2N$ in the Wright–Fisher model is large and the time is measured in units of $M = 2N$ generations, the distribution of the time (counted backward) when there are j ancestors of the given sample is approximately exponential with parameter $j(j-1)/2$. In particular, the time to the most recent common ancestors of a pair of individuals is approximately exponential with parameter $M^{-1} = (2N)^{-1}$.

In the related **Moran-type model** with mutations, time is continuous, individuals are exchangeable, and the distribution of the time to the common ancestor of two individuals is exponential with parameter N^{-1}, where N is the so-called effective population size (in general, the whole of the population cannot mate randomly, hence, to correct for this, the actual size of the population must be reduced to its effective size N; N does not have to be an integer).

Consider a locus on a chromosome of an individual in such a model, and suppose that the **alleles** (variants) at this locus may be numbered by positive integers so that the mutation process is a Markov chain with values in $\mathbb{N}$. This is the case for example with micro-satellite loci, where the Durret-Kruglyak Markov chain describes the number of short tandem repeats [120]. Let $\{P(t), t \geq 0\}$ be the related semigroup of Markov operators in l^1, the space of absolutely summable sequences; if $x \in l^1$ is the initial distribution of the chain, then $P(t)x$ is its distribution at time t. Operators $P(t)$ may be identified with transition matrices of the chain and the value of $P(t)x$ is obtained by multiplying the corresponding matrix by x from the left. Analogously, if the intensity matrix Q involved is nonexplosive, Q may be identified with the generator of $\{P(t), t \geq 0\}$ (see Chapter 13 or [49, 180, 278]).

Figure 27.1 Extremely strong genetic drift by Marek Bobrowski

Let $\mathcal{M}$ be the space of matrices $(m_{ij})_{i,j\in\mathbb{N}}$ with absolutely summable entries, and let $\mathcal{M}_s$ be its subspace where $m_{ij} = m_{ji}$ ("s" is for "symmetric"). Because the members of our population are exchangeable, the joint distribution of allelic types of two individuals sampled from the population is a member of $\mathcal{M}_s$. Our first goal is to study the time evolution of such a joint distribution.

To this end, suppose that we have observed the population in the time interval $[0, t]$ and that the joint allelic distribution of two individuals sampled from this population at time 0 was $m \in \mathcal{M}_s$. Consider two individuals living at time t; with probability $e^{-t/N}$ their common ancestor died before time 0, when we started to observe the population. In such a case, since during the time interval $[0, t]$ the individuals evolved independently, the joint allelic distribution is $T(t)m = P(t)mP^*(t)$, where $*$ denotes the transpose. (We note in passing that $T(t)$ is the tensor product of two copies of $P(t)$, i.e., $T(t) = P(t) \otimes P(t)$. Here, $\mathcal{M}$ is viewed as the **projective tensor product** [108, 304] of two copies of l^1; $\mathcal{M}_s$ is invariant for $T(t)$. More specifically $\mathcal{M} = l^1 \hat{\otimes}_\pi l^1$, and $T(t) = P(t) \otimes_\pi P(t)$.) This is just saying that if $X(t)$ and $Y(t)$ are processes representing allelic types of our individuals, then, by independence assumption, $\Pr(X(t) = i, Y(t) = j) = \sum_{k,l\geq 1} p_{ki}(t)p_{lj}(t) \Pr(X(0) = k, Y(0) = l)$, where $p_{ki}(t)$ are the entries of $P(t)$, that is, the transition probabilities of the Markov chain involved.

Suppose next that the time to the most recent common ancestor equals $s \in [0, t]$, so that the split of genealogical lines of our individuals occurred at $t - s$. The distribution of allelic type of the common ancestor at 0 is the marginal distribution of m, and at time $t - s$, this distribution is $P(t - s)Km$, where the operator $K : \mathcal{M}_s \to l^1$:

$$Km = \left(\sum_{i=1}^{\infty} m_{ij}\right)_{j\geq 1} = \left(\sum_{j=1}^{\infty} m_{ij}\right)_{i\geq 1}$$

calculates the marginal distribution of a symmetric joint probability matrix. Immediately after the split, the joint probability distribution of the pair of processes formed of the common ancestor was concentrated on the main diagonal and was equal to $\Theta P(t - s)Km$ where $\Theta(\xi_i)_{i\geq 1}$ is the diagonalization of $(\xi)_{i\geq 1}$, that is, the entries of $\Theta(\xi_i)_{i\geq 1}$ are all zero except on the diagonal where the (i, i) entry is ξ_i. Hence, at time t the joint distribution of our two individuals is $T(s)\Theta P(t - s)Km$.

Gathering all the information, we obtain that the joint distribution of allelic types of two individuals sampled at time t from the Moran population is:

$$S_N(t)m = \mathrm{e}^{-\frac{t}{N}}T(t)m + \frac{1}{N}\int_0^t \mathrm{e}^{-\frac{s}{N}}T(s)\Theta P(t - s)Km\,\mathrm{d}s. \tag{27.1}$$

Using the facts that $K\Theta = I_{l^1}$ and:

$$KT(t) = P(t)K$$

(the latter formula says that the operator K **intertwines** semigroups $\{T(t), t \geq 0\}$ and $\{P(t), t \geq 0\}$, [127, 287]), we check that $\{S_N(t), t \geq 0\}$ is a strongly continuous semigroup in $\mathcal{M}_s$ generated by $A + \frac{1}{N}\Theta K - \frac{1}{N}I_{\mathcal{M}_s}$, where A is the generator of $\{T(t), t \geq 0\}$. (For the characterization of A in terms of Q see, e.g., [269].)

Clearly, for extremely small N, the Markov chains involved here are strongly correlated: the genetic drift is so strong that the individuals are almost identical. Hence, as $N \to 0$, we should expect that the semigroups $\{S_N(t), t \geq 0\}$ converge in a sense to $\{P(t), t \geq 0\}$, which describes the evolution of a single individual.

This may be achieved as follows. The subspace $\mathcal{D}$ of $\mathcal{M}_s$ formed by matrices with off-diagonal entries equal to zero, is isometrically isomorphic to l^1; the isomorphism is $\Theta : l^1 \to \mathcal{M}_s$ with the inverse K restricted to the range of Θ. Hence, the semigroup $\{P^\sharp(t), t \geq 0\}$ given by $P^\sharp(t) = \Theta P(t)K$ is an isometrically isomorphic copy of $\{P(t), t \geq 0\}$ in $\mathcal{D}$. Clearly, $P^\sharp(t)$ has an extension to the whole of $\mathcal{M}_s$, which we denote by the same symbol. We remark that although $t \mapsto P^\sharp(t)m$ is continuous for $m \in \mathcal{M}_s$, $\{P^\sharp(t), t \geq 0\}$ is not a strongly continuous semigroup, since $\lim_{t\to 0} P^\sharp(t)m = \Theta K m \neq m$ unless $m \in \mathcal{D}$.

We claim that $\lim_{N\to 0} S_N(t)m = P^\sharp(t)m, t > 0$. For $m \in \mathcal{D}$ this may be proved using the the Trotter–Kato Theorem (see Exercise 27.2). However, formula (27.1) provides a simple proof of the general result. To show this, we note that as $N \to 0$ the measures on $[0, \infty)$ with densities $t \mapsto \frac{1}{N}\mathrm{e}^{-t/N}$ tend weakly to the Dirac measure at zero (the point mass at 0). Therefore, since $T(t)$ and $P(t)$ are contractions, the integral in (27.1) converges as $N \to 0$ to the value of the integrand at 0, that is, to $\Theta P(t)Km = P^\sharp(t)m$. Since the first term in (27.1) converges to 0 (for $t > 0$), the claim is proved.

Exercise 27.1 Check directly that, for $S_N(t)$ defined in (27.1),

$$\lim_{N\to\infty} S_N(t) = T(t),$$

uniformly in compact intervals.

Exercise 27.2 Let $\mathbb{X}$ and $\mathbb{Y}$ be two Banach spaces, and let $K : \mathbb{X} \to \mathbb{Y}$ and $\Theta : \mathbb{Y} \to \mathbb{X}$ be two bounded linear operators such that $K\Theta = I_{\mathbb{Y}}$; in the nontrivial case, though, we *do not* have $\Theta K = I_{\mathbb{X}}$. Also, let $(\mathrm{e}^{tA})_{t\geq 0}$ and $(\mathrm{e}^{tB})_{t\geq 0}$ be two strongly continuous semigroups of equibounded operators in $\mathbb{X}$ and $\mathbb{Y}$, respectively, and assume that K intertwines $(\mathrm{e}^{tA})_{t\geq 0}$ and $(\mathrm{e}^{tB})_{t\geq 0}$: $\mathrm{e}^{tB}K = K\mathrm{e}^{tA}$, $t \geq 0$.

- Using the Phillips Perturbation Theorem or checking directly prove that for any $\epsilon > 0$ the operator $A_\epsilon = A + \frac{1}{\epsilon}\Theta K - \frac{1}{\epsilon} I_{\mathbb{X}}$ (with domain $D(A)$) generates a strongly continuous semigroup $\{S_\epsilon(t), t \geq 0\}$ given by:

$$S_\epsilon(t) = \mathrm{e}^{-\frac{t}{\epsilon}}\mathrm{e}^{tA} + \frac{1}{\epsilon}\int_0^t \mathrm{e}^{-\frac{s}{\epsilon}}\mathrm{e}^{sA}\Theta\mathrm{e}^{(t-s)B}K\,\mathrm{d}s. \tag{27.2}$$

- Check the following facts. K is onto $\mathbb{Y}$ since it has the right inverse Θ. Hence, the range of Θ is equal to the range of ΘK. Since ΘK is idempotent ($(\Theta K)^2 = \Theta K$), its range $\mathbb{X}'$ is a closed linear subspace of $\mathbb{X}$. Thus, Θ being injective, it establishes an isomorphism between $\mathbb{Y}$ and $\mathbb{X}'$, with inverse $K_{|\mathbb{X}'}$. The semigroup composed of $S(t) = \Theta\mathrm{e}^{tB}K$ (as restricted to $\mathbb{X}'$) is thus the isomorphic copy in $\mathbb{X}'$ of the semigroup $(\mathrm{e}^{tB})_{t\geq 0}$ in $\mathbb{Y}$.
- Check that $(\lambda - A_\epsilon)^{-1} = (\lambda + \frac{1}{\epsilon} - A)^{-1} + \frac{1}{\epsilon}(\lambda + \frac{1}{\epsilon} - A)^{-1}\Theta(\lambda - B)^{-1}K$ for $\lambda > 0$. Use this to prove that $\mathbb{X}' = \mathbb{X}_0$ (the regularity space), and $\lim_{\epsilon\to 0} S_\epsilon(t)x = S(t)x, x \in \mathbb{X}', t \geq 0$ from the Trotter–Kato Theorem.
- Use (27.2) to show that $\lim_{\epsilon\to 0} S_\epsilon(t)x = S(t)x, x \in \mathbb{X}, t > 0$.

☞ Chapter's summary

As another illustration of the fact that semigroups may converge outside of their regularity space, we consider the limit, as $N \to 0$, of a family of semigroups modeling dynamics of joint distribution of allelic types of two individuals sampled from a Moran population of effective size N. The limit semigroup is "degenerated" in the sense that it describes a pair of two identical individuals, so that while the approximating semigroups are defined in the space of summable matrices, the regularity space is composed of its subspace of matrices that are diagonal.

28

The Nature of Irregular Convergence

By the Trotter–Kato Theorem, on the regularity space $\mathbb{X}_0$ the semigroups converge almost uniformly in $t \in \mathbb{R}^+$. The situation outside of $\mathbb{X}_0$ is somewhat similar: the limit, if it exists, is almost uniform in $t \in \mathbb{R}^+_*$ [48]. This chapter is devoted to the proof of this result.

We start with the connection between irregular convergence and generation of noncontinuous semigroups. To this end, suppose that A is a Hille–Yosida operator. We know that the part A_p of A in $\mathbb{X}_0 = cl\, D(A)$ generates a strongly continuous semigroup there. In particular, $(\lambda - A_p)^{-1}x = \int_0^\infty \mathrm{e}^{-\lambda t}\mathrm{e}^{tA_p}x\, \mathrm{d}t$, $x \in \mathbb{X}_0$. The operator A itself generates a once-integrated semigroup $(U(t))_{t\geq 0}$ and we have:

$$(\lambda - A)^{-1}x = \lambda \int_0^\infty \mathrm{e}^{-\lambda t}U(t)x\, \mathrm{d}t, \qquad x \in \mathbb{X}. \tag{28.1}$$

However, sometimes there exists a semigroup $(T(t))_{t\geq 0}$ of bounded operators such that:

$$(\lambda - A)^{-1}x = \int_0^\infty \mathrm{e}^{-\lambda t}T(t)x\, \mathrm{d}t, \qquad x \in \mathbb{X}, \tag{28.2}$$

and then we say that A generates a noncontinuous semigroup, see, for example, [9]. (This is the case, e.g., with the operator in Example 26.3 of Chapter 26 and the semigroup (26.4).) The adjective *noncontinuous* refers to the fact that this semigroup is not strongly continuous except for $x \in \mathbb{X}_0$. Instead, the trajectories $t \mapsto T(t)x$ are assumed to be merely (Bochner) measurable. By [180] p. 305, Thm. 10.2.3, this implies that they are continuous in $t \in (0, \infty)$. Since the Laplace transform restricted to continuous functions is injective, this implies that the semigroup, if it exists, is unique. We note the following criterion for existence of such a semigroup.

Proposition 28.1 *The following conditions are equivalent:*

a) $\mathrm{e}^{tA_p}(\lambda - A)^{-1}f$ *belongs to* $D(A_p)$, *for all* $x \in \mathbb{X}$, $\lambda > 0$, *and* $t > 0$;
b) for all $x \in \mathbb{X}$ *and* $t > 0$, *the limit* $\lim_{\mu\to\infty} \mathrm{e}^{tA_p}\mu(\mu - A)^{-1}x$ *exists;*
c) A *generates a noncontinuous semigroup* $(T(t))_{t\ge 0}$.

If these conditions are satisfied, $T(t)x = \lim_{\mu\to\infty} \mathrm{e}^{tA_p}\mu(\mu - A)^{-1}x$.

Proof For $x \in \mathbb{X}$, $t \ge 0$ and $\mu > 0$ let:

$$U_\mu(t)x = \mu \int_0^t \mathrm{e}^{sA_p}(\mu - A)^{-1}x\,\mathrm{d}s - \mathrm{e}^{tA_p}(\mu - A)^{-1}x + (\mu - A)^{-1}x. \quad (28.3)$$

Clearly, $t \mapsto U_\mu(t)$ is strongly continuous with $\|U_\mu(t)\| \le M^2t + (M^2 + M)\mu^{-1}$, and:

$$\begin{aligned}
\lambda \int_0^\infty \mathrm{e}^{-\lambda t} U_\mu(t)x\,\mathrm{d}t &= \lambda\mu \int_0^\infty \mathrm{e}^{-\lambda t} \int_0^t \mathrm{e}^{sA_p}(\mu - A)^{-1}x\,\mathrm{d}s\,\mathrm{d}t \\
&\quad - \lambda\,(\lambda - A)^{-1}\,(\mu - A)^{-1}x + (\mu - A)^{-1}x \\
&= \mu\,(\lambda - A)^{-1}\,(\mu - A)^{-1}x \\
&\quad - \lambda\,(\lambda - A)^{-1}\,(\mu - A)^{-1}x + (\mu - A)^{-1}x \\
&= (\lambda - A)^{-1}\,x.
\end{aligned}$$

Hence, by (28.1) and the injectivity of the Laplace transform for continuous functions, $U_\mu(t)$ does not depend on $\mu > 0$ and coincides with the integrated semigroup generated by A.

If (c) holds, then integrating (28.2) by parts and comparing with (28.1), by injectivity of the Laplace transform for continuous functions, we see that $U(t)x = \int_0^t T(s)x\,\mathrm{d}s$, $t > 0$, $x \in \mathbb{X}$. Hence, $(0, \infty) \ni t \mapsto U(t)$ is strongly differentiable. Because U coincides with U_μ, so is $t \mapsto \mathrm{e}^{tA_p}(\mu - A)^{-1}$ or, which is the same, a) holds.

By Exercise 8.4, (a) implies that $\lim_{\mu\to\infty} \mu(\mu(\mu - A)^{-1} - I)\mathrm{e}^{tA_p}(\lambda - A)^{-1}x$ exists for all $x \in \mathbb{X}$, $t > 0$ and $\lambda > 0$. Since e^{tA_p} commutes with $(\lambda - A)^{-1}$ and because of the Hilbert Equation, the expression above is $\mathrm{e}^{tA_p}\mu\lambda(\mu - A)^{-1}(\lambda - A)^{-1}x - \mathrm{e}^{tA_p}\mu(\mu - A)^{-1}x$. Since the limit of the first term here exists also, b) follows.

Finally, if (c) holds, we may define $T(t) = \lim_{\lambda\to\infty} \mathrm{e}^{tA_p}\lambda(\lambda - A)^{-1}$, $t > 0$. Since the range of $(\lambda - A)^{-1}$ belongs to $\mathbb{X}_0$, $T(t)$ is an extension of e^{tA_p} and

$T(t)x \in \mathbb{X}_0, x \in \mathbb{X}$. Hence, by (8.4),

$$T(s)T(t)x = \lim_{\mu\to\infty} e^{sA_p}e^{tA_p}\mu(\mu - A)^{-1}x = \lim_{\mu\to\infty} e^{(s+t)A_p}\mu(\mu - A)^{-1}x$$
$$= T(t+s)x$$

proving the semigroup property. Finally, the trajectories of $(T(t))_{t\geq 0}$ are bounded and measurable, as limits of continuous functions, and the calculation:

$$\int_0^\infty e^{-\lambda t}T(t)x\,dt = \lim_{\mu\to\infty}\int_0^\infty e^{-\lambda t}\mu e^{tA_p}(\mu - A)^{-1}x\,dt$$
$$= \lim_{\mu\to\infty}(\lambda - A_0)^{-1}\mu(\mu - A)^{-1}x$$
$$= \lim_{\mu\to\infty}\mu(\mu - A)^{-1}(\lambda - A)^{-1}x = (\lambda - A)^{-1}x,$$

shows (28.2). □

Example 28.2 A well-known example of a nondensely defined operator generating a noncontinuous semigroup is that of a sectorial operator. Apparently the first to consider this example where Da Prato and Sinestrari [291, 314], who noted that with an operator A satisfying (15.4), but not densely defined, one may associate the semigroup defined by the Dunford integral (15.5). Except for $x \in \mathbb{X}_0$ this semigroup no longer has the property that $\lim_{t\to 0, t\in\Sigma_\delta} e^{tA}x = x$ for $\delta \in (0, \delta_0)$. See also [246].

For $t > 0$ this semigroup coincides with the one described in Proposition 28.1. For A_p is a sectorial operator and generates the strongly continuous semigroup given by (15.5). It is well known that $e^{tA_p}x \in D(A_p)$ for all $x \in \mathbb{X}_0$ and $t > 0$ ([9] p.160, [128] p. 101, [201] p. 488, or [284] p. 61). This implies that condition a) in Proposition 28.1 is satisfied, and the semigroup generated by A exists. Moreover,

$$T(t)x = \lim_{\mu\to\infty} e^{tA_p}\mu(\mu - A)^{-1}x$$
$$= \lim_{\mu\to\infty}\frac{1}{2\pi i}\int_\gamma e^{t\lambda}\mu(\mu - A)^{-1}(\lambda - A)^{-1}x\,d\lambda$$
$$= \frac{1}{2\pi i}\int_\gamma e^{t\lambda}(\lambda - A)^{-1}x\,d\lambda, \qquad x \in \mathbb{X}, t > 0$$

as claimed.

Proposition 28.3 *Assume that the semigroups* $(e^{tA_n})_{t\geq 0}$ *satisfy* (2.1). *The strong limit* $\lim_{n\to\infty} e^{tA_n}x$ *exists for all* $x \in \mathbb{X}$, *iff the operator* $\mathcal{A}$ *defined by* (14.1) *generates a noncontinuous semigroup in* $c(\mathbb{X})$.

Proof For necessity, suppose that $(\mathcal{T}(t))_{t\geq 0}$ is the semigroup generated by $\mathcal{A}$. Then, for $(x_n)_{n\geq 1} \in c(\mathbb{X})$,

$$\int_0^\infty \mathrm{e}^{-\lambda t}\mathcal{T}(t)(x_n)_{n\geq 1}\,\mathrm{d}t = (\lambda - \mathcal{A})^{-1}(x_n)_{n\geq 1} = \left((\lambda - A_n)^{-1}x_n\right)_{n\geq 1}$$
$$= \left(\int_0^\infty \mathrm{e}^{-\lambda t}\mathrm{e}^{tA_n}x_n\,\mathrm{d}t\right)_{n\geq 1}.$$

Since the Laplace transform is injective for continuous functions, we must have $\mathcal{T}(t)(x_n)_{n\geq 1} = (\mathrm{e}^{tA_n}x_n)_{n\geq 1}$. In particular, for all $x \in \mathbb{X}$, $(\mathrm{e}^{tA_n}x)_{n\geq 1}$ belongs to $c(\mathbb{X})$, that is, $\lim_{n\to\infty}\mathrm{e}^{tA_n}x$ exists.

Conversely, note that if $(\mathrm{e}^{tA_n}x)_{n\geq 1}$ belongs to $c(\mathbb{X})$ for all $x \in \mathbb{X}$, then $(\mathrm{e}^{tA_n}x_n)_{n\geq 1}$ belongs to $c(\mathbb{X})$ for all $(x_n)_{n\geq 1} \in c(\mathbb{X})$, too. Define $(\mathcal{T}(t))_{t\geq 0}$, $\mathcal{T}(t) : c(\mathbb{X}) \to c(\mathbb{X})$, and $(T_0(t))_{t\geq 0}$, $T_0(t) : \mathbb{X} \to \mathbb{X}$ by:

$$\mathcal{T}(t)(x_n)_{n\geq 1} = \left(\mathrm{e}^{tA_n}x_n\right)_{n\geq 1}, \qquad T_0(t)x = \lim_{n\to\infty}\mathrm{e}^{tA_n}x.$$

It is evident that $(\mathcal{T}(t))_{t\geq 0}$ and $T_0(t)$ are semigroups. Furthermore, the trajectories $t \mapsto T_0(t)x$ being pointwise limits of continuous functions are strongly measurable ([180] p. 72, Thm 3.5.4), hence strongly continuous in $t > 0$.

We proceed to show that the same is true for $(\mathcal{T}(t))_{t\geq 0}$. Let us recall that by the B. J. Pettis Theorem ([180] p.72, Thm 3.5.3), a vector-valued function is strongly measurable iff it is weakly measurable and almost separably valued. Fix $(x_n)_{n\geq 1} \in c(\mathbb{X})$ and let $x = \lim_{n\to\infty}x_n$. The vectors $T_0(w)x$ where w is a rational number form a dense set in $Z = \{y \in \mathbb{X}; y = T_0(t)x, t > 0\}$. Let $\mathcal{Z} \subset c(\mathbb{X})$ be the set of all sequences $(y_n)_{n\geq 1}$ that are ultimately one of $T_0(w)x$ and initially are of the form $y_n = T_n(w_n)x_n$, where w_n are rational numbers. Then, $\mathcal{Z}$ is countable and its closure contains the set $\mathcal{W} = \{(z_n)_{n\geq 1}, (z_n)_{n\geq 1} = \mathcal{T}(t)(x_n)_{n\geq 1}, t > 0\}$. Indeed, for $t > 0$ and $\epsilon > 0$ there exists an $n_0 \in N$ such that for $n > n_0$, $\|\mathrm{e}^{tA_n}x_n - T_0(t)x\| < \epsilon$. Consequently, there is a rational number w such that $\|\mathrm{e}^{tA_n}x_n - T_0(w)x\| < \epsilon$ for such n. Next, we find rational numbers $w_1, w_2, \ldots w_{n_0}$ such that $\|\mathrm{e}^{tA_n}x_n - \mathrm{e}^{w_nA_n}x_n\| < \epsilon$, for $n \leq n_0$. Then, $(z_n)_{n\geq 1}$ defined as $z_n = \mathrm{e}^{w_nA_n}x_n$ for $n \leq n_0$ and $z_n = T_0(w)x$ for $n > n_0$, belongs to $\mathcal{Z}$ and $\|\mathcal{T}(t)(x_n)_{n\geq 1} - (z_n)_{n\geq 1}\| < \epsilon$, as desired.

Furthermore, $t \to \mathcal{T}(t)$ is weakly measurable. Indeed, any functional $\Psi \in c(\mathbb{X})^*$ may be represented in the form:

$$\Psi(x_n)_{n\geq 1} = \psi_0(x) + \sum_{n=1}^{\infty}\psi_n(x_n)$$

where $x = \lim_{n\to\infty} x_n$, $\psi_i \in \mathbb{X}^*$, and $\sum_{n=0}^{\infty} \|\psi_n\|_{\mathbb{X}^*} < \infty$, which implies that $t \to \Psi[(\mathrm{e}^{tA_n}x_n)_{n\geq 1}]$ is a pointwise limit of measurable functions. Thus, our claim on strong measurability of $(\mathcal{T}(t))_{t\geq 0}$ follows from Pettis's Theorem.

Finally,

$$\begin{aligned}\int_0^\infty \mathrm{e}^{-\lambda t}\mathcal{T}(t)(x_n)_{n\geq 1}\,\mathrm{d}t &= \left(\int_0^\infty \mathrm{e}^{-\lambda t}\mathrm{e}^{tA_n}x_n\,\mathrm{d}t\right)_{n\geq 1}\\ &= \left((\lambda - A_n)^{-1}x_n\right)_{n\geq 1}\\ &= (\lambda - \mathcal{A})^{-1}(x_n)_{n\geq 1},\end{aligned}$$

as desired. □

Theorem 28.4 *Assume that the semigroups* $(\mathrm{e}^{tA_n})_{t\geq 0}$ *satisfy* (2.1) *and converge on the whole of* $\mathbb{X}$. *Then the limit is almost uniform in* $t \in (0, \infty)$.

Proof The semigroup $(\mathcal{T}(t))_{t\geq 0}$, being measurable, is strongly continuous for $t > 0$, and, thus, uniformly continuous on any compact subinterval of $(0, \infty)$. The same remarks apply to $(T_0(t))_{t\geq 0}$. Fix $x \in \mathbb{X}$ and $r > 1$. For any $\epsilon > 0$, there exists a δ such that:

$$\|\mathcal{T}(t)\,(x)_{n\geq 1} - \mathcal{T}(s)\,(x)_{n\geq 1}\| < \frac{\epsilon}{3} \quad \text{and} \quad \|T_0(t)f - T_0(s)f\| < \frac{\epsilon}{3},$$

whenever $|s - t| < \delta$ and $r^{-1} \leq s, t \leq r$. Therefore, for such s, t, and all $n \geq 1$ we get, recalling that $\mathcal{T}(t)(x)_{n\geq 1} = (\mathrm{e}^{tA_n}x)_{n\geq 1}$,

$$\begin{aligned}\|\mathrm{e}^{tA_n}x - T_0(t)x\| &\leq \|\mathrm{e}^{tA_n}x - \mathrm{e}^{sA_n}x\| + \|\mathrm{e}^{sA_n}x - T_0(s)x\|\\ &\quad + \|T_0(s)x - T_0(t)x\| \leq \frac{2}{3}\epsilon + \|\mathrm{e}^{sA_n}x - T_0(s)x\| \qquad (28.4)\end{aligned}$$

Set $s_i = r^{-1} + i\frac{\delta}{2}$, $i = 1, 2, \ldots, [\frac{2}{\delta}(r - r^{-1})]$ where $[\cdot]$ is integer part. For any $t \in [r^{-1}, r]$, there exists an i such that $|s_i - t| < \delta$. Since one can choose an n_0 such that for $n \geq n_0$, $\sup_i \|\mathrm{e}^{s_iA_n}f - T_0(s_i)\| < \frac{\epsilon}{3}$, by (28.4) we get:

$$\sup_{t\in[r^{-1},r]} \|\mathrm{e}^{tA_n}x - T_0(t)x\| < \epsilon, \qquad \text{for } n \geq n_0,$$

as desired. □

Corollary 28.5 *For all* $t > 0$ *and* $x \in \mathbb{X}$, $T_0(t)x$ *belongs to* $\mathbb{X}_0$.

Proof By Proposition 28.3, operator $\mathcal{A}$ generates a noncontinuous semigroup $(\mathcal{T}(t))_{t\geq 0}$. By Proposition 28.1, $\mathcal{T}(t)(x_n)_{n\geq 1} \in clD(\mathcal{A})$. Since $clD(\mathcal{A})$ is composed of sequences converging to elements of $\mathbb{X}_0$, we are done. □

☞ Chapter's summary

The irregular convergence is somewhat regular: if the limit of semigroups exists outside of $\mathbb{X}_0$, it is almost uniform in $t \in (0, \infty)$. Moreover, by nature, the ranges of the operators forming the limit semigroup are contained in $\mathbb{X}_0$.

29

Irregular Convergence Is Preserved Under Bounded Perturbations

A useful property of irregular convergence is that it is preserved under bounded perturbations. To explain this in more detail, suppose that $\{\mathrm{e}^{tA_n}, t \geq 0\}$ are equibounded, strongly continuous semigroups converging to a semigroup $\{T(t), t \geq 0\}$, which is strongly continuous only on the regularity space $\mathbb{X}_0 \subset \mathbb{X}$. Assume also that B_n are bounded linear operators converging strongly to a B. Then, by the Phillips Perturbation Theorem:

$$\mathrm{e}^{t(A_n+B_n)} = \sum_{k=0}^{\infty} S_{n,k}(t), \tag{29.1}$$

where:

$$S_{n,0} = \mathrm{e}^{tA_n}, \qquad S_{n,k+1}(t) = \int_0^t \mathrm{e}^{(t-s)A_n} B_n S_{n,k}(s)\,\mathrm{d}s, \quad k \geq 0.$$

We note that $\|S_{n,k}(t)\| \leq M\frac{(MKt)^k}{k!}$, where $M := \sup_{n\geq 1, t\geq 0} \|\mathrm{e}^{tA_n}\|$ and $K := \sup_{n\geq 1} \|B_n\|$. Since $S_{n,0}$ converges by assumption, and the Lebesgue Dominated Convergence Theorem together with $\lim_{n\to\infty} S_{n,k}(t) =: S_k(t)$ implies:

$$\lim_{n\to\infty} S_{n,k+1}(t) = \int_0^t T(t-s)BS_k(s)\,\mathrm{d}s,$$

all summands in (29.1) converge. Using the Lebesgue Dominated Convergence Theorem again, we see that:

$$\lim_{n\to\infty} \mathrm{e}^{t(A_n+B_n)} = \sum_{k=0}^{\infty} S_k(t),$$

where:

$$S_0(t) = T(t), \quad S_{k+1}(t) = \int_0^t T(t-s)BS_k(s)\,\mathrm{d}s, \; k \geq 0.$$

This is what we mean by saying that irregular convergence is preserved under bounded perturbations. We note that the regularity spaces of $\{e^{tA_n}, t \ge 0\}, n \ge 1$ and $\{e^{t(A_n+B_n)}, t \ge 0\}, n \ge 1$ coincide.

A special case is obtained for:

$$T(t) = e^{tA}P, \tag{29.2}$$

where A is the generator of a strongly continuous semigroup in $\mathbb{X}_0$ and P is an idempotent operator projecting $\mathbb{X}$ onto $\mathbb{X}_0$. Then,

$$S_0(t) = e^{tA}P, \qquad S_{k+1}(t) = \int_0^t e^{(t-s)A}PBS_k(s)\,ds\,P, \quad k \ge 0.$$

In other words,

$$\lim_{n\to\infty} e^{t(A_n+B_n)} = e^{t(A+PB)}P \tag{29.3}$$

where PB is treated as a bounded operator in $\mathbb{X}_0$. As we shall see, this explicit formula comes in handy in many situations.

Here is an example. Let $\{S_{i,r}(t), t \ge 0\}, i \in \{0, 1\}, r > 0$ be the strongly continuous semigroups in $C([0, 1]^2)$, introduced in Example 26.6, and let $\mathbb{X}$ be the Cartesian product of two copies of $C([0, 1]^2)$. Also, let A_r be the generator of the Cartesian product semigroup in $\mathbb{X}$ given by $S_r(t)(f, g) = (S_{0,r}(t)f, S_{1,r}(t)g)$, $f, g \in C([0, 1]^2)$ and let $B \in \mathcal{L}(\mathbb{X})$ be given in the matrix form: $B = \left(\begin{smallmatrix} -\beta & \beta \\ \alpha & -\alpha \end{smallmatrix}\right)$, where α and β are given non-negative members of $C([0, 1]^2)$. By Example 26.6,

$$\lim_{r\to\infty} S_r(t) = e^{tA}P,$$

where:

$$P(f, g)(x, y) = (f(x, x), g(x, x)),$$
$$e^{tA}(f, g)(x, y) = (f(x_0(t), x_0(t)), g(x_1(t), x_1(t))),$$

for all $(x, y) \in [0, 1]^2$, $(f, g) \in \mathbb{X}$, $t \ge 0$. The operator P is a projection onto the subspace $\mathbb{X}_0$ of pairs $(f, g) \in \mathbb{X}$ of functions that do not depend on the second coordinate y, and $(e^{tA})_{t\ge0}$ is a semigroup in $\mathbb{X}_0$. Since convergence is preserved under bounded convergence,

$$\lim_{r\to\infty} e^{t(A_r+B)} = e^{t(A+PB)}P, \tag{29.4}$$

and to obtain the matrix form for PB it suffices to replace α and β in the definition of B by $\tilde{\alpha}(x, y) = \alpha(x, x)$ and $\tilde{\beta}(x, y) = \beta(x, x)$.

To interpret this result we note that the semigroups $(e^{tA_r+B})_{t\ge0}$ describe a stochastic process $(x(t), y(t), \gamma(t)), t \ge 0$ on two copies of the unit square

$[0, 1]^2$, indexed by 0 and 1, respectively. While on the zeroth copy of the square, $\gamma = 0$ and the process moves along the integral curves of the system (26.8) with $i = 0$. At a random time, the process jumps to the other copy of the square, changing γ to 1 but without changing the x and y coordinates; the intensity of the jumps' β depends on the position of the process. Similarly, while at the first copy the process moves along the trajectories of the system (26.8) with $i = 1$, and at a random time jumps back to the zeroth copy with intensity α. In other words, $(x(t), y(t), \gamma(t)), t \geq 0$ is a solution of the stochastic differential equation:

$$\begin{aligned} \frac{\mathrm{d}x}{\mathrm{d}t} &= \gamma - x, \qquad & \gamma = 1 \overset{\alpha}{\to} \gamma = 0, \\ \frac{\mathrm{d}y}{\mathrm{d}t} &= r(x - y), \qquad & \gamma = 0 \overset{\beta}{\to} \gamma = 1. \end{aligned} \tag{29.5}$$

This is a single-gene-copy-version of Lipniacki et al.'s model of gene expression to be discussed later in Chapter 46. As $r \to \infty$, y 'becomes' x and the model reduces to:

$$\frac{\mathrm{d}x}{\mathrm{d}t} = \gamma - x, \qquad \gamma = 1 \overset{\tilde{\alpha}}{\to} \gamma = 0, \quad \gamma = 0 \overset{\tilde{\beta}}{\to} \gamma = 1. \tag{29.6}$$

This is exactly what (29.4) says, when we identify $\mathbb{X}_0$ with the Cartesian product of two copies of $C[0, 1]$, and the semigroup $(\mathrm{e}^{tA})_{t\geq 0}$ with its isomorphic image $\mathrm{e}^{tA}(f, g)(x) = (f(x_0(t)), g(x_1(t))), f, g \in C[0, 1], x \in [0, 1], t \geq 0$, in the latter space.

☞ Chapter's summary

By perturbing an irregularly converging sequence of semigroups by a sequence of bounded operators having a strong limit, we obtain another sequence of irregularly converging semigroups. If the limit of the first sequence is of special form (29.2), the other limit is of the form (29.3).

30

Stein's Model

In the Stein's model of **neuronal variability** – classical today, and quite generally "accepted as acceptable" – the state of a neuron is characterized by the difference V between its actual potential and potential "at rest" [322, 323].

This difference, termed **depolarization**, increases at **excitation impulses**, which form a Poisson process with parameter λ, and decreases in a similar way at **inhibitory impulses** according to an independent Poisson process with parameter μ.[1] A characteristic feature of the model is a threshold value $\Theta > 0$, above which the neuron "fires," and V immediately resets to zero. Then a **refractory period** begins, when excitation and inhibitory impulses have no effect on V.

Because of this special period, a natural state-space for the model is not the half-line $(-\infty, \Theta)$ but rather the set depicted at Figure 30.1, with horizontal interval $(-1, 0) - (0, 0)$ corresponding to this period, here assumed to be of unit length (see [303] for a slightly different approach). While at a point $(0, p)$ of the vertical half-line, the underlying Markov process is nearly of pure jump type: it waits for an exponential time with parameter $\lambda + \mu$ and then either jumps to $(0, p + a)$, with probability $\frac{\lambda}{\lambda+\mu}$ or to $(0, p - a)$, with probability $\frac{\mu}{\lambda+\mu}$, where a and b are excitation and inhibitory magnitudes; in between the jumps the absolute value of the potential decreases along one of the curves $t \mapsto pe^{-\nu t}$ (cf., e.g., [146, Section 10.2]). Once V exceeds Θ, the process jumps to the point $(-1, 0)$ and proceeds through the interval $(-1, 0) - (0, 0)$ with constant speed: after reaching $(0, 0)$ it resumes jumping (and decreasing).

Notwithstanding the fact that the refractory period, from the mathematical point of view, makes the model nontrivial, sometimes its existence may be

[1] In the simplest version of the model (see [322, p. 175] or [346, p. 107]), the latter possibility is excluded – see, however, [322, p. 181]. Of course, for the model to work properly, parameters λ, μ, a, b, and ν (see further on) must be chosen so that the general tendency for V to increase remains.

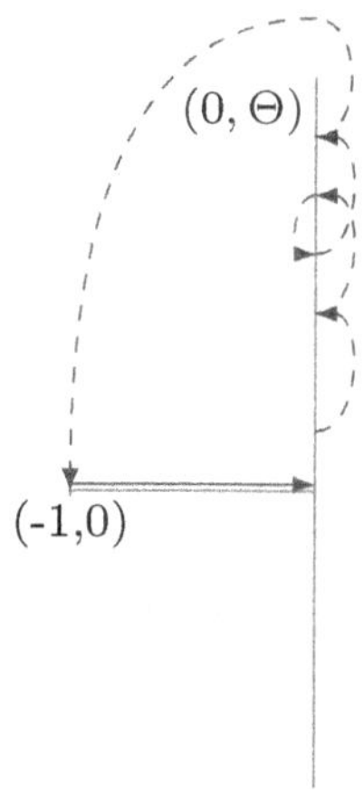

Figure 30.1 State-space of the Stein model with dashed arrows depicting jumps and the solid line denoting motion with constant speed. Exponential decay between jumps is neglected.

disregarded (see, e.g., [346]). In such a case, the state-space shrinks back to the vertical half-line $(-\infty, \Theta)$, and the entire process is really simple. In this chapter, we would like to obtain this simplified model as a limit of the full model with the length of refractory period decreasing to zero or, equivalently, with the speed υ with which the process proceeds through the refractory period increasing to infinity (see Figure 30.2). In our analysis, for simplicity of exposition, we assume that the exponential decay parameter ν is zero, that is, that the effect of decay between jumps is negligible. (Incorporating $\nu > 0$ to the model requires changing the strong topology to the topology of bounded pointwise convergence; besides this technical inconvenience, the analysis is practically the same.)

The choice of the Banach space for the semigroup describing Stein's process is a nontrivial matter. For, while the motion through the refractory period is conveniently put in terms of a translation semigroup in the space $C[-1, 0]$ of continuous functions on $[-1, 0]$, the jump-type process on the vertical half-line is not of Feller type – the space of continuous functions is not a good choice here.

For that reason we will work in the space $\mathbb{X}$ of pairs (f, g) such that $f \in C[-1, 0]$, $g \in BM(-\infty, \Theta)$ and $f(0) = g(0)$; here $BM(-\infty, \Theta)$ is the space of bounded (Lebesgue or Borel) measurable functions on $(-\infty, \Theta)$. The norm in $\mathbb{X}$ is inherited from the Cartesian product $C[-1, 0] \times BM(-\infty, \Theta)$.

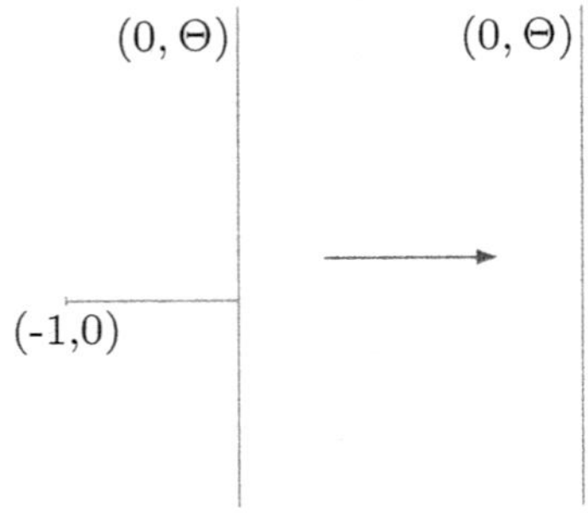

Figure 30.2 State-space collapse in the Stein model

The generator of the jump part of the process is the following bounded linear operator:

$$B(f, g) = (f, h) \tag{30.1}$$

where:

$$h(x) = \begin{cases} \lambda[g(x+a) - g(x)] + \mu[g(x-b) - g(x)], & x \in (-\infty, \Theta - a), \\ \lambda[f(-1) - g(x)] + \mu[g(x-b) - g(x)], & x \in [\Theta - a, \Theta]; \end{cases}$$

we assume throughout that $\Theta > a$. (Note that h need not be continuous if g is: this is why the process of jumps is not of Feller type.) The other part of the process (related to the refractory period behavior) is also quite easy to describe. In fact, this part of the semigroup is given explicitly by the formula:

$$T_{v,t}(f, g) = (f_{v,t}, g), \tag{30.2}$$

where:

$$f_{v,t}(x) = f\left((x + vt) \wedge 0\right), x \in [-1, 0]$$

and v, as previously explained, is the speed with which the process proceeds through the refractory period.

It is easy to check that $\{T_{v,t}, t \geq 0\}$ is a strongly continuous semigroup in $\mathbb{X}$. The domain of its generator is composed of pairs $(f, g) \in \mathbb{X}$ such that $f \in C^1[-1, 0]$ with $f'(0) = 0$, and for such a pair:

$$A_v(f, g) = (vf', 0).$$

Now, as already remarked, operator B of (30.1) is bounded, and the Phillips Perturbation Theorem implies that $A_v + B$ is a generator as well. Since both the semigroup generated by A_v and the semigroup generated by B are contraction semigroups, so is the semigroup generated by $A_v + B$ (by the Trotter Product Formula). This semigroup describes the full Stein process.

Our task is therefore to find the strong limit:

$$\lim_{\upsilon\to\infty} e^{t(A_\upsilon+B)}.$$

To this end, we will use the main result of Chapter 29, and therefore, start by considering:

$$\lim_{\upsilon\to\infty} e^{tA_\upsilon}.$$

This is very simple, since e^{tA_υ} is given explicitly by (30.2), that is $e^{tA_\upsilon} = T_{\upsilon,t}$. It follows that for any $t > 0$, choosing $\upsilon > t^{-1}$, we obtain:

$$f_{\upsilon,t}(x) = f(1), \qquad x \in [-1,0].$$

Let $\mathbb{X}_0$ be the subspace $\mathbb{X}_0$ of $\mathbb{X}$ composed of pairs (f,g) where $f(x) = g(0)$, $x \in [-1,0]$ (hence, $\mathbb{X}_0$ may be identified with $BM(-\infty,\Theta)$). Also, let A be the zero operator in $\mathbb{X}_0$, and let P be the projection of $\mathbb{X}$ on $\mathbb{X}_0$ given by:

$$P(f,g) = (g(0), g).$$

In these notations, we have:

$$\lim_{\upsilon\to\infty} e^{tA_\upsilon} = e^{tA}P, \tag{30.3}$$

that is, the limit semigroup is of the form (29.2). Since:

$$G(g(0),g) := PB(g(0),g) = (h(0),h)$$

where:

$$h(x) = \begin{cases} \lambda[g(x+a) - g(x)] + \mu[g(x-b) - g(x)], & x \in (-\infty, \Theta - a),\\ \lambda[g(0) - g(x)] + \mu[g(x-b) - g(x)], & x \in [\Theta - a, \Theta]; \end{cases}$$

the result established in Chapter 29 shows that:

$$\lim_{\upsilon\to\infty} e^{t(A_\upsilon+B)} = e^{tG}P.$$

As already mentioned, $\mathbb{X}_0$ is isometrically isomorphic to $BM(-\infty,\Theta)$. The natural isomorphism maps the semigroup $(e^{tG})_{t\ge 0}$ into the semigroup $(e^{t\mathfrak{G}})_{t\ge 0}$, where $\mathfrak{G}$ is defined by $\mathfrak{G}g = h$ with h defined earlier. The latter semigroup describes the pure jump process in which a particle starting at $x < \Theta - a$ jumps after exponential time with parameter λ to $x+a$ (with probability $\frac{\lambda}{\lambda+\mu}$) or to $x-b$ (with probability $\frac{\mu}{\lambda+\mu}$); for $x > \Theta - a$, the jump to $x+a$ is replaced by a jump to 0. This is exactly the process from the Stein model without refractory period. Hence, our task is completed.

Exercise 30.1 Consider the space $\mathbb{X} = C([0,1] \cup \{2\})$ of continuous functions on the union of the unit interval and the one-point set $\{2\}$. Show that, for $\lambda > 0$

and $v > 0$, the formula:

$$T_v(t)f(x) = \begin{cases} f(x+vt), & x \in [0, 1-vt], \\ e^{-\lambda(t-\frac{1-x}{v})}f(1) + (1 - e^{\lambda(t-\frac{1-x}{v})})f(2), & x \in (1-vt, 1], \\ f(2), & x = 2, \end{cases}$$

defines the strongly continuous semigroup in $C([0, 1] \cup \{2\})$ generated by $A_v f(x) = vf'(x), x \in [1, 2], A_v f(2) = 0$ with domain composed of f with the property that $f_{|[0,1]}$ is of class C^1 and $vf(1) = \lambda(f(2) - f(1))$.

Exercise 30.2 Let P be the projection of $\mathbb{X} = C([0, 1] \cup \{2\})$ onto the subspace $\mathbb{X}_0$ of functions f that are constant on $[0, 1]$, given by $Pf(x) = f(1), x \in [0, 1], Pf(2) = 2$. The subspace may be identified with $\mathbb{R}^2$. Let $T(t)$ be the isomorphic copy in $\mathbb{X}_0$ of the semigroup $S(t)(x, y) = (e^{-\lambda t}x + (1 - e^{-\lambda t})y, y)$ in $\mathbb{R}^2$. Show that, in the notations of the previous exercise,

$$\lim_{v\to\infty} T_v(t)f = T(t)Pf, \qquad f \in \mathbb{X}, t > 0.$$

☞ Chapter's summary

As refractory period in the Stein model of neural variability shrinks to zero, the state-space of the related Markov process collapses to a smaller set, and the limit process significantly simplifies. Using the principle established in the previous chapter, we prove pointwise convergence of the related semigroups of operators.

31

Uniformly Holomorphic Semigroups

How can we prove irregular convergence in practice? As the following lemma shows, if convergence of the resolvents is already established (and this is a necessary condition for convergence), the problem boils down to showing that the trajectories of the semigroups are equicontinuous in $t \in (0, \infty)$. We omit the proof of the lemma since it is a simple modification of the proof of Lemma 2.1; see Exercise 31.1.

Lemma 31.1 *Uniformly bounded functions $f_n : (0, \infty) \to \mathbb{X}$ converge almost uniformly in $t \in (0, \infty)$ iff a) for each $t > 0$ and $\epsilon > 0$ there exists a $\delta \in (0, t)$ such that $\|f_n(t) - f_n(s)\| < \epsilon$ provided $|s - t| < \delta$, and b) their Laplace transforms converge (pointwise).*

One of the ways of obtaining such equicontinuity in a hidden way is assuming uniform holomorphicity of the semigroups involved [44, 48]. This approach is especially useful because uniform holomorphicity may be conveniently expressed in terms of the related resolvents. Semigroups generated by a sequence A_n, $n \ge 1$ of sectorial operators are said to be **uniformly holomorphic** if the choice of δ_0 and M_δ's in (15.4) does not depend on n. In other words, we require that there exists $\delta_0 \in (0, \frac{\pi}{2}]$ such that the sector $\Sigma_{\frac{\pi}{2}+\delta_0} \subset \mathbb{C}$ is contained in the resolvent sets $\rho(A_n)$ of A_n, and for all $\delta \in (0, \delta_0)$ there exists $M_\delta > 0$ such that:

$$\|(\lambda - A_n)^{-1}\| \le \frac{M_\delta}{|\lambda|}, \qquad \lambda \in \Sigma_{\frac{\pi}{2}+\delta}. \tag{31.1}$$

Note that this estimate combined with the Dunford Integral Formula (15.5) implies that e^{tA_n} are equibounded in each sector $\Sigma_{\frac{\pi}{2}+\delta}$, $\delta < \delta_0$.

Theorem 31.2 *Uniformly holomorphic semigroups converge iff their resolvents do. Specifically, the following are equivalent:*

(a) $\lim_{n\to\infty} e^{tA_n}$ *exists for all* $t \in \Sigma_{\frac{\pi}{2}+\delta_0}$,
(b) $\lim_{n\to\infty} e^{tA_n}$ *exists for all* $t \ge 0$,
(c) $\lim_{n\to\infty}(\lambda - A_n)^{-1}$ *exists for some* λ *in the open right half-plane,*
(d) operator $\mathcal{A}$ *defined in (14.1) is sectorial with angle* δ_0,
(e) $\lim_{n\to\infty}(\lambda - A_n)^{-1}$ *exists for all* $\lambda \in \Sigma_{\frac{\pi}{2}+\delta_0}$.

Proof Our plan is to prove implications (a) $\Rightarrow$ (b) $\Rightarrow$ (c) $\Rightarrow$ (d) first, and then to establish (e) $\Rightarrow$ (d) and (d) $\Rightarrow$ [(a) & (e)].

The first of these is trivial. Condition (b) implies convergence in (c) for all λ in the open right half-plane, because for such λ, $(\lambda - A_n)^{-1} = \int_0^\infty e^{-\lambda t} e^{tA_n}\,dt$.

Assuming (c), we define the operator $\mathcal{R}_\lambda$ in $c(\mathbb{X})$, the space of convergent $\mathbb{X}$-valued sequences, by:

$$\mathcal{R}_\lambda (x_n)_{n\ge1} = \left((\lambda - A_n)^{-1}x_n\right)_{n\ge1}. \tag{31.2}$$

Then $\mathcal{R}_\lambda$ is a right and left inverse to $\lambda - \mathcal{A}$, where $\mathcal{A}$ was defined in (14.1). Moreover, by (31.1), $\|\mathcal{R}_\lambda\| \le \frac{M_\delta}{|\lambda|}$ for appropriate $\delta < \delta_0$. Hence $\lambda \in \rho(\mathcal{A})$. Since $\Sigma_{\frac{\pi}{2}+\delta_0}$ is connected, to show that:

$$\Sigma_{\frac{\pi}{2}+\delta_0} \subset \rho(\mathcal{A}), \tag{31.3}$$

it suffices to prove that $\Sigma_{\frac{\pi}{2}+\delta_0} \cap \rho(\mathcal{A})$ is both open and closed in $\Sigma_{\frac{\pi}{2}+\delta_0}$. It is open, since $\rho(\mathcal{A})$ is open in $\mathbb{C}$. To check that it is closed, we consider $(\lambda_n)_{n\ge1} \subset \rho(\mathcal{A}) \cap \Sigma_{\frac{\pi}{2}+\delta_0}$ converging to $\lambda \in \Sigma_{\frac{\pi}{2}+\delta_0}$. Then for some $\delta < \delta_0$, λ and all λ_n are members of $\Sigma_{\frac{\pi}{2}+\delta}$, and (31.1) implies that:

$$\|(\lambda - \mathcal{A})X\| \ge \frac{M_\delta}{|\lambda|}\|X\|, \qquad X \in D(\mathcal{A}). \tag{31.4}$$

It follows that the range of $\lambda - \mathcal{A}$ is closed: if $\lim_{n\to\infty}(\lambda X_n - \mathcal{A}X_n) = Y \in c(\mathbb{X})$, then $(X_n)_{n\ge1}$ is a Cauchy sequence in $c(\mathbb{X})$, and since $\mathcal{A}$ is closed, $X := \lim_{n\to\infty} X_n$ belongs to $D(\mathcal{A})$ and $\lim_{n\to\infty} \lambda X_n - \mathcal{A}X_n = \lambda X - \mathcal{A}X$. Moreover, the range of $\lambda - \mathcal{A}$ is dense in $c(\mathbb{X})$: for $Y \in c(\mathbb{X})$, we take $X_n = (\lambda_n - \mathcal{A})^{-1}Y$ to see that:

$$\|Y - (\lambda - \mathcal{A})X_n\| = \|(\lambda - \lambda_n)X_n\| \le \frac{M_\delta}{|\lambda_n|}|\lambda - \lambda_n|\|Y\| \tag{31.5}$$

converges to zero. Hence, the range of $\lambda - \mathcal{A}$ is $c(\mathbb{X})$. Relation (31.4) shows now that $\lambda - \mathcal{A}$ is injective, and then that $(\lambda - \mathcal{A})^{-1}$ is bounded with norm not exceeding $\frac{M_\delta}{|\lambda|}$, that is, $\lambda \in \rho(\mathcal{A})$, completing the proof of (31.3). Finally, if λ belongs to $\Sigma_{\frac{\pi}{2}+\delta}$, then (31.4) shows that $\|(\lambda - \mathcal{A})^{-1}\| \le \frac{M_\delta}{|\lambda|}$, establishing (d).

If (e) holds, then $\mathcal{R}_\lambda$ in (31.2) is well defined for all $\lambda \in \Sigma_{\frac{\pi}{2}+\delta_0}$, and is a right and left inverse of $\lambda - \mathcal{A}$. Also, (31.1) shows that $\|(\lambda - \mathcal{A})^{-1}\| \le \frac{M_\delta}{|\lambda|}$ for $\lambda \in \Sigma_{\frac{\pi}{2}+\delta}$, proving (d).

Finally, assume (d). Fix z with $|z| = 1$ and $|\arg z| < \delta_0$. The operators zA_n generate strongly continuous, equibounded semigroups $\{T_{z,n}(t), t \geq 0\}$ given by $T_{z,n}(t) = \mathrm{e}^{tzA_n}, t \geq 0$. Since $\mathcal{A}$ is sectorial with angle δ_0, $z\mathcal{A}$ generates a (non-continuous) semigroup $\{\mathrm{e}^{tz\mathcal{A}}, t \geq 0\}$. Proposition 28.3 shows that $\lim_{n\to\infty} \mathrm{e}^{tzA_n}$ exists for $t \geq 0$. Since z is arbitrary, this proves (a). Moreover, for λ in the open right half-plane there exists the limit of:

$$\int_0^\infty \mathrm{e}^{-\lambda t}\mathrm{e}^{tzA_n}\,\mathrm{d}t = (\lambda - zA_n)^{-1} = z^{-1}(\lambda z^{-1} - A_n)^{-1}.$$

Since all members of $\Sigma_{\frac{\pi}{2}+\delta_0}$ are of the form λz^{-1} where λ is in the open right half-plane and z is as above, this shows (e). □

For our first example, let A_ϵ be the operators of Chapter 10 related to Brownian motions with elastic barrier, this time considered in the space $C[0,\infty]$ of *complex* functions. For $\lambda \in \Sigma_\pi$, let $\sqrt{\lambda}$ denote the unique square root of λ lying in the right half-plane. Then a calculation similar to that presented in Chapter 3 shows that the resolvent of A_ϵ is still given by (10.1). Introducing $H(z) = \frac{z-1}{z+1}$, we rewrite the latter formula as:

$$(\lambda - A_\epsilon)^{-1} f(x) = \frac{1}{\sqrt{2\lambda}} \int_{-\infty}^\infty \mathrm{e}^{-\sqrt{2\lambda}|x-y|} f_\epsilon(y)\,\mathrm{d}y \tag{31.6}$$

where $f_\epsilon(x) = f(x)$ for $x \geq 0$ and $f_\epsilon(x) = H(\epsilon\sqrt{2\lambda})f(-x)$ for $x < 0$. Since the linear fractional function H maps the imaginary axis into the unit circle and the right half-plane into the open unit ball, we have $|H(\epsilon\sqrt{2\lambda})| < 1$ and $\sup_{x\in\mathbb{R}} |f_\epsilon(x)| = \sup_{x\geq 0} |f(x)|$. Therefore, for $|\arg\lambda| \leq \frac{\pi}{2} + \delta < \pi$:

$$\|(\lambda - A_\epsilon)^{-1}\| \leq \frac{2}{|\sqrt{2\lambda}|}\frac{1}{\Re e\sqrt{2\lambda}} \leq \frac{1}{|\lambda|\cos(\frac{\pi}{4} + \frac{\delta}{2})}$$

the last inequality following from $\frac{\Re e\sqrt{2\lambda}}{|\sqrt{2\lambda}|} = \cos(\arg\sqrt{2\lambda}) \geq \cos(\frac{\pi}{4} + \frac{\delta}{2})$. This shows that A_ϵ's generate uniformly holomorphic semigroups.

Moreover, using (10.1) again,

$$(\lambda - A_\epsilon)^{-1} f(x) - (\lambda - A_0)^{-1} f(x) = \frac{2\epsilon}{\epsilon\sqrt{2\lambda} + 1} \int_0^\infty \mathrm{e}^{-\sqrt{2\lambda}y} f(y)\,\mathrm{d}y \mathrm{e}^{-\sqrt{2\lambda}x}.$$

Thus for $f \in C[0,\infty]$ and $\lambda \in \Sigma_\pi$,

$$\|(\lambda - A_\epsilon)^{-1} f - (\lambda - A_0)^{-1} f\| \leq \left|\frac{2\epsilon}{\epsilon\sqrt{2\lambda} + 1}\right| \frac{1}{\Re e\sqrt{2\lambda}} \|f\|,$$

which tends to 0, as $\epsilon \to 0$. It follows that the semigroups generated by A_ϵ converge on a space much larger than $C_0(0,\infty]$ (see Chapter 10), in fact they converge on the whole of $C[0, \infty]$, but the limit semigroup is continuous merely on $C_0(0, \infty]$.

Turning to the second example, we consider a nondensely defined, sectorial operator A of angle δ_0, that is, assume that (15.4) is satisfied. Our aim is to show that the semigroups generated by the related Yosida approximation: $A_\mu = \mu^2 R_\mu - \mu$, $\mu > 0$ where $R_\mu = (\mu - A)^{-1}$, are uniformly holomorphic and converge to the noncontinuous semigroup generated by A. We start with the following lemma.

Lemma 31.3 *For any $r, \mu > 0$ and $\pi > |\alpha| \geq 0$,*

$$\left| \frac{\mu + r}{\mu + re^{i\alpha}} \right| \leq \sqrt{\frac{2}{1 + \cos \alpha}}. \tag{31.7}$$

Proof Fix $\mu, r > 0$ and define $f(\alpha) = 2(\mu^2 + 2\mu r \cos \alpha + r^2) - (1 + \cos \alpha)(\mu + r)^2$, for $\alpha \in [0, \pi)$. We have $f(0) = 0$, $f'(\alpha) = (\mu - r)^2 \sin \alpha \geq 0$. Thus $f(\alpha) \geq 0$, for all $\alpha \in [0, \pi)$. Hence, $\frac{2}{1+\cos\alpha} \geq \frac{(\mu+r)^2}{\mu^2+2\mu r\cos\alpha+r^2}$, $\alpha \in [0, \pi)$, which is equivalent to (31.7) because $|\mu + re^{i\alpha}| = \sqrt{\mu^2 + 2\mu r \cos \alpha + r^2}$, and cos is an even function. □

Returning to our example, observe that, by the Hilbert Equation,

$$(I - \nu R_\mu)(I + \nu R_{\mu-\nu}) = (I + \nu R_{\mu-\nu})(I - \nu R_\mu) = I, \tag{31.8}$$

provided $\mu, \mu - \nu \in \Sigma_{\frac{\pi}{2}+\delta}$. It is easy to check that if $\mu > 0$ and $\lambda \in \Sigma_{\frac{\pi}{2}+\delta}$ then $\frac{\lambda\mu}{\lambda+\mu} \in \Sigma_{\frac{\pi}{2}+\delta}$. Thus, since $\frac{\lambda\mu}{\lambda+\mu} = \mu - \frac{\mu^2}{\lambda+\mu}$ we may use (31.8) with $\nu = \frac{\mu^2}{\lambda+\mu}$ to obtain:

$$\left(I - \frac{\mu^2}{\lambda + \mu} R_\mu \right)^{-1} = I + \frac{\mu^2}{\lambda + \mu} R_{\frac{\lambda\mu}{\lambda+\mu}}.$$

It follows that for any $\mu > 0$, the resolvent set of A_μ contains $\Sigma_{\frac{\pi}{2}+\delta}$ and (compare (8.4)):

$$\begin{aligned}
(\lambda - A_\mu)^{-1} &= \left(\mu + \lambda - \mu^2 R_\mu\right)^{-1} \\
&= \frac{1}{\lambda + \mu} \left(I - \frac{\mu^2}{\lambda + \mu} R_\mu \right)^{-1} \\
&= \frac{1}{\lambda + \mu} + \left(\frac{\mu}{\lambda + \mu} \right)^2 R_{\frac{\lambda\mu}{\lambda+\mu}}, \qquad \mu > 0, \lambda \in \Sigma_{\frac{\pi}{2}+\delta}.
\end{aligned}$$

This combined with (15.4) yields:

$$\begin{aligned}\|(\lambda - A_\mu)^{-1}\| &\le \frac{1}{|\lambda+\mu|} + \frac{\mu^2}{|\lambda+\mu|^2}\|R_{\frac{\lambda\mu}{\lambda+\mu}}\| \\ &\le \frac{1}{|\lambda+\mu|} + \frac{\mu}{|\lambda+\mu|}\frac{M_\delta}{|\lambda|} \\ &\le \max\{M_\delta, 1\}\frac{\mu+|\lambda|}{|\mu+\lambda|}\frac{1}{|\lambda|}.\end{aligned}$$

Writing $\lambda = |\lambda|e^{i\alpha}$ and using (31.7), we obtain:

$$\|(\lambda - A_\mu)^{-1}\| \le \frac{M'_\delta}{|\lambda|}, \qquad \mu > 0, \lambda \in \Sigma_{\frac{\pi}{2}+\delta},$$

where $M'_\delta = \max\{M_\delta, 1\}\sqrt{\frac{2}{1-\sin\delta}}$, since $\cos\alpha \ge \cos(\frac{\pi}{2}+\delta) = -\sin\delta$, establishing the first claim.

Hence, there exists the limit:

$$S(t)x = \lim_{\mu\to\infty} e^{tA_\mu}x, \qquad t \in \Sigma_{\frac{\pi}{2}+\delta}, x \in \mathbb{X}.$$

By the Lebesgue Dominated Convergence Theorem and (5.4):

$$\begin{aligned}\int_0^\infty e^{-\lambda t}S(t)x\,dt &= \lim_{\mu\to\infty}\int_0^\infty e^{-\lambda t}e^{tA_\mu}x\,dt = \lim_{\mu\to\infty}(\lambda - A_\mu)^{-1}x \\ &= \lim_{\mu\to\infty}\left\{\frac{1}{\lambda+\mu}x + \left(\frac{\mu}{\lambda+\mu}\right)^2 R_{\frac{\lambda\mu}{\lambda+\mu}}x\right\} \\ &= (\lambda - A)^{-1}x.\end{aligned}$$

Thus, $\{S(t), t \ge 0\}$ coincides with the semigroup generated by A. The same is true for $t \in \Sigma_{\frac{\pi}{2}+\delta}$ by uniqueness of holomorphic extension (Exercise 31.3).

Exercise 31.1 Prove Lemma 31.1.

Exercise 31.2 Proof (e) $\Rightarrow$ (a) in Theorem 31.2 using the Dunford integral and the Dominated Convergence Theorem.

Exercise 31.3 Prove the last claim of this chapter. *Hint:* use linear functionals to reduce the problem to scalar-valued functions. Recall that zeros of an analytic function are isolated points [329].

Exercise 31.4 For $a < b$, let $C[a, b]$ be the space of all continuous functions $f : [a, b] \to \mathbb{C}$. For any non-negative numbers μ, ν define the operator $A_{\mu,\nu}$, by $D(A_{\mu,\nu}) = \{f \in C^2[a, b]; f(a) - \mu f'(a) = 0, f(b) + \nu f'(b) = 0\}$, $A_{\mu,\nu}f = \frac{1}{2}\frac{d^2 f}{dx^2}$. Check that $A_{\mu,\nu}$ is densely defined iff $\mu, \nu > 0$. Show that for

$f \in C[a,b]$ there exists $f_{\mu,\nu}: \mathbb{R} \to \mathbb{C}$ such that $\sup_{x\in\mathbb{R}} |f_{\mu,\nu}(x)| = \|f\|_{C[a,b]}$ and the resolvent of $A_{\mu,\nu}$ has the form:

$$(\lambda - A_{\mu,\nu})^{-1} f(x) = \frac{1}{\sqrt{2\lambda}} \int_{-\infty}^{\infty} \mathrm{e}^{-\sqrt{2\lambda}|x-y|} f_{\mu,\nu}(y)\,\mathrm{d}y, \qquad \lambda \in \Sigma_\pi.$$

Conclude that $A_{\mu,\nu}$ generate uniformly holomorphic semigroups (and Feller semigroups at the same time), and that:

$$\lim_{\mu,\nu\to 0} \mathrm{e}^{tA_{\mu,\nu}} f = \mathrm{e}^{tA_{0,0}} f, \qquad f \in C[a,b]$$

even though the limit semigroup is strongly continuous merely on the subspace $C_0(a,b)$ where $f(a) = f(b) = 0$. Beware, calculations are lengthy.

☞ Chapter's summary

Uniformly holomorphic semigroups converge iff their resolvents do, regardless of what the range of the limit pseudoresolvent is. Besides proving this fundamental result, we give two examples of such convergence. In the first of these, we come back to the elastic Brownian motions of Chapter 10, and in the second we study the Yosida approximation of a sectorial operator.

32

Asymptotic Behavior of Semigroups

We start with a simple corollary to Theorem 31.2; see for example [126], p. 105 for an alternative proof.

Corollary 32.1 *Assume A is sectorial. Then the limit* $\lim_{t\to\infty} \mathrm{e}^{tA}$ *exists iff so does* $\lim_{\lambda\to 0} \lambda(\lambda - A)^{-1}$.

Proof (Necessity) Let $(\lambda_n)_{n\ge 1}$ be a sequence of positive numbers such that $\lim_{n\to\infty} \lambda_n = 0$ and define $\kappa_n = \lambda_n^{-1}$. Since $\lim_{n\to\infty} \mathrm{e}^{t\kappa_n A}$ exists, there exists the limit of $(1 - \kappa_n A)^{-1} = \lambda_n(\lambda_n - A)^{-1}$. Since $(\lambda_n)_{n\ge 1}$ is arbitrary, this proves the claim.

(Sufficiency) Let $(\kappa_n)_{n\ge 1}$ be a sequence of positive numbers with infinite limit. The operators $A_n = \kappa_n A$ are then sectorial with the same angle as A : for $\lambda \in \Sigma_{\frac{\pi}{2}+\delta}, \delta < \delta_0$ we have $\|(\lambda - \kappa_n A)^{-1}\| = \|\frac{\lambda}{\kappa_n}(\frac{\lambda}{\kappa_n} - A)^{-1}\| \le \frac{M_\delta}{|\lambda|}$. Moreover,

$$\lim_{n\to\infty} (\lambda - \kappa_n A)^{-1} = \lim_{n\to\infty} \frac{\lambda}{\kappa_n} \left(\frac{\lambda}{\kappa_n} - A \right)^{-1}, \lambda > 0$$

exists. Therefore, $\lim_{n\to\infty} \mathrm{e}^{t\kappa_n A}, t \ge 0$ exists. Since $(\kappa_n)_{n\ge 1}$ is arbitrary, this completes the proof. □

For a particular instance of Corollary 32.1, let $a < b$ and A in $C[a, b]$ be given by:

$$Af = f'', \qquad D(A) = \{f \in C^2[0, 1] : f'(a) = f'(b) = 0\}. \tag{32.1}$$

By Exercise 15.3, A is sectorial of angle $\frac{\pi}{2}$ and generates a conservative, bounded holomorphic Feller semigroup.

Corollary 32.2 *For A defined above,*

$$\lim_{t\to\infty} \mathrm{e}^{tA} f = \lim_{\lambda\to 0} \lambda(\lambda - A)^{-1} f = \frac{1}{b-a}\int_a^b f, \qquad f \in C[a,b]$$

where $\int_a^b f$ is identified with the appropriate constant function on $[a, b]$.

Proof To establish the convergence of the resolvents, it suffices to show that for any positive sequence $(\kappa_n)_{n\ge 1}$ with $\lim_{n\to\infty} \kappa_n = \infty$, we have $\lim_{n\to\infty} \lambda(\lambda - \kappa_n A)^{-1} f = \frac{1}{b-a}\int_a^b f$, $\lambda > 0$. It is clear that the domain of the extended limit A_{ex} of $\kappa_n A$ is contained in the kernel of A, and the latter set is composed of constant functions. On the other hand, given a constant u and $f \in C[a, b]$, we consider $g(x) = \int_x^b \int_y^b f(z)\,\mathrm{d}z\,\mathrm{d}y$ and $\chi(x) = \frac{\cosh(b-x)}{\sinh(b-a)}$, $x \in [a, b]$, to define $f_n = u + \frac{1}{\kappa_n} g - \frac{1}{\kappa_n}(\int_a^b f)\chi$. Then $f_n \in D(A)$, $\lim_{n\to\infty} f_n = u$ and $\kappa_n A f_n = f - (\int_a^b f)\chi$, proving that $f - (\int_a^b f)\chi$ is a possible value of A_{ex} on u:

$$A_{ex} u = f - \left(\int_a^b f\right)\chi. \tag{32.2}$$

Therefore, given $\lambda > 0$ and $h \in C[a, b]$ we may choose $u = \lambda^{-1}(b-a)^{-1}\int_a^b h$ and $f = \lambda u - h$ to see that $\int_a^b f = 0$ and so $\lambda u - (f - (\int_a^b f)\chi) = h$. Since h is arbitrary, this shows that the range of $\lambda - A_{ex}$ equals $C[a, b]$, proving that the resolvents converge.

Furthermore, (32.2) implies that all f with $\int_a^b f = 0$ are values of A_{ex} on 0, that is, that for such f, $\lim_{n\to\infty}(\lambda - \kappa_n A)^{-1} f = 0$. Since $\{\mathrm{e}^{tA}, t \ge 0\}$ is conservative, it leaves constant functions invariant, and so must $\lambda(\lambda - \kappa_n A)^{-1}$. Combining the previous and writing $f = (f - \frac{1}{b-a}\int_a^b f) + \frac{1}{b-a}\int_a^b f$ we obtain our claim.

We are left with establishing the part concerning semigroups. Let $(\kappa_n)_{n\ge 1}$ be as earlier. Operators $\kappa_n A$ are sectorial with the same angle as A. Hence, convergence of resolvents implies convergence of the semigroups. For f with $\int_a^b f = 0$, the semigroups must converge to 0, since their resolvents do. Arguing as for the resolvents, we obtain $\lim_{n\to\infty} \mathrm{e}^{t\kappa_n A} f = \frac{1}{b-a}\int_a^b f$, $t > 0$. This completes the proof because $(\kappa_n)_{n\ge 1}$ is arbitrary. □

Corollary 32.2 is well known, and may be proved in many ways. Its physical interpretation is that as time passes the temperature distribution in an isolated finite rod "averages out" and becomes constant throughout the rod. Another proof of this result may be deduced, for example, from the explicit form of $(\mathrm{e}^{tA})_{t\ge 0}$ given in [128], p. 68. In fact, a much stronger[1] result than Corollary 32.2

[1] The result involves a change from strong topology to operator topology. Hence, perhaps, we should say, "yet *operatorer* result."

is true: there is $K > 0$ and $\epsilon > 0$ such that:

$$\left\| e^{tA} - P \right\|_{\mathcal{L}(C[a,b])} \leq K e^{-\epsilon t}, \tag{32.3}$$

where $Pf = \frac{1}{b-a} \int_a^b f$, and the scalar $\int_a^b f$ is identified with a constant function in $C[a, b]$. (See also Chapter 38.)

Estimate (32.3) is a typical application of the theory of asymptotic behavior of semigroups, as developed, for example, in [7, 9, 128, 269, 330] (compare [127]). Here we recall a result from [7, Section 3.5.1] that seems to be most suitable for establishing (32.3).

To begin with, for a semigroup $(e^{tA})_{t \geq 0}$, by:

$$s = s(A) = \sup\{\Re\lambda : \lambda \in \sigma(A)\},$$

where $\sigma(A)$ is the spectrum of A, we denote the **spectral bound** of A. Let S be a locally compact Hausdorf space, and let $C_0(S)$ be the space of continuous functions on S, vanishing at infinity. An $f \in C_0(S)$ is said to be **strictly positive** if $f(x) > 0$ for all $x \in S$. A functional $F \in [C_0(S)]^*$ is termed strictly positive if $Ff > 0$ for all nonzero $f \geq 0$. A semigroup $(e^{tA})_{t \geq 0}$ is said to be irreducible if for some (all) $\lambda > s(A)$, $(\lambda - A)^{-1} f$ is strictly positive for all nonzero $f \geq 0$.

Theorem 32.3 *Let $(e^{tA})_{t \geq 0}$ be a positive, irreducible C_0-semigroup on $C_0(S)$. Assume that $e^{t_0 A}$ is compact for some $t_0 > 0$. Then, $s(A) > -\infty$, and there is a strictly positive $f_* \in C_0(S)$ and a strictly positive $F \in [C_0(S)]^*$ such that $F f_* = 1$ and:*

$$\|e^{-s(A)t} e^{tA} - P\|_{\mathcal{L}(C_0(S))} \leq K e^{-\epsilon t}, \tag{32.4}$$

for some $K \geq 0$ and $\epsilon > 0$, where P is the projection given by $Pf = (Ff) f_*$.

In this theorem, the assumption that is arguably hardest to check is that of compactness of $e^{t_0 A}$, because the semigroup is rarely given explicitly. A useful criterion for automatic compactness of all e^{tA}, $t > 0$ is for the semigroup to be norm-continuous for all $t > 0$ (i.e., $\lim_{s \to t} \|e^{sA} - e^{tA}\| = 0, t > 0$) and for the resolvent operator $(\lambda - A)^{-1}$ to be compact for some λ (see, e.g., [128, p. 119]). Finally (see [128, p. 117]), $(\lambda - A)^{-1}$ is compact iff $D(A)$ with graph norm embeds compactly into $C_0(S)$, that is, iff a unit ball in $D(A)$ is relatively compact as a set in $C_0(S)$. Since all holomorphic semigroups are norm continuous for $t > 0$, we obtain the following corollary.

Corollary 32.4 *Let an operator A generate a positive, irreducible, holomorphic semigroup in $C_0(S)$. Assume that $D(A)$ embeds compactly into $C_0(S)$. Then the thesis of Theorem 32.3 holds.*

Before diving into examples, we recall a well-known criterion for exponential decay to zero (see, e.g., [7, p. 13]).

Proposition 32.5 *Let $(e^{tA})_{t\geq 0}$ be norm-continuous for $t > 0$ and assume that $s(A) < 0$. Then, constants $K \geq 0$ and $\epsilon > 0$ may be chosen so that:*

$$\|e^{tA}\| \leq Ke^{-\epsilon t}, \qquad t > 0.$$

Assumptions of this proposition are particularly easy to check for positive semigroups, because for such semigroups $s(A)$ is a member of $\sigma(A)$, see [9, Prop. 3.11.2]. Therefore, conditions $s(A) \leq 0$ and $0 \in \rho(A)$ ($\rho(A)$ is the resolvent set of A) imply $s(A) < 0$.

Example 32.6 We start by showing (32.3); for simplicity we restrict ourselves to the case where $a = 0, b = 1$ (the natural isometric isomorphism of $C[a, b]$ with $C[0, 1]$ allows recovering the general case). Defining, given $f \in C[0, 1]$, its extension $\tilde{f}$ to the whole real line by:

$$\tilde{f}(2n + x) = f(x), \quad \tilde{f}(2n + 1 + x) = f(1 - x), \qquad n \in \mathbb{Z}, x \in [0, 1],$$

we easily check that:

$$C(t)f(x) = \frac{1}{2}(\tilde{f}(x + t) + \tilde{f}(x - t)), \qquad t \in \mathbb{R}, x \in [0, 1] \tag{32.5}$$

is a cosine family of contractions in $C[0, 1]$ with generator (32.1) (where $a = 0$ and $b = 1$). This allows calculating $(\lambda - A)^{-1}$ explicitly:

$$\begin{aligned}
(\lambda - A)^{-1} f(x) &= \frac{1}{\sqrt{\lambda}} \int_0^\infty e^{-\sqrt{\lambda}t} C(t)f(x)\,dt \\
&= \frac{1}{2\sqrt{\lambda}} \int_0^\infty e^{-\sqrt{\lambda}t}[\tilde{f}(x + t) + \tilde{f}(x - t)]\,dt \\
&= \frac{1}{2\sqrt{\lambda}} \int_{-\infty}^\infty e^{-\sqrt{\lambda}|t-x|}\tilde{f}(t)\,dt \\
&= \frac{1}{2\sqrt{\lambda}} \sum_{n=-\infty}^{\infty} \left(\int_{2n-1}^{2n} + \int_{2n}^{2n+1}\right) e^{-\sqrt{\lambda}|t-x|}\tilde{f}(t)\,dt \\
&= \frac{1}{2\sqrt{\lambda}} \sum_{n=-\infty}^{\infty} [\int_0^1 e^{-\sqrt{\lambda}|y+2n-1-x|} f(1 - y)\,dy \\
&\quad + \int_0^1 e^{-\sqrt{\lambda}|y+2n-x|} f(y)\,dy] \\
&= \int_0^1 k_\lambda(x, y)f(y)\,dy,
\end{aligned}$$

where (compare [146], pp. 341–343 and p. 478):

$$k_\lambda(x, y) = \sum_{n=-\infty}^{\infty} [\mathrm{e}^{-\sqrt{\lambda}|2n-y-x|} + \mathrm{e}^{-\sqrt{\lambda}|2n+y-x|}].$$

Since $k_\lambda(x, y) > 0$ for all $x, y \in [0, 1]$, the related semigroup is not only positive but also irreducible. Since the semigroup generated by A is holomorphic, we are left with showing the compact embedding assumption. Now, members f of the unit ball in $D(A)$ satisfy:

$$\|f\| + \|f''\| \le 1 \quad \text{and} \quad f(x) = f(0) + \int_0^x \int_0^y f''(y)\,\mathrm{d}y, x \in [0, 1].$$

It follows that $\|f\| \le 1$ and $|f(x) - f(y)| \le |x - y|, x, y \in [0, 1]$. The Arzelà–Ascoli Theorem shows now that such f form a relatively compact set in $C[0, 1]$.

Hence, (32.4) holds for some projection P. Since e^{tA} leaves constant functions invariant, $s(A)$ must equal 0. Finally, Corollary 32.2 forces $Pf = \int_0^1 f$, and completes the proof.

Example 32.7 Let $C_0(0, 1)$ be the subspace of $f \in C[0, 1]$ satisfying $f(0) = f(1) = 0$. Given $f \in C_0(0, 1)$, we define its extension $\tilde{f}$ to the whole real line by:

$$\tilde{f}(2n + x) = f(x), \quad \tilde{f}(2n + 1 + x) = -f(1 - x), \qquad n \in \mathbb{Z}, x \in [0, 1].$$

Then,

$$C(t)f(x) = \frac{1}{2}(\tilde{f}(x + t) + \tilde{f}(x - t)), \qquad t \in \mathbb{R}, x \in [0, 1] \tag{32.6}$$

is a cosine family of contractions in $C_0(0, 1)$ with generator:

$$Af = f'', D(A) = \{f \in C^2[0, 1]; f(0) = f''(0) = f''(1) = f(1) = 0\}. \tag{32.7}$$

In particular, $(\mathrm{e}^{tA})_{t\ge 0}$ is norm-continuous for $t > 0$, and since A satisfies the positive maximum principle, $(\mathrm{e}^{tA})_{t\ge 0}$ is a Feller semigroup.

Given $g \in C_0(0, 1)$, we note that the function f given by $f(x) = ax + \int_0^x \int_0^y g(z)\,\mathrm{d}z\,\mathrm{d}y$, with $a = -\int_0^1 \int_0^y g(z)\,\mathrm{d}z\,\mathrm{d}y$, belongs to $D(A)$ and $Af = g$. Since $\ker A = \{0\}$, we conclude that $0 \in \rho(A)$. Proposition 32.5 is now in force and we obtain that there are constants $K \ge$ and $\epsilon > 0$ such that:

$$\|\mathrm{e}^{tA}\| \le K\mathrm{e}^{-\epsilon t}. \tag{32.8}$$

This agrees with our intuition (except for, perhaps, the speed of decay): In the related stochastic process, particles touching either of the boundaries are killed and removed from the state-space. Hence, expected number of particles in the interval diminishes with time.

Example 32.8 Let $C_0(0,1]$ be the space of continuous functions on $[0,1]$ vanishing at $x=0$. We consider the family of operators in $C_0(0,1]$, given by $A_\nu f = f''$, on the domain:

$$D(A_\nu) = \{f \in C^2[0,1] : f(0) = f''(0) = 0,\ f(1) + \nu f''(1) = 0\}.$$

Operators A_ν, $\nu > 0$ are densely defined in $C_0(0,1]$ and satisfy the positive maximum principle. Since, as we shall see soon, the range condition is also satisfied, they are generators of Feller semigroups in $C_0(0,1]$. These semigroups describe Brownian motions on the unit interval in which a particle touching $x=0$ is killed and removed from the state-space. When touching $x=1$, it is trapped there for an exponential time with parameter ν^{-1}, so that the expected time spent at $x=1$ is ν: after this time elapses, the particle is also killed and removed from the state-space. It is clear that in such a process all particles will eventually be killed, and the particles' expected survival time grows with ν. In other words, we should have:

$$\lim_{t\to\infty} \mathrm{e}^{tA_\nu} f = 0, \qquad f \in C_0(0,1] \tag{32.9}$$

and:

$$\mathrm{e}^{tA_{\nu_1}} \le \mathrm{e}^{tA_{\nu_2}}, \qquad \text{for } \nu_1 \le \nu_2. \tag{32.10}$$

To prove the second of these assertions, we note that a solution to the resolvent equation: $\lambda f - A_\nu f = g$, where $g \in C_0(0,1]$ and $\lambda > 0$, may be found by searching for $f \in D(A_\nu)$ of the form:

$$f(x) = C_1 \sinh\sqrt{\lambda}x + C_2 \cosh\sqrt{\lambda}x + f_{g,\lambda}(x), \tag{32.11}$$

where $f_{g,\lambda}(x) = \frac{1}{2\sqrt{\lambda}} \int_0^1 \mathrm{e}^{-\sqrt{\lambda}|x-y|} g(y)\,\mathrm{d}y$. Such an f belongs to $D(A_\nu)$ iff:

$$C_2 = -f_{g,\lambda}(0) \quad \text{and} \quad (\sinh\sqrt{\lambda})C_1 = \frac{\nu g(1)}{1+\nu\lambda} + h(0)\cosh\sqrt{\lambda} + h(1). \tag{32.12}$$

In particular, as claimed, the range condition is satisfied. Since C_1 is an increasing function of ν (provided $g(1) \ge 0$), we obtain $(\lambda - A_{\nu_1})^{-1} \le (\lambda - A_{\nu_2})^{-1}$ for $\nu_1 \le \nu_2$ and (32.10) follows.

Also, as in the previous example, given $g \in C_0(0,1]$, we take $f(x) = ax + \int_0^x \int_0^y g(z)\,\mathrm{d}z\,\mathrm{d}y$. For $a = -\int_0^1 \int_0^y g(y)\,\mathrm{d}y - \nu g(1)$, $f \in D(A_\nu)$ and $A_\nu f = g$. Again, since $\ker A_\nu$ is trivial, $0 \in \rho(A_\nu)$ and (32.9) follows by Proposition 32.5.

Example 32.9 A more interesting asymptotic behavior is obtained in the limit case $\nu \to \infty$ of the previous example, that is, for the operator:

$$A_\infty f = f'', \qquad D(A_\infty) = \{f \in C^2[0,1] : f(0) = f''(0) = 0,\ f''(1) = 0\}.$$

Again, A_∞ is the generator of a Feller semigroup, but here the boundary condition describes a trap at $x = 1$: a particle reaching $x = 1$ stays there forever. An elementary potential theory [124, p. 75] or [146, p. 478] tells us that the probability that a Brownian traveler starting at $x \in (0, 1)$ will reach $x = 1$ before reaching $x = 0$ equals x. This is the probability the traveller will be trapped at $x = 1$; with probability $1 - x$ the traveller will eventually be killed and removed from the state-space. Hence, we expect:

$$\lim_{t\to\infty} \mathrm{e}^{tA_\infty} f(x) = f(1)x, \qquad f \in C_0(0, 1], x \in [0, 1].$$

(Recall $\mathrm{e}^{tA_\infty} f(x) = E_x f(w(t))$, where $w(t), t \geq 0$ is the stochastic process described above, and E_x is the expected value, conditional on the process starting at x).

One is tempted to deduce this claim from (32.4), for everything seems to be prepared for the application of Theorem 32.3. Members of the unit ball in $D(A_\infty)$ are of the form:

$$f(x) = a_f x + \int_0^x \int_0^y f''(y)\,\mathrm{d}y, x \in [0, 1].$$

where $|a_f|$ may be estimated by 2, and the Arzelà–Ascoli Theorem can be used again to prove compactness of the embedding $D(A) \hookrightarrow C_0(0, 1]$. Irreducibility is also within our reach: $(\lambda - A_\infty)^{-1}$ is an increasing limit of $(\lambda - A_\nu)^{-1}$ (see (32.12)). Since e^{tA_0} is irreducible (see Exercise 32.7), so is e^{tA_∞}.

Or, is it? Well, C_2 in (32.12) is an increasing function of ν iff $g(1) > 0$. For $g(1) = 0$, C_2 does not change with ν and $(\lambda - A_\nu)^{-1}$ coincides with $(\lambda - A_0)^{-1}$, which is zero at both interval's ends. Hence, for such g, the irreducibility condition is violated (at *one* point!). By the way, the fact that for all $\nu \geq 0$ (including $\nu = \infty$):

$$(\lambda - A_\nu)^{-1} g = (\lambda - A_0)^{-1} g, \text{ provided } g(1) = 0 \tag{32.13}$$

is, on the second thought, not surprising at all, and could have been predicted.

Fortunately, the very reason that does not allow applying Theorem 32.3, allows applying Proposition 32.5.[2] For (32.13) implies that the semigroups $(\mathrm{e}^{tA_\nu})_{t\geq 0}$ leave the subspace $C_0(0, 1) \subset C_0(0, 1]$ invariant, and coincide with $(\mathrm{e}^{tA_0})_{t\geq 0}$ there. Therefore, introducing $f_*(x) = x$, $Ff = f(1)$ and $Pf = (Ff)f_*$, we obtain $A_\infty f_* = 0$ and, by (32.8),

$$\begin{aligned}\|\mathrm{e}^{tA_\infty} f - Pf\|_{C_0(0,1]} &= \|\mathrm{e}^{tA_\infty}(f - Pf)\|_{C_0(0,1]} = \|\mathrm{e}^{tA_0}(f - Pf)\|_{C_0(0,1)} \\ &\leq K\mathrm{e}^{-\epsilon t}\|f - Pf\|_{C_0(0,1)} \leq 2K\mathrm{e}^{-\epsilon t}\|f\|_{C_0(0,1]}\end{aligned}$$

[2] "Wiles realized that exactly what was making the Euler System fail is what would make the Horizontal Iwasawa Theory approach he had abandoned three years earlier *work*" [1, p. 132].

since $f - Pf$ belongs to $C_0(0, 1)$. In other words,

$$\|e^{tA_\infty} f - Pf\| \le 2Ke^{-\epsilon t},$$

as desired.

Example 32.10 Let A be the Laplace operator in $C[0, 1]$ defined on $D(A) \subset C^2[0, 1]$ where:

$$f'(0) = f'(1) = f(1) - f(0). \tag{32.14}$$

We have encountered this boundary condition in (5.5). As shown in [71], A is the generator of a unique semigroup (even of a cosine family) that preserves the first two moments:

$$F_i = \int_0^1 x^i f(x)\,\mathrm{d}x, \qquad i = 0, 1,$$

that is, such that $t \mapsto F_i e^{tA} f$ is constant for each $f \in C[0, 1]$. Moreover (see [71, Theorem 4.3]), there is $K \ge 0$ and $\epsilon > 0$ such that:

$$\|e^{tA} - P\| \le Ke^{-\epsilon t}, \tag{32.15}$$

where:

$$Pf = (F_0 f) f_0 + (F_1 f) f_1,$$

$f_0(x) = 1$ and $f_1(x) = 12x - 6, x \in [0, 1]$. (We note in passing that $F_1 f_1 = F_0 f_0 = 1$.) Interpreted, (32.15) means that the smoothing property of diffusion combined with preservation of moments forces the trajectories of the semigroup to forget, as time passes, the shape of the initial condition and remember merely its zeroth and first moments.

It is worth stressing that this result is not a particular instance of (32.4): P is not a projection on a one-dimensional space. Interestingly, however, two cases of convergence of type (32.4) may be derived from (32.15). To this end, we note that the spaces:

$$C_{\text{odd}}[0, 1] \text{ and } C_{\text{even}}[0, 1]$$

of odd and even functions, respectively, are left invariant by the semigroup $(e^{tA})_{t \ge 0}$. (By definition $f \in C_{\text{odd}}[0, 1]$ if $f(1 - x) = -f(x), x \in [0, 1]$, while $f \in C_{\text{even}}[0, 1]$ if $f(x) = f(1 - x), x \in [0, 1]$.) As it transpires, on the latter space, e^{tA} coincides with the Neumann Laplace semigroup (see [71]), and our result reduces to Example 32.6.

An arguably more interesting conclusion may be obtained from the part in the other space: We start by noting that $C_{\text{odd}}[0, 1]$ is isometrically isomorphic

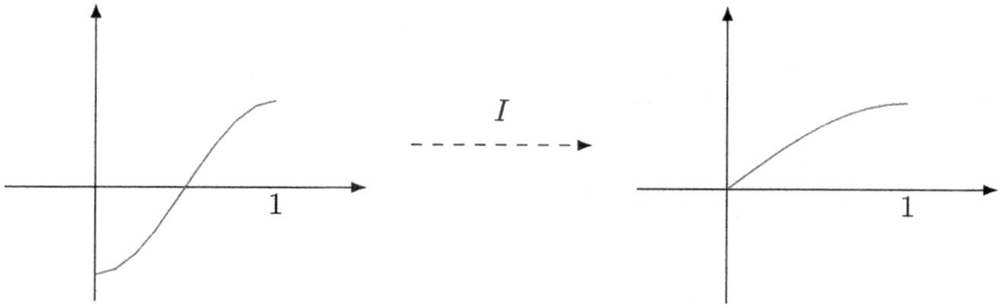

Figure 32.1 Isomorphism $I : C_{\mathrm{odd}}[0, 1] \to C_0(0, 1]$

to $C_0(0, 1]$. The isomorphism is given by (see Figure 32.1):

$$I : C_{\mathrm{odd}}[0, 1] \to C_0(0, 1], \qquad If(x) := f\left(\frac{1+x}{2}\right), \quad x \in [0, 1]$$

with inverse given by:

$$I^{-1}f(x) = \begin{cases} -f(1-2x), & x \in \left[0, \frac{1}{2}\right), \\ f(2x-1), & x \in \left[\frac{1}{2}, 1\right]. \end{cases}$$

Let A_{odd} be the generator of e^{tA} as restricted to $C_{\mathrm{odd}}[0, 1]$; we have $A_{\mathrm{odd}} f = f''$ with domain composed of twice continuously differentiable odd functions satisfying $f'(0) = -2f(0)$. The isomorphic image of A_{odd} in $C_0(0, 1]$ is given by $A_{\mathrm{image}} f = IA_{\mathrm{odd}} I^{-1} f$ with domain equal to the image of the domain of $D(A_{\mathrm{odd}})$, that is,

$$D(A_{\mathrm{image}}) = \{f \in C^2[0, 1] : f(0) = f''(0) = 0 \text{ and } f'(1) = f(1)\},$$
$$A_{\mathrm{image}} f = 4f''.$$

By (32.15), since $F_0 f = 0$ for $f \in C_{\mathrm{odd}}[0, 1]$,

$$\|\mathrm{e}^{tA_{\mathrm{odd}}} f - (F_1 f) f_1\| \le M\mathrm{e}^{-\epsilon t}\|f\|, \qquad f \in C_{\mathrm{odd}}[0, 1].$$

It follows that:

$$\|\mathrm{e}^{tA_{\mathrm{image}}} f - (F_1 I^{-1} f) I f_1\| \le M\mathrm{e}^{-\epsilon t}\|f\|, \qquad f \in C_0(0, 1].$$

Introducing $f_*(x) = x, x \in [0, 1]$, we obtain $If_1 = 6f_*$ and $F_1 I^{-1} f = \frac{1}{2} F_1$. Therefore,

$$\|\mathrm{e}^{tA_{\mathrm{image}}} f - 3(F_1 f) f_*\| \le M\mathrm{e}^{-\epsilon t}\|f\|. \tag{32.16}$$

Since $F_1(3f_*) = 1$, we obtained a result of the type (32.4).

The strongly continuous semigroup generated by A_{image}, and the related cosine family, were considered in [54]. This semigroup describes chaotic movement of particles in the interval $[0, 1]$ with constant inflow of particles from the boundary at $x = 1$ and outflow at the boundary $x = 0$ (see Chapter 40). Condition (32.16) says that the rates of inflow and outflow are so tuned here that in the limit a nontrivial equilibrium is attained.

In Chapter 40 we will have handy estimates of $(\lambda - A_{\text{image}})^{-1}$ at our disposal, and will be able to show (32.16) directly from Theorem 32.3 without alluding to the results of [71].

Exercise 32.1 For the elastic Brownian motion semigroups $(\mathrm{e}^{tA_\epsilon})_{t\ge 0}$, $\epsilon > 0$ show that $\lim_{t\to\infty} \mathrm{e}^{tA_\epsilon} = 0$. *Hint:* use (10.1); consider constant functions and those in $L^1(\mathbb{R}^+) \cap C[0, \infty]$ separately.

Exercise 32.2 Let A be the generator of the reflected Brownian motion semigroup in $C[0, \infty]$: $Af = \frac{1}{2}f''$, $D(A) = \{f \in C^2[0, \infty]; f'(0) = 0\}$. Show that A is sectorial of angle $\frac{\pi}{2}$ and that $\lim_{t\to\infty} \mathrm{e}^{tA} f = f(\infty)$, where $f(\infty)$ is identified with the appropriate constant function in $C[0, \infty]$.

Exercise 32.3 Let A be the generator of the absorbed (stopped) Brownian motion semigroup in $C_0(0, 1]$ with a trap at $x = 1$. More specifically, $Af = \frac{1}{2}f''$, $D(A) = \{f \in C_0^2(0, 1]; f'' \in C_0(0, 1], f''(1) = 0\}$. Show that A is sectorial of angle $\frac{\pi}{2}$ and that $\lim_{t\to\infty} \mathrm{e}^{tA} f = Pf$, where $Pf(x) = xf(1)$, $x \in (0, 1]$. (Probabilistically, this result is interpreted as follows: In the absorbed Brownian motion a particle starting at $x \in [0, 1]$ will eventually be trapped (stopped) at $x = 1$ or killed and removed from the state-space at $x = 0$. x is the probability that the particle will be trapped at 1 and $1 - x$ is the probability that it will be killed at 0, and we have $Pf(x) = (1 - x)f(0) + xf(1)$.)

Exercise 32.4 Let, as in Exercise 1.12, $C_{\mathrm{p}}[0, 1]$ be the space of continuous functions f on $[0, 1]$ such that $f(0) = f(1)$. As a generator of a bounded cosine family, the operator $Af = f''$ with domain composed of twice continuously differentiable members of $C_{\mathrm{p}}[0, 1]$ such that $f', f'' \in C_{\mathrm{p}}[0, 1]$, generates a holomorphic semigroup of angle $\frac{\pi}{2}$ (by the Weierstrass Formula). Show that $\lim_{t\to\infty} \mathrm{e}^{tA} f = \int_0^1 f$, $f \in C_{\mathrm{p}}[0, 1]$.

Exercise 32.5 Show that formula (32.5) defines a cosine family of contractions in $C[0, 1]$ with generator (32.1) (where $a = 0$ and $b = 1$).

Exercise 32.6 Show that formula (32.6) defines a cosine family of contractions with generator (32.7). Calculating as in Example 32.6, prove that:

$$(\lambda - A)^{-1} f(x) = \int_0^1 k_\lambda(x, y) f(y)\, \mathrm{d}y$$

where (compare [146], p. 478):

$$k_\lambda(x,y) = \sum_{n=-\infty}^{\infty} [\mathrm{e}^{-\sqrt{\lambda}|2n+y-x|} - \mathrm{e}^{-\sqrt{\lambda}|2n-y-x|}]. \tag{32.17}$$

Exercise 32.7 Prove that for $\alpha \in [-2, 2]$ and $\mu > 0$,

$$\sum_{n=-\infty}^{\infty} \mathrm{e}^{-\mu|2n+\alpha|} = \frac{\mathrm{e}^{-\mu|\alpha|} + \mathrm{e}^{-2\mu}\mathrm{e}^{\mu|\alpha|}}{1-\mathrm{e}^{-2\mu}};$$

(check this for $\alpha \in [0, 2]$ first, and then deduce the general case). Conclude that, since $|x-y| \le x+y$, $k_\lambda(x,y)$ in (32.17) is strictly positive for $x, y \in (0,1)$.

Exercise 32.8 Let $C_0(0,1]$ be the space of continuous functions on $[0,1]$ vanishing at $x=0$. Consider the family of operators in $C_0(0,1]$, given by:

$$A_\nu f = f'', D(A_\nu) = \{f \in C^2[0,1] : f(0) = f''(0) = 0, f(1) + \nu f'(1) = 0\}.$$

(a) Check that operators A_ν, $\nu > 0$ are densely defined in $C_0(0,1]$ and satisfy the positive maximum principle.
(b) Show that (32.11) is a solution to the resolvent equation for A_ν iff $C_2 = -f_{g,\lambda}(0)$ and

$$C_1 = \frac{\nu\sqrt{\lambda}(f_{g,\lambda}(1) + f_{g,\lambda}(0)\sinh\sqrt{\lambda}) + f_{g,\lambda}(0)\cosh\sqrt{\lambda} - f_{g,\lambda}(1)}{\sinh\sqrt{\lambda} + \nu\sqrt{\lambda}\cosh\sqrt{\lambda}}. \tag{32.18}$$

(c) Recall that a linear-fractional function $x \mapsto \frac{a_{11}x+a_{12}}{a_{21}x+a_{22}}$ (strictly) increases iff the determinant $a_{11}a_{22} - a_{12}a_{21}$ is (strictly) positive. Check that in the case considered the determinant is $\sqrt{\lambda}(\mathrm{e}^{\sqrt{\lambda}} f_{g,\lambda}(1) - f_{g,\lambda}(0)] > 0$, so that C_1 is a strictly increasing function of ν.
(d) Describe heuristically stochastic processes governed by $(\mathrm{e}^{tA_\nu})_{t\ge 0}$ and interpret point (c).

☞ Chapter's summary

The holomorphic semigroups convergence theorem of the previous chapter has led us naturally into the world of asymptotic behavior of semigroups. We recalled a typical criterion for existence of the limit, as $t \to \infty$, of semigroups' trajectories, in terms of positivity and compactness. Also, we provided a few examples of such limits for Brownian motions on a unit interval with various boundary conditions. One result we will use often in what follows (and greatly generalize, see Chapter 38) says that diffusion on a compact interval with Neumann boundary conditions averages everything out as time goes to infinity.

As we shall see in the following chapters, this fundamental property of diffusion, when coupled with other mechanisms, leads to interesting, often surprising, singular limits. In fact, all the following chapters, up to Chapter 40, are a long detour from the main subject, aimed at one goal: to show variety of phenomena with fast diffusion acting behind the scene. (The only exception is Chapter 37, where we will be completely led astray; see Figure 37.1.) We will continue investigating the main subject (which, to recall for those who have been lost by the abundance of distractions, is: how do we prove irregular convergence?) in Chapter 41.

33

Fast Neurotransmitters

Our next two chapters are devoted to an application of the theoretical results concerning uniformly holomorphic semigroups to modeling fast neurotransmitters. We start by drawing the necessary bio-mathematical background.

In an attempt to understand phenomena behind synaptic depression, Aristizabal and Glavinovič introduced a simple ODE model of dynamics of levels of neurotransmitters [12]. They adopted the following widely accepted, simplified but sufficiently accurate description (see [273] and other papers cited in [12]): neurotransmitters are localized in three compartments, or pools: the large pool, where also their synthesis takes place, the small intermediate pool, and the immediately available pool, from which they are released during stimulus. Moreover, they assumed that the dynamics of levels U_i, $i = 1, 2, 3$ of vesicles with neurotransmitters in the pools is analogous to that of voltages across the capacitors in the electric circuit reproduced (with minor changes) as our Figure 33.1. This results in the following system of ordinary differential equations for U_i:

$$\begin{pmatrix} U_1' \\ U_2' \\ U_3' \end{pmatrix} = Q \begin{pmatrix} U_1 \\ U_2 \\ U_3 \end{pmatrix} + \begin{pmatrix} 0 \\ 0 \\ \frac{1}{R_3C_3}(E - U_3) \end{pmatrix}, \tag{33.1}$$

where:

$$Q = \begin{pmatrix} -\frac{1}{R_0C_1} - \frac{1}{R_1C_1} & \frac{1}{R_1C_1} & 0 \\ \frac{1}{R_1C_2} & -\frac{1}{R_2C_2} - \frac{1}{R_1C_2} & \frac{1}{R_2C_2} \\ 0 & \frac{1}{R_2C_3} & -\frac{1}{R_2C_3} \end{pmatrix}.$$

(The reason why $\frac{U_3}{R_3C_3}$ is included in the second and not in the first summand in (33.1) is explained in Corollary 34.5.) For the electric circuit, E denotes the electromagnetic source, the constants C_is are capacitors' sizes, while R_is

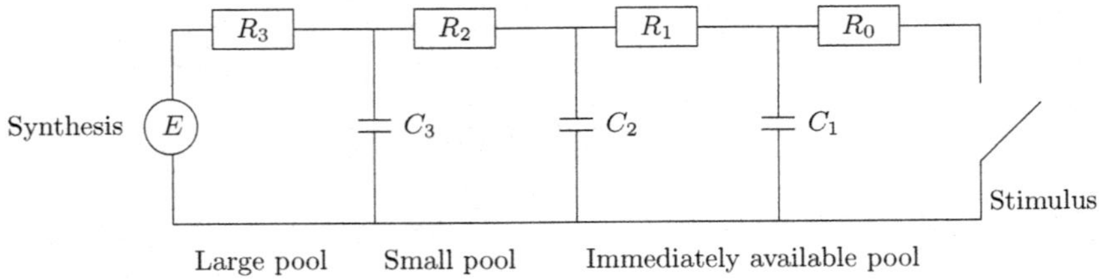

Figure 33.1 The ODE model of Aristizabal and Glavinovič (redrawn with insignificant changes from [12])

characterize the resistors. Biologically, E represents synthesis and C_is are the capacities to store vesicles, but R_is do not have a clear meaning. Merely the compounds $\frac{1}{R_iC_j}$ are interpreted as the pools' replenishment rates.

A more recent PDE model of Bielecki and Kalita [35] zooms in on vesicles with neurotransmitters, and assumes that they move according to a diffusion process in a three-dimensional domain Ω. As a result, in the linear version of the model, the (unknown) concentration ρ of vesicles in the cytoplasm satisfies a Fokker–Planck-type equation:

$$\frac{\partial \rho}{\partial t} = A\rho + \beta(\overline{\rho} - \rho) \tag{33.2}$$

where A is a second order, elliptic partial differential operator, $\beta : \Omega \to \mathbb{R}$ is the rate of neurotransmitters' production, and $\overline{\rho}$ is a balance concentration of vesicles.

Our goal is to show a connection between these models. To this end, following [70], first we note that equations (33.1) and (33.2) are of quite a different nature: while (in the absence of stimulus and production) the latter is conservative, the former is not. However, the dual to the matrix Q in (33.1) is an intensity matrix (it has non-negative off-diagonal entries, and the sums of its entries in each column are zero), and the dual equation to (33.1) is conservative. This suggests that to find a link between the two models, one must first pass from the description of dynamics of densities (concentrations), to that of expected values, that is, instead of considering the Fokker–Planck-type equation (33.2) one should pass to the dual Kolmogorov backward equation. Moreover, to find such a link, one needs to specify the way vesicles move from one pool to another, that is, to specify the transmission conditions [100], which are missing in Bielecki and Kalita's model. (In proving formula (33.3) below, instead of introducing appropriate transmission conditions, Bielecki and Kalita use what they call "technical conditions," without showing that these technical conditions are satisfied.) These transmission conditions, describing communication between pools, are of crucial importance in the analysis.

We introduce a one-dimensional version of the Bielecki and Kalita model, where vesicles perform a Brownian motion on three adjacent intervals (corresponding to pools) with diffusion coefficients varying from pool to pool, and where the mechanism of passing from one pool to another is specified by means of transmission conditions. Our main result (Theorem 34.4) says that as the diffusion coefficients in the model tend to infinity and the boundary and transmission conditions are scaled in an appropriate way, the solutions to the related Cauchy problems converge to those of the model of Aristizabal and Glavinovič. Roughly speaking, if diffusion in three separate pools is large and communication between pools is slow, the ODE model is a good approximation of the PDE model.

We note that to show a connection between the two models, Bielecki and Kalita also divide Ω into three subregions Ω_3, Ω_2, and Ω_1, corresponding to the three pools. They assume that the diffusion process the vesicles perform is a three-dimensional Brownian motion and the diffusion coefficients, say $\sigma_1, \sigma_2, \sigma_3$ vary from region to region, and suggest (see [35, Thm. 2]) that the quantities:

$$U_i = \frac{\int_{\Omega_i} \rho}{\text{volume } \Omega_i}, \qquad i = 1, 2, 3, \tag{33.3}$$

satisfy the ODE system (33.1) of Aristizabal and Glavinovič with $C_i = \frac{\text{volume } \Omega_i}{\sigma_i}$. However, this formula is at least doubtful: the proof of (33.3) given in [35] contains a number of errors. Moreover, in the absence of stimulus and neurotransmitter's production, the total number of vesicles should remain constant. Hence, $U_1 + U_2 + U_3 = const$, provided volume $\Omega_i = 1, i = 1, 2, 3$. However, system (33.1) is not conservative, that is, $(U_1 + U_2 + U_3)' \neq 0$ unless all C_i' are the same (no stimulus case is obtained by letting $R_0 \to \infty$, and no production results in removing the second summand in (33.1)). Hence, formula (33.3) cannot hold unless all σ_i' are the same.

In our model, we imagine the three pools as three adjacent intervals $[0, r_3]$, $[r_3, r_2]$ and $[r_2, r_1]$ of the real line, corresponding to the large, the small, and the immediately available pools, respectively. As in the model of Bielecki and Kalita, in each of those intervals, vesicles perform Brownian motions with respective diffusion coefficients σ_3, σ_2, and σ_1. The pools are separated by semipermeable membranes located at $x = r_3$ and $x = r_2$. Therefore, it is convenient to think of the actual state-space Ω of the process performed by the

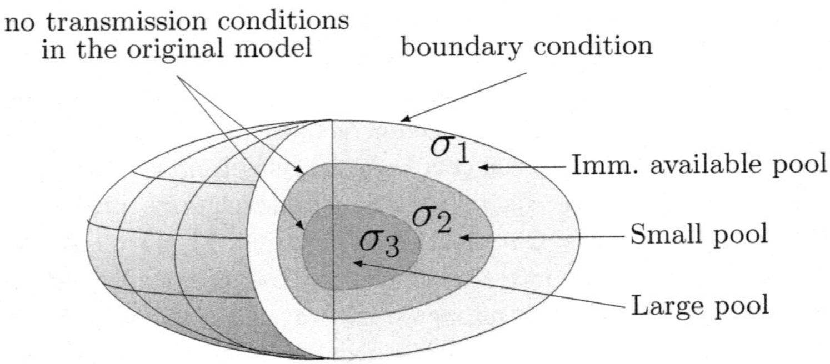

Figure 33.2 The PDE model of Bielecki and Kalita

vesicles as the union of three intervals:

$$\Omega := \Omega_3 \cup \Omega_2 \cup \Omega_1 := \left[0, r_3^-\right] \cup \left[r_3^+, r_2^-\right] \cup \left[r_2^+, r_1\right].$$

(In order to keep our notations consistent with those of [12] and [35], the intervals are numbered "from the right to the left.") Note that r_3 is now split into two points: r_3^- and r_3^+, representing positions to the immediate left and to the immediate right from the first membrane; a similar remark concerns r_2. Vesicles in all pools may permeate through the membrane(s) to the adjacent pool(s), and their ability to filter from the ith into the jth pool is characterized by permeability coefficients $k_{ij} \geq 0$, $i, j = 1, 2, 3$, $|i - j| = 1$. The left end-point $x = 0$ is a reflecting boundary for the process, and the right end-point $x = r_1$ is an elastic boundary with elasticity coefficient $k_{10} \geq 0$. The case $k_{10} > 0$ characterizes the boundary during stimulus, and $k_{10} = 0$ describes it in between stimuli (i.e., when there is no stimulus, $x = r_1$ is a reflecting boundary). Hence, k_{10} characterizes vesicles' ability to be released from the terminal bouton.

To describe our model more formally, we note that Ω is a (disconnected) compact space and the function σ defined on Ω by:

$$\sigma(x) = \sigma_i, \quad x \in \Omega_i,\ i = 1, 2, 3,$$

is continuous. A typical member of the Banach space:

$$\mathbb{X} = C(\Omega)$$

of complex continuous functions on Ω is depicted in Figure 33.3. We note that the space is isometrically isomorphic to the Cartesian product $C(\Omega_1) \times C(\Omega_2) \times C(\Omega_3)$ of the spaces of continuous functions on the three intervals. In other words, a member of $C(\Omega)$ may be identified with three continuous

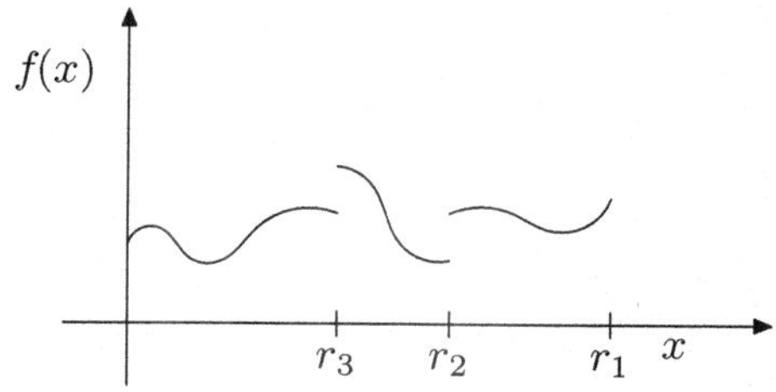

Figure 33.3 A typical member of $C(\Omega)$ (redrawn from [70])

functions f_1, f_2, f_3 being restrictions of f to the three intervals $\Omega_1, \Omega_2, \Omega_3$, respectively.

We prove (see Theorem 34.4) that the operator A in $C(\Omega)$ defined as:

$$Af = \sigma f'' \tag{33.4}$$

for twice continuously differentiable functions f on Ω satisfying the conditions:

$$\begin{aligned} f'(0) &= 0, & f'(r_3^-) &= k_{32}[f(r_3^+) - f(r_3^-)], \\ f'(r_3^+) &= k_{23}[f(r_3^+) - f(r_3^-)], & f'(r_2^-) &= k_{21}[f(r_2^+) - f(r_2^-)], \\ f'(r_2^+) &= k_{12}[f(r_2^+) - f(r_2^-)], & f'(r_1) &= -k_{10} f(r_1), \end{aligned} \tag{33.5}$$

generates a Feller semigroup $\{e^{tA_\kappa}, t \geq 0\}$ in $C(\Omega)$. The semigroup is conservative iff $k_{10} = 0$, and describes the dynamics of expected values of neurotransmitters' levels in the three pools.

Using this semigroup, we study the case when the diffusion coefficients tend to infinity while the permeability coefficients tend to zero. To this end, we consider operators $A_{\kappa_n}, n \geq 1$ defined by (33.4) with σ replaced by $\kappa_n \sigma$ and all permeability coefficients in (33.5) divided by κ_n, where $(\kappa_n)_{n \geq 1}$ is a sequence of positive numbers tending to infinity. We will show that:

$$\lim_{n\to\infty} e^{tA_{\kappa_n}} f = e^{tQ} P f, \qquad t > 0, f \in C(\Omega) \tag{33.6}$$

where:

$$Q = \begin{pmatrix} -k'_{10} - k'_{12} & k'_{12} & 0 \\ k'_{21} & -k'_{21} - k'_{23} & k'_{23} \\ 0 & k'_{32} & -k'_{32} \end{pmatrix}, \qquad k'_{ij} = \frac{\sigma_i k_{ij}}{|\Omega_i|},$$

$|\Omega_i|$ is the length of the ith interval, and the operator P given by $Pf = (|\Omega_i|^{-1} \int_{\Omega_i} f)_{i=1,2,3}$ is a projection on the subspace $\mathbb{X}_0$ of $C(\Omega)$ of functions that are constant on each of the three subintervals separately; the subspace may

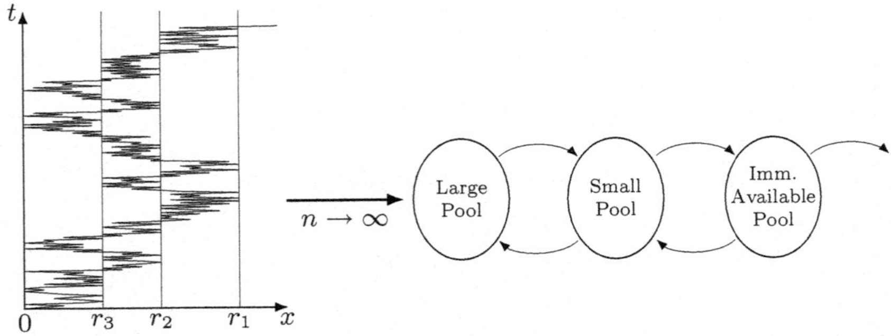

Figure 33.4 Approximating a Markov chain by diffusion processes (redrawn from [70])

be identified with $\mathbb{R}^3$, and its members may be identified with triples of real numbers.

Intuitively, as the diffusion coefficients increase, the transition probabilities between points in each interval separately tend to 1. As a result, the points become indistinguishable and may be lumped together. Points from different intervals may not be lumped together since as $n \to \infty$ the permeability coefficients $\frac{k_{ij}}{\kappa_n}$ tend to zero and in the limit the membranes become reflecting boundaries, separating the intervals. However, because of the intimate relation between diffusion and permeability coefficients, the three states of the limit process, that is, the three intervals contracted to three separate points, communicate as the states of a Markov chain with intensity matrix Q (see Figure 33.4).

Comparing this intensity matrix with the Q of (33.1), we obtain the following relations between parameters in the two models:

$$\begin{aligned} \frac{1}{R_0C_1} &= \frac{\sigma_1 k_{10}}{|\Omega_1|}, \frac{1}{R_1C_1} = \frac{\sigma_1 k_{12}}{|\Omega_1|}, \frac{1}{R_1C_2} = \frac{\sigma_2 k_{21}}{|\Omega_2|}, \\ \frac{1}{R_2C_2} &= \frac{\sigma_2 k_{23}}{|\Omega_2|}, \frac{1}{R_2C_3} = \frac{\sigma_3 k_{32}}{|\Omega_3|}. \end{aligned} \tag{33.7}$$

These relations agree with the intuition that the replenishment rate from the ith to the jth pool is directly proportional to permeability of the membrane separating them and to the speed of diffusion in the ith interval, and inversely proportional to the length of this interval. Moreover, see Corollary 34.5,

$$\frac{1}{R_3C_3} = \frac{1}{r_3}\int_0^{r_3} \beta \quad \text{and} \quad \frac{E}{R_3C_3} = \frac{1}{r_3}\int_0^{r_3} \beta\overline{\rho}, \tag{33.8}$$

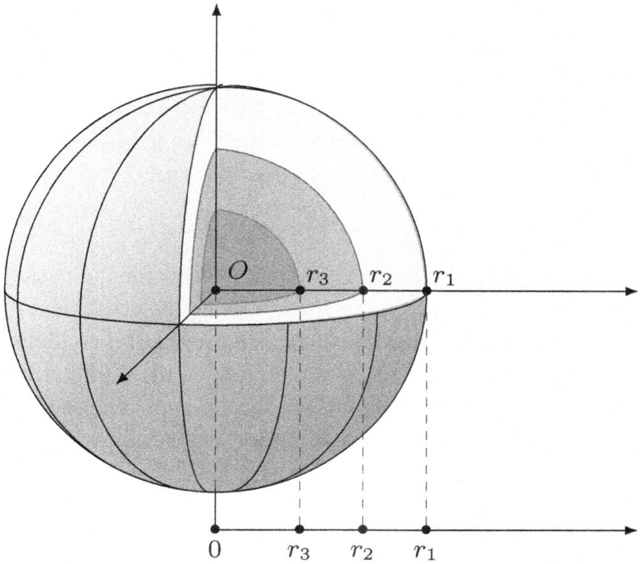

Figure 33.5 Reducing the number of dimensions by spherical symmetry

where $\overline{\rho}$ and β were introduced in (33.2). In other words, we are able to interpret the compound parameters of the ODE model (33.1) without alluding to their electrical analogues.

The model presented here is one-dimensional for simplicity of exposition. It may be readily generalized to the full three-dimensional model, but the proof of the related convergence result is much more involved and technical. One may also reduce the three-dimensional model to a variant of the model considered here as follows (see Figure 33.5). Suppose that the domains in the Bielecki and Kalita model are balls centered at the origin, with radiuses $r_3 < r_2 < r_1$. Assuming spherical symmetry of the process, we obtain that the neurotransmitter's distribution depends merely on the distance $x \in [0, r_3]$ from the origin. Equivalently, vesicles' distances from the origin may be modeled as a Bessel process [296] on Ω. This leads (compare [54]) to the model presented above with $Af(x) = \sigma f''(x)$ replaced by $\sigma \frac{1}{x}(xf(x))''$ and the reflecting boundary at $x = 0$ removed. (This is because the three-dimensional Brownian motion with probability 1 does not reach the origin in a finite time, and so the Bessel process never reaches $x = 0$. In other words, in Feller's classification [134, 143], $x = 0$ is a natural boundary for the latter process and there is no need for specifying boundary condition there.)

Finally, a word of explanation of transmission conditions is in order. Considering, for example, the third condition in (33.5), we note that it has the form of the elastic barrier condition (3.8) with 0 replaced by r_3^+, $b = 1$, $c = d = k_{23}$, and μ equal to the Dirac measure at r_3^-. Hence, it describes the process in which the vesicles in Ω_2 bounce from the membrane separating it from Ω_3 to filter into the latter interval at a random time T with distribution:

$$P(T > t) = \mathrm{e}^{-k_{23}t^+}, \quad t \geq 0.$$

Clearly, the larger k_{23} is, the shorter the time needed for the vesicle to filter through the membrane. Hence, k_{23} is truly a permeability coefficient for passing from Ω_2 to Ω_3. On the other hand, as indicated previously, dividing k_{23} by κ_n and letting $n \to \infty$ we obtain $P(T > t) = 1$, that is, the time to filter through the membrane is infinite and the boundary is reflecting. The interpretation of the other boundary and transmission conditions in (33.5) is analogous.

☞ Chapter's summary

We describe two models of fast neurotransmitters, a simpler one based on a system of ODEs, and a more involved one, coming down to a PDE, and discuss ways of connecting them. We anticipate that fast diffusion (which, we know from the previous chapter, averages things out) combined with appropriately tuned permeability conditions should lump an interval with semipermeable membranes at two midpoints and reflecting barriers at the ends, into three separate points communicating as the states of a Markov chain.

34

Fast Neurotransmitters II

Continuing the subject of neurotransmitters, we turn to proving the main convergence result announced in the previous chapter; we keep the notations introduced there.

Corollary 34.1 *Let A be sectorial and assume that B is A-bounded in the sense that $D(A) \subset D(B)$ and $\|Bf\| \le a\|Af\|$, $f \in D(A)$, for some constant a. If $\lim_{t\to\infty} \mathrm{e}^{tA}$ exists, then for any sequence $(\kappa_n)_{n\ge 1}$ with $\lim_{n\to\infty} \kappa_n = \infty$, we have:*

$$\lim_{n\to\infty} \mathrm{e}^{t(\kappa_n A + B)} = \lim_{s\to\infty} \mathrm{e}^{sA}, \qquad t > 0.$$

Proof Let δ_0 be the angle of A. For $\lambda \in \Sigma_{\frac{\pi}{2}+\delta}$, $\delta < \delta_0$,

$$\begin{aligned}\|B(\lambda - \kappa_n A)^{-1}\| &\le a\kappa_n^{-1}\|\kappa_n A(\lambda - \kappa_n A)^{-1}\| \\ &\le a\kappa_n^{-1}\|\lambda(\lambda - \kappa_n A)^{-1} - I\| \\ &\le a\kappa_n^{-1}(M_\delta + 1) =: q_n = q_n(\delta). \qquad (34.1)\end{aligned}$$

Therefore, for $n \ge n_0$ where n_0 is sufficiently large, $\|B(\lambda - \kappa_n A)^{-1}\| \le q_n < q_{n_0} < 1$. It follows that $\lambda \in \rho(\kappa_n A + B)$,

$$(\lambda - \kappa_n A - B)^{-1} = (\lambda - \kappa_n A)^{-1} \sum_{k=0}^{\infty} [B(\lambda - \kappa_n A)^{-1}]^k \qquad (34.2)$$

and:

$$\begin{aligned}\|(\lambda - \kappa_n A - B)^{-1}\| &= \|(\lambda - \kappa_n A)^{-1} \sum_{k=0}^{\infty} [B(\lambda - \kappa_n A)^{-1}]^k\| \\ &\le \|(\lambda - \kappa_n A)^{-1}\| \frac{1}{1 - q_n} \\ &\le \frac{M_\delta}{1 - q_{n_0}} |\lambda|^{-1}. \qquad (34.3)\end{aligned}$$

This implies that the operators $\kappa_n A + B$, $n \ge n_0$ are sectorial with angle δ. On the other hand, using (34.1), (34.2), and the Lebesgue Dominated Convergence Theorem, $\lim_{n\to\infty}(\lambda - \kappa_n A - B)^{-1} = \lim_{n\to\infty}(\lambda - \kappa_n A)^{-1}$. Therefore, $\lim_{n\to\infty} \mathrm{e}^{t(\kappa_n A+B)}$ exists and equals $\lim_{n\to\infty} \mathrm{e}^{t\kappa_n A} = \lim_{s\to\infty} \mathrm{e}^{sA}$, $t > 0$. □

Let G_i in $C(\Omega_i)$ be the version of the operator A of Corollary 32.2, and let B in $\mathbb{X}$ be defined by $Bf = (\sigma_i G_i f_i)_{i\in\mathbb{I}} = \sigma f''$ on the domain $D(G_1) \times D(G_2) \times D(G_3)$. Then B is sectorial with angle $\frac{\pi}{2}$ because G_i are, and by Corollary 32.2 we have:

$$\lim_{t\to\infty} \mathrm{e}^{tB} f = Pf, \tag{34.4}$$

where P is as defined after (33.6).

Lemma 34.2 *A satisfies the positive maximum principle.*

Proof Let $x_0 \in \Omega$ be the coordinate where the positive maximum of f is attained. If x_0 belongs to the interior of any of the intervals in Ω, the claim is obvious. If $x_0 = 0$, then by the first condition in (33.5), $f'(x_0) = 0$ and f may be uniquely extended to the even, twice continuously differentiable function on $[-r_3, r_3]$. Since the maximum of this extension is attained at $x_0 = 0$, $f''(x_0) \le 0$, proving the claim. If $x_0 = r_3^-$, then $f'(r_3^-) \ge 0$ and, on the other hand, by the second relation in (33.5), $f'(r_3^-) \le 0$. Therefore, $f'(r_3^-) = 0$ and we may proceed as earlier to show that $f''(r_3^-) \le 0$. The remaining cases are treated similarly. We note that it is only at $x = r_1$ that we use the fact that the maximum is positive. □

We introduce the functionals $F_{ij} \in \mathbb{X}^*$ describing permeability from the ith to the jth pool:

$$\begin{aligned} F_{32}(f) &= k_{32}[f(r_3^+) - f(r_3^-)], & F_{23}(f) &= k_{23}[f(r_3^+) - f(r_3^-)], \\ F_{21}(f) &= k_{21}[f(r_2^+) - f(r_2^-)], & F_{12}(f) &= k_{12}[f(r_2^+) - f(r_2^-)], \\ F_{10}(f) &= -k_{10} f(r_1), \end{aligned} \tag{34.5}$$

the 0th pool representing the space outside of the terminal bouton. Then the domain of A_{κ_n} may be equivalently described as composed of twice continuously differentiable functions f on Ω satisfying:

$$\begin{aligned} f'(0) &= 0, & \kappa_n f'(r_3^-) &= F_{32}(f), \\ \kappa_n f'(r_3^+) &= F_{23}(f), & \kappa_n f'(r_2^-) &= F_{21}(f), \\ \kappa_n f'(r_2^+) &= F_{12}(f), & \kappa_n f'(r_1) &= F_{10}(f). \end{aligned} \tag{34.6}$$

Given $a \in \mathbb{R}$ we may choose $\psi \in C^2[0, 1]$ with arbitrarily small supremum norm so that $\psi(0) = \psi(1) = \psi'(0) = 0$ and $\psi'(1) = a$; for example

$\psi(x) = \frac{a\epsilon}{\pi}x^2 \sin\frac{\pi(x-1)}{\epsilon}$. Using translations, symmetries, and stretching transformations, we may find ψ_{ij} with norms not exceeding arbitrary ϵ such that:

$$\begin{array}{lll} \psi_{10} \in C^2[r_2^+, r_1], & \psi_{10}'(r_2^+) = 0, & \psi_{10}'(r_1) = 1, \\ \psi_{12} \in C^2[r_2^+, r_1], & \psi_{12}'(r_2^+) = 1, & \psi_{12}'(r_1) = 0, \\ \psi_{21} \in C^2[r_3^+, r_2^-], & \psi_{21}'(r_3^+) = 0, & \psi_{21}'(r_2^-) = 1, \\ \psi_{23} \in C^2[r_3^+, r_2^-], & \psi_{23}'(r_3^+) = 1, & \psi_{23}'(r_2^-) = 0, \\ \psi_{32} \in C^2[0, r_3^-], & \psi_{32}'(0) = 0, & \psi_{32}'(r_3^-) = 1, \end{array} \tag{34.7}$$

and ψ_{ij} vanish at the ends of intervals where they are defined.

Lemma 34.3 *Fix n. The map $I_n = I_{\mathbb{X}} - \kappa_n^{-1}J$, where:*

$$Jf = \begin{pmatrix} (F_{12}f)\psi_{12} + (F_{10}f)\psi_{10} \\ (F_{23}f)\psi_{23} + (F_{21}f)\psi_{21} \\ (F_{32}f)\psi_{32} \end{pmatrix} \quad \text{and} \quad I_{\mathbb{X}}f = f, \tag{34.8}$$

is an isomorphism of $\mathbb{X}$ with $I_n^{-1} = I_{\mathbb{X}} + \kappa_n^{-1}J$.

Proof Since all ψ_{ij}'s vanish at the ends of the intervals where they are defined,

$$F_{ij} \circ J = 0. \tag{34.9}$$

Clearly, I_n is linear and bounded. To show that I_n is injective, we assume that $I_n f = I_n g$ for some $f, g \in \mathbb{X}$. Applying F_{ij} to both sides of this relation, by (34.9), we obtain $F_{ij}(f) = F_{ij}(g)$. Therefore, $Jf = Jg$, and this coupled with $I_n f = I_n g$ implies $f = g$.

Finally, given $g \in \mathbb{X}$ we define $f = g + \kappa_n^{-1}Jg$ to see that, by (34.9), $F_{ij}(f) = F_{ij}(g)$ and so $Jf = Jg$. Then, $I_n f = (g + \kappa_n^{-1}Jg) - \kappa_n^{-1}Jf = g$, proving that I_n is onto and establishing the formula for I_n^{-1}. □

Theorem 34.4 *A is the generator of a holomorphic Feller semigroup, which is conservative iff $k_{10} = 0$. Moreover,* (33.6) *holds.*

Proof The key step is to pass to similar (or: isomorphic) semigroups [49, 129]. Let $B_n = I_n A_{\kappa_n} I_n^{-1}$. For $f \in D(A_{\kappa_n})$, $I_n f$ is twice continuously differentiable, since ψ_{ij} are, and conditions (34.7) imply $I_n f \in D(B)$. Since the converse implication is also true, $f \in D(A_{\kappa_n})$ iff $I_n f \in D(B)$. This shows that $D(B_n) = D(B)$.

Moreover, for $f \in D(B)$,

$$A_{\kappa_n} I_n^{-1} f = \kappa_n \sigma f'' + \sigma (Jf)'' = \kappa_n \sigma f'' + \sigma K f,$$

where Kf is given by the right-hand side of (34.8) with ψ_{ij}'s replaced by their second derivatives. Thus,

$$B_n f = \kappa_n Bf + Cf + D_n f,$$

where $Cf = -J(\sigma f'') = -JBf$ and $D_n f = \sigma Kf - \frac{1}{\kappa_n} J(\sigma Kf)$.

Since $\|\psi_{ij}\| < \epsilon$, for $l := \max\{\|F_{32}\|, \|F_{23}\| + \|F_{21}\|, \|F_{12}\| + \|F_{10}\|\}$, we have $\|J\| < \epsilon l$. Therefore:

$$\|Cf\| \le \epsilon l \|Bf\| \quad f \in D(B),$$

that is, C is B-bounded with bound $l\epsilon$. Calculating as in (34.1), for $\lambda \in \Sigma_{\pi/2+\delta}$,

$$\begin{aligned} \|C(\lambda - \kappa_n B)^{-1}\| &\le l\epsilon\kappa_n^{-1}\|\kappa_n B(\lambda - \kappa_n B)^{-1}\| \\ &= l\epsilon\kappa_n^{-1}\|\lambda(\lambda - \kappa_n B)^{-1} - I\| \\ &\le l\epsilon\kappa_n^{-1}(M_\delta + 1). \end{aligned} \tag{34.10}$$

Taking sufficiently small ϵ we see that $\|C(\lambda - \kappa_n B)^{-1}\| \le q$ for some $q < 1$ and all $n \ge 1$. It follows that $\lambda \in \rho(B_n + C)$, that is, $\kappa_n B + C$ satisfies the range condition. Since D_n are bounded, the same is true for B_n, and for A_{κ_n} (for sufficiently large λ). In particular, choosing $\kappa_n = 1$ for some n we see, in view of Lemma 34.2 that A generates a holomorphic, Feller semigroup; this semigroup is conservative iff $1_\Omega \in D(A_\kappa)$, that is, iff $k_{10} = 0$. We are left with showing (33.6).

By Corollary 34.1 and (34.4) $\lim_{n\to\infty} \mathrm{e}^{t(\kappa_n B + C)} = P, t > 0$, and since irregular convergence is preserved under bounded perturbations (see (29.3)):

$$\lim_{n\to\infty} \mathrm{e}^{tB_n} = \mathrm{e}^{t\sigma PK}P = \mathrm{e}^{tQ}P, t > 0$$

because, calculating:

$$\sigma PKf = \begin{pmatrix} \frac{\sigma_1}{|\Omega_1|}(F_{10} - F_{12})f \\ \frac{\sigma_2}{|\Omega_2|}(F_{21} - F_{23})f \\ \frac{\sigma_3}{|\Omega_3|}F_{32}f \end{pmatrix}$$

we see that σPK as restricted to $\mathbb{X}_0$ equals Q (for $f \in \mathbb{X}_0$, $f(r_2^+) = f(r_1)$ and $f(r_2^-) = f(r_3^+)$). Since $\mathrm{e}^{tA_{\kappa_n}} = I_n^{-1}\mathrm{e}^{tB_n}I_n$ and $\lim_{n\to\infty} I_n = \lim_{n\to\infty} I_n^{-1} = I_{\mathbb{X}}$, this completes the proof. □

As a corollary, we obtain the following result. Given a non-negative $\beta \in C(\Omega_3)$, we may identify it with $(0, 0, \beta) \in C(\Omega)$ and interpret it as a neurotransmitters' production rate in the large pool. β may also be identified with the bounded multiplication operator in $C(\Omega)$ mapping $f \in C(\Omega)$ to βf.

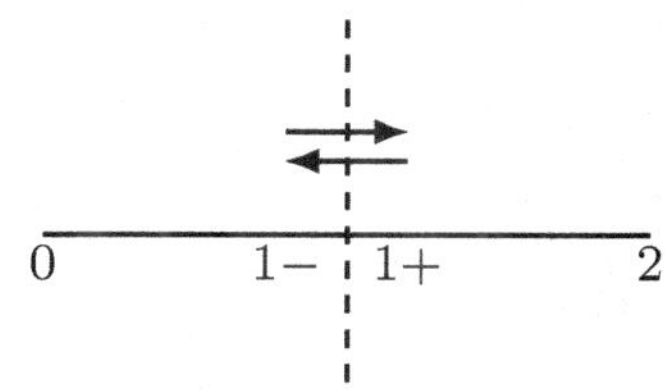

Figure 34.1 Two pools with one separating membrane

Corollary 34.5 *Let $\overline{\rho} \in C(\Omega)$ be given, and fix $f \in C(\Omega)$. The mild solutions to the inhomogeneous Cauchy problems:*

$$\frac{\mathrm{d}u_n}{\mathrm{d}t} = (A_{\kappa_n} - \beta)u_n + \beta\overline{\rho}, \qquad u(0) = f, \tag{34.11}$$

converge to those of the Cauchy problem related to (33.1) *with initial conditions* $U_i(0) = |\Omega_i|^{-1}\int_{\Omega_i} f$ *and the constants in* Q *satisfying* (33.7) *and* (33.8)*. The convergence is uniform in* t *in compact subintervals of* $\mathbb{R}^+_*$.

Proof By (33.6) and (29.3),

$$\lim_{n\to\infty} \mathrm{e}^{t(A_{\kappa_n}-\beta)} = \mathrm{e}^{t(Q-P\beta)}P \tag{34.12}$$

strongly and almost uniformly in $t \in R^+_*$. The mild solution to (34.11) is (see [128]):

$$u_n(t) = \mathrm{e}^{t(A_{\kappa_n}-\beta)}f + \int_0^t \mathrm{e}^{(t-s)(A_{\kappa_n}-P\beta)}\beta\overline{\rho}\,\mathrm{d}s.$$

By (34.12), this converges to:

$$U(t) = \mathrm{e}^{t(Q-\beta)}Pf + \int_0^t \mathrm{e}^{(t-s)(Q-P\beta)}P\beta\overline{\rho}\,\mathrm{d}s.$$

Since $P\beta = (0, 0, \frac{1}{r_3}\int_0^{r_3}\beta)$ and $P\beta\overline{\rho} = (0, 0, \frac{1}{r_3}\int_0^{r_3}\beta\overline{\rho})$, U solves (33.1) with the specified initial conditions. □

Exercise 34.1 This exercise clarifies (hopefully) the mechanism discussed in this chapter, by considering the case of two pools. For further simplicity, assume, as in Figure 34.1, that two intervals representing pools are of the same unit length, and consider the operator $Af = f''$ defined for twice continuously differentiable functions in $\mathbb{B} = C[0,1] \times C[1,2]$ (each pair of functions in $\mathbb{B}$ is identified with a single function on $[0,2]$, continuous in this interval except perhaps at $x = 1$ where it has limits from the left and from the right), satisfying boundary and transmission conditions:

$$f'(0) = f'(2) = 0,$$
$$f'(1-) = \alpha[f(1+) - f(1-)],$$
$$f'(1+) = \beta[f(1+) - f(1-)],$$

where α and β play the role of permeability coefficients of the membrane located at $x = 1$ (see Figure 34.1).

(a) Prove directly that A is a Feller generator and find its resolvent.

(b) Using (a), prove that, if $Af = f''$ is replaced by $A_n f = nf''$ and the permeability coefficients are divided by n, the related semigroups e^{tA_n} converge, as $n \to \infty$, to $\mathrm{e}^{tQ}P$ where:

$$Pf = \left(\int_0^1 f, \int_1^2 f\right)$$

is a map from $\mathbb{X}$ to $\mathbb{R}^2$ (identified with the subspace $\subset \mathbb{X}$ of functions that are constant on each of the intervals) and:

$$Q = \begin{pmatrix} -\alpha & \alpha \\ \beta & -\beta \end{pmatrix}, \tag{34.13}$$

is the intensity matrix of the simplest Markov chain.

(c) Let A be the operator in $L^1[0, 2]$ defined by $A\phi = \phi''$ for all functions ϕ with the following properties. (a) ϕ restricted to $[0, 1]$ is differentiable, its derivative is absolutely continuous and the second derivative belongs to $L^1[0, 1]$, (b) similar conditions are met for the restriction of ϕ to $[1, 2]$, (c) transmission conditions:

$$\phi'(1+) = \phi'(1-) = \beta\phi(1+) - \alpha\phi(1-).$$

are satisfied. Show that, in the sense described in Exercise 11.2, this operator is dual to the one of point (a). Imitating point (b), prove a convergence theorem in this space; note that it involves the transpose of the matrix (34.13).

☞ Chapter's summary

We provide a rigorous proof of the convergence theorem for semigroups modelling activity of neurotransmitters, anticipated in the previous chapter. The argument relies on the fact that the semigroups are uniformly holomorphic.

35

From Diffusions on Graphs to Markov Chains and Back Again

The result from the previous chapter may be readily generalized to a nice abstract setup [55] (see also [19, 20] for further developments). To this end, imagine a finite graph $\mathcal{G}$ without loops, and a Markov process on $\mathcal{G}$ obeying the following rules:

- While on the ith edge, imagined as a C^1 curve in $\mathbb{R}^3$, the process behaves like a one-dimensional Brownian motion with variance $\sigma_i > 0$.
- Graph's vertices are semipermeable membranes, allowing communication between the edges; permeability coefficients p_{ij}, describing the possibility to filter through the membrane from the ith to the jth edge, depend on the edges. In particular, p_{ij} is in general different from p_{ji}. At each vertice, the process may also be killed and removed from the state-space.

Now, suppose the diffusion's speed increases while membranes' permeability decreases (i.e., $\sigma_i \to \infty$ and $p_{ij} \to 0$). As a result points in each edge communicate almost immediately and in the limit are lumped together, but the membranes prevent lumping of points from different edges. Again, an appropriate choice of scaling of permeability coefficients leads to a limit process in which communication between lumped edges is possible. The lumped edges form then vertices in the so-called **line graph of** $\mathcal{G}$ (see [112]) and communicate as the states of a Markov chain with jumps' intensities directly proportional to permeability coefficients p_{ij} and the diffusion coefficients σ_i, and inversely proportional to the edges' lengths (see Figure 35.1 and Theorem 35.2).

This procedure may also be reversed: given a finite-state Markov chain, we may find a graph $\mathcal{G}$ and construct a fast diffusion on $\mathcal{G}$ approximating the chain.

Let us be more specific. Let $\mathcal{G} = (\mathcal{V}, \mathcal{E})$ be a finite geometric graph [265] without loops, where $\mathcal{V} \subset \mathbb{R}^3$ is the set of vertices and $\mathcal{E}$ is the set of edges of finite length. The number of edges is N, and the edges are seen as C^1 curves connecting vertices. For $i \in \mathcal{N} := \{1, \dots, N\}$, $L_i, R_i \in \mathcal{V}$ denote the left and

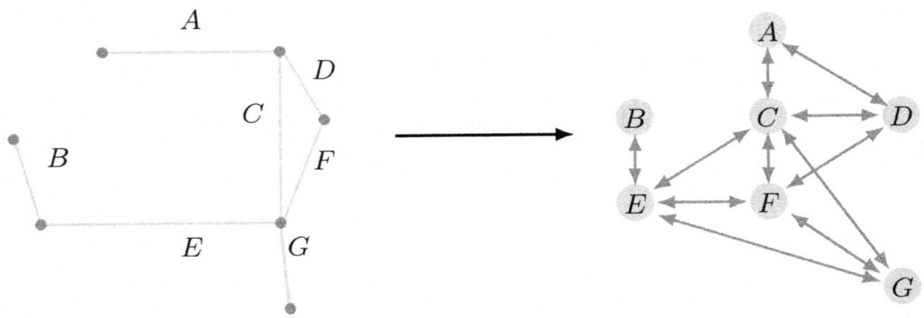

Figure 35.1 From diffusion on $\mathcal{G}$ to a Markov chain on the vertices of the line graph of $\mathcal{G}$; edges "shrink" to vertices, vertices "split" into edges (redrawn from [55])

right endpoints of the ith edge E_i, respectively; the decision which is the left point and which is the right point is made arbitrarily.

Since each vertex V is a semipermeable membrane, the future behavior of the related diffusion process starting at V depends on which side of the membrane it lies. Hence, if $k \le N$ edges are incident in V, V splits naturally into k points. By $V_i \in \mathcal{N}$ we denote V as an endpoint of the ith edge (if V is not an endpoint of this edge, we leave V_i undefined).

The state-space of the diffusion process is the union $S = \cup_{i\in\mathcal{N}} E_i$ of *disjoint* edges; as earlier, even though endpoints of many edges coincide, we treat them as distinct. S is a (disconnected) compact topological space, and the related space of continuous complex functions $C(S)$ is isometrically isomorphic to the Cartesian product $C(E_1) \times \cdots \times C(E_N)$ of spaces of continuous functions on the edges (equipped with the norm $\|(f_i)_{i\in\mathcal{N}}\| = \max_{i\in\mathcal{N}} \|f_i\|_{C(E_i)}$). On the other hand, each $C(E_i)$ is isometrically isomorphic to the space $C[0, d_i]$ of continuous functions on the interval $[0, d_i]$, where d_i is the length of the ith edge: a function $f \in C(E_i)$ is then identified with its image $If \in C[0, d_i]$, given by $If(t) = f(p)$, where p is the unique point on E_i, whose distance from the left endpoint of E_i (along the edge) is $t \in [0, d_i]$. In particular, we may meaningfully speak of continuously differentiable functions on E_i; these are simply the images of continuously differentiable functions on $[0, d_i]$.

Let $\sigma \in C(S)$ be defined by $\sigma(p) = \sigma_i$, for $p \in E_i$, where σ_i are given positive numbers. We define the operator A in $C(S)$ by:

$$Af = \sigma f'' \tag{35.1}$$

for twice continuously differentiable functions on $C(S)$, satisfying the transmission conditions described later.

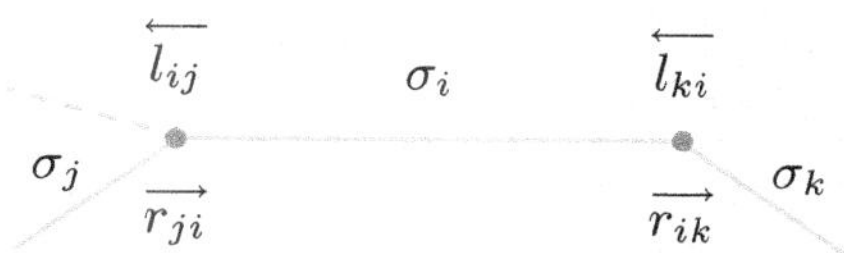

Figure 35.2 l_{ij}s and r_{ij}s are permeability coefficients of the membranes at vertices, and determine probabilities of passing from one edge of the graph to another (redrawn from [55])

For each i, let l_i and r_i be non-negative numbers describing the possibility of passing through the membrane from the ith edge to the neighboring edges at the left and right endpoints, respectively. Also, let l_{ij} and r_{ij}, $j \neq i$ be non-negative numbers satisfying $\sum_{i \neq j} l_{ij} \leq l_i$ and $\sum_{i \neq j} r_{ij} \leq r_i$. These numbers determine the probability that after filtering through the membrane from the ith edge, a particle will enter the jth edge (see Figure 35.2). By default, if E_j is not incident in L_i, we put $l_{ij} = 0$. In particular, by convention $l_{ij} f(V_j) = 0$ for $f \in C(S)$, if V_j is not defined; the same remark concerns r_{ij}. In these notations, the transmission conditions mentioned earlier are as follows: if $L_i = V$, then:

$$f'(V_i) = l_i f(V_i) - \sum_{j \neq i} l_{ij} f(V_j); \tag{35.2}$$

and if $R_i = V$, then:

$$-f'(V_i) = r_i f(V_i) - \sum_{j \neq i} r_{ij} f(V_j). \tag{35.3}$$

Again, their interpretation is that the diffusion process starting at the ith edge "bounces" from the membrane at the left end, and the time it spends at the membrane is measured by the related Lévy local time t^+. Then, at a random time T distributed according to $P(T > t) = \mathrm{e}^{-l_i t^+}, t \geq 0$, the process filters through the membrane; the probability that it will filter to the jth edge is $\frac{l_{ij}}{l_i}$; $1 - l_i^{-1} \sum_{j \neq i} l_{ij}$ is the probability that the particle is removed from the state-space. The behavior at the right end is analogous.

Proposition 35.1 *A generates a Feller semigroup in* $C(S)$, *denoted* $\{\mathrm{e}^{tA}, t \geq 0\}$. *The semigroup is conservative iff:*

$$\sum_{j \neq i} l_{ij} = l_i \text{ and } \sum_{j \neq i} r_{ij} = r_i, \qquad i \in \mathcal{N}. \tag{35.4}$$

Theorem 35.2 *Let* $(\kappa_n)_{n \geq 1}$ *be a sequence of positive numbers converging to infinity, and let operators* A_{κ_n} *be defined by* (35.1) *with* σ *replaced by* $\kappa_n \sigma$ *and* *with domain composed of $C^2(S)$ functions satisfying the transmission conditions* (35.2) *and* (35.3) *with all permeability coefficients divided by κ_n, respectively. Then:*

$$\lim_{n\to\infty} \mathrm{e}^{tA_{\kappa_n}} f = \mathrm{e}^{tQ} Pf, \qquad f \in C(S), t > 0, \tag{35.5}$$

where P is the projection of $C(S)$ on the space $C_0(S)$ of functions that are constant on each edge, given by:

$$Pf = \left(d_i^{-1} \int_{E_i} f \right)_{i\in\mathcal{N}}, \tag{35.6}$$

while Q is the operator in $C_0(S)$ represented by the matrix $(q_{ij})_{i,j\in\mathcal{N}}$ with $q_{ij} = \sigma_i d_i^{-1}(l_{ij} + r_{ij})$ for $i \neq j$ and $q_{ij} = -\sigma_i d_i^{-1}(l_i + r_i)$ for $i = j$.

As always, the limit here is strong and almost uniform in $t \in (0, \infty)$; for $f \in C_0(S)$ the formula holds for $t = 0$, as well, and the limit is almost uniform in $t \in [0, \infty)$. Q is the intensity matrix of the limit Markov chain; the limit semigroup is in general substochastic, because the rows of Q do not necessarily add up to 0.

The proof is a straightforward generalization of the one presented in the previous chapter, where $\mathcal{G}$ was composed of three adjacent intervals. Instead of discussing all the details, we restrict ourselves to presenting its main steps.

First for $i \in \mathcal{N}$ we define $F_{\mathrm{l},i}, F_{\mathrm{r},i} \in [C(S)]^*$ by:

$$F_{\mathrm{l},i} f = [l_i f(V_i) - \sum_{j\neq i} l_{ij} f(V_j)]; \qquad F_{\mathrm{r},i} f = -[r_i f(V_i) - \sum_{j\neq i} r_{ij} f(V_j)], \tag{35.7}$$

"l" and "r" standing for "left" and "right," respectively. Also, given $\epsilon > 0$ we find $\psi_{\mathrm{l},i}, \psi_{\mathrm{r},i} \in C^2(E_i)$ such that $\|\psi_{\mathrm{l},i}\|, \|\psi_{\mathrm{r},i}\| < \epsilon$, $\psi_{\mathrm{l},i}$ and $\psi_{\mathrm{r},i}$ vanish at both ends of E_i while:

$$\psi'_{\mathrm{l},i}(L_i) = \psi'_{\mathrm{r},i}(R_i) = 1, \qquad \psi'_{\mathrm{l},i}(R_i) = \psi'_{\mathrm{r},i}(L_i) = 0. \tag{35.8}$$

Then, we define $J \in \mathcal{L}(C(S))$ by:

$$Jf = \left((F_{\mathrm{l},i} f)\psi_{\mathrm{l},i} + (F_{\mathrm{r},i} f)\psi_{\mathrm{r},i} \right)_{i\in\mathcal{N}}. \tag{35.9}$$

Then, for $\kappa \neq 0$, the map $I_\kappa = I_{C(S)} - \kappa^{-1} J$ turns out to be an isomorphism of $C(S)$ with $I_\kappa^{-1} = I_{C(S)} + \kappa^{-1} J$, where $I_{C(S)}$ is the identity operator in $C(S)$.

Let G_i in $C(E_i)$ be the version of the operator A of Corollary 32.2, and let B in $C(S)$ be defined by:

$$Bf = (\sigma_i G_i f_i)_{i\in\mathcal{N}} \tag{35.10}$$

on the Cartesian product of domains of G_i. Equivalently, $Bf = \sigma f''$ for all $f \in C^2(S)$ satisfying the Neumann boundary conditions $f'(L_i) = f'(R_i) = 0$, $i \in \mathcal{N}$. Then B is sectorial with angle $\frac{\pi}{2}$ because G_i are, and by Corollary 32.2 we have:

$$\lim_{n\to\infty} \lambda(\lambda - \kappa_n B)^{-1} f = Pf,$$

where P was defined in Theorem 35.2.

Turning to isomorphic images $B_n = I_{\kappa_n} A_{\kappa_n} I_{\kappa_n}^{-1}$ of A_{κ_n} we check that $D(B_n)$ coincides with $D(B)$ and:

$$B_n f = \kappa_n Bf + Cf + D_n f, \qquad f \in D(B)$$

where $Cf = -J(\sigma f'') = -JBf$, $D_n f = \sigma Kf - \frac{1}{\kappa_n} J(\sigma Kf)$, and Kf is given by the right-hand side of (35.9) with $\psi_{\mathrm{l},i}$'s and $\psi_{\mathrm{r},i}$'s replaced by their second derivatives.

Since B is sectorial and C is B-bounded, B_n generate semigroups, and so do A_{κ_n}. Since the latter operators satisfy the maximum principle, they are generators of Feller semigroups. These semigroups are conservative iff $I_S \in D(A_{\kappa_n})$, that is, iff (35.4) holds. This proves Proposition 35.1. Moreover, $\kappa_n B + C$ are equiholomorphic.

Since the semigroups generated by $\kappa_n B + C$ are uniformly holomorphic, convergence of their resolvents implies convergence of the semigroups, and we check that $\lim_{n\to\infty}(\lambda - \kappa_n B - C)^{-1} = \lim_{n\to\infty}(\lambda - \kappa_n B)^{-1}$, which we know to be equal to $\lambda^{-1}P$, $\lambda > 0$. Thus,

$$\lim_{n\to\infty} \mathrm{e}^{(\kappa_n B + C)t} = P, \qquad t > 0$$

and since bounded perturbations preserve convergence,

$$\lim_{n\to\infty} \mathrm{e}^{tB_n} = \mathrm{e}^{t\sigma PK} P, \qquad t > 0.$$

Finally, calculating:

$$\begin{aligned} \sigma PKf &= \left(\sigma_i d_i^{-1}(F_{\mathrm{r},i} f - F_{\mathrm{l},i} f)\right)_{i\in\mathbb{I}} \qquad (35.11) \\ &= \left(\sigma_i d_i^{-1} \sum_{j\neq i} (l_{ij} + r_{ij}) f(V_j) - \sigma_i d_i^{-1} (l_i + r_i) f(V_i)\right)_{i\in\mathbb{I}} \end{aligned}$$

we see that σPK as restricted to $C_0(S)$ equals Q, and the result follows as before.

Interestingly, the procedure described in Theorem 35.2 may be reversed: any finite-state (possibly not honest) Markov chain is a limit of fast diffusions on a graph in the following sense. Let Q be a (substochastic) $N \times N$ intensity

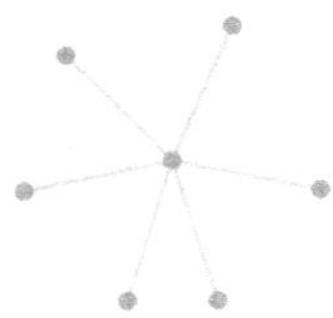

Figure 35.3 Star-shaped graph with six edges (redrawn from [55])

matrix, and let $\mathcal{G}$ be the star-shaped graph with N edges, all vertices but one of degree 1, the special vertex with degree N (see Figure 35.3), and all the edges of length 1. We number the edges clockwise, from 1 to N, and agree that the special vertex is the left end of all the edges. Finally, we take $\sigma_i = 1$, $r_i = 0$, and $l_{ij} = q_{ij}$. Then, by Theorem 35.2 the related diffusions on $\mathcal{G}$ with arbitrary choice of κ_n converge to the Markov chain on the vertices of the related line graph (these vertices may be identified with natural numbers $1, \ldots, N$) with intensity matrix Q.

An $L^1(\mathbb{R})$ analogue of our main Theorem 35.2 may be found in [166] – a step toward this result is contained in Exercise 35.2.

Exercise 35.1 Show that A defined in this chapter satisfies the maximum principle. Assuming (35.4), use the same argument to check that if $\mathcal{G}$ is connected, then the kernel of A contains merely constant functions. *Hint:* a function belonging to the kernel must be linear on each edge, and its maximal value must be attained at the end of an edge. Use (35.2) and (35.3) to persuade yourself that the derivative at this edge must be zero, and hence, that the maximal value is attained at the entire edge. Use the transmission conditions again to check that the function must have the same maximal value at all neighboring edges.

Exercise 35.2 Let $W^{2,1}(S)$ be the space of functions on S that, as restricted to any edge, have absolutely continuous derivatives and integrable second derivative. Let A^* be the operator $A^*\phi = \sigma\phi''$ with domain composed of members of $W^{2,1}(S)$ satisfying the transition conditions:

$$\begin{aligned}
\sigma_j\phi'(L_j) &= \sigma_j l_j \phi(L_j) - \sum_{i\in I_j^L}{}^{*} (\sigma_i l_{ij}\phi(L_i) + \sigma_i r_{ij}\phi(R_i)), \\
\sigma_j\phi'(R_j) &= \sum_{i\in I_j^R}{}^{*} (\sigma_i l_{ij}\phi(L_i) + \sigma_i r_{ij}\phi(R_i)) - \sigma_j r_j \phi(R_j), \qquad j \in \mathcal{N}.
\end{aligned} \tag{35.12}$$

Here, I_j^L and I_j^R are the sets of indexes $i \neq j$ of edges incident in L_j and R_j, respectively, and the asterisk in the sums denotes the fact that, since there are no loops, at most one of the terms $\sigma_i\phi(L_i)l_{ij}$ and $\sigma_i\phi(R_i)r_{ij}$ is taken into account.

Check that $\int_S (Af)\phi = \int_S (A^*\phi) f$ for all $f \in D(A)$ and $\phi \in D(A^*)$. Compare Exercise 11.2. Calculations are intelligent and somewhat lengthy, but not as lengthy as in Exercise 31.4.

☞ Chapter's summary

As proved in this chapter, fast diffusions on finite graphs with semipermeable membranes on vertices may be approximated by finite-state Markov chains provided the related permeability coefficients are appropriately small.

36

Semilinear Equations, Early Cancer Modeling

This chapter provides background material for the homogenization theorem of Chapter 38. We start by recalling basic theory of semilinear Cauchy problems (see [177, 284] and [318] for more information on the subject). Let A be the generator of a semigroup of equibounded operators in a Banach space $\mathbb{X}$, and let $F : \mathbb{X} \to \mathbb{X}$ be a continuous function. For $f \in D(A)$, the Cauchy problem of finding a continuously differentiable function u such that $u(0) = f, u(t) \in D(A)$ and:

$$\frac{\mathrm{d}u(t)}{\mathrm{d}t} = Au(t) + F(u(t)) \tag{36.1}$$

for t in an interval $[0, \delta)$, $\delta > 0$ is said to be *semilinear*. Replacing t in (36.1) by s, applying $\mathrm{e}^{(t-s)A}$ to both sides, and integrating from 0 to $t < \delta$, we see that u satisfies:

$$u(t) = \mathrm{e}^{tA} f + \int_0^t \mathrm{e}^{(t-s)A} F(u(s))\, \mathrm{d}s \tag{36.2}$$

in the same interval. For $f \in \mathbb{X}$, a u satisfying (36.2) is said to be a **mild solution** of (36.1), and of course such a u may but needs not satisfy (36.1). A sufficient condition for a mild solution to solve (36.1) is that $f \in D(A)$ and F is continuously differentiable, see [284].

If F is (globally) Lipschitz continuous, that is, if there is a constant $L > 0$ such that

$$\|F(f) - F(g)\| \le L\|f - g\|, \qquad f, g \in \mathbb{X}, \tag{36.3}$$

then for each $f \in \mathbb{X}$ the solution of (36.2) exists and is unique on the whole of $\mathbb{R}^+$. To show this, we argue as in Chapter 22. Let $\mathcal{X}_\mu$ be the Banach space of continuous functions on $\mathbb{R}^+$, with values in $\mathbb{X}$, and of at most exponential growth with exponent $\mu > 0$; we equip $\mathcal{X}_\mu$ with the Bielecki-type norm

[34, 125]:

$$\|u\|_\mu = \sup_{t\geq 0} \mathrm{e}^{-\mu t} \|u(t)\|.$$

Then, the operator K given by $Ku(t) = \mathrm{e}^{tA}f + \int_0^t \mathrm{e}^{(t-s)A}F(u(s))\,\mathrm{d}s$ is a contraction in $\mathcal{X}_\mu$ provided $\mu > LM$, where M is the bound for the semigroup $(\mathrm{e}^{tA})_{t\geq 0}$:

$$\begin{aligned}
\|Ku - Kv\|_\mu &\leq ML \sup_{t\geq 0} \mathrm{e}^{-\mu t} \int_0^t \|u(s) - v(s)\|\,\mathrm{d}s \\
&\leq ML \sup_{t\geq 0} \int_0^t \mathrm{e}^{-\mu(t-s)}\,\mathrm{d}s\, \|u - v\|_\mu \\
&\leq \frac{ML}{\mu}\|u - v\|_\mu, \qquad u, v \in \mathcal{X}_\mu. \qquad (36.4)
\end{aligned}$$

Hence, our claim follows by Banach's Fixed Point Theorem. (Again, to show local uniqueness of solutions one needs to invoke the Gronwall Lemma.)

As a by-product of the proof, we see that mild solutions are of at most exponential growth with parameter $\mu > ML$. (Also, bearing in mind that existence and uniqueness of classical solutions of a Cauchy problem for initial values from a dense set combined with continuous dependence on those initial values is equivalent to existence of the related semigroup, and taking $F(f) = Bf$, $f \in \mathbb{X}$ for a bounded linear operator B we obtain an elegant proof of the Phillips Perturbation Theorem.)

However, for many applications, condition (36.3) is too restrictive. A more realistic assumption is that for all $r > 0$ there exists an $L(r)$ such that:

$$\|F(f) - F(g)\| \leq L(r)\|f - g\|, \qquad \|f\|, \|g\| \leq r. \qquad (36.5)$$

This assumption of local Lipschitz continuity does not guarantee existence of solutions of (36.1) on the whole half-line. A more careful analysis involving Banach's Fixed Point Theorem shows that in this case for all $f \in \mathbb{X}$ the mild solution of (36.1) is unique and exists on a maximal interval of the form $[0, t_{\max}(f))$, see [284], p. 186.

The reason for nonexistence of global solutions is that the vector field F may cause the solutions to reach infinity in finite time, which is amply expressed in the statement that if $t_{\max}(f) < \infty$ then:

$$\lim_{t\to t_{\max}(f)} \|u(t)\| = \infty,$$

and in general there is no remedy for this phenomenon even if $A = 0$. Fortunately, in some cases the form of the vector field forces solutions of (36.1) starting from specific f to stay bounded in time. In view of the above discussion this guarantees that for such f the solutions exist and are unique on the whole half-line.

To describe a particular situation of this type, we specialize to the case where $\mathbb{X}$ is the space $C(S_N)$ of continuous functions on the union S_N of N copies of a compact metric space S. We will write:

$$\mathcal{N} = \{1, \dots, N\}.$$

Note that a member f of $C(S_N)$ may be naturally identified with a member of $C(S \times \mathcal{N})$: $f(x, i)$ is the value of f at x in the ith copy of S. Equivalently, f may be identified with an N-tuple of members of $C(S)$:

$$f = (f_1, \dots, f_N), \qquad f_i(\cdot) = f(\cdot, i).$$

We assume that we are given operators A_i, $i \in \mathcal{N}$ being the generators of conservative Feller semigroups in $C(S)$, and that the domain of A is composed of f such that $f_i \in D(A_i)$, and we have $(Af)_i = A_i f_i$. Furthermore, we assume that (36.1) takes the form:

$$\frac{\mathrm{d}u_i(t, x)}{\mathrm{d}t} = A_i u_i(t, x) + F_i(u(t, x)), \qquad x \in S, i \in \mathcal{N}, \tag{36.6}$$

where $u(t, x) \in \mathbb{R}^N$ is the vector $(u_1(t, x), \dots, u_N(t, x))$, $F : \mathbb{R}^N \to \mathbb{R}^N$ is of class C^1, and $F_i : \mathbb{R}^N \to \mathbb{R}$ is the ith coordinate of F. (Note that F defines, in a natural way, a function $\mathbb{X} \to \mathbb{X}$, denoted in what follows by the same letter, see also Exercises 36.2 and 36.3.) F being smooth, it is locally Lipschitz continuous in the sense that for $r > 0$ there exists $L(r)$ such that $|F(\alpha) - F(\beta)| \le L(r)|\alpha - \beta|$ provided the norms of $\alpha \in \mathbb{R}^N$ and $\beta \in \mathbb{R}^N$ do not exceed r. This guarantees existence of local solutions to (36.6). Moreover, for $f \in D(A)$, the mild solutions are classical solutions as well.

Proposition 36.1 *Given constants $m_i < M_i$, $i = 1, \dots, N$, consider a 'rectangle' $\mathcal{R}$ composed of $\alpha \in R^N$ with ith coordinate in the interval $[m_i, M_i]$, $i = 1, \dots, N$. Suppose that F is of class C^1 and that for all $\alpha \in \mathcal{R}$ with the ith coordinate equal to M_i, and for all $\beta \in \mathcal{R}$ with ith coordinate equal to m_i we have:*

$$F_i(\alpha) < 0 \text{ and } F_i(\beta) > 0. \tag{36.7}$$

Let $\mathcal{M}$ be the set of $f \in C(S_N)$ such that:

$$m_i \le f_i \le M_i, \qquad i \in \mathcal{N}.$$

Then, solutions of (36.6) starting at $f \in D(A) \cap \mathcal{M}$ never leave $\mathcal{M}$.

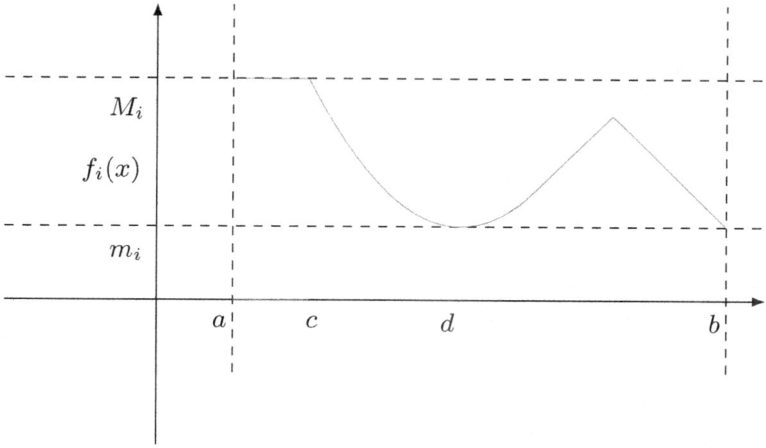

Figure 36.1 Sets B_i and C_i. Here, S_i is the interval $[a, b]$, $B_i = [a, c]$ and $C_i = \{b, d\}$

Proof Suppose that our thesis is not true for some $f \in D(A) \cap \mathcal{M}$. Then there is a $t_0 \geq 0$ such that $u(t) \in \mathcal{M}$ for $t \leq t_0$ and for any $\epsilon > 0$ there is a $t' \in (t_0, t_0 + \epsilon)$ with $u(t') \notin \mathcal{M}$. Without loss of generality we may assume $t_0 = 0$ (for otherwise we may replace f by $u(t_0)$). Let $u'(t)$ denote $\frac{\mathrm{d}u(t)}{\mathrm{d}t}$. Note that f cannot belong to the interior of $\mathcal{M}$, for this would imply that the solution stays in $\mathcal{M}$ for at least some short time interval. It follows that one of the sets $B_i, C_i, i \in \mathcal{N}$ defined by (see Figure 36.1):

$$(x, i) \in B_i \text{ if } f_i(x) = M_i, \qquad (x, i) \in C_i \text{ if } f_i(x) = m_i$$

is nonempty. We claim that so is one of the sets B_i':

$$(x, i) \in B_i' \text{ if } f_i(x) = M_i \text{ and } u_i'(0, x) \geq 0,$$

or C_i':

$$(x, i) \in C_i' \text{ if } f_i(x) = m_i \text{ and } u_i'(0, x) \leq 0.$$

For the proof, suppose without loss of generality (this assumption merely allows focusing our attention) that both $B = \bigcup_{i \in I} B_i$ and $C = \bigcup_{i \in J} C_i$ are nonempty, where:

$$I = \{i \in \mathcal{N} \,|\, B_i \text{ is nonempty}\} \text{ and } J = \{i \in \mathcal{N} \,|\, C_i \text{ is nonempty}\}.$$

If the claim is not true, B and C being compact, there is an $\epsilon > 0$ such that $u'(0, x) < -\epsilon$ for $x \in B$ and $u'(0, x) > \epsilon$ for $x \in C$. Let:

$$D_n = \bigcup_{i \in I} D_{i,n} \text{ and } E_n = \bigcup_{i \in J} E_{i,n}$$

be unions of compact sets:

$$D_{i,n} = \left\{(x,i) | f_i(x) \ge M_i - \frac{1}{n}\right\} \text{ and } E_{i,n} = \left\{(x,i) | f_i(x) \le m_i + \frac{1}{n}\right\},$$

and let:

$$\delta_n = \max_{i\in I} \max_{(x,i)\in D_{n,i}} \min_{(y,i)\in B_i} d(x,y), \quad \delta_n' = \max_{i\in J} \max_{(x,i)\in E_{n,i}} \min_{(y,i)\in C_i} d(x,y),$$

where d is the metric in S. Since $D_{i,n+1} \subset D_{i,n}$, $\delta_{n+1} \le \delta_n$; analogously $\delta_{n+1}' \le \delta_n'$. If $\lim_{n\to\infty} \delta_n > 0$, there exists a $\delta > 0$ and $y_n \in D_{i(n),n}$, such that for each $x \in B_{i(n)}$, $d(x, y_n) \ge \delta$. Since S is compact, it follows that there is an i_0 and $y \in \bigcap_{n\ge1} D_{i_0,n}$ (in particular: $f(y, i_0) = M_{i_0}$) such that for $x \in B_{i_0}$ we have $d(x, y) \ge \delta$. This being clearly impossible, we must have $\lim_{n\to\infty} \delta_n = 0$. Analogously, we check that $\lim_{n\to\infty} \delta_n' = 0$. By continuity, there is an n_0 such that $u'(0, x) < -\epsilon$ for $x \in D_{n_0}$ and $u'(0, x) > \epsilon$ for $x \in E_{n_0}$, and then there is $t_1 > 0$ such that $u'(t, x) < -\epsilon$ for $x \in D_{n_0}$ and $t \in [0, t_1]$, and $u'(t, x) > \epsilon$ for $x \in E_{n_0}$ and $t \in [0, t_1]$. Clearly, for $x \in D_{n_0}$ and $t \le t_1$, $u_i(t, x) \le u_i(0, x) = f_i(x) \le M_i$. Let $\tilde{M} = \sup_{t\in[0,t_1]} \|u'(t)\|$ and $\tilde{m} = \min_{i\notin I} \min_{x\in S}[M_i - f_i(x)] > 0$. Then, for $t \le t_2 := \min(\frac{1}{n_0\tilde{M}}, \frac{\tilde{m}}{\tilde{M}}, t_1)$ and $(x, i) \notin D_{n_0}$, $u_i(t, x) \le u_i(0, x) + t\tilde{M} \le M_i$, implying that $u_i(t, x) \le M_i$ for all $x \in S$, $i \in \mathcal{N}$ and $t \in [0, t_2]$. Analogously, we show that $u_i(t, x) \ge m_i$ for all i and x, provided t is in a short interval. This contradicts the way f was chosen, proving the claim.

Let i be such that B_i' is nonempty. At $(x, i) \in B_i'$, f_i attains its maximum and $f_i \in D(A_i)$. Hence, $A_i f_i(x) \le 0$ (since A_i is the generator of a conservative Feller semigroup, it satisfies the maximum principle). Moreover, $F_i(u(0, x)) < 0$, since the ith coordinate of $u_i(0, x)$ is M_i. This shows that $u_i'(0, x)$ cannot be equal to $A_i u_i(0, x) + F_i(u(0, x))$, contradicting (36.6). Analogously, if a C_i' is non-empty, f_i attains its minimum at $(x, i) \in C_i'$. Hence, $A_i f_i(x) \ge 0$, and $u_i'(0, x) \le 0$ cannot be equal to the sum of two terms: $A_i u_i(0, x) \ge 0$ and $F_i(u(0, x)) > 0$. This contradiction completes the proof. □

Coupled with the previous discussion, Proposition 36.1 shows that under the stated conditions, for initial data in $\mathcal{M}$, the semilinear Cauchy problem is well posed on the whole half-line.

Taking for S the space composed of N distinct points, we may identify members of $C(S)$ with N-dimensional vectors, that is, members of $\mathbb{R}^N$. Then, (36.6) with $A = 0$ is a system of N ordinary differential equations of N variables. In this context, relations (36.7) (and their relatives (36.8)) are known as the **Müller conditions** [337, 338]. They guarantee that a solution of such a system starting in the 'rectangle' $\mathcal{R}$ never leaves this rectangle. In the context of reaction-diffusion equations, reasonings akin to that presented in our proposition are referred to as the **method of (invariant) rectangles**, see [318], and in particular Thm. 14.7, p. 200 there.

Before turning to an example, we need to weaken condition (36.7).

Proposition 36.2 *Suppose that in Proposition 36.1,* (36.7) *is changed to read:*

$$F_i(\alpha) \le 0 \text{ and } F_i(\beta) \ge 0. \tag{36.8}$$

Then, the thesis remains the same.

Proof The idea is of course to approximate solutions of (36.1) with F satisfying (36.8) by solutions of (36.1) with F satisfying (36.7). Let us equip $\mathbb{R}^N$ with the supremum norm. Let $r > 0$ be so large that $\mathcal{R}$ is contained in the ball $K(0, r) \subset \mathbb{R}^N$, and let $L_0 (= L(r))$ be a common Lipschitz constant for all F_i's, as restricted to $K(0, r)$. Since the F_i's may be extended to the whole of $\mathbb{R}^N$ without increasing their Lipschitz constants ([117] p. 189), we may find a globally Lipschitz continuous function $\tilde{F}$ on $\mathbb{R}^N$ with Lipschitz constant L_0 that agrees with F on $K(0, r)$.

Let $H : \mathbb{R}^N \to \mathbb{R}^N$ be given by $H = (H_1(\cdot), \ldots, H_N(\cdot))$ and:

$$H_i(\alpha) = (0, \ldots, \underbrace{\cos(\frac{\alpha_i - m_i}{M_i - m_i}\pi)}_{i\text{th coordinate}}, \ldots, 0), \qquad i = 1, \ldots, N,$$

where α_i is the ith coordinate of α. Clearly $H_i(\alpha) = 1$ provided $\alpha_i = m_i$, and $H_i(\alpha) = -1$ provided $\alpha_i = M_i$. It follows that for all n, the $F_n := \tilde{F} + \frac{1}{n}H$ satisfies condition (36.7) and is globally Lipschitz continuous with Lipschitz constant $L_1 = L_0 + \max_{i=1,\ldots,N} \frac{\pi}{M_i - m_i}$.

Let u^n be the solution to (36.1) with F replaced by F_n:

$$u^n(t) = e^{tA} f + \int_0^t e^{(t-s)A} F_n(u^n(s))\, ds. \tag{36.9}$$

For $\mu > L_1$ we have, as in (36.4):

$$\begin{aligned}
\|u^n - u^m\|_\mu &\le \sup_{t\ge 0} \|e^{-\mu t} \int_0^t e^{-(t-s)A} [F_n(u^n(s)) - F_m(u^m(s))]\, ds\| \\
&\le \sup_{t\ge 0} e^{-\mu t} \int_0^t [\|F_n(u^n(s)) - F_n(u^m(s))\| \\
&\qquad + \|F_n(u^m(s)) - F_m(u^m(s))\|]\, ds \\
&\le \frac{L_1}{\mu} \|u^n - u^m\|_\mu + \frac{1}{\mu}\left|\frac{1}{n} - \frac{1}{m}\right|.
\end{aligned}$$

Hence,

$$\|u^n - u^m\|_\mu \le \left|\frac{1}{n} - \frac{1}{m}\right| \frac{1}{\mu - L_1}, \qquad n, m \ge 1,$$

and we see that $(u^n)_{n\ge 1}$ is a Cauchy sequence in $\mathcal{X}_\mu$. Denoting by u its limit, and passing with n to infinity in (36.9), we see that u satisfies (36.2) with F replaced by $\tilde{F}$. On the other hand, by Proposition 36.1, all u^n's belong to $\mathcal{M}$. Since $\mathcal{M}$ is closed, so does u. In particular, since $\tilde{F}$ coincides with F on $K(0, r)$, we recognize u as the solution of (36.2) and of (36.6). □

As an example, we consider the following (rescaled) model of early lung cancer due to Marciniak-Czochra and Kimmel, see [249, 250]:

$$\begin{aligned}\frac{\partial c}{\partial t} &= ((2p-1)a(b,c) - d_c)c,\\ \frac{\partial b}{\partial t} &= \alpha(c)g - d_b b - db,\\ \frac{\partial g}{\partial t} &= \frac{1}{\gamma}\frac{\partial^2 g}{\partial x^2} - \alpha(c)g - d_g g + \kappa(c) + db,\end{aligned} \tag{36.10}$$

where c, b, and g are densities of cells, and of bound and free growth factor molecules, respectively, distributed across the unit interval $[0, 1]$ (we keep the notations of [249]). The first equation expresses the fact that cells may divide with efficiency $p \in [0, 1]$, and that the rate at which they enter mitosis depends on the number of cells and on the number of bound receptors via the Hill-type function (see, e.g., [3]):

$$a(b, c) = \frac{a_0 b^m}{b^m + c^m}; \tag{36.11}$$

($a_0 > 0$ and an integer m are given constants); the constant $d_c > 0$ is the rate at which the cells die. The third equation tells us that free growth factors diffuse freely on the interval with diffusion constant γ^{-1}, and we assume that the interval endpoints are reflecting barriers:

$$\frac{\partial g(t,x)}{\partial x}_{|x=0} = \frac{\partial g(t,x)}{\partial x}_{|x=1} = 0.$$

These molecules may also bind to the cells on the cell's membranes, and the rate of binding depends on the cells density: $\alpha(c) = \alpha c^2$, for some $\alpha > 0$. Such free growth factors are produced by external medium at the rate $\kappa_0 > 0$ and by the cells at the rate $\frac{\kappa_1 c}{1+c}$ (with $\kappa_1 > 0$), so that the total production rate is $\kappa(c) = \kappa_0 + \frac{\kappa_1 c}{1+c}$; the constant $d_g > 0$ is the rate at which free growth factors are eliminated. Finally, the first term in the second equation describes influx of bound growth factors, and the two remaining terms are responsible for dissociation (at the rate $d > 0$) and elimination of these molecules (at the rate $d_b > 0$).

Let S_3 be the union of three copies of the unit interval. The members of the space $C(S_3)$ may be identified with triplets of continuous functions on $[0, 1]$. Using this identification, let A be the operator in $C(S_3)$ given by $A(c, b, g) = (0, 0, \gamma^{-1}g'')$ with domain composed of the triplets where $g \in C^2[0, 1]$ and $g'(0) = g'(1) = 0$. Clearly, A generates the Feller semigroup $e^{tA}(c, b, g) = (c, b, T(t)g)$ where $\{T(t), t \geq 0\}$ is the semigroup describing the Brownian motion on $[0, 1]$ with reflecting barriers at both endpoints. Hence, (36.10) is a special case of (36.6) with:

$$F_1(c, b, g) = ((2p-1)a(b, c) - d_c)c,$$
$$F_2(c, b, g) = \alpha(c)g - d_b b - db,$$
$$F_3(c, b, g) = -\alpha(c)g - d_g g + \kappa(c) + db.$$

We proceed to show that assumptions of Proposition 36.2 are satisfied with $m_1 = m_2 = m_3 = 0$. Since $F_1(0, b, g) = 0$, $F_2(c, 0, g) = \alpha(c)g \geq 0$ and $F_3(c, b, 0) = \kappa(c) + db \geq 0$ (provided $b, c, g \geq 0$), we see that the second condition in (36.8) is fulfilled. It remains to find $M_1, M_2, M_3 > 0$ such that:

$$(2p-1)a(b, M_1) - d_c \leq 0,$$
$$\alpha(c)g - d_b M_2 - dM_2 \leq 0,$$
$$-\alpha(c)M_3 - d_g M_3 + \kappa(c) + db \leq 0,$$

as long as $0 \leq c \leq M_1$, $0 \leq b \leq M_2$ and $0 \leq g \leq M_3$. The first two inequalities are satisfied iff $\frac{(2p-1)a_0 M_2^m}{M_2^m + M_1^m} \leq d_c$ and $M_1 \leq \sqrt{\frac{(d_b+d)M_2}{\alpha M_3}}$, and the third is satisfied iff $M_3 \geq \frac{\kappa(c)+dM_2}{\alpha c^2 + d_g}$ for $c \in [0, M_1]$ which clearly holds if $M_3 \geq \frac{\kappa_0+\kappa_1+dM_2}{d_g}$. If $(2p-1)a_0 > d_c$, we define M_1 and M_3 as functions of M_2:

$$M_1 = M_2 \left(\frac{(2p-1)a_0 - d_c}{d_c} \right)^{\frac{1}{m}}, \quad M_3 = \frac{\kappa_0 + \kappa_1 + dM_2}{d_g}.$$

Then all the three conditions are satisfied if:

$$(\kappa_0 + \kappa_1 + dM_2)M_2 \leq \frac{(d_b + d)d_g}{\alpha} \left(\frac{(2p-1)a_0 - d_c}{d_c} \right)^{-\frac{2}{m}},$$

and it is clear that such an M_2 may be chosen. In the case $(2p-1)a_0 \leq d_c$, the first condition is satisfied for all M_1 and M_2. Hence, we may take an arbitrary $M_2 > 0$ and define $M_1 = \sqrt{\frac{(d_b+d)M_2}{\alpha M_3}}$, $M_3 = \frac{\kappa_0+\kappa_1+dM_2}{d_g}$.

It follows that, provided initial conditions are non-negative and uniformly bounded by M_1, M_2 and M_3, respectively, the solutions of (36.10) are well defined, unique, bounded, and non-negative on the whole half-line.

Exercise 36.1 Under conditions of Proposition 36.2 and arguing as in its proof, check that if u_1 and u_2 are mild solutions of (36.6) with initial conditions f_1 and f_2, respectively, then $\|u_1 - u_2\|_\mu \le \frac{\mu\|f_1 - f_2\|}{\mu - L_1}$. Conclude that mild solutions starting in $\mathcal{M}$ never leave $\mathcal{M}$, and are defined on the whole half-line.

Exercise 36.2 Suppose $F : \mathbb{R} \to \mathbb{R}$ is of class C^1, and let $F_C : C(S) \to C(S)$, where S is a compact topological space, be given by $[F_C(f)](x) = F(f(x))$, $f \in C(S)$, $x \in S$. Prove that F_C is of class C^1, that is, that for each $f \in C(S)$ there exists a bounded linear operator A_f in $C(S)$ such that:

$$\lim_{h \in C(S), \|h\| \to 0} \frac{F_C(f+h) - F_C(f) - A_f h}{\|h\|} = 0,$$

and A_f depends continuously on f (A_f is the **Fréchet differential** of F_C at f). *Hint:* take $(A_f h)(x) = F'(f(x))h(x)$; in other words, A_f is the operator of multiplication by the function $x \mapsto F'(f(x))$.

Exercise 36.3 Given a natural k, repeat the previous exercise for $F : \mathbb{R}^k \to \mathbb{R}^k$ and F_C mapping $[C(S)]^k$ into itself via the formula:

$$F_C(f_1, \ldots, f_k) = (g_1, \ldots, g_k)$$

where $g_i(x) = F_i(f_1(x), \ldots, f_k(x))$, $i = 1, \ldots, k$.

☞ Chapter's summary

As an application of the method of invariant rectangles (or the Müller Theorem) for semilinear equations, we establish global existence and non-negativity of solutions of an early cancer model due to A. Marciniak-Czochra and M. Kimmel.

37

Coagulation-Fragmentation Equation

On another detour from the main subject we consider a discrete model of coagulation and fragmentation. To explain the reason for leaving the main road, we recall from the previous chapter that the Müller conditions serve as a means to guarantee global existence (in time) of solutions of a semilinear equation in the case where the nonlinear part is merely locally Lipschitz continuous. They simply force solutions to stay in a bounded region and hence not blow up. A similar idea may be used to show existence of solutions of other equations, of which the system this chapter is devoted to is a nice example; under certain conditions, the norm is preserved along solutions of the equation involved, and hence there can be no blowup.

The model is taken from McBride et al. [257] (see [16, Chapter 8] for the continuous analogue). A population we are interested in is composed of clusters (or: mers) that can fragment, or break into smaller pieces, and coagulate, that is, join to other clusters to form larger ones. We assume there is a fundamental unit for all mers, and each mer is composed of an integer number of such units; a mer of n fundamental units is referred to as an n-mer. Hence, a population is characterized by a sequence $(\xi_n)_{n\geq 1}$ where ξ_n is the number (concentration) of n-mers. The total "mass" or "size" of the population is $\sum_{n=1}^{\infty} n\xi_n$. The master equation for the concentrations $u_n(t)$ of n-mers at time $t \geq 0$ reads:

$$\frac{\mathrm{d}u_n(t)}{\mathrm{d}t} = -a_n u_n(t) + \sum_{m=n+1}^{\infty} a_m b_{m,n} u_m(t) + \frac{1}{2}\sum_{m=1}^{n-1} k_{n-m,m} u_{n-m}(t) u_m(t) - \sum_{m=1}^{\infty} k_{n,m} u_n(t) u_m(t), \qquad (37.1)$$

for $n \geq 1, t \geq 0$. Here, the first two terms correspond to fragmentation events: $a_n \geq 0$ is the breakdown rate for n-mers, and $b_{m,n} \geq 0$ is the fraction of n-mers produced upon the breakdown of an m-mer. The remaining two terms describe

Figure 37.1 Coagulation and fragmentation by Radek Bobrowski

coagulation: if an n-mer coagulates with an m-mer with rate $k_{n,m} \geq 0$, then the first sum is the total number of new n-mers, and the second is the number of n-mers that become larger mers by coagulating with other mers. We assume:

$$a_1 = 0, \qquad k_{n,m} = k_{m,n}, \quad n, m \geq 1, \qquad \sum_{n=1}^{m-1} n b_{m,n} = m, \quad m \geq 2,$$

that is, that 1-mer cannot break up any further and that the mass of a fragmented m-mer is the same as before its fragmentation.

It will be convenient to consider the master equation (37.1) as a semi-linear equation in the space $\mathbb{X}$ of sequences $x = (\xi_n)_{n\geq 1}$ such that $\|x\| := \sum_{n=1}^{\infty} n|\xi_n| < \infty$. Of course, $\mathbb{X}$ is a Banach space (this space is isometrically isomorphic to l^1 via isomorphism $I : \mathbb{X} \to l^1$ given by $I(\xi_n)_{n\geq 1} = (n\xi_n)_{n\geq 1}$). To this end, we write (for the time being, merely formally):

$$\frac{\mathrm{d}u(t)}{\mathrm{d}t} = Au(t) + Bu(t) + K(u(t)), \qquad t \geq 0, \tag{37.2}$$

where:

$$A\,(\xi_n)_{n\geq 1} = (-a_n\xi_n)_{n\geq 1}, \quad B\,(\xi_n)_{n\geq 1} = \left(\sum_{m=n+1}^{\infty} a_m b_{m,n} \xi_m\right)_{n\geq 1},$$

with common domain $D = \{(\xi_n)_{n\geq 1} \in \mathbb{X} | A(\xi_n)_{n\geq 1} \in \mathbb{X}\}$ (see later), and:

$$K((\xi_n)_{n\geq 1}) = \left(\frac{1}{2}\sum_{m=1}^{n-1} k_{n-m,m}\xi_{n-m}\xi_m - \sum_{m=1}^{\infty} k_{n,m}\xi_n\xi_m\right)_{n\geq 1};$$

by convention $\sum_{m=1}^{0} = 0$. The function K is defined for all $x \in \mathbb{X}$ provided we assume, as we do, that there is a $k \ge 0$ such that:

$$k_{n,m} \le k, \qquad n, m \ge 1. \tag{37.3}$$

(For the case of unbounded $k_{n,m}$ see [15, 26].)

Except for a special choice of coefficients, the Cauchy problem related to (37.2) is not well posed. However, we will show now that there is a generator G of a positive semigroup, such that G extends $A + B$. To this end, we use the main theorem of Chapter 13 (Thm. 13.3): it is clear that $\mathbb{X}$ is a KB-space and the first two assumptions in Theorem 13.3 are satisfied. Next, for $x = (\xi_n)_{n\ge1} \in D$,

$$\begin{aligned} \|Bx\| &\le \sum_{n=1}^{\infty} n \sum_{m=n+1}^{\infty} a_m b_{m,n} |\xi_m| = \sum_{m=2}^{\infty} a_m |\xi_m| \sum_{n=1}^{m-1} n b_{m,n} \\ &= \sum_{m=1}^{\infty} m a_m |\xi_m| = \|Ax\| \qquad \text{(recall } a_1 = 0\text{)}, \end{aligned}$$

so that the third assumption is satisfied as well. Since a similar calculation shows that for any $x \ge 0, x \in D$, and for the functional M defined by $M(\xi_n)_{n\ge1} = \sum_{n=1}^{\infty} n\xi_n$,

$$M(Ax + Bx) = 0, \quad Mx = \|x\|, \text{ and } \quad \|M\| = 1, \tag{37.4}$$

existence of the generator is secured by Theorem 13.3.

We proceed to show that $(e^{tG})_{t\ge0}$ is a Markov semigroup, that is, that it leaves $\mathcal{D} = \{x \ge 0, x \in \mathbb{X} \,|\, Mx = 1\}$ invariant. For this, we need the following lemma.

Lemma 37.1

(a) For $\lambda > 0$, $(\lambda - G)^{-1} = \sum_{n=0}^{\infty} (\lambda - A)^{-1} B_\lambda^n$ where by definition $B_\lambda = B(\lambda - A)^{-1}$.

(b) Operators $\lambda(\lambda - G)^{-1}, \lambda > 0$ are Markov iff $\lim_{n\to\infty} B_\lambda^n = 0$.

Proof The proof of Theorem 13.3 makes it clear that we have:

$$(\lambda - G)^{-1} = \lim_{t\uparrow1} R(r, \lambda)$$

where, see (13.5), $R(r, \lambda) = (\lambda - A)^{-1} \sum_{n=0}^{\infty} r^n B_\lambda^n$, and the limit is monotone. Since B_λ and $(\lambda - A)^{-1}$ are non-negative,

$$\sum_{n=0}^{N} (\lambda - A)^{-1} r^n B_\lambda^n \le R(r, \lambda) \le (\lambda - G)^{-1}, \qquad \lambda > 0, N \ge 1.$$

This implies:

$$\sum_{n=0}^{N} (\lambda - A)^{-1} B_\lambda^n \le (\lambda - G)^{-1}, \qquad \lambda > 0, N \ge 1. \tag{37.5}$$

Hence, for $x \ge 0$, $\sum_{n=0}^{N}(\lambda - A)^{-1}B_\lambda^n x \le \sum_{n=0}^{N+1}(\lambda - A)^{-1}B_\lambda^n x, \lambda > 0$, and $\| \sum_{n=0}^{N}(\lambda - A)^{-1}B_\lambda^n \| \le \|(\lambda - G)^{-1}\| \|x\|, \lambda > 0$. Since $\mathbb{X}$ is a KB-space, the series $\sum_{n=0}^{\infty}(\lambda - A)^{-1}B_\lambda^n x$ exists for such x. Using decomposition $x = x^+ - x^-$ (see Chapter 13) we see that:

$$S_\lambda := \sum_{n=0}^{\infty} (\lambda - A)^{-1} B_\lambda^n$$

converges in the strong topology, and (37.5) implies $S_\lambda \le (\lambda - G)^{-1}$. On the other hand, $R(r, \lambda) \le S_\lambda$ and taking the limit as $r \uparrow 1$ we obtain $(\lambda - G)^{-1} \le S_\lambda$. This completes the proof of (a).

For the proof of (b) we note first that:

$$I_{\mathbb{X}} + B \sum_{n=0}^{N} (\lambda - A)^{-1} B_\lambda^n = (\lambda - A) \sum_{n=0}^{N} (\lambda - A)^{-1} B_\lambda^n + B_\lambda^{N+1}, \qquad \lambda > 0,$$

both sides being equal to $\sum_{n=0}^{N+1} B_\lambda^n$. Hence, for $x \in \mathcal{D}$,

$$Mx + M(A + B) \sum_{n=0}^{N} (\lambda - A)^{-1} B_\lambda^n x = M\lambda \sum_{n=0}^{N} (\lambda - A)^{-1} B_\lambda^n + MB_\lambda^{N+1} x.$$

Using (a) and (37.4), and letting $N \to \infty$,

$$1 = M\lambda\, (\lambda - G)^{-1}\, x + \lim_{N \to \infty} \|B_\lambda^{N+1} x\|.$$

Thus, $M\lambda(\lambda - G)^{-1}x = 1$ iff $\lim_{n\to\infty} B_\lambda^n x = 0$ for $x \in \mathcal{D}$. The latter condition holds iff $\lim_{n\to\infty} B_\lambda^n = 0$ strongly, completing the proof. □

We return to the proof of the fact that $e^{tG}, t \ge 0$ are Markov operators, or that $\lambda(\lambda - G)^{-1}, \lambda > 0$ are Markov operators (see Chapter 13 for this equivalence). Using (37.4) and the definition of A, we show first that $B_\lambda, \lambda > 0$ are contractions (see Lemma 13.1):

$$\begin{aligned} \|B_\lambda x\| &= MB\,(\lambda - A)^{-1}\, x = -MA\,(\lambda - A)^{-1}\, x = \|A\,(\lambda - A)^{-1}\, x\| \\ &\le \|x\|, \qquad x \ge 0. \end{aligned}$$

Next, let $e_i = (\delta_{i,n})_{n\ge 1} \in \mathbb{X}, i \ge 1$ where $\delta_{i,n}$ is the Kronecker symbol. Since $\{e_i, i \ge 1\}$ is a **Schauder basis** for $\mathbb{X}$ (see, e.g., [78]), in view of Lemma 37.1, it suffices to show:

$$\lim_{n\to\infty} B_\lambda^n e_i = 0, \qquad \lambda > 0, i \ge 1. \tag{37.6}$$

To this end, we fix $\lambda > 0$, and for $i \ge 1$ let:

$$q_i = q(\lambda, i) = \max_{k=1,\dots,i} \frac{a_k}{\lambda + a_k} < 1.$$

By the definition of the operators A and B,

$$(\lambda - A)^{-1} e_i = \frac{1}{\lambda + a_i} e_i \qquad \text{and} \qquad Be_i = a_i \sum_{j=1}^{i-1} b_{ij} e_j.$$

We claim that:

$$\|B_\lambda^n e_i\| \le q_i^n \|e_i\| = q_i^n i, \qquad n \ge 0, i \ge 1. \tag{37.7}$$

This is clear for $n = 0$ and the induction step is proved as follows:

$$\begin{aligned}\|B_\lambda^{n+1} e_i\| &= \frac{a_i}{\lambda + a_i} \left\| \sum_{j=1}^{i-1} b_{ij} B_\lambda^n e_j \right\| \le q_i \left\| \sum_{j=1}^{i-1} b_{ij} B_\lambda^n e_j \right\| \le q_i^{n+1} \sum_{j=1}^{i-1} b_{ij} j \\ &= q_i^{n+1} i,\end{aligned}$$

completing the proof of the claim. Since (37.7) implies (37.6), we are done.

This analysis allows replacing the ill-posed Cauchy problem related to equation (37.2) by the well-posed one:

$$\frac{\mathrm{d}u(t)}{\mathrm{d}t} = Gu(t) + K(u(t)), \qquad t \ge 0, u(0) = x \ge 0, \tag{37.8}$$

where G extends $A + B$ and is the generator of a Markov semigroup. (In fact, see Exercise 37.3, G is the closure of $A + B$.) However, to see that this problem is indeed well posed, we need to exhibit basic properties of K.

We begin by introducing two bilinear forms in $\mathbb{X}$:

$$\mathcal{K}_1(x, y) = \left(\frac{1}{2} \sum_{m=1}^{n-1} k_{n-m,m} \xi_{n-m} \eta_m \right)_{n\ge1}$$

$$\mathcal{K}_2(x, y) = \left(\xi_n \sum_{m=1}^{\infty} k_{m,n} \eta_m \right)_{n\ge1}$$

where $x = (\xi_n)_{n\geq 1}, y = (\eta_n)_{n\geq 1} \in \mathbb{X}$. To show that these forms map $\mathbb{X} \times \mathbb{X}$ into $\mathbb{X}$, we estimate (recall (37.3)):

$$\begin{aligned}\|\mathcal{K}_1(x,y)\| &\leq \sum_{n=1}^{\infty} \frac{n}{2} \sum_{m=1}^{n-1} k_{n-m,m} |\xi_{n-m}||\eta_m| \\ &\leq k \sum_{m=1}^{\infty} \frac{n}{2} \sum_{n=m+1}^{\infty} |\xi_{n-m}||\eta_m| \\ &\leq k \sum_{m=1}^{\infty} \sum_{n=1}^{\infty} \frac{n+m}{2} |\xi_n||\eta_m| \\ &\leq k \sum_{m=1}^{\infty} \sum_{n=1}^{\infty} nm |\xi_n||\eta_m| = k\|x\| \, \|y\|,\end{aligned}$$

and $\|\mathcal{K}_2(x,y)\| \leq k \sum_{n=1}^{\infty} n|\xi_n| \sum_{j=1}^{\infty} |\eta_j| \leq k\|x\| \, \|y\|$. As a by-product of this calculation, both forms are seen to be continuous. Hence, $\mathcal{K} := \mathcal{K}_1 - \mathcal{K}_2$ is also a continuous bilinear form, and $\|\mathcal{K}(x,y)\| \leq 2k\|x\| \, \|y\|$. Since K is a quadratic form associated with $\mathcal{K}$, that is, $Kx = \mathcal{K}(x,x)$, K is locally Lipschitz continuous (see Exercise 37.1). Hence, (37.8) has local solutions in time.

For the global existence we need the following property of K:

$$MK(x) = 0, \qquad x \in \mathbb{X}, \tag{37.9}$$

or, expanded,

$$\sum_{n=1}^{\infty} \frac{n}{2} \sum_{m=1}^{n-1} k_{n-m,m} \xi_{n-m} \xi_m = \sum_{n=1}^{\infty} n\xi_n \sum_{m=1}^{\infty} k_{n,m} \xi_m,$$

where, as before, $x = (\xi_n)_{n\geq 1}$. To explain, note that:

$$\frac{1}{2} \sum_{m=1}^{n-1} k_{n-m,m} \xi_{n-m} \xi_m$$

is the fraction of mers of size smaller than n that coagulated into mers of size n, or the fraction of "gained" mers of size n. Thus the left-hand side is the mass of all mers "gained" by coagulation. Similarly, $\xi_n \sum_{m=1}^{\infty} k_{n,m}\xi_m$ is the fraction of n-mers that coagulated with other mers, thus becoming mers of larger size; the right-hand side is therefore the mass of all mers "lost" by coagulation. Hence, for non-negative x, the formula simply expresses the fact that by coagulation no mass is really lost or gained. For the formal proof, we change the order of summation on the left-hand side to obtain, as in the estimate for $\|\mathcal{K}_1(x,y)\|$, that it equals $\sum_{m=1}^{\infty} \sum_{n=1}^{\infty} \frac{n+m}{2} k_{n,m} \xi_m \xi_n$. By $k_{m,n} = k_{n,m}$ this can be written as the half of:

$$\sum_{m=1}^{\infty}\sum_{n=1}^{\infty} nk_{n,m}\xi_m\xi_n + \sum_{m=1}^{\infty}\sum_{n=1}^{\infty} mk_{m,n}\xi_m\xi_n,$$

and thus it equals the right-hand side.

Had we established (as in [257, Theorem 4.7]) that mild solutions of (37.8) are non-negative, provided initial condition is, with (37.9) at hand, the proof of global existence would be easy. We would simply use the relation $u(t) = \mathrm{e}^{tG}x + \int_0^t \mathrm{e}^{(t-s)G}K(u(s))\,\mathrm{d}s$ to see that:

$$\|u(t)\| = Mu(t) = M\mathrm{e}^{tG}x + 0 = Mx = \|x\|, \qquad x \geq 0, \tag{37.10}$$

because for $y \in \ker M$, and any $s \geq 0$, $M\mathrm{e}^{sG}y = M\mathrm{e}^{sG}y^+ - M\mathrm{e}^{sG}y^- = My^+ - My^- = My = 0$, by the Markov property of e^{sG}; in other words, e^{sG} leaves $\ker M$ invariant. (Recall that by (37.9) K maps $\mathbb{X}$ into $\ker M$.) This would imply that solutions cannot blow up, and the result would follow.

However, because we do not have the result mentioned earlier at our disposal, we will take a slightly different route and prove global existence directly (of course, with the help of (37.9)). To this end, for fixed real number $\rho > 0$ and non-negative $x \in \mathbb{X}$, we consider the set $S = \{y \in \mathbb{X} \,|\, y \geq 0,\ \|y\| = \|x\|\}$, and the complete metric space C_ρ of continuous functions $v : \mathbb{R}^+ \to S$, with the Bielecki-type distance [34, 125]:

$$d_\rho(v_1, v_2) = \sup_{t\geq 0} \mathrm{e}^{-\rho t}\|v_1(t) - v_2(t)\|.$$

Also, let L be a Lipschitz constant for K restricted to S, so that:

$$\|K(y_1) - K(y_2)\| \leq L\|y_1 - y_2\|, \qquad y_1, y_2 \in S,$$

and let bounded linear functionals $F_i \in \mathbb{X}^*$ be given by:

$$F_i\,(\eta_n)_{n\geq 1} = \eta_i, \qquad i \geq 1.$$

For $y = (\eta_n)_{n\geq 1} \in S$ and $\alpha := k\|x\|$, $F_i(\mathcal{K}_2(y, y)) = \eta_i \sum_{m=1}^{\infty} k_{m,i}\eta_m \leq k\|y\|$ $F_i y = \alpha F_i y$, or $F_i[\alpha y - \mathcal{K}_2(y, y)] \geq 0$. Since $\mathcal{K}_1(y, y) \geq 0$, it follows that:

$$K(y) + \alpha y \geq 0, \qquad y \in S. \tag{37.11}$$

Next, for $v \in C_\rho$, let:

$$w(t) = \mathrm{e}^{-\alpha t}\mathrm{e}^{tG}x + \int_0^t \mathrm{e}^{-\alpha(t-s)}\mathrm{e}^{(t-s)G}[K(v(s)) + \alpha v(s)]\,\mathrm{d}s, \qquad t \geq 0.$$

By (37.11), $w(t) \geq 0$. As in (37.10),

$$\|w(t)\| = Mw(t) = \mathrm{e}^{-\alpha t}Mx + \alpha \int_0^t \mathrm{e}^{-\alpha(t-s)}Mv(s)\,\mathrm{d}s$$
$$= \mathrm{e}^{-\alpha t}Mx + \alpha \int_0^t \mathrm{e}^{-\alpha(t-s)}Mx\,\mathrm{d}s = Mx = \|x\|.$$

It follows that the transformation $u \mapsto Uu := w$ maps C_ρ into C_ρ. Moreover, for $v_1, v_2 \in C_\rho$, arguing similarly as in (36.4),

$$\begin{aligned}
&d_\rho(Uv_1, Uv_2)\\
&\quad\leq \sup_{t\geq 0} \mathrm{e}^{-\rho t} \int_0^t \mathrm{e}^{-\alpha(t-s)} \|K(v_1(s) - v_2(s)) + \alpha(v_1(s) - v_2(s)]\|\,\mathrm{d}s\\
&\quad\leq \sup_{t\geq 0}(L+\alpha) \int_0^t \mathrm{e}^{-\rho(t-s)}\mathrm{e}^{-\alpha(t-s)}\mathrm{e}^{-\rho s}\|v_1(s) - v_2(s)\|\,\mathrm{d}s\\
&\quad\leq \sup_{t\geq 0}(L+\alpha) \int_0^t \mathrm{e}^{-(\rho+\alpha)s}\,\mathrm{d}s\, d_\rho(v_1, v_2)\\
&\quad= \frac{L+\alpha}{\rho+\alpha} d_\rho(v_1, v_2).
\end{aligned}$$

For $\rho > L$, the fraction here is less than 1. By the Banach Fixed Point Theorem, this proves existence of $u \in C_\rho$ such that $Uu = u$, that is,

$$u(t) = \mathrm{e}^{-\alpha t}\mathrm{e}^{tG}x + \int_0^t \mathrm{e}^{-\alpha(t-s)}\mathrm{e}^{(t-s)G}[K(u(s)) + \alpha u(s)]\,\mathrm{d}s, \tag{37.12}$$

for $t \geq 0$.

It remains to show that u is a mild solution to (37.8). This can be deduced as follows. First, we note that all the three terms in (37.12) are bounded, and their Laplace transforms are well defined for $\lambda > 0$. Writing:

$$\hat{u}(\lambda) = \int_0^\infty \mathrm{e}^{-\lambda t}u(t)\,\mathrm{d}t \quad \text{and} \quad \widehat{Ku}(\lambda) = \int_0^\infty \mathrm{e}^{-\lambda t}K(u(t))\,\mathrm{d}t, \quad \lambda > 0,$$

we have, by (37.12),

$$\hat{u}(\lambda) = (\lambda+\alpha - G)^{-1}x + (\lambda+\alpha - G)^{-1}\widehat{Ku}(\lambda) + \alpha(\lambda+\alpha - G)^{-1}\hat{u}(\lambda), \quad \lambda > 0.$$

Therefore, $\hat{u}(\lambda)$ belongs to the domain of G, and for all $\lambda > 0$, we have $(\lambda + \alpha - G)\hat{u}(\lambda) = x + \widehat{Ku}(\lambda) + \alpha\hat{u}(\lambda)$. This reduces to $(\lambda - G)\hat{u}(\lambda) = x + \widehat{Ku}(\lambda)$, or:

$$\hat{u}(\lambda) = (\lambda - G)^{-1}x + (\lambda - G)^{-1}\,\widehat{Ku}(\lambda).$$

This, however, means that the Laplace transforms of continuous functions $t \mapsto u(t)$ and $t \mapsto \mathrm{e}^{tG}x + \int_0^t \mathrm{e}^{(t-s)G}(K(u(s)))\,\mathrm{d}s$ coincide. It follows that these functions are equal to each other, completing the proof.

Exercise 37.1 Check that K is locally Lipschitz continuous.

Exercise 37.2 Check that K is Fréchet differentiable, and find its derivative. Deduce that for $x \in D(G)$ mild solutions of (37.8) are strong solutions of this equation.

Exercise 37.3 (See [16, p. 120]) Using (37.6) and:

$$(\lambda - A - B)\sum_{n=0}^{N}(\lambda - A)^{-1}B_\lambda^n x = x - B_\lambda^{N+1}x, \qquad x \in \mathbb{X}$$

$$\sum_{n=0}^{N}(\lambda - A)^{-1}B_\lambda^n(\lambda - A - B)x = x - (\lambda - A)^{-1}B_\lambda^N Bx, \qquad x \in D,$$

show that $(\lambda - \overline{A+B})^{-1} = \sum_{n=0}^{\infty}(\lambda - A)^{-1}B_\lambda^n$. Deduce that $G = \overline{A+B}$.

☞ Chapter's summary

On a little detour from the main subject, we prove well-posedness of the coagulation-fragmentation equation from the discrete model of McBride et al. [257]. This is a particular case of a semilinear equation in an l^1-type space, with locally Lipschitz-continuous nonlinearity. For non-negative initial conditions, the solutions preserve the norm and hence do not blow up, forcing existence of global solutions in time.

38

Homogenization Theorem

Our main goal in this chapter is (a simple version of) the homogenization theorem of Conway, Hoff, and Smoller [95, 318]. In the case we are interested in, the theorem deals with equation (36.6) with S being a compact interval $[a, b]$, and all generators A_i being scalar multiples of the one-dimensional Laplace operator with Neumann boundary conditions. In other words, we are considering:

$$\frac{\partial u_i(t,x)}{\partial t} = \sigma_i \frac{\partial^2 u_i(t,x)}{\partial x^2} + F_i(u(t,x)), \qquad x \in [a,b],\ i \in \mathcal{N}, \tag{38.1}$$

where σ_i's are given positive constants, with boundary conditions:

$$\frac{\partial u_i(t,x)}{\partial x}_{|x=a} = \frac{\partial u_i(t,x)}{\partial x}_{|x=b} = 0, \qquad i \in \mathcal{N}. \tag{38.2}$$

We start with the following Poincaré's and Sobolev's inequalities (comp. [2, 84, 318]). By D we denote the derivative with respect to $x \in [a, b]$ (one-sided, where necessary). Also, $L^2[a, b]$ is the Hilbert space of square integrable functions on $[a, b]$, with the usual norm $\|f\|_2^2 = \int_a^b [f(x)]^2\,\mathrm{d}x$.

Lemma 38.1 *Let $f \in C^1[a, b]$ and suppose for some $x_0 \in [a, b]$, $f(x_0) = 0$. Then:*

(a) (Poincaré's inequality) there is a $\lambda > 0$ (independent of f) such that:

$$\lambda \|f\|_2^2 \le \|Df\|_2^2,$$

(b) (Sobolev's inequality) and:

$$\|f\|_{\sup} := \sup_{x\in[a,b]} |f(x)| \le \|f\|_{W^{2,1}} := \sqrt{\|f\|_2^2 + \|Df\|_2^2}.$$

Proof We have $f(x) = \int_{x_0}^{x} Df(y)\,\mathrm{d}y$, $x \in [a, b]$, where, to recall, for $x \le x_0$, $\int_{x_0}^{x} Df(y)\,\mathrm{d}y$ equals by definition $-\int_{x}^{x_0} Df(y)\,\mathrm{d}y$. Hence, by Schwarz's inequality applied to the inner integral:

$$\|f\|_2^2 \le \int_a^b \left| \int_{x_0}^{x} Df(y)\,\mathrm{d}y \right|^2 \mathrm{d}x \le \int_a^b \left(\int_a^b |Df(y)|\,\mathrm{d}y \right)^2 \mathrm{d}x$$
$$\le (b-a) \int_a^b \int_a^b |Df(y)|^2\,\mathrm{d}y\,\mathrm{d}x = (b-a)^2 \int_a^b |Df(y)|^2\,\mathrm{d}y.$$

This implies (a) with $\lambda = (b-a)^{-2}$.

Next, introducing $g(x) = (f(x))^2 \ge 0$, we have for any $x \in [a, b]$,

$$g(x) = \int_{x_0}^{x} g'(y)\,\mathrm{d}y = \int_{x_0}^{x} 2f(y)f'(y)\,\mathrm{d}y \le \int_a^b 2|f(y)|\,|f'(y)|\,\mathrm{d}y$$
$$\le \int_a^b [f(y)]^2\,\mathrm{d}y + \int_a^b [Df(y)]^2\,\mathrm{d}y = \|f\|^2_{W^{2,1}}.$$

Hence, $\|f\|_{\sup} = \sup_{x\in[a,b]} \sqrt{g(x)} \le \|f\|_{W^{2,1}}$. □

Suppose that assumptions of Proposition 36.2 concerning F are satisfied, so that $\mathcal{M}$ is an invariant set for equation (38.1). Let $\mathcal{R} \subset \mathbb{R}^N$ be the rectangle related to $\mathcal{M}$, and let:

$$\sigma := \min_{i\in\mathcal{N}} \sigma_i \qquad \text{and} \qquad M = \sup_{\alpha\in\mathcal{R}} \sqrt{\sum_{i=1}^{N}\sum_{j=1}^{N} \left(\frac{\partial F_i(\alpha)}{\partial \alpha_j} \right)^2}. \tag{38.3}$$

Also, let $L^2(S_N)$ be the Cartesian product of N copies of $L^2[a, b]$; the norm in the latter space, denoted in what follows $\|\cdot\|_{L^2(S_N)}$, is related to the scalar product:

$$< f, g >= \sum_{i=1}^{N} \int_a^b f_i g_i, \qquad f = (f_i)_{i\in\mathcal{N}}, g = (g_i)_{i\in\mathcal{N}} \in L^2(S_N).$$

If all coordinates of $f = (f_i)_{i\in\mathcal{N}}$ are differentiable, we write Df for $(Df_i)_{i\in\mathcal{N}}$.

Theorem 38.2 (Conway–Hoff–Smoller) *Assume $F \in C^3$ and:*

$$d := \sigma\lambda - M > 0,$$

where λ is the constant from Poincaré Lemma. Let u be the solution of the Cauchy problem related to (38.1) with initial condition $f \in D(A^2) \cap \mathcal{M}$ and let $c := \|Df\|_{L^2(S_N)}$.

(a) *We have* $\|Du(t)\|_{L^2(S_N)} \le c\mathrm{e}^{-dt}, t \ge 0$.
(b) *Denoting* $\overline{u}(t) = (\overline{u}_i(t))_{i\in\mathcal{N}}$, *where* $\overline{u}_i(t) = \frac{1}{b-a}\int_a^b u_i(t,x)\,\mathrm{d}x$, *we have* $\|u(t) - \overline{u}(t)\|_{L^2(S_N)} \le \frac{c}{\sqrt{\lambda}}\mathrm{e}^{-dt}, t \ge 0$.
(c) *Moreover,*

$$\|u(t) - \overline{u}(t)\|_{\sup} \le \sqrt{\frac{\lambda+1}{\lambda}}c\mathrm{e}^{-dt}, t \ge 0.$$

(d) *There are: a constant* c_1 *and functions* $G_i : [0,\infty) \to \mathbb{R}, i \in \mathcal{N}$ *such that:*

$$\overline{u}_i'(t) = F_i(\overline{u}(t)) + G_i(t),$$

while $|G_i(t)| \le c_1\mathrm{e}^{-dt}, t \ge 0$.

For the proof of the theorem we need a (not too) technical lemma.

Lemma 38.3 *Suppose* $F \in C^3$, $f \in D(A^2) \cap \mathcal{M}$, *and let u be the solution of* (38.1). *Then, treated as a function of two variables, u possesses the mixed partial derivative:*

$$\frac{\partial^2 u(t,x)}{\partial t\,\partial x}, \qquad t \ge 0, x \in [a,b],$$

and the latter is continuous.

Proof Recalling notation of Chapter 36, we let:

$$A(f_i)_{i\in\mathcal{N}} = (\sigma_i D^2 f_i)_{i\in\mathcal{N}}$$

on the domain:

$$D(A) = \{(f_i)_{i\in\mathcal{N}} \in C(S_N);\ f_i \in C^2[a,b], Df_i(a) = Df_i(b) = 0, i \in \mathcal{N}\}.$$

When equipped with the graph norm:

$$\|f\|_A := \|f\| + \|Af\| := \sup_{i\in\mathcal{N}} \|f_i\|_{C(S)} + \sup_{i\in\mathcal{N}} \|D^2 f_i\|_{C(S)},$$

$D(A)$ is a Banach space, denoted in what follows by $[D(A)]$. Moreover, $(\mathrm{e}^{tA})_{t\ge0}$ leaves $D(A)$ invariant, and its restriction to this subspace is a strongly continuous semigroup in $[D(A)]$ with generator being the restriction of A do $D(A^2)$.

Let $\mathcal{R} \subset \mathbb{R}^N$ be the rectangle related to $\mathcal{M} \subset C(S_N)$, and $\tilde{F} : \mathbb{R}^N \to \mathbb{R}^N$ be a C^3 function that agrees with F on $\mathcal{R}$ and vanishes outside of a ball containing $\mathcal{R}$. It may be checked that for $f \in D(A)$, the function $F_{D(A)}f := \tilde{F} \circ f$ belongs to $D(A)$. In fact, since $\tilde{F}$ and all its three derivatives vanish outside of a ball, $F_{D(A)} : [D(A)] \to [D(A)]$ is globally Lipschitz. A calculation similar to Exercise 36.2, but slightly more tedious, shows that $F_{D(A)} : [D(A)] \to [D(A)]$ has a continuous Fréchet differential.

It follows that for $f \in D(A^2)$, there is a classical, unique, global solution u to (38.1) with F replaced by $\tilde{F}$, in the space $[D(A)]$ (in particular, derivatives and continuity are understood in the graph norm). Since the norm in $[D(A)]$ is stronger than that of $C(S_N)$, u is a solution to (38.1) with F replaced by $\tilde{F}$, in $C(S_N)$ as well. If $f \in D(A^2) \cap \mathcal{M}$, the latter solution never leaves $\mathcal{M}$, and hence is a solution of the original equation (38.1). This implies, by uniqueness of solutions, that for $f \in D(A^2) \cap \mathcal{M}$, the solution to (38.1) has values in $D(A)$ (even in $D(A^2)$), and is continuously differentiable as a function $\mathbb{R}^+ \to [D(A)]$. In particular, $Du'(t)$ exists for all $t \geq 0$ and $t \mapsto Du'(t) \in C(S_N)$ is continuous. This implies existence and continuity of the partial derivative:

$$\frac{\partial^2 u(t,x)}{\partial x\, \partial t}, \qquad t \geq 0, x \in [a,b],$$

and so the lemma follows by the well-known result of multivariate calculus. □

Proof of Theorem 38.2. (a) Lemma 38.3, combined with compactness of the interval $[a,b]$, implies that $t \mapsto Du(t) \in C(S_N)$ is continuously differentiable. Because the supremum norm is stronger than the Hilbert space norm, the same is true if we consider this function as having values in $L^2(S_N)$. Let:

$$\phi(t) = \|Du(t)\|^2_{L^2(S_N)} =< Du(t), Du(t) >, \qquad t \geq 0.$$

Then ϕ is differentiable and:

$$\begin{aligned}\frac{\mathrm{d}}{\mathrm{d}t}\phi(t) &= 2 < \frac{\mathrm{d}}{\mathrm{d}t}Du(t), Du(t) >= 2\sum_{i=1}^{N}\int_a^b \frac{\partial^2 u_i(t,x)}{\partial x \partial t}\frac{\partial u_i(t,x)}{\partial x}\,\mathrm{d}x\\ &= -2\sum_{i=1}^{N}\int_a^b \frac{\partial u_i(t,x)}{\partial t}\frac{\partial^2 u_i(t,x)}{\partial x^2}\,\mathrm{d}x,\end{aligned}$$

where in the last step we have used integration by parts formula plus the Neumann boundary conditions.

On the other hand, multiplying the ith equality in (38.1) by $D^2 u_i(t,x)$ and integrating over the interval $[a,b]$, we obtain:

$$\begin{aligned}\int_a^b \frac{\partial u_i(t,x)}{\partial t}\frac{\partial^2 u_i(t,x)}{\partial x^2}\,\mathrm{d}x &= \sigma_i \|D^2 u_i(t,\cdot)\|_2^2\\ &\quad + \int_a^b F_i(u(t,x))\frac{\partial^2 u_i(t,x)}{\partial x^2}\,\mathrm{d}x.\end{aligned}$$

Summing over $i \in \mathcal{N}$ renders now:

$$\frac{\mathrm{d}}{\mathrm{d}t}\phi(t) + 2\sigma\lambda\phi(t) \leq -2\sum_{i=1}^{N}\int_a^b F_i(u(t,x))\frac{\partial^2 u_i(t,x)}{\partial x^2}\,\mathrm{d}x,$$

where we have used the Poincaré Lemma. Integrating the right-hand side by parts, using the boundary conditions, and applying Schwarz inequality twice, we see that it does not exceed $2M\|Du(t)\|^2_{L^2(S_N)} = 2M\phi(t)$. (Alternatively, we use the following estimate:

$$\left|\sum_{i=1}^{N}\sum_{j=1}^{N} a_{i,j}v_i v_j\right| \le \sqrt{\sum_{i=1}^{N}\sum_{j=1}^{N} a_{i,j}^2 \sum_{k=1}^{N} v_k^2},$$

that holds for all numbers $a_{i,j}, v_i, i, j = 1, \ldots, N$. This inequality simply means that $|(Av, v)| \le \|A\|\|v\|^2$, where Av is the matrix product of $(a_{i,j})_{i,j=1,\ldots N}$ and $(v_i)_{i=1,\ldots,N}$, and $(\cdot,\cdot)$ denotes the standard scalar product in $\mathbb{R}^N$.) Therefore, $\frac{\mathrm{d}}{\mathrm{d}t}\phi(t) + 2d\phi(t) \le 0$, or, which is the same, $\frac{\mathrm{d}}{\mathrm{d}t}(\mathrm{e}^{2dt}\phi(t)) \le 0$. This implies $\mathrm{e}^{2dt}\phi(t) \le \phi(0)$, and (a) follows since $\|Du(t)\|_{L^2(S_N)} = \sqrt{\phi(t)}$.

(b) Fix $i \in \mathcal{N}$ and $t \ge 0$. Since $u_i(t,\cdot)$ is a continuous function, for some $x \in [a,b]$ we have $u_i(t,x) = \frac{1}{b-a}\int_a^b u_i(t,y)\,\mathrm{d}y$. Hence, $u_i(t) - \overline{u}_i(t)$ satisfies the assumptions of Poincaré's Lemma. Since $D[u_i(t) - \overline{u}_i(t)] = Du_i(t)$, (b) is an immediate consequence of (a).

(c) follows immediately by (a), (b), and the Sobolev inequality.

(d) We have:

$$\begin{aligned}\overline{u}_i'(t) &= \frac{1}{b-a}\int_a^b \frac{\partial u_i(t,x)}{\partial t}\,\mathrm{d}x \\ &= \frac{1}{b-a}\int_a^b \sigma_i \frac{\partial^2 u_i(t,x)}{\partial x^2}\,\mathrm{d}x + \frac{1}{b-a}\int_a^b F_i(u(t,x))\,\mathrm{d}x.\end{aligned}$$

The Neumann boundary conditions imply that the first term on the right-hand side vanishes. Hence, by (c), (d) follows with:

$$G_i(t) = \frac{1}{b-a}\int_a^b F_i(u(t,x))\,\mathrm{d}x - F_i(\overline{u}(t)),$$

because:

$$|G_i(t)| \le \frac{1}{b-a}\int_a^b |F_i(u(t,x)) - F_i(\overline{u}(t))|\,\mathrm{d}x \le L\|u(t) - \overline{u}(t)\|_{\sup},$$

where L is a Lipschitz constant for F_i's. □

The primary significance of the theorem is that it describes long-time behavior of solutions to (38.1): if diffusion coefficients are sufficiently large as compared to the size of the interval (which is hidden in the constant from the Poincaré Lemma) and the forcing nonlinear terms F_i's, solutions to (38.1) homogenize along N copies of the interval $[a,b]$ as time tends to infinity, and when identified with N real-valued functions, for large times they are very close

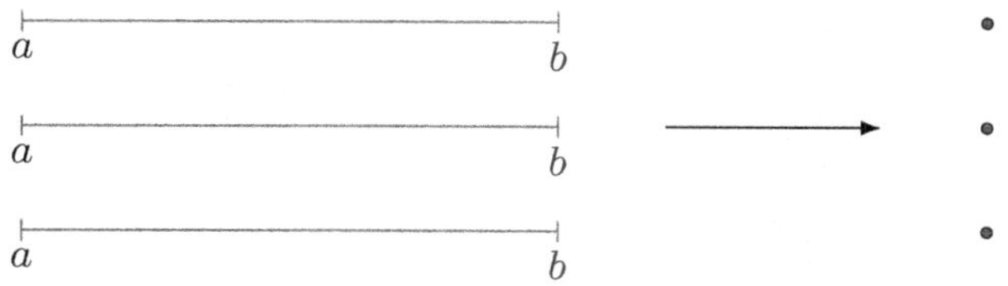

Figure 38.1 State-space collapse for reaction-diffusion processes with fast diffusion

to solutions of

$$\overline{u}'(t) = F(\overline{u}(t)), \qquad t \geq 0. \tag{38.4}$$

Point (a) says that as $t \to \infty$, the derivative of the solution decays exponentially, suggesting that – and (b) makes it precise – the solution becomes "flat." Point (c) says that the latter statement is true in the supremum norm as well, and point (d) specifies the dynamics of the "flattened" solution.

However, one could also look at the theorem from a different angle: suppose $\sigma > 0$ and we multiply all σ_i's in (38.1) by $\kappa_n > 0$, where $\lim_{n\to\infty} \kappa_n = \infty$. Then, for sufficiently large n, we have $d > 0$ and the theorem's estimates are in force. Letting $n \to \infty$, we obtain that solutions of (38.1) converge to those of (38.4). It is the latter point of view that prevails in this book. The intuitive reason for this result is that fast diffusion "merges" or "lumps" points in each interval into a single point (see Figure 38.1).

For example, let us consider a (one-dimensional variant of a) model of **kinase** phosphorylation and dephosphorylation, due to Żuk et al. [356]. In the model, kinase molecules diffusing in a compartment, modeled as an interval $[a, b]$, are divided into three classes: unphosphorylated, $\mathcal{K}_0$, singly phosphorylated, $\mathcal{K}_1$, and doubly phosphorylated, $\mathcal{K}_2$. Any kinase, whether phosphorylated or not, is an enzyme, leading to phosphorylation of other kinases, and the higher the phosphorylation level is, the higher the intensity c_i of such reaction, $i = 0, 1, 2$. In particular,

$$\mathcal{K}_i + \mathcal{K}_1 \to \mathcal{K}_i + \mathcal{K}_2, \text{ with intensity } c_i, \quad i = 0, 1, 2,$$

where $c_0 < c_1 < c_2$ are given. Similarly,

$$\mathcal{K}_i + \mathcal{K}_0 \to \mathcal{K}_i + \mathcal{K}_1, \text{ with intensity } 2c_i, \quad i = 0, 1, 2,$$

the factor 2 reflecting the fact that a kinase of type $\mathcal{K}_0$ may be phosphorylated at any of its two residues.

A phosphorylated kinase may also be dephosphorylated by a phosphatase $\mathcal{P}$:

$$\mathcal{P} + \mathcal{K}_i \to P + \mathcal{K}_{i-1}, \text{ with intensity } id, \quad i = 1, 2;$$

(for doubly phosphorylated kinase, the intensity of dephosphorylation is twice larger, because there are two residues).

These assumptions lead to the following system of equations for kinase concentrations, $u_i(t, x)$ denoting concentration of kinase $\mathcal{K}_i$ at time t and spatial location x:

$$\begin{aligned}
\frac{\partial u_0}{\partial t} &= \sigma \frac{\partial^2 u_0}{\partial x^2} + du_1C - 2u_0 \sum_{i=0}^{2} c_i u_i, \\
\frac{\partial u_1}{\partial t} &= \sigma \frac{\partial^2 u_1}{\partial x^2} - du_1C + 2u_0 \sum_{i=0}^{2} c_i u_i - u_1 \sum_{i=0}^{2} c_i u_i + 2dCu_2, \\
\frac{\partial u_2}{\partial t} &= \sigma \frac{\partial^2 u_2}{\partial x^2} + u_1 \sum_{i=0}^{2} c_i u_i - 2dCu_2,
\end{aligned} \tag{38.5}$$

with "no-flux" Neumann boundary conditions (38.2). Here, C is the (constant) number, or concentration, of phosphatases and $\sigma > 0$ is the common diffusion coefficient for all kinases. To explain, the term $2u_0 \sum_{i=0}^{2} c_i u_i$ in the nonlinear part in the first equation is the number of kinase molecules that changed the class $\mathcal{K}_0$ to $\mathcal{K}_1$ by being phosphorylated at one residue, and du_1C is the number of singly phosphorylated kinase molecules that underwent a process of dephosphorylation. Similarly, the term $u_1 \sum_{i=0}^{2} c_i u_i$ is the number of singly phosphorylated molecules that became doubly phosphorylated, and $2dCu_2$ is the number of doubly phosphorylated molecules that became singly phosphorylated.

If σ is very large, concentrations of kinase molecules do not depend on the spatial position x, and hence are constant along the interval $[a, b]$. Then, $t \mapsto u_i(t, \cdot)$ may be identified with real functions $t \mapsto u_i(t)$, and the Conway–Hoff–Smoller Theorem says that the system is well-approximated by the differential equations:

$$\begin{aligned}
\frac{du_0}{dt} &= du_1C - 2u_0 \sum_{i=0}^{2} c_i u_i, \\
\frac{du_1}{dt} &= -du_1C + 2u_0 \sum_{i=0}^{2} c_i u_i - u_1 \sum_{i=0}^{2} c_i u_i + 2dCu_2, \\
\frac{du_2}{dt} &= u_1 \sum_{i=0}^{2} c_i u_i - 2dCu_2.
\end{aligned} \tag{38.6}$$

Exercise 38.1 Check that the nonlinear part in (38.5) is locally Lipschitz-continuous, and that solutions starting at non-negative initial conditions stay non-negative all the time. Also, $\|u_0 + u_1 + u_2\|_{\sup}$ does not vary with time. In particular, solutions starting at non-negative initial conditions never blow up.

☞ Chapter's summary

If no mass is lost at the boundary (which is expressed in the Neumann boundary conditions), diffusion has a property of homogenizing solutions along bounded intervals: as diffusion coefficients tend to infinity, solutions to (38.1) converge to those of (38.4).[1]

[1] For an ingenious, probabilist's look at a homogenization theorem (with periodic boundary condition instead of Neumann boundary condition) consult the works of Pardoux, Pang, and Veretennikov (see, e.g., [283] and references given there).

39

Shadow Systems

In modeling, one often needs to combine diffusion equations with ordinary differential equations – see, for example, the model of early cancer, presented in Chapter 36. In such cases σ defined in (38.3) equals 0, and the main assumption of the Conway–Hoff–Smoller Theorem is not satisfied. On the other hand, the latter theorem speaks about convergence of solutions of systems of equations involving diffusion to solutions of ordinary differential equations. Hence, intuitively, lack of diffusion in some of the equations of the approximating systems should not spoil convergence. As we shall see, and this is the subject of this chapter, solutions do in fact converge to those of a **shadow system** (this notion seems to be originally introduced by J. P. Keener, see [203]). On our way to this goal, we will generalize the Conway–Hoff–Smoller Theorem in its part concerning convergence as diffusion coefficients tend to infinity. We follow [59].

We begin by noting that in the case where $F \equiv 0$ and $N = 1$, combined points (c) and (d) of the Conway–Hoff–Smoller Theorem have already been

Figure 39.1 Shadow(s) system(s) by Radek Bobrowski

encountered in Chapter 32. For, in this case G_i in (d) equals zero (see the proof), so that $\overline{u}'(t) = 0$, and $\overline{u}(t) = const. = \frac{1}{b-a}\int_a^b f$. Hence, (c) says that:

$$\left\|e^{tA}f - Pf\right\|_{\sup} \le \sqrt{\frac{\lambda+1}{\lambda}}ce^{-\lambda t}, \tag{39.1}$$

where $c = c(f)$, P is the projection $Pf = \frac{1}{b-a}\int_a^b f$, and A is the Neumann Laplace operator. This is Corollary 32.2 supplied with the rate of convergence, or a version of (32.3) – (c depends on f, but instead of unknown, perhaps quite small ϵ, we have $\lambda = \frac{1}{b-a}$).

We will show that the main result of the previous chapter concerning convergence as diffusion coefficients converge to infinity may be recovered from (32.3). Moreover, we will generalize the result to all semigroups satisfying (32.3) and to systems where some "diffusion coefficients" are zero.

More specifically, we come back to the setup of Chapter 36: S is a compact metric space, N is a natural number, and A_i, $i \in \mathcal{N}$ are generators of conservative Feller semigroups in $C(S)$. Additionally, we assume that there are $N_0 \le N$, $\epsilon > 0$, $M > 0$ and rank-one projections P_i, $i \in \mathcal{N}_0 := \{1, \dots, N_0\}$ such that:

$$\|e^{tA_i} - P_i\| \le Me^{-\epsilon t}, \qquad t > 0, i \in \mathcal{N}_0, \tag{39.2}$$

while:

$$A_i = 0, \qquad i \in \mathcal{N} \setminus \mathcal{N}_0.$$

(A rank-one projection is a projection of the form $Pf = (Ff)f_*$, where the vector f_* and the functional F are given.) In other words, only $N_0 \le N$ first equations in (36.6) are partial differential equations; the remaining ones are ordinary differential equations, although all u_i's depend on x and t (see, for instance, (36.10)).

It will be convenient to rewrite (36.6) in the following form:

$$\begin{aligned} \mathfrak{u}'(t) &= \mathfrak{A}\mathfrak{u}(t) + F(\mathfrak{u}(t), \mathfrak{v}(t)), & \mathfrak{u}(0) &= f \in [C(S)]^{N_0}, \\ \mathfrak{v}'(t) &= G(\mathfrak{u}(t), \mathfrak{v}(t)), & \mathfrak{v}(0) &= g \in [C(S)]^{N-N_0}, \end{aligned}$$

where $\mathfrak{u}(t) = (u_1(t), \dots, u_{N_0}(t))$, $\mathfrak{v}(t) = (u_{N_0+1}, \dots, u_N(t))$, and $\mathfrak{A}$ is the generator of the Cartesian product semigroup:

$$e^{t\mathfrak{A}}(f_i)_{i=1,\dots,N_0} = (e^{tA_i}f_i)_{i=1,\dots,N_0}.$$

In studying the fast-diffusion analogue of this system:

$$\begin{aligned} \mathfrak{u}_n'(t) &= \kappa_n\mathfrak{A}\mathfrak{u}_n(t) + F(\mathfrak{u}_n(t), \mathfrak{v}_n(t)), & \mathfrak{u}_n(0) &= f \in [C(S)]^{N_0}, \\ \mathfrak{v}_n'(t) &= G(\mathfrak{u}_n(t), \mathfrak{v}_n(t)), & \mathfrak{v}_n(0) &= g \in [C(S)]^{N-N_0}, \quad n \ge 1 \end{aligned} \tag{39.3}$$

where $\lim_{n\to\infty}\kappa_n = \infty$, we assume that either the Müller conditions are satisfied or for some other reason (see Chapter 37 or Exercise 38.1) solutions to these equations, starting in a bounded subset $\mathcal{M}$ of $C(S_N)$ never leave $\mathcal{M}$.

Theorem 39.1 *Under the stated conditions, (the mild) solutions to the system* (39.3) *starting in $\mathcal{M}$, converge to those of the shadow system:*

$$\begin{aligned} u'(t) &= PF(u(t), v(t,\cdot)), && u(0) = Pf \in \mathbb{R}^{N_0},\\ v'(t) &= G(u(t), v(t)), && v(0) = g \in [C(S)]^{N-N_0}, \end{aligned} \tag{39.4}$$

almost uniformly in $t \in (0,\infty)$, where $P(f_i)_{i=1,\dots,N_0} = (P_i f_i)_{i=1,\dots,N_0}$.

To explain: $F(u(t), v(t,\cdot))$ is a member of $[C(S)]^{N_0}$; P on the right-hand side of the first equation is applied to this member. Here and in what follows, N_0-tuples of real numbers will be alternately treated as elements of $\mathbb{R}^{N_0}$, and elements of $[C(S)]^{N_0}$, that is, identified with an N_0-tuple of constant functions.

For example, solutions to the early cancer model (36.10) converge as $\gamma \to 0$, to those of:

$$\begin{aligned} \frac{\partial c}{\partial t} &= ((2p-1)a(b,c) - d_c)c,\\ \frac{\partial b}{\partial t} &= \alpha(c)g - d_b b - db,\\ \frac{\partial g}{\partial t} &= -\int_0^1 [\alpha(c)g + d_g g - \kappa(c) - db]. \end{aligned} \tag{39.5}$$

(The integral here is with respect to the variable x.)

For the proof of the theorem we recall **Gronwall's Lemma**.

Lemma 39.2 *Suppose $\alpha : \mathbb{R}^+ \to \mathbb{R}^+$ is a measurable function that is bounded on each subinterval of $\mathbb{R}^+$. If there are non-negative constants a and b such that:*

$$\alpha(t) \le a + b\int_0^t \alpha(s)\,\mathrm{d}s, \qquad t \ge 0, \tag{39.6}$$

then $\alpha(t) \le a e^{bt}, t \ge 0$.

Proof By (39.6) and induction argument, it follows that:

$$\alpha(t) \le a\sum_{k=0}^{n-1}\frac{b^k t^k}{k!} + b^n\int_0^t \frac{(t-s)^{n-1}}{(n-1)!}\alpha(s)\,\mathrm{d}s, \qquad t \ge 0, n \ge 1. \tag{39.7}$$

Since the first term on the right-hand side converges, as $n \to \infty$, to ae^{bt}, it suffices to show that:

$$\lim_{n\to\infty} b^n \int_0^t \frac{(t-s)^{n-1}}{(n-1)!}\alpha(s)\,\mathrm{d}s = 0.$$

However, α is bounded on $[0, t]$. Hence, the integral here does not exceed $const.\frac{t^n}{n!}$, and we are done. □

Proof of Theorem 39.1 Let:

$$\alpha_n(t) = \|\mathfrak{u}_n(t) - u(t)\| + \|\mathfrak{v}_n(t) - v(t)\|, \qquad t \geq 0,$$

where norms are those of $[C(S)]^{N_0}$ and $[C(S)]^{N-N_0}$, respectively. Since:

$$\mathfrak{u}_n(t) = \mathrm{e}^{\kappa_n t \mathfrak{A}} f + \int_0^t \mathrm{e}^{\kappa_n (t-s)\mathfrak{A}} F(\mathfrak{u}_n(s), \mathfrak{v}_n(s))\, \mathrm{d}s,$$
$$\mathfrak{v}_n(t) = g + \int_0^t G(\mathfrak{u}_n(s), \mathfrak{v}_n(s))\, \mathrm{d}s,$$

while:

$$u(t) = Pf + \int_0^t PF(u(s), v(s, \cdot))\, \mathrm{d}s,$$
$$v(t) = g + \int_0^t G(u(s), v(s))\, \mathrm{d}s,$$

we have:

$$\begin{aligned}\alpha_n(t) \leq\ & \|\mathrm{e}^{\kappa_n t \mathfrak{A}} f - Pf\| \\ &+ \int_0^t \|\mathrm{e}^{\kappa_n (t-s)\mathfrak{A}} F(\mathfrak{u}_n(s), \mathfrak{v}_n(s)) - PF(\mathfrak{u}_n(s), \mathfrak{v}_n(s, \cdot))\|\, \mathrm{d}s \\ &+ \int_0^t \|PF(\mathfrak{u}_n(s), \mathfrak{v}_n(s, \cdot)) - PF(u(s), v(s, \cdot))\|\, \mathrm{d}s \\ &+ \int_0^t \|G(\mathfrak{u}_n(s), \mathfrak{v}_n(s)) - G(u(s), v(s))\|\, \mathrm{d}s.\end{aligned}$$

However, solutions stay in a bounded region. It follows that we can choose a Lipschitz constant L such that the last two integrands do not exceed $L\alpha_n(s)$, $s \in [0, t]$. For the same reason, using (39.2), we see that the first integrand does not exceed $C\mathrm{e}^{-\epsilon\kappa_n(t-s)}$, where $C = M \sup_{(\mathfrak{u}, \mathfrak{v}) \in \mathcal{M}} \|F(\mathfrak{u}, \mathfrak{v})\|$.

Now, consider $t \in [t_0, t_1]$ where $t_0 > 0$. Since the first term above may also be estimated using (39.2), for the non-negative constants:

$$a_n = M\mathrm{e}^{-\epsilon\kappa_n t_0} \|f\|_{[C(S)]^{N_0}} + \frac{C}{\epsilon\kappa_n}$$

we obtain:

$$\alpha_n(t) \leq a_n + 2L \int_0^t \alpha_n(s)\, \mathrm{d}s,$$

while $\lim_{n\to\infty} a_n = 0$. Therefore, by Gronwall's Lemma,

$$\alpha_n(t) \le a_n e^{Lt} \le a_n e^{2Lt_1}.$$

This completes the proof. □

We note that in the case where A is the Neumann Laplace operator, Theorem 39.1 has been established in [248]. The latter paper also gives more delicate information on the speed of convergence based on heat semigroup estimates to be found, for example, in [300, p. 25] or [349, Lemma 1.3].

As already mentioned, Theorem 39.1 recovers the main result of the previous chapter on convergence of solutions of reaction-diffusion equations with fast diffusion to those of (38.4): it suffices to take A as the Neumann Laplacian and $N_0 = N$. We note however that the theorem does not include the Conway–Hoff–Smoller Theorem in its estimates and information on the limit behavior as $t \to \infty$. Remarkably, for $N_0 < N$, the limit behavior of solutions to a reaction-diffusion equation with some diffusion coefficients equalling zero may be quite different from that of the solution to the related shadow system [248].

On the other hand, Theorem 39.1 shows that "homogenization" is not a phenomenon related directly to diffusion equation with reflecting boundary conditions but, rather, with asymptotic behavior of the involved semigroup. In particular, the class of semigroups satisfying (39.2) contains many more objects than just the Neumann Laplacian semigroup. As we have seen in Chapter 32, even if we restrict ourselves to the Laplace operator in $C[0, 1]$, there is a number of boundary conditions that may replace the reflecting boundary conditions. (For yet another boundary condition suitable for Theorem 39.1, see Exercise 32.3.)

☞ Chapter's summary

We discuss convergence, as diffusion coefficients tend to infinity, of solutions to reaction-diffusion equations coupled with ordinary differential equations, to shadow systems. We show in particular that it is not the relation to Laplace operator and Neumann boundary conditions that is crucial here but, rather, asymptotic behavior of the semigroup involved – it is assumption (39.2) that does the trick.

40

Kinases

In the last chapter devoted to various models involving fast diffusion, we consider a model of **kinase** activity due to Kaźmierczak and Lipniacki [202] (see also [97]). Kinases are enzymes that transport phosphate groups: in doing this, protein kinases transmit signals and control complex processes in cells. In [202], following [75], a cell is modeled as a unit $3D$ ball. All kinases, whether active (i.e., phosphorylated) or inactive, are diffusing inside the ball. Binding a receptor located at the cell membrane (the sphere) by an extracellular is a signal that is to be conveyed to the cell: this is done by the kinases that, when touching the boundary (the sphere), become activated; such active kinases are randomly inactivated when meeting phosphatases that are uniformly distributed over the cell (see Figure 40.1).

In the no feedback case, where all receptors at the membrane are ligand-bound almost simultaneously, reaching uniform stable concentration R, the master equation for the concentration of active kinases (after suitable rescaling) is a diffusion-degradation equation:

$$\frac{\partial K^*}{\partial t} = d \triangle K^* - K^*, \qquad t \geq 0,$$

with boundary condition (this b.c. is missing in [75] and was introduced in [202]):

$$aR(1 - K_b^*) = dn(\nabla K^*)_b.$$

Here, $d > 0$ and $a >$ are a diffusion coefficient and a reaction coefficient, respectively, K_b^* is the value of K^* at the boundary, $n(\nabla K^*)_b$ is the normal derivative at the boundary, and the term $-K^*$ describes random dephosphorylation of active kinases.

Under the assumption of spherical symmetry (made in both papers cited earlier), that is, under the assumption that K^* depends merely on the distance r

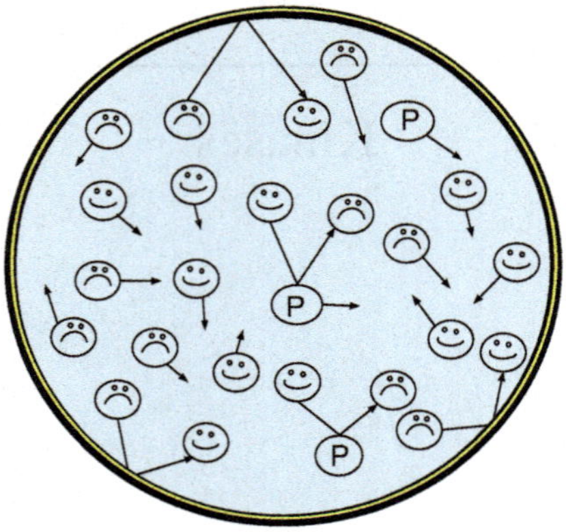

Figure 40.1 Kinases' activation and inactivation by Marek Bobrowski

from the ball's center, the master equation becomes,

$$\frac{\partial K^*}{\partial t}(t, r) = \frac{d}{r}\frac{\partial^2}{\partial r^2}\big(rK^*(t, r)\big) - K^*(t, r), \qquad t \geq 0, r \in (0, 1], \tag{40.1}$$

with boundary condition:

$$d\frac{\partial K^*}{\partial r}(t, 1) = aR(1 - K^*(t, 1)), \qquad t \geq 0. \tag{40.2}$$

We note that, in agreement with the intuitions presented earlier, the first part of the operator on the right-hand side of (40.1) is the generator of the 3*D* Bessel process, modeling the distance of a 3*D* Brownian motion from the origin (see [296] p. 251). Condition (40.2) in turn describes an inflow of active kinases from the boundary (see also later in this chapter).

One of the aims of both [75] and [202] is to show that (perhaps somewhat surprisingly) *slow* diffusion is necessary for effective signal transmission. Hence, Kaźmierczak and Lipniacki study the case of infinitely fast diffusion and compare the properties of solutions of the limit equation with those of (40.1)–(40.2). In doing this, they derive the form of the limit equation to be:

$$\frac{\mathrm{d}K^*}{\mathrm{d}t} = 3aR(1 - K^*) - K^*, \qquad t \geq 0. \tag{40.3}$$

To explain, as diffusion coefficients increase to infinity, the active kinases' distribution becomes uniform over the interval and may be identified with a real function of time: its dynamics is then described by (40.3). Nevertheless, the form of the limit equation is quite curious, with a particularly intriguing factor 3. In [202], equation (40.3) is obtained by integrating the master equation over the ball and using the Gauss Theorem.

Here, following [59], we would like to present a semigroup-theoretical proof of convergence of solutions to (40.1)–(40.2) with $d \to \infty$, to those of (40.3). (In [202], no proof of such convergence is given.)

Convergence of related semigroups

Let $C_0(0, 1]$ be the space of $f \in C[0, 1]$ satisfying $f(0) = 0$. Given $c \leq 1$, let:

$$G_c f = f'' \tag{40.4}$$

with domain:

$$D(G_c) = \{f \in C^2[0, 1]; f(0) = f''(0) = 0 \text{ and } f'(1) = cf(1)\}.$$

For $c \leq 0$, we know from Exercise 32.8 that G_c generates a Feller semigroup in $C_0(0, 1]$ – in the notations of that exercise, $G_c = A_{-c}$. In the related stochastic process (a Brownian motion), all particles touching the boundary $x = 0$ are instantaneously killed and removed from the state-space, while those touching $x = 1$ are partially killed and removed, and partially reflected for $c < 0$, and reflected for $c = 0$ (see Chapter 3). However, for $c \in (0, 1]$, the boundary condition $f'(1) = cf(1)$ describes inflow of particles through the boundary $x = 1$, and the related semigroup cannot be of Feller type. The operator G_1 has already been encountered in Chapter 32: condition (32.16) involves a scalar multiple of G_1 and forces:

$$\lim_{t\to\infty} \mathrm{e}^{tG_1} f = Pf, \qquad f \in C_0(0, 1], \tag{40.5}$$

where $P : f \mapsto 3(F_1 f)h$, where $h(x) = x$, $x \in [0, 1]$; in particular, $(\mathrm{e}^{tG_1})_{t\geq 0}$ is a bounded semigroup. To repeat, this condition says that the rates of inflow and outflow are so tuned here that in the limit a nontrivial equilibrium is attained. On the other hand, in the process related to $c = 1$, the inflow of particles through the boundary $x = 1$ is faster than for $0 < c < 1$, and obviously for $c = 0$ there is no inflow at all. Therefore, expected numbers of particles in the interval in the case $0 < c < 1$ should be larger than in the case $c = 0$ and smaller than in the case $c = 1$. In other words, we should have:

$$\mathrm{e}^{tG_0} \leq \mathrm{e}^{tG_c} \leq \mathrm{e}^{tG_1}, \qquad c \in [0, 1], t \geq 0. \tag{40.6}$$

We make these intuitions precise in the following lemma.

Lemma 40.1 *The operators G_c, $c \leq 1$ generate strongly continuous, equibounded semigroups in $C_0(0, 1]$: there is $M > 0$ such that:*

$$\|\mathrm{e}^{tG_c}\| \leq M, \qquad t \geq 0, c \leq 1. \tag{40.7}$$

Moreover, (40.6) *holds.*

Proof In view of the remarks made earlier, we may restrict ourselves to $c \in (0, 1]$; for other c, (40.7) holds with $M = 1$. Given $\lambda > 0$ and $g \in C_0(0, 1]$, we want to solve the resolvent equation for G_c: $\lambda f - G_c f = g$. Any solution must be of the form:

$$f(x) = C_1 \sinh(\sqrt{\lambda}x) + C_2 \cosh(\sqrt{\lambda}x) + f_{g,\lambda}(x), \tag{40.8}$$

where:

$$f_{g,\lambda}(x) = \frac{1}{2\sqrt{\lambda}} \int_0^1 \mathrm{e}^{-\sqrt{\lambda}|x-y|} g(y)\,\mathrm{d}y.$$

We see that $f \in D(G_c)$ iff $C_2 = -f_{g,\lambda}(0)$ and:

$$\begin{aligned}\sqrt{\lambda}C_1 \cosh\sqrt{\lambda} + \sqrt{\lambda}C_2 \sinh\sqrt{\lambda} &= cC_1 \sinh\sqrt{\lambda} + cC_1 \cosh\sqrt{\lambda} \\ &\quad + (c + \sqrt{\lambda}) f_{g,\lambda}(1),\end{aligned} \tag{40.9}$$

where we used $f'_{g,\lambda}(1) = -\sqrt{\lambda} f_{g,\lambda}(1)$ and $f''_{g,\lambda}(0) = \lambda f_{g,\lambda}(0)$. It follows that the resolvent equation has a unique solution, and C_1 is a linear-fractional function of c: $C_1(c) = \frac{\alpha c+\beta}{\gamma c+\delta}$ where:

$$\begin{aligned}\alpha\delta - \beta\gamma &= \sqrt{\lambda}\left(f_{g,\lambda}(1) + f_{g,\lambda}(0)\cosh\sqrt{\lambda}\right)\cosh\sqrt{\lambda} \\ &\quad + \sqrt{\lambda}\left(f_{g,\lambda}(1) - f_{g,\lambda}(0)\sinh\sqrt{\lambda}\right)\sinh\sqrt{\lambda} \\ &= \sqrt{\lambda}\left(f_{g,\lambda}(1)\mathrm{e}^{\sqrt{\lambda}} + f_{g,\lambda}(0)\right) > 0,\end{aligned}$$

provided $g \geq 0$. It follows that, for such g, C_1 increases with c (while C_2 is constant). This shows in particular that:

$$(\lambda - G_0)^{-1} \leq (\lambda - G_c)^{-1} \leq (\lambda - G_1)^{-1}, c \in [0, 1]. \tag{40.10}$$

Since $(\lambda - G_0)^{-1}$ is the resolvent of a Feller generator, $(\lambda - G_0)^{-1}$, $\lambda > 0$ are positive operators in $C_0(0, 1]$. Therefore, $(\lambda - G_c)^{-1}$, $\lambda > 0$ are positive operators too. On the other hand, as we have already seen, G_1 is the generator of a bounded, positive semigroup: there is $M > 0$ such that $\|\mathrm{e}^{tG_1}\| \leq M$. It follows that for $\lambda > 0$ and integers $k \geq 0$,

$$\|\lambda^k (\lambda - G_1)^{-k}\| \leq M,$$

and (40.10) implies:

$$\|\lambda^k (\lambda - G_c)^{-k}\| \leq M.$$

Since the domains $D(G_c)$ are clearly dense in $C_0(0, 1]$, the Hille–Yosida Theorem completes the proof of (40.7). Condition (40.6) is now a direct consequence of (40.10). □

Our main step in proving Kaźmierczak and Lipniacki's claim is the following theorem.

Proposition 40.2 *Given real numbers $d_n > aR$ with $\lim_{n\to\infty} d_n = \infty$, we consider operators:*

$$A_n = d_n G_{c_n}$$

where $c_n = 1 - \frac{aR}{d_n}$ (with domain $D(A_n) = D(G_{c_n})$). Then:

$$\lim_{n\to\infty} \mathrm{e}^{tA_n} f = \mathrm{e}^{-3aRt} Pf, \qquad f \in C_0(0, \infty)$$

almost uniformly in $t \in (0, \infty)$, where P is defined after (40.5).

The easiest road to this result leads through Kurtz's Theorem of Chapter 42. Hence, we postpone the proof to that chapter. We note that the curious factor 3 is already here, and it turns out to be a direct consequence of the definition of P!

The physical phenomenon described by Proposition 40.2 is quite similar to that of the main result of Chapter 35. To explain, in the particular case where the graph contains only one edge identified with the interval $[0, 1]$, Theorem 35.2 states that for the semigroups generated by:

$$A_n f = \kappa_n f'', \quad D(A_n) = \{f \in C^2[0, 1] \colon f'(0) = 0, f'(1) = -\frac{a}{\kappa_n} f(1)\},$$

where $a > 0$ is given, we have:

$$\lim_{n\to\infty} \mathrm{e}^{tA_n} f = \mathrm{e}^{-at} \int_0^1 f.$$

This is interpreted as follows: Condition $f'(1) = -\frac{a}{\kappa_n} f(1)$ describes a partial outflow of particles through the boundary $x = 1$. When $\kappa \to \infty$, the boundary $x = 1$ becomes reflecting, but at the same time solutions homogenize over the interval. Because of the tuning of speed of homogenization and outflow, in the limit we still have a possibility of probability mass' loss. The limit process is a single-state Markov chain in which a particle after an exponential time with parameter $a > 0$, disappears from the state-space.

Proposition 40.2 is similar: here conditions $f'(1) = f(1)$ and $f(0) = 0$ describe a balance: inflow and outflow rates cancel out. By considering $f'(1) = (1 - \frac{aR}{d_n}) f(1)$, we decrease the rate of inflow through $x = 1$. Again letting

$d_n \to \infty$, on the one hand, we "homogenize" solutions and, on the other, gradually increase the rate of inflow to the balance state. However, the tuning of the rates of outflow and inflow is so that in the limit, probability mass still decays exponentially.

Proof of the Main Convergence Result

We start by noting that the boundary condition in (40.2) is arguably easier to comprehend, when we turn to the equations for the concentration of inactive kinases:

$$K^\diamond = 1 - K^*$$

(the total concentration of kinases is normalized to 1). The master equation then becomes:

$$\frac{\partial K^\diamond}{\partial t}(t, r) = \frac{d}{r}\frac{\partial^2}{\partial r^2}\big(rK^\diamond(t, r)\big) + 1 - K^\diamond(t, r), \tag{40.11}$$

for $t \geq 0$, $r \in (0, 1]$, while the boundary condition:

$$\frac{\partial K^\diamond}{\partial r}(t, 1) = -\frac{aR}{d}K^\diamond(t, 1), \qquad t \geq 0, \tag{40.12}$$

turns out to be the familiar Robin boundary condition, describing partial outflow of inactive kinases at the boundary. Hence, (40.2) indeed describes an inflow of active kinases from the boundary.

Next, we rewrite (40.11) and (40.12) in terms of:

$$L(t, r) = rK^\diamond(t, r).$$

This gives:

$$\frac{\partial L}{\partial t} = d\frac{\partial^2 L}{\partial r^2} + h - L, \qquad t \geq 0, \tag{40.13}$$

where $h(r) = r$, $r \in [0, 1]$, with boundary condition:

$$\frac{\partial L}{\partial r}(t, 1) = \left(1 - \frac{aR}{d}\right)L(t, 1). \tag{40.14}$$

Proposition 40.3 *Solutions to the system* (40.13)–(40.14) *with initial condition $f \in C_0(0, 1]$ converge, as $d \to \infty$, to those of*

$$\frac{\mathrm{d}L}{\mathrm{d}t} = -(3aR + 1)L(t) + h \tag{40.15}$$

with initial condition Pf, for all $t > 0$.

Proof Let $d_n, n \ge 1$ be as in Proposition 40.2. Equations (40.13) with d replaced by d_n, may be seen as semilinear, and their solutions are given by:

$$L_n(t) = \mathrm{e}^{-t}\mathrm{e}^{tA_n} f + \int_0^t \mathrm{e}^{-(t-s)}\mathrm{e}^{(t-s)A_n} h \, \mathrm{d}s.$$

Proposition 40.2 yields now:

$$\lim_{n\to\infty} L_n(t) = \mathrm{e}^{-t}\mathrm{e}^{-3aRt} Pf + \int_0^t \mathrm{e}^{-(t-s)}\mathrm{e}^{-3aR(t-s)} Ph \, \mathrm{d}s.$$

Since $Ph = h$, the right-hand side is the mild solution of (40.15). □

Equation (40.15) describes dynamics in the one-dimensional subspace spanned by h. In particular, all $L(t)$' are scalar multiples of h. Since $L = hK^\diamond L$, solutions of (40.11)–(40.12) converge, as $d \to \infty$, to scalars evolving according to:

$$\frac{\mathrm{d}K^\diamond}{\mathrm{d}t} = -(3aR + 1)K^\diamond(t) + 1.$$

Hence, the limit equation for K^* is:

$$\frac{\mathrm{d}K^*}{\mathrm{d}t} = -\frac{\mathrm{d}K^\diamond}{\mathrm{d}t} = (3aR + 1)K^\diamond(t) - 1 = 3aR(1 - K^*) - K^*,$$

which was our goal.

Alternative Proof of (32.16)

We complete this chapter by giving an alternative proof of (32.16) – this relation was crucial for the proof of the main result in this chapter (see (40.5)): we will show that $K \ge 0$ and $\epsilon > 0$ may be chosen so that:

$$\|\mathrm{e}^{tG_1} - P\| \le K\mathrm{e}^{-\epsilon t}, \qquad t > 0, \tag{40.16}$$

where $Pf = 3(F_1 f)h$.

Exercise 32.8 (c) combined with (40.10) shows that for operators G_c defined in (40.4), we have:

$$(\lambda - A_0)^{-1} \le (\lambda - G_c)^{-1} \le (\lambda - G_1)^{-1}, \qquad c \le 1;$$

(we recall that A_ν in Exercise 32.8 (c) coincides with $G_{-\nu}$, $\nu \ge 0$). Moreover, see Exercise 32.7, $(\lambda - A_0)^{-1} f(x) = \int_0^1 k_\lambda(x, y) f(y) \, \mathrm{d}y > 0$ for all $x \in (0, 1)$, provided $f \ge 0$ is nonzero. Using Exercise 32.8 (c) again, we see that:

$$(\lambda - G_1)^{-1} f(1) \ge (\lambda - G_0)^{-1} f(1) \ge (\lambda - A_\nu)^{-1} f(1)$$
$$> (\lambda - A_0)^{-1} f(1) = 0,$$

for $f \geq 0$, $\nu > 0$. It follows that $(\mathrm{e}^{tG_1})_{t\geq 0}$ is irreducible. Moreover, by the proof of Theorem 4.1 in [54], G_1 is the generator of a cosine family. Therefore, the semigroup generated by G_1 is holomorphic.

To apply Corollary 32.4, we need to show that $D(G_1)$ embeds compactly in $C_0(0, 1]$. To this end, assume $f \in C^2[0, 1]$ satisfies $f'(1) = f(1)$. Then, for $g(x) = f(1 - x)$ we have $g'(0) = -g(0)$, and so $g(x) = g(0)(1 - x) + \int_0^x \int_0^y g''(z)\,\mathrm{d}z\,\mathrm{d}y$. It follows that members of the unit ball in $D(G_1)$ satisfy:

$$\|f\| + \|f''\| \leq 1 \quad \text{and} \quad f(x) = f(1)x + \int_0^{1-x} \int_0^y f''(z)\,\mathrm{d}z\,\mathrm{d}y,\ x \in [0, 1].$$

Since $|f(1)| \leq 1$, all f in the ball are Lipschitz-continuous with constant 2. (Alternatively, use a version of Landau's inequality.) Therefore, the Arzelà–Ascoli Theorem implies the claim.

Thus, by Corollary 32.4, there is a strictly positive f_* and a functional F such that $Ff_* = 1$ and:

$$\|\mathrm{e}^{-s(G_1)t}\mathrm{e}^{tG_1}f - (Ff)f_*\| \leq K\mathrm{e}^{-\epsilon t}\|f\|, \qquad f \in C_0(0, 1], t > 0 \tag{40.17}$$

for some constants $K \geq 0$ and $\epsilon > 0$. Since $h \in D(G_1)$ and $G_1 h = 0$, we have $\mathrm{e}^{tG_1}h = h$. Therefore, $s(G_1)$ in this formula must be equal to 0. Next, (40.17) with $s(G_1) = 0$, implies $\lim_{t\to\infty} \mathrm{e}^{tG_1} f_* = f_*$. This in turn shows that f_* is a fixed point of the semigroup, and hence belongs to the kernel of G_1. Since this kernel is spanned by h, without loss of generality we may assume $f_* = 3h$. We are left with showing that F in (40.17) coincides with F_1. We note that our remarks combined with (40.17) yield:

$$\lim_{t\to\infty} \mathrm{e}^{tG_1} f = 3(Ff)h. \tag{40.18}$$

Integrating by parts and using boundary conditions, we check that $F_1(G_1 f) = 0$ for all $f \in D(G_1)$. It follows that $t \mapsto F_1(\mathrm{e}^{tG_1})$ is constant, that is, that $F_1(\mathrm{e}^{tG_1}f) = F_1 f$ for all $t \geq 0$ and $f \in D(G_1)$. A density argument shows that the same is true for all $f \in C_0(0, 1]$. Therefore, using (40.17), $F_1 f = \lim_{t\to\infty} F_1(\mathrm{e}^{tG_1} f) = F_1(3(Ff)h) = Ff$ for all $f \in C_0(0, 1]$, since $F_1 3h = 1$, completing the proof.

Notes

The model of Kazmierczak and Lipniacki [202] has two variants: the no-feedback case described earlier, and the case with feedback, in which kinases interact with receptors. More specifically, initially, only a part of receptors is activated by the extracellular ligands: subsequently inactive receptors are activated by meeting active kinases, which in turn were previously activated by active receptors. In the latter variant, dynamics is described by the system:

$$\begin{aligned} \frac{\partial K^*}{\partial t}(t,r) &= \frac{d}{r}\frac{\partial^2}{\partial r^2}\big(rK^*(t,r)\big) - K^*(t,r), \\ d\frac{\partial K^*}{\partial r}(t,1) &= aR(t)(1-K^*(t,1)), \\ \frac{\mathrm{d}R(t)}{\mathrm{d}t} &= qK^*(t,1)[R_{max}-R(t)] - bR(t), \qquad t \geq 0, r \in (0,1], \end{aligned} \tag{40.19}$$

where $b, q, R_{max} > 0$ are constants, and $R(t) \in [0, R_{max}]$ is the number of active receptors. As in the no-feedback case, the model's authors predict the limit equations, as $d \to \infty$, to be:

$$\begin{aligned} \frac{\mathrm{d}K^*(t)}{\mathrm{d}t} &= 3aR(t)(1-K^*(t)) - K^*(t), \\ \frac{\mathrm{d}R(t)}{\mathrm{d}t} &= qK^*(t)[R_{max}-R(t)] - bR(t), \qquad t \geq 0. \end{aligned} \tag{40.20}$$

Clearly, the first equation here may be obtained by replacing the constant R in (40.3) by the function $R(\cdot)$. However, system (40.19), while natural biologically, is quite complicated mathematically. In particular, (40.19) is an example of a dynamic boundary value problem, with interactions between the "interior" and the boundary. Hence, even the question of existence of solutions is nontrivial and the nonlinearity in the boundary makes the problem even more difficult. In particular, (40.19) does not seem to come within reach of the elegant, linear theory developed in [80, 276]. Hence, analysis of asymptotic behavior of this systems, as $d \to \infty$, must be the subject of further research.

☞ Chapter's summary

In considering a signaling pathways model of activity of kinases we show that solutions to (40.1) and (40.2) converge to those of (40.3). In other words, if diffusion inside the cell is fast, a probabilistic model involving diffusion becomes a deterministic one. Although the idea seems to be self-explanatory, the limit model is of curious form because of the influence of boundary conditions involved.

41

Uniformly Differentiable Semigroups

After a long detour leading through various convergence theorems involving fast diffusions, we come back to our main question: "How can we prove irregular convergence of semigroups?". In Theorem 31.2, we covered the case where the semigroups are uniformly holomorphic, but this condition is quite restrictive, and may be weakened considerably. To this end, recall that a semigroup $(e^{tA})_{t\geq 0}$ is said to be differentiable iff $e^{tA}x \in \mathcal{D}(A)$ for all $x \in \mathbb{X}$ and $t > 0$. Holomorphic semigroups are differentiable, and a differentiable semigroup of equibounded operators is holomorphic iff there exists a constant $C > 0$ such that $\|Ae^{tA}\| \leq \frac{C}{t}$ for all $t > 0$ ([9] p. 160, [128] p. 101, [201] p. 488, or [284] p. 61).

The necessary and sufficient conditions for an operator A to generate a differentiable semigroup, due to A. Pazy, are presented in the following theorem.

Theorem 41.1 *Suppose $\|e^{tA}\| \leq M, t \geq 0$. The semigroup is differentiable iff for all $b > 0$ there exist constants $a_b \in \mathbb{R}$, and $C_b > 0$ such that:*

$$\rho(A) \supset \Sigma_b := \{\lambda \in \mathbb{C}; \Re e\, \lambda > a_b - b \log |\Im m\, \lambda|\} \tag{41.1}$$

and:

$$\|(\lambda - A)^{-1}\| \leq C_b |\Im m\, \lambda| \qquad \text{for } \lambda \in \Sigma_b,\ \Re e\, \lambda \leq 0. \tag{41.2}$$

Corollary 41.2 *Let A be a (nondensely defined) Hille–Yosida operator. If the semigroup $(e^{tA_p})_{t\geq 0}$ is differentiable then A generates a noncontinuous semigroup (see Chapter 28).*

Proof Since the range of $(\lambda - A)^{-1}$ is contained in $\mathbb{X}$, our assumption implies a) in Proposition 28.1, and the corollary follows. □

Here is the main theorem of this chapter, taken from [48].

Proposition 41.3 *Suppose that equibounded semigroups* $(e^{tA_n})_{t\geq 0}$ *are equidifferentiable in the sense that for all* $b > 0$ *there exist constants* $a_b \in R$, *and* $C_b > 0$ *such that for all* $n \in N$:

$$\rho(A_n) \supset \Sigma_b := \{\lambda;\, \Re e\, \lambda > a_b - b\, log|\Im m\lambda|\}$$

and:

$$\|(\lambda - A_n)^{-1}\| \leq C_b|\Im m\lambda| \qquad for\ \lambda \in \Sigma_b,\ Re\lambda \leq 0,\ n \geq 1. \tag{41.3}$$

Then the following are equivalent,

(a) $\lim_{n\to\infty} e^{tA_n}$ *exists for all* $t \geq 0$,
(b) $\lim_{n\to\infty}(\lambda - A_n)^{-1}$ *exists for all* λ *in the open right half-plane,*
(c) $\lim_{n\to\infty}(\lambda - A_n)^{-1}$ *exists for some* λ *in the open right half-plane.*

Proof Implication (a) $\Rightarrow$ (b) was explained in Theorem 31.2, and the fact that (c) implies (b) is left as an exercise (Exercise 41.1). Since (b) $\Rightarrow$ (c) is trivial we are left with showing (b) $\Rightarrow$ (a). The proof consists of three steps.

Step 1. Fix $b > 0$. Consider a purely imaginary $\lambda \in \Sigma_b$; we want to show that $\lambda \in \rho(\mathcal{A})$, where $\mathcal{A}$ was defined in (14.1). For $\alpha := \frac{1}{4|\Im m\lambda|C_b}$ and $\mu := \lambda + \alpha$, we have $(\mu - A_n)^{-1} = \sum_{k=0}^{\infty}(-\alpha)^k(\lambda - A_n)^{-(k+1)}$. Therefore:

$$\|(\mu - A_n)^{-1}\| \leq \|\,(\lambda - A_n)^{-1}\,\| \sum_{k=0}^{\infty}\frac{1}{4^k} \leq \frac{4}{3}|\Im m\lambda|C_b.$$

It follows that $(\lambda - A_n)^{-1} = \sum_{k=0}^{\infty} s^k(\mu - A_n)^{-(k+1)}$, and this implies that $\lim_{n\to\infty}(\lambda - A_n)^{-1}$ exists by the Dominated Convergence Theorem. Hence, $\mathcal{R}_\lambda$ defined in (31.2) is both a right and left inverse of $\lambda - \mathcal{A}$, and $\|\mathcal{R}_\lambda\| \leq C_b|\Im m\lambda|$, showing that $\lambda \in \rho(\mathcal{A})$.

Step 2. We show that $\Sigma_b \cap \{\lambda;\, \Re e\lambda \leq 0\} \subset \rho(\mathcal{A})$. To this end we note that $\Sigma_b \cap \{\lambda;\, \Re e\lambda \leq 0\} \cap \rho(\mathcal{A})$ is open in $\Sigma_b \cap \{\lambda;\, \Re e\lambda \leq 0\}$ since $\rho(\mathcal{A})$ is open in $\mathbb{C}$. To show that the former set is closed in the latter we argue as in the proof of (c) $\Rightarrow$ (d) in Theorem 31.2: it is enough to replace $\frac{M_\delta}{|\lambda_n|}$ by $C_b|\Im m\lambda_n|$ in (31.5). Since $\Sigma_b \cap \{\lambda;\, \Re e\lambda \leq 0\} \cap \rho(\mathcal{A})$ is nonempty by Step 1, and $\Sigma_b \cap \{\lambda;\, \Re e\lambda \leq 0\}$ is connected, $\Sigma_b \cap \{\lambda;\, \Re e\lambda \leq 0\} \cap \rho(\mathcal{A}) = \Sigma_b \cap \{\lambda;\, \Re e\lambda \leq 0\}$ and we are done.

Step 3. By Step 2, $\mathcal{A}$ satisfies (41.1). Moreover, for $\lambda \in \Sigma_b$, (41.3) implies $\|(\lambda - \mathcal{A})(x_n)_{n\geq 1}\| \geq C_b|\Im m\lambda|\|(x_n)_{n\geq 1}\|$, proving that $\mathcal{A}$ satisfies (41.2). Hence, the semigroup generated by $\mathcal{A}_p$ is differentiable. By Corollary 41.2, $\mathcal{A}$ generates a noncontinuous semigroup, and Proposition 28.3 completes the proof. □

Exercise 41.1 Use the Hille–Yosida estimates for $(\lambda - A_n)^{-1}$ to show that, in Proposition 41.3, (c) implies (b).

☞ Chapter's summary

The main theorem of Chapter 31 is generalized: its main assumption of uniform holomorphicity is dropped and replaced by that of uniform differentiability.

42

Kurtz's Singular Perturbation Theorem

Figure 42.1 A singular theorem by Radek Bobrowski

Theorem 41.1 is a dead end. First of all, conditions (41.1) and (41.2) are quite hard to check in practice and I am not aware of interesting examples of equidifferentiable semigroups (that are not equiholomorphic). There is no hope for further generalizations either, such as to equi-normcontinuous semigroups (comp. [87, 253]). Last but not least, in dealing with "hyperbolic" problems we should not expect additional regularity of solutions, and the method of the previous chapters simply fails.

Luckily, a surprising number of examples of degenerate convergence falls into the following scheme, devised by T. G. Kurtz [132, 232, 233]. Let $(\epsilon_n)_{n\geq 1}$ be a sequence of positive numbers converging to 0. Suppose A_n, $n \geq 1$ are generators of equibounded semigroups $\{e^{tA_n}, t \geq 0\}$ in a Banach space $\mathbb{X}$, and Q generates a strongly continuous semigroup $(e^{tQ})_{t\geq 0}$ such that:

$$\lim_{t\to\infty} e^{tQ}x =: Px, \qquad x \in \mathbb{X} \tag{42.1}$$

exists. Then (see Exercise 8.7) P is a bounded idempotent operator and:

$$Ker\, Q = Range\, P, \qquad \overline{Range\, Q} = Ker\, P. \tag{42.2}$$

Lemma 42.1 (Kurtz) *Suppose (42.1) is satisfied and* $\lim_{n\to\infty} e^{\epsilon_n A_n t} = e^{tQ}$. *Then:*

$$\lim_{n\to\infty} e^{tA_n}x = 0, \qquad x \in Ker\, P, t > 0 \tag{42.3}$$

uniformly in intervals of the form $[t_0, \infty)$, $t_0 > 0$.

Proof Let $x \in Ker\, P$ and suppose (42.3) does not hold uniformly in one of the specified intervals. Then there exists a sequence $(t_n)_{n\geq 1}$, separated from 0 and such that:

$$\|e^{t_n A_n}x\| \geq \delta, \qquad n \geq 1 \tag{42.4}$$

for some positive $\delta > 0$. On the other hand, there exists an $s > 0$ such that $\|e^{sQ}x\| < \frac{\delta}{2M}$ where M is a constant larger than all $\|e^{tA_n}\|$. Also, for large n, $\|e^{sQ}x - e^{s\epsilon_n A_n}x\| < \frac{\delta}{2M}$ and $\frac{t_n}{\epsilon_n} > s$. For such n,

$$\begin{aligned}\|e^{t_n A_n}x\| = \|e^{\frac{t_n}{\epsilon_n}(\epsilon_n A_n)}x\| &\leq M\|e^{s\epsilon_n A_n}x\| \\ &\leq M\|e^{s\epsilon_n A_n}x - e^{sQ}x\| + M\|e^{sQ}x\| < \delta,\end{aligned}$$

contradicting (42.4). □

Let, as before, A_n's be the generators of equibounded semigroups and suppose (42.1) holds for a Q. Denote:

$$\mathbb{X}' = Range\, P.$$

Theorem 42.2 (Kurtz) *Let A be an operator in* $\mathbb{X}$, *D be a subset of its domain, and assume that:*

(a) for $x \in D$, $(x, Ax) \in A_{ex}$ *where* A_{ex} *is the extended limit of* A_n, $n \geq 1$,
(b) for y in a core D' *of* Q, $(y, Qy) \in B_{ex}$ *where* B_{ex} *is the extended limit of* $\epsilon_n A_n$, $n \geq 1$,
(c) the operator PA with domain $D \cap \mathbb{X}'$ *is closable and its closure* $\overline{PA}$ *generates a strongly continuous semigroup in* $\mathbb{X}'$.

Then:

$$\lim_{n\to\infty} e^{A_n t}x = e^{t\overline{PA}}Px, \qquad x \in \mathbb{X}, t > 0. \tag{42.5}$$

Proof For $x \in D(A)$ and $y \in D'$, we take $z_n = x_n + \epsilon_n y_n$ where $x_n \in D(A_n)$ and $y_n \in D(A_n)$ are such that $\lim_{n\to\infty} y_n = y$, $\lim_{n\to\infty} \epsilon_n A_n y_n = Qy$, $\lim_{n\to\infty} x_n = x$ and $\lim_{n\to\infty} A_n x_n = Ax$. Clearly, $\lim_{n\to\infty} z_n = x$ and $\lim_{n\to\infty} A_n z_n = Ax + Qy$, that is, $A_{ex}x = Ax + Qy$. Since A_{ex} is closed, and D' is a core for Q, Qy here may be replaced by any member of the closure of the range of Q. Moreover, by (42.2), Qy may be replaced by any member of the kernel of P. Taking $PAx - Ax$ instead of Qy we obtain that $PAx \in \mathbb{X}'$ is a value of A_{ex} on x; $PAx - Ax$ belongs to the kernel, since P is idempotent.

By (c) and Corollary 8.5 it follows that (42.5) holds for $x \in \mathbb{X}'$ and the convergence is almost uniform in $t \in [0, \infty)$. By the same result, (b) implies $\lim_{n\to\infty} \mathrm{e}^{t\epsilon_n A_n} = \mathrm{e}^{tQ}$. Hence, by Kurtz's Lemma, $\lim_{t\to\infty} \mathrm{e}^{tA_n} x = 0$ for $x \in$ *KerP*, and the convergence is uniform in intervals of the form $[t_0, \infty)$. Writing $x \in \mathbb{X}$ as $(x - Px) + Px$ and noting that $x - Px$ belongs to *KerP* since P is idempotent, we complete the proof. □

As a by-product of the proof we obtain that $\mathbb{X}' = \mathbb{X}_0$, that is, that $\mathbb{X}'$ coincides with the regularity space. For, we have seen that for $x \in \mathbb{X}'$, the limit (42.5) is almost uniform in $[0, \infty)$. On the other hand, for $x \notin \mathbb{X}'$, the limit cannot be uniform in a neighborhood of 0, since the limit function is discontinuous at 0 (for $t = 0$, the left-hand side obviously converges to x). We also note that Kurtz's Theorem does not require checking directly the range condition – this is one of its biggest advantages. However, a kind of range condition for the space $\mathbb{X}'$ is, certainly, involved in requirement (c).

For a more general version of Kurtz's Theorem, see [132].

Armed with this magnificent tool, we may give a simple proof of the main theorem of Chapter 35 (assuming Proposition 35.1). In fact, the biggest obstacle in the analysis is the apparent conflict of notations we need to overcome. To begin with, the role of Q of Kurtz's Theorem will be played by B defined in (35.10). By Corollary 32.2, we have:

$$\lim_{t\to\infty} \mathrm{e}^{tB} = P,$$

with P defined in (35.6) (here, happily, notation for P of Kurtz's Theorem and that used in Chapter 35 agree well). Moreover, we have $\mathbb{X}' :=$ *Range* $P = C_0(S)$, the space of functions that are constant on each edge. The semigroups we want to consider are $(\mathrm{e}^{tA_{\kappa_n}})_{t\geq 0}$ defined in Theorem 35.2, which by Proposition 35.1 are composed of contractions. To comply with notations of Kurtz's Theorem, these semigroups will be denoted $(\mathrm{e}^{tA_n})_{t\geq 0}$. Also, the role of ϵ_n is played by κ_n^{-1}. Next, we define $Af := \sigma K f$ and take $D = C_0(S)$.

Lemma 42.3 *We have* $D = C_0(S) = \mathbb{X}' \subset D(A_{\mathrm{ex}})$ *and:*

$$A_{\mathrm{ex}} u = \sigma K u = Au, \qquad u \in D,$$

so that condition (a) in Kurtz's Theorem is satisfied.

Proof As the proof of Corollary 32.2 reveals, there are $f_n \in D(B)$ such that $\lim_{n\to\infty} f_n = u$ and, $\lim_{n\to\infty} \kappa_n f_n'' = 0$ (take $f = 0$ in relation (32.2)). Let $g_n = I_{\kappa_n}^{-1} f_n$. Clearly, $\lim_{n\to\infty} g_n = u$. Moreover,

$$A_n g_n = \sigma\kappa_n f_n'' + \sigma(Jf_n)'' = \sigma\kappa_n f_n'' + \sigma K f_n \underset{n\to\infty}{\longrightarrow} \sigma K u,$$

as claimed. □

Lemma 42.4 *For all* $f \in D(B)$,

$$B_{\text{ex}} f = Bf,$$

so that (b) in Kurtz's Theorem is satisfied.

Proof We take $f_n = I_{\kappa_n}^{-1} f \in D(A_n)$. Certainly, $\lim_{n\to\infty} f_n = f$ and:

$$\kappa_n^{-1} A_n f_n = \sigma f'' + \kappa_n^{-1}\sigma(Jf)'' \underset{n\to\infty}{\longrightarrow} \sigma f'' = Bf,$$

as desired. □

To complete the proof, we note that by (35.11), *PA* restricted to $C_0(S) = D = \mathbb{X}'$ coincides with Q of Theorem 35.2:

$$PAu = P(\sigma K u) = \sigma P K u = Qu.$$

Since this Q is the generator of a semigroup in $C_0(S)$, we do not need to close it, (c) in Kurtz's Theorem is satisfied, and (42.5) with $\overline{PA}$ replaced by Q yields (35.5).

We close this chapter with the promised proof of Proposition 40.2.

Proof of Proposition 40.2 We check the conditions of Kurtz's Theorem. The semigroups $(\mathrm{e}^{tA_n})_{t\geq 0}, n \geq 1$ are equibounded by Lemma 40.1. Also, (40.5) shows that (42.1) is satisfied with G_1 playing the role of Q. Next, for $f \in D(G_1)$, we consider:

$$f_n = f + a_n\chi$$

where $\chi(x) := \frac{\sinh x}{\sinh 1}, x \in [0,1]$, and $a_n := \frac{f(1)(1-c_n)}{c_n - \coth 1}$. We have $f_n \in D(G_{c_n})$ and $\lim_{n\to\infty} f_n = f$, while:

$$G_{c_n} f_n = f'' + a_n\chi'' \to f'' = G_1 f,$$

proving condition (b) in Kurtz's Theorem. Moreover, taking f from the subspace $\mathbb{X}' = \mathit{Range}\, P$, that is f of the form $f = \alpha h$ for some $\alpha \in \mathbb{R}$, we have for the same f_n,

$$A_n f_n = d_n a_n \chi'' = d_n a_n \chi \to \frac{\alpha a R}{1 - \coth 1}\chi.$$

Let A_{ex} be the extended limit of A_n, $n \geq 1$. It follows that for $f = \alpha h$, we have $(f, Af) \in A_{ex}$ where $Af = \frac{\alpha a R}{1-\coth 1}\chi$. This proves condition (b).

Finally, $F_1\chi = \coth 1 - 1$. Therefore, $PAf = -3\alpha a R h = -3aRf$ and PA is clearly the generator of the semigroup $e^{tPA} f = e^{-3aRt} f$ on $\mathbb{X}'$. This proves condition (c) and completes the proof. □

Exercise 42.1 Show that Example 26.2 fits into the setup of Kurtz's Theorem and deduce (26.3) without alluding to (26.2).

Exercise 42.2 (This generalizes Example 26.2.) Consider a Banach space $\mathbb{X}$ and let A_0 be the generator of a contraction semigroup in $\mathbb{X} \times \mathbb{X}$. Show that the operators $A_n = A_0 - \left(\begin{smallmatrix} 0 & 0 \\ 0 & nI_{\mathbb{X}} \end{smallmatrix}\right)$ with domain $\mathcal{D}(A_0)$ generate contraction semigroups in $\mathbb{X} \times \mathbb{X}$ with the regularity space $\mathbb{X} \times \{0\}$. Prove also:

$$\lim_{n\to\infty} e^{tA_n} \begin{pmatrix} 0 \\ x \end{pmatrix} = 0, \qquad \text{for } x \in \mathbb{X}. \tag{42.6}$$

Exercise 42.3 Assume that generators A_n, $n \geq 1$ satisfy the assumptions of Kurtz's Theorem (in fact, it suffices to assume (a) and (b)). Let B be a bounded linear operator such that BA_n, $n \geq 1$ are generators of equibounded semigroups and so is BQ. Also, assume that $\lim_{t\to\infty} e^{tBQ} =: P_B$ exists. Prove that if $P_B BA$ is closable and and its closure generates a strongly continuous semigroup, then:

$$\lim_{n\to\infty} e^{tB_n} = e^{t\overline{P_B BA}} P_B.$$

☞ Chapter's summary

A fundamental (and very useful) theorem of T. G. Kurtz is established. Many chapters to follow are devoted to its applications. Here, we present two of such applications: an alternative proof of Theorem 35.2 and a proof of Proposition 40.2.

43

A Singularly Perturbed Markov Chain

To begin with applications of Kurtz's Theorem, we consider a singularly perturbed Markov chain. This example is a piece of a bigger puzzle presented in [21], and at the same time provides probabilistic interpretation of Lemma 5.1 there. For a systematic treatment of singularly perturbed Markov chains, including inhomogeneity in time and infinite state-space, see [352].

Let $N \in \mathbb{N}$, and two $N \times N$ Kolmogorov matrices A and Q be given; as in Chapter 13, a Kolmogorov matrix is identified with an operator in l^1_N (the space of sequences of length N, equipped with the l^1-type norm). Given a sequence $(\epsilon_n)_{n\geq 1}$ of positive numbers converging to zero, we consider the operators:

$$A_n = A + \frac{1}{\epsilon_n} Q.$$

Each A_n describes a Markov chain being a combination of two: the "slow" one (A), and the "fast" one $(\epsilon_n^{-1} Q)$. We are interested in the limit behavior of the related Markov semigroups $(\mathrm{e}^{tA_n})_{t\geq 0}$, $n \geq 1$.

Before proceeding, we need more information on the structure of Q. To recall (see [90, 278]), states of a Markov chain are of two types: the transient and the recurrent ones. The state-space (which may be identified with $\{1, \dots, N\}$) may be divided into closed recurrent classes, say $C_1, C_2, \dots, C_m$, where $m \leq N$, with all states communicating within each class but not between the classes, and the remaining states being transient. The process starting at one of the latter states will after some time reach one of the recurrent classes never to return to its starting point. A process starting at a state of the former type, visits this state infinitely many times (see Figure 43.1).

This fact has its reflection in the structure of the null space of Q: its basis is formed by m non-negative eigenvectors $v_1, v_2, \dots, v_m \in \mathbb{R}^N$, corresponding to the stationary, limit distributions of the closed, recurrent classes. These eigenvectors have disjoint supports (and in particular are linearly

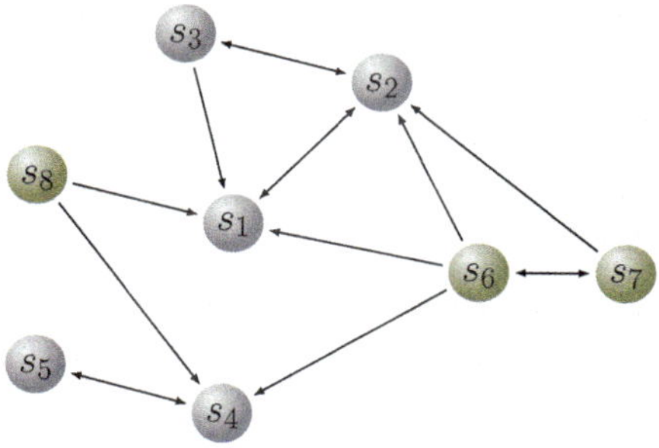

Figure 43.1 A Markov chain example: out of eight points, five are recurrent and form two clusters (in blue), and the remaining three (in olive) are transient. Two basic eigenvectors for this chain have supports $\{s_1, s_2, s_3\}$ and $\{s_4, s_5\}$, respectively.

independent), and by suitable rearrangement of the states, we may assume that these supports are $\{n_1 = 1, \ldots, N_1\}, \ldots, \{n_m, \ldots, N_m\}$, where $n_{j+1} = N_j + 1$, $j \in \{1, \ldots, m-1\}$, and $N_m \leq N$. In the terminology of Chapter 13, these vectors are distributions.

Moreover, the limits $\lim_{t\to\infty} p_{i,j}(t) = p_{i,j}$ exist, where $p_{i,j}(t)$ are coordinates of the matrix $P(t) = \mathrm{e}^{tQ}$. Since the matrices involved here are finite-dimensional, it follows that (42.1) is fulfilled. If i and j belong to the same class of communicating states, $p_{i,j}$ in fact does not depend on i. More specifically, for $i \in \{n_k, \ldots, N_k\}$, the ith row of $P = (p_{i,j})_{i,j=1,\ldots,N}$ is composed of the coordinates of the eigenvector v_k, the fact that the process never leaves the kth class being expressed in all the coordinates $v_{k,l}$, $l \notin \{n_k, \ldots, N_k\}$, of v_k being equal zero. Finally, the remaining rows are convex combinations of the eigenvectors, say the ith row is:

$$\sum_{k=1}^{m} \alpha_{i,k} v_k, \qquad i > N_m. \tag{43.1}$$

(For example, in the case of the Markov chain depicted at Figure 43.1,

$$P = \begin{pmatrix} v_{1,1} & v_{1,2} & v_{1,3} & 0 & 0 & 0 & 0 & 0 \\ v_{1,1} & v_{1,2} & v_{1,3} & 0 & 0 & 0 & 0 & 0 \\ v_{1,1} & v_{1,2} & v_{1,3} & 0 & 0 & 0 & 0 & 0 \\ 0 & 0 & 0 & v_{2,4} & v_{2,5} & 0 & 0 & 0 \\ 0 & 0 & 0 & v_{2,4} & v_{2,5} & 0 & 0 & 0 \\ \alpha_{6,1}v_{1,1} & \alpha_{6,1}v_{1,2} & \alpha_{6,1}v_{1,3} & \alpha_{6,2}v_{2,4} & \alpha_{6,2}v_{2,5} & 0 & 0 & 0 \\ \alpha_{7,1}v_{1,1} & \alpha_{7,1}v_{1,2} & \alpha_{7,1}v_{1,3} & \alpha_{7,2}v_{2,4} & \alpha_{7,2}v_{2,5} & 0 & 0 & 0 \\ \alpha_{8,1}v_{1,1} & \alpha_{8,1}v_{1,2} & \alpha_{8,1}v_{1,3} & \alpha_{8,2}v_{2,4} & \alpha_{8,2}v_{2,5} & 0 & 0 & 0 \end{pmatrix},$$

where $v_{i,k}$ is the kth coordinate of v_i.)

In particular, the related operator P (mapping x to $x \cdot P$) is a projection on the kernel of Q, in agreement with (42.2). Writing:

$$P\,(\xi_i)_{i=1,\dots,N} = \sum_{j=1}^{m}\left(\sum_{i=n_j}^{N_j}\xi_i\right)\upsilon_j + \sum_{j=N_m+1}^{N}\xi_j\sum_{k=1}^{m}\alpha_{j,k}\upsilon_k,$$
$$= \sum_{j=1}^{m}\left(\sum_{i=n_j}^{N_j}\xi_i + \sum_{i=N_m+1}^{N}\xi_i\alpha_{i,j}\right)\upsilon_j,$$

we see that, as a map from $\mathbb{R}^n$ with standard basis to $\mathbb{X}_0$ with the basis $\{\upsilon_1, \dots, \upsilon_m\}$, P may be identified with the $N \times m$ matrix P^* given by (see also Exercise 43.1):

$$p^*_{i,j} = \begin{cases} 1, & i \in \{n_j, \dots, N_j\}, \\ 0, & i \in \{1, \dots, N_m\} \setminus \{n_j, \dots, N_j\}, \\ \alpha_{i,j}, & i > N_m. \end{cases}$$

The quantity $p^*_{i,k}$ is the probability that a chain starting at $i \in \{1, \dots, N\}$ will end up in the kth class, $k = 1, \dots, m$.

Returning to the limit behavior of the semigroups: by Exercise 7.5, the related regularity space is contained in the kernel of Q. Hence, we expect Kurtz's Theorem to work with $\mathbb{X}_0 = \mathbb{X}' = Ker\,Q$. In notations of Kurtz's Theorem, $(x, Ax) \in A_{\text{ex}}$ and $(x, Qx) \in B_{\text{ex}}$ for all $x \in \mathbb{X}_0$. Also, $(PA)_{|\mathbb{X}_0}$, being a bounded operator, generates a semigroup. Applying the theorem:

$$\lim_{n\to\infty} \mathrm{e}^{A_n t}x = \mathrm{e}^{tPA}Px, \qquad x \in \mathbb{R}^n, t > 0. \tag{43.2}$$

To interpret this result, we note that the transformation matrix for $(PA)_{|\mathbb{X}_0}$, in the basis $\{\upsilon_1, \upsilon_2, \dots, \upsilon_m\}$, is the (matrix) product of three matrices:

$$Q^* = C \cdot A \cdot P^*,$$

where C is the change of basis matrix, composed of m rows being the m eigenvectors (in the normal order). We claim that Q^* is a Kolmogorov matrix, so that the limit process is a Markov chain with m states. To this end, for $i = 1, \dots, N$, let $1_i = (1, \dots, 1) \in \mathbb{R}^i$. Then, since the coordinates in each row of P^* add up to 1, $P^* \cdot 1_m = 1_n$ and, since A is a Kolmogorov matrix, $Q \cdot 1_n = 0$. This implies that $Q^* \cdot 1_m = 0$, that is, that the coordinates in each row of Q^* add up to 1. To see that the off-diagonal entries in Q^* are non-negative, we calculate, recalling that the support of υ_i's is $\{n_i, \dots, N_i\}$, in self-explanatory notations:

$$q^*_{ij} = \sum_{k=1}^{M}\sum_{l=n_i}^{N_i}\upsilon_{i,l}a_{l,k}p^*_{k,j} = \sum_{k=n_j}^{N_j}\sum_{l=n_i}^{N_i}\upsilon_{i,l}a_{l,k} + \sum_{k=N_m+1}^{N}\sum_{l=n_i}^{N_i}\upsilon_{i,l}a_{l,k}\alpha_{k,j}. \tag{43.3}$$

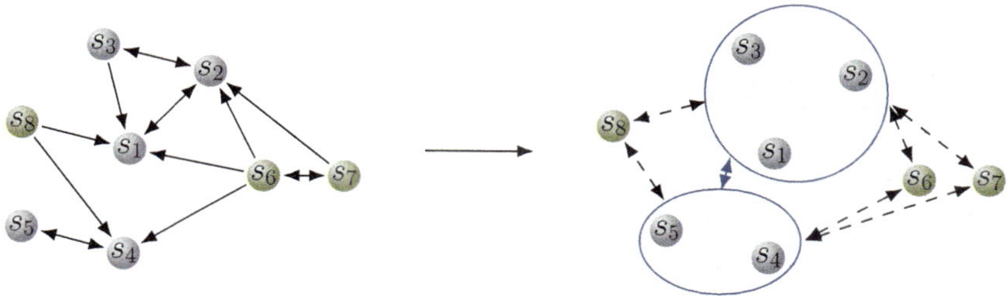

Figure 43.2 State-space lumping for a Markov chain example: two recurrent classes become two "aggregate" points of the limit Markov chain, communicating directly (blue arrows), or via transient states (which are not states of the limit chain; that is the reason for black arrows to be dashed).

If $i \neq j$, since $\{n_i, \ldots, N_j\} \cap \{n_i, \ldots, N_j\} = \emptyset$, the indexes k and l in the first sum on the right are never equal. Similarly, in the second sum we have $k > N_m \geq l$. Hence, for such i and j, the sum does not involve the diagonal elements of the matrix A, proving that the sum is non-negative, because matrices P^* and C are non-negative.

Probabilistically, when transitions in the chain Q are "very fast," the communicating states in each closed class become lumped together to form a single point (each point corresponds to one class, that is, to one basic eigenvector). Since the remaining, transient states are disregarded, the limit Markov chain's state-space has m points, which may be identified with m classes of Q. In itself, the Markov chain described by Q does not allow communication between these classes, but transitions are possible through the slow process related to A. As seen in (43.3), there are two ways for the process starting in class C_i to jump to class C_j. At Q-equilibrium, a particle starting in C_i will be at $l \in C_i$ with probability $v_{i,l}$, but the slow process may transfer it to one of the states $k \in C_j$ – the first sum on the right is the intensity of such a jump. The other way is via one of the transient states: from $l \in C_i$ the particle may be transferred to one of $k > N_m$ and then go to C_j forced by the sweeping Q-chain – the second sum is the intensity of such an indirect jump (see Figure 43.2). Hence, while Q lumps points in communicating classes, A allows transitions between the classes (interestingly, both A and Q influence transition rates).

A "biological" interpretation could read as follows. Imagine a colony of birds living on a number of islands, and think of the fast chain (Q) as representing transitions between various feeding grounds located at those islands, occurring on a daily basis. While on an island, birds may change feeding places but generally do not leave the island; hence, each island with various feeding grounds corresponds to a cluster of communicating states for Q. Transient states may then represent deserted islands, where there is not enough food, forcing birds to look for a better place to live. Also, think of A as describing "slow" migration processes, occurring on yearly basis; such transitions may be due to nesting, seasons, and so on. Then, (43.3) may be explained as follows: if the number of birds is large, their "fast" daily migrations between feeding grounds on each island do not change their "stationary" distribution, and, as remarked earlier, no transitions between the islands are possible on daily basis. However, on a yearly basis, such transitions are possible: in search for a companion, a nest, or for other reasons, a bird may be transferred from one island to another via a deserted place or directly, and – on the scale of years – intensities of such transitions are given by (43.3).

Yet another interpretation, a psychological one, is that the states of thoughtless happiness seem to be recurrent in humans, while the states of intellectual activity are apparently painful and hence transient. The vision of stationary distributions related to recurrent, closed classes is not really encouraging. We need another independent intensity matrix to lead us from such a stationary distribution to a transient, temporary state of mind called thinking, quickly to return to another stationary distribution. Nevertheless, sometimes in making such a forced jump from one closed set of preoccupations to another, we manage to avoid going through dangerous places described earlier.

A number of examples involving singularly perturbed Markov chains can be found in [352].

Exercise 43.1 Check that for the Markov chain depicted at Figure 43.1,

$$P^* = \begin{pmatrix} 1 & 0 \\ 1 & 0 \\ 1 & 0 \\ 0 & 1 \\ 0 & 1 \\ \alpha_{6,1} & \alpha_{6,2} \\ \alpha_{7,1} & \alpha_{7,2} \\ \alpha_{8,1} & \alpha_{8,2} \end{pmatrix}.$$

☞ Chapter's summary

We present one of the simplest of tons of applications of Kurtz's Singular Perturbation Theorem: a limit of singularly perturbed Markov chains. Applications are "birdy"-like [344].

44

A Tikhonov-Type Theorem

As another application of Kurtz's Singular Perturbation Theorem we show a Tikhonov-type theorem on dynamical systems with fast motion along chosen axes. Our approach is based on [61, 233].

We start with a description of semigroups related to systems of ordinary differential equations in an M-dimensional rectangle J, defined as:

$$J := \{\mathsf{x} \in \mathbb{R}^M; \mathsf{v} \leq \mathsf{x} \leq \mathsf{w}\},$$

where $\mathsf{v} = (v_i)_{i\in\mathcal{M}} \leq \mathsf{w} = (w_i)_{i\in\mathcal{M}}$ are vectors in $\mathbb{R}^M$, $\mathcal{M} := \{1, \dots, M\}$, and inequalities are understood coordinate-wise.

Let $\mathsf{a} = (a_j)_{j\in\mathcal{M}}$, be a vector of Lipschitz continuous functions $a_j : J \to \mathbb{R}$. The functions a_j may be extended to functions on $\mathbb{R}^M$ without increasing their Lipschitz constants (see [117] p. 189). Hence, for each extension, the initial value problem for the system of ODEs in $\mathbb{R}^M$:

$$\frac{\mathrm{d}\mathsf{x}(t)}{\mathrm{d}t} = \mathsf{a}(\mathsf{x}(t)), \qquad t \geq 0, \tag{44.1}$$

has a unique global solution. Moreover, the Müller conditions:

$$\begin{aligned} a_j(\mathsf{u}) &\geq 0 \quad \text{for} \quad \mathsf{u} \in J \text{ with } u_j = v_j, \\ a_j(\mathsf{u}) &\leq 0 \quad \text{for} \quad \mathsf{u} \in J \text{ with } u_j = w_j, \end{aligned} \tag{44.2}$$

imply that the trajectories of (44.1) starting in J remain there forever and do not depend on the extension chosen. This condition simply says that on the faces of J, the vector field a points toward the interior of J, forcing the trajectories to remain in J; see Chapter 36 or [337, 338].

By $s(t, \cdot)$ we denote the function mapping the initial value $\mathsf{x} \in J$ into the solution $\mathsf{x}(t)$ of (44.1) at time t; $\mathsf{x} \mapsto \mathsf{x}(t) = s(t, \mathsf{x})$. By the theorem on continuous dependence of solutions on initial conditions ([339], p. 145), $s(t, \mathsf{x})$ is continuous in x.

Let $C = C(J)$ be the space of continuous functions on J, and let C^1 denote the set of $f \in C$ that have continuous partial derivatives in J. Equivalently, $f \in C^1$ iff it may be extended to a continuously differentiable function on $\mathbb{R}^M$. Such an extension is not unique, but the partial derivatives of all extensions of f are the same in J.

Proposition 44.1 *The formula $S(t)f(\mathsf{x}) = f(s(t, \mathsf{x}))$ defines a c_0 semigroup of contractions in C. For the infinitesimal generator A of $\{S(t), t \geq 0\}$, we have $C^1 \subset \mathcal{D}(A)$ and:*

$$Af(\mathsf{x}) = \sum_{j=1}^{M} a_j(\mathsf{x}) \frac{\partial f}{\partial x_j}(\mathsf{x}), \qquad f \in C^1, \mathsf{x} \in J. \tag{44.3}$$

Moreover, C^1 is a core for A.

Lemma 44.2 *There exists a sequence $\{\mathsf{a}_n\}_{n\geq 1}$ of continuously differentiable functions $\mathsf{a}_n : J \to \mathbb{R}^M$ with the same Lipschitz constants, satisfying the Müller condition and such that* $\lim_{n\to\infty} \sup_{\mathsf{x}\in J} |\mathsf{a}(\mathsf{x}) - \mathsf{a}_n(\mathsf{x})|_{\mathbb{R}^M} = 0$.

Proof Rectangles J and $[0, 1]^M$ are diffeomorphic with diffeomorphism $\rho : \mathbb{R}^M \to \mathbb{R}^M$, defined by:

$$\rho\,(u_i)_{i=1,\dots,M} = \left(\frac{u_i - v_i}{w_i - v_i}\right)_{i=1,\dots,M},$$

mapping faces of J onto corresponding faces of $[0, 1]^M$. Hence, without loss of generality, we may assume that $J = [0, 1]^M$. Let a_n be the multivariate Bernstein polynomials [245]:

$$\mathsf{a}_n(\mathsf{x}) = \mathbb{E}\,\mathsf{a}\left(\frac{1}{n}X\right), \quad \mathsf{x} = (x_1, \dots, x_M) \in J, \quad n \geq 1, \tag{44.4}$$

where $X = (X_1, \dots, X_M)$ is a random vector with independent, binomially distributed coordinates; the probability of success for X_i being x_i and the number of trials being n. The Müller condition says that on the face where the ith coordinate of x is 0 (or 1, respectively) the ith coordinate of a is non-negative (or nonpositive, respectively). Since for x on such a face, $\frac{1}{n}X$ admits values from this face, a_n satisfy the Müller condition as well. Since it is well known that a_n converge uniformly to a (for the one-dimensional case, see e.g., [49]; the generalization to the multivariate case is straightforward), we are left with showing that they are Lipschitz continuous with the Lipschitz constant Λ. To this end we rephrase the proof from [74] (devoted to one-dimensional case) in probabilistic terms to make it suitable for the M-dimensional case we are interested in here.

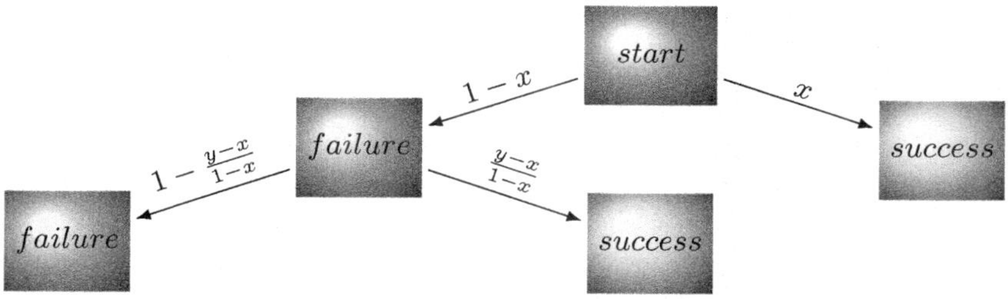

Figure 44.1 A Bernoulli trial with a chance to make up

Let $x \in [0, 1)$ and $y \in [x, 1]$ be given. Suppose that in a Bernoulli trial the probability of success is x, but in the case of failure we obtain a second, independent chance with probability of success equal $\frac{y-x}{1-x}$. Clearly, the resulting probability of success in such combined experiment is y (see Figure 44.1). Therefore, if X is a binomial variable with parameters n and x, and conditionally on $X = k \in \{0, \dots, n\}$, Z is binomial with parameters $n-k$ and $\frac{y-x}{1-x}$, then $X + Z$ is binomial with parameters n and y. (Z is the number of makeup trials.) Moreover, $\mathbb{E} Z = EY - EX = n(y-x)$.

Now, let X be as in (44.4) and Y be a similar random vector with parameters $\mathsf{y} \in J$. In estimating $|\mathbb{E}\,\mathsf{a}(X/n) - \mathbb{E}\,\mathsf{a}(Y/n)|$ we may replace Y_i by $X_i + Z_i$ if $x_i \le y_i$, and X_i by $Y_i + Z_i$ in the other case, where Z_i is as described earlier; in particular $\mathbb{E} Z_i = n|x_i - y_i|$. Hence, assuming $|\mathsf{a}(\mathsf{x}) - \mathsf{a}(\mathsf{y})| \le \Lambda \|\mathsf{x} - \mathsf{y}\| := \Lambda \sum_{i=1}^M |x_i - y_i|$, we obtain $|\mathbb{E}\,\mathsf{a}(X/n) - \mathbb{E}\,\mathsf{a}(Y/n)| \le \frac{\Lambda}{n} \sum_{i=1}^M \mathbb{E} Z_i = \Lambda \|\mathsf{x} - \mathsf{y}\|$.

Proof of Proposition 44.1 Without loss of generality, here we also assume $J = [0, 1]^M$. The Müller condition implies that $s(t, \mathsf{x}) \in J$, provided $\mathsf{x} \in J$, so $S(t)$ is an operator in C ($S(t)f \in C$ since $s(t, \cdot)$ is continuous). Since $s(t, \cdot), t \ge 0$, being related to an autonomous differential equation, is a semiflow, $\{S(t),\ t \ge 0\}$ is a semigroup. Moreover, using $\sup_{\mathsf{x}(t) \in J} |f(\mathsf{x}(t))| \le \sup_{\mathsf{x} \in J} |f(\mathsf{x})|$, we have $\|S(t)f\| \le \|f\|$. For the jth coordinate of $\mathsf{x}(t)$ we have $|x_j(t) - x_j| \le |\int_0^t a_j(\mathsf{x}(s))\,\mathrm{d}s| \le t\,\|a_j\|$, where $\|a_j\| = \sup_{\mathsf{x} \in J} |a_j(\mathsf{x})|$. It follows that $\mathsf{x}(t)$ tends to x uniformly in $\mathsf{x} \in J$ as $t \to 0$. Hence, f being uniformly continuous, $\{S(t), t \ge 0\}$ is a c_0 contraction semigroup in C.

For $f \in C^1$, let $\hat{f}$ be a continuously differentiable extension of f to $\mathbb{R}^M$. Fix $\mathsf{x} \in J$ and define $\varphi(t) := \hat{f}(\mathsf{x}(t))$. We have $\frac{S(t)f(\mathsf{x}) - f(\mathsf{x})}{t} = \frac{\varphi(t) - \varphi(0)}{t}$. By the Lagrange Theorem, for any $t \ge 0$ there exists $\theta \in (0, 1)$ such that:

$$\frac{\varphi(t) - \varphi(0)}{t} = \varphi'(\theta t) = \sum_{j=1}^M \frac{\partial \hat{f}}{\partial x_j}(\mathsf{x}(\theta t))\, a_j(\mathsf{x}(\theta t)). \tag{44.5}$$

The derivative of $\hat{f}$ is continuous, hence uniformly continuous when restricted to the compact set J. Thus, for $\mathsf{x} \in J$, the right-hand side of (44.5) tends as $t \to 0$ uniformly to the right-hand side in (44.3).

In order to prove that C^1 is a core for A, it suffices to show that the operator defined by the right-hand side of (44.3), say A_1, with domain equal to C^1, is closable and its closure generates a strongly continuous semigroup in C. To this end, by Proposition 3.7, p. 19 in [132], it suffices to show that there exists a norm $|\|\cdot\||$ on C^1, a constant $\omega > 0$, a sequence $\{\varepsilon_n\} \subset (0, \infty)$, $n \ge 1$, tending to 0, and a sequence $\{U_n(t), t \ge 0\}$, $n \ge 1$, of contraction semigroups in C, with generators G_n such that:

(a) $C^1 \subset \mathcal{D}(G_n)$,
(b) $\|A_1 f - G_n f\| \le \varepsilon_n |\|f\||$, for $f \in C^1$, and
(c) $U_n(t)C^1 \subset C^1$ and $|\|U_n(t)f\|| \le \mathrm{e}^{\omega t}|\|f\||$, for $t \ge 0,\ f \in C^1$.

We define $|\|f\|| = \|f\| + \sum_{j=1}^M \|\frac{\partial f}{\partial x_j}\|$, where $\|\cdot\|$ is the supremum norm in C.

For a_n defined in Lemma 44.2, let $\hat{\mathsf{a}}_n = (\hat{a}_{jn})_{j\in\mathcal{M}}$, be its continuously differentiable extension to $\mathbb{R}^M$. Since $\hat{\mathsf{a}}_n$ satisfy the Müller condition, the integral curves $\mathsf{x}_n(t)$ of (44.1) with a replaced by a_n, never leave J. Let $s_n(t,\cdot), t \ge 0$ be the functions mapping the initial value $\mathsf{x} \in J$ into the related solution $\mathsf{x}_n(t)$, and let $\{U_n(t), t \ge 0\}$ be the corresponding semigroups defined by $U_n(t)f(\mathsf{x}) = f(s_n(t,\mathsf{x}))$, with the generators G_n. By the first part of the proof, $C^1 \subset \mathcal{D}(G_n)$ and, for $f \in C^1$, G_n is given by the right-hand side of (44.3) with a replaced by a_n. Next, for $f \in C^1$ we have $A_1 f - G_n f = \sum_{j=1}^M (a_j - a_{jn})\frac{\partial f}{\partial x_j}$, hence (b) follows with $\varepsilon_n := \max_{j\in\mathcal{M}}\{\|a_j - a_{jn}\|\}$. For the proof of (c), let $f \in C^1$ and let $\hat{f}$ be its continuously differentiable extension to $\mathbb{R}^M$. Since $\hat{a}_{jn}$ are continuously differentiable, we have $s_n(t,\cdot) \in C^1$, $t \ge 0$, by the theorem on smooth dependence of solutions on the initial data (Chapter 1 in [196]). Hence, $U_n(t)f \in C^1$ for all $t \ge 0$, being restriction of a continuously differentiable function $\hat{f} \circ s_n(t,\cdot)$. We are left with proving that $|\|U_n(t)f\|| \le \mathrm{e}^{\omega t}|\|f\||$, for $t \ge 0,\ f \in C^1$. Defining $D_n(t) = (\frac{\partial s_{kn}(t,\cdot)}{\partial x_j})_{k,j\in\mathcal{M}}$, where s_{kn} are coordinates of s_n, we obtain by the chain rule $\nabla(U_n(t)f) = (\nabla f)D_n(t)$. By [196], p. 28, we have $\frac{\mathrm{d}}{\mathrm{d}t}D_n(t) = M_n D_n(t)$, where $M_n = (\frac{\partial a_{kn}}{\partial x_j})_{k,j\in\mathcal{M}}$ and $D_n(0)$ is the identity matrix, hence $D_n(t) = \exp(tM_n)$. Since $a_{kn} \in C^1$ are Lipschitz continuous with Lipschitz constants Λ, $\|\frac{\partial a_{kn}}{\partial x_j}\| \le \Lambda$ for $k, j \in \mathcal{M}$. Using $\nabla(U_n(t)f) = (\nabla f)\exp(tM_n)$ we obtain $|\|U_n(t)f\|| \le \mathrm{e}^{\omega t}|\|f\||$, for $\omega = M\Lambda$, $t \ge 0$ and $f \in C^1$. □

Armed with Proposition 44.1 we turn to studying convergence of semigroups related to fast motions along some axes. To this end, we choose a natural $K \in [0, M)$ and split a into two parts, $\mathsf{a} = \binom{\bar{\mathsf{a}}}{\underline{\mathsf{a}}}$, composed of its first K, and the remaining $M-K$ coordinates, respectively. The last $M-K$ coordinates are then multiplied by $\kappa_n > 0$, where $\lim_{n\to\infty}\kappa_n = \infty$. As a result, we consider the sequence of initial value problems:

$$\frac{\mathsf{d}\mathsf{y}_n(t)}{\mathsf{d}t} = \overline{\mathsf{a}}(\mathsf{y}_n(t),\mathsf{z}_n(t)), \qquad \mathsf{y}_n(0)=\tilde{\mathsf{y}},$$
$$\frac{\mathsf{d}\mathsf{z}_n(t)}{\mathsf{d}t} = \kappa_n\,\underline{\mathsf{a}}(\mathsf{y}_n(t),\mathsf{z}_n(t)), \quad \mathsf{z}_n(0)=\tilde{\mathsf{z}}; \tag{44.6}$$

we write $\mathsf{x}_n(t) = \binom{\mathsf{y}_n(t)}{\mathsf{z}_n(t)}$ for their solutions. Following Tikhonov [328, 332], we will show that these solutions converge to a certain limit. However, in contrast to Tikhonov, we want the limit to be uniform in initial data, and thus will assume more than he did. Namely, we will assume that problems (44.6) are of *uniformly Tikhonov-type*, that is, introduce the following Assumptions A and B.

Assumption A *For any* $\mathsf{y}\in J_\mathsf{y}$ *there exists a unique* $\varphi(\mathsf{y})\in J_\mathsf{z}$ *such that* $\underline{\mathsf{a}}(\mathsf{y},\varphi(\mathsf{y}))=0$, *where* φ *is Lipschitz continuous. For* $K=0$, *there is a unique* $\varphi\in J$ *solving* $\mathsf{a}(\varphi)=0$.

Here, and in what follows, J_z and J_y are multidimensional 'faces' of J defined as:

$$J_\mathsf{z} = \{\mathsf{z}\in\mathbb{R}^{M-K}|\ \text{there exist } \mathsf{y}\in\mathbb{R}^K \text{ such that } \begin{pmatrix}\mathsf{y}\\ \mathsf{z}\end{pmatrix}\in J\},$$
$$J_\mathsf{y} = \{\mathsf{y}\in\mathbb{R}^{K}|\ \text{there exist } \mathsf{z}\in\mathbb{R}^{M-K} \text{ such that } \begin{pmatrix}\mathsf{y}\\ \mathsf{z}\end{pmatrix}\in J\}.$$

The initial value problem for the *boundary layer system* corresponding to (44.6) is:

$$\frac{\mathsf{d}\mathsf{z}(t)}{\mathsf{d}t} = \underline{\mathsf{a}}(\tilde{\mathsf{y}},\mathsf{z}(t)), \quad \mathsf{z}(0)=\tilde{\mathsf{z}}, \quad t\ge 0; \tag{44.7}$$

$\tilde{\mathsf{y}}\in J_\mathsf{y}$ is treated as a fixed parameter. The solution to (44.7) will be denoted $\mathsf{z}(t,\tilde{\mathsf{y}},\tilde{\mathsf{z}})$. The point $\varphi(\tilde{\mathsf{y}})\in J_\mathsf{z}$ is called the *singular point* or the *equilibrium* of (44.7). The equilibrium is called *(Lyapunov) stable* if for every $\epsilon>0$ there exists $\delta(\epsilon)$ such that $|\mathsf{z}(t,\tilde{\mathsf{y}},\tilde{\mathsf{z}})-\varphi(\tilde{\mathsf{y}})|<\epsilon$ for $t>0$ as long as $|\tilde{\mathsf{z}}-\varphi(\tilde{\mathsf{y}})|<\delta(\epsilon)$.

The point $\varphi(\tilde{\mathsf{y}})$ is called *attractive* if there exists δ_0 such that:

$$\lim_{t\to\infty}\mathsf{z}(t,\tilde{\mathsf{y}},\tilde{\mathsf{z}}) = \varphi(\tilde{\mathsf{y}}), \tag{44.8}$$

provided $|\tilde{\mathsf{z}}-\varphi(\tilde{\mathsf{y}})|<\delta_0$. A stable equilibrium that is also attractive is called *asymptotically stable*. The initial value $\tilde{\mathsf{z}}$ for which (44.8) holds is said to lie in the *basin of attraction* of $\varphi(\tilde{\mathsf{y}})$.

Assumption B *For any* $\tilde{\mathsf{y}}\in J_\mathsf{y}$, *the point* $\varphi(\tilde{\mathsf{y}})$, *is an asymptotically stable equilibrium of (44.7) and the whole of* J_z *is a subset of its basin of attraction.*

The main result (Theorem 1) of [182] says that in such a case $\varphi(\tilde{\mathsf{y}})$ is stable uniformly with respect to $\tilde{\mathsf{y}}$:

B′ for each $\epsilon > 0$ there exists $\delta(\epsilon) > 0$ such that, for any $\tilde{\mathsf{y}}$, $|\tilde{\mathsf{z}} - \varphi(\tilde{\mathsf{y}})| < \delta(\epsilon)$ implies $|\mathsf{z}(t, \tilde{\mathsf{y}}, \tilde{\mathsf{z}}) - \varphi(\tilde{\mathsf{y}})| < \epsilon$ for all $t > 0$.

As already mentioned, our assumptions are stronger than those of Tikhonov [328, 332]. First of all, the ODEs we consider are autonomous and their solutions never leave the compact set J. Moreover, we require that the entire J lies in the domain of attraction of a unique singular point. Originally it was assumed that in a sufficiently small neighborhood of any singular point there were no other singular points.

Under our assumptions the solutions to problems (44.6) converge to $\binom{\mathsf{y}(t)}{\varphi(\mathsf{y}(t))}$ where:

$$\frac{\mathrm{d}\mathsf{y}(t)}{\mathrm{d}t} = \overline{\mathsf{a}}(\mathsf{y}(t), \varphi(\mathsf{y}(t))), \ \mathsf{y}(0) = \tilde{\mathsf{y}}. \tag{44.9}$$

The limit here is uniform in initial data $\tilde{\mathsf{y}}$ and $\tilde{\mathsf{z}}$ and almost uniform in $t \in (0, \infty)$. If $K = 0$, all equations in (44.6) are singularly perturbed and the y coordinate does not appear there. In this case, the limit here is a constant function equal to φ of Assumption A, and the result follows in fact by Lemma 44.3.

For the proof of our Tikhonov-type theorem we will need the following form of Gronwall's Lemma. Suppose $A > 0$ and $B \geq 0$ are constants, and v is a continuous, non-negative function on an interval $[t_0, t_0 + h]$. Then, the inequality:

$$v(t) \leq A \int_{t_0}^{t} v(s)\,\mathrm{d}s + B, \qquad t \in [t_0, t_0 + h] \tag{44.10}$$

implies:

$$v(t) \leq B\mathrm{e}^{A(t-t_0)}, \quad t \in [t_0, t_0 + h].$$

For the proof, it suffices to show inductively that (44.10) implies:

$$v(t) \leq A^n \int_{t_0}^{t} \frac{(t-s)^{n-1}}{(n-1)!} v(s)\,\mathrm{d}s + B \sum_{k=0}^{n-1} \frac{A^k (t-t_0)^k}{k!},$$

for $n \geq 1$ and $t \in [t_0, t_0 + h]$, and then pass with n to infinity.

Lemma 44.3 *We have:*

$$\lim_{t \to \infty} \mathsf{z}(t, \tilde{\mathsf{y}}, \tilde{\mathsf{z}}) = \varphi(\tilde{\mathsf{y}}), \tag{44.11}$$

uniformly in $\binom{\tilde{\mathsf{y}}}{\tilde{\mathsf{z}}} \in J$.

Proof We argue as in the proof of Theorem 7 in [252]. Suppose (44.11) is not uniform in $\binom{\tilde{\mathsf{y}}}{\tilde{\mathsf{z}}}$. Then, there is $\epsilon > 0$ such that for all $n \geq 1$ there exists $\binom{\tilde{\mathsf{y}}_n}{\tilde{\mathsf{z}}_n} \in J$ and $t_n \geq n$ with:

$$|\mathsf{z}(t_n, \tilde{\mathsf{y}}_n, \tilde{\mathsf{z}}_n) - \varphi(\tilde{\mathsf{y}}_n)| \geq \epsilon. \tag{44.12}$$

By compactness of J, without loss of generality we may assume that the sequence $\binom{\tilde{\mathsf{y}}_n}{\tilde{\mathsf{z}}_n}_{n\geq 1}$ converges to a $\binom{\tilde{\mathsf{y}}^*}{\tilde{\mathsf{z}}^*} \in J$. Since $\upsilon_n(t) := |\mathsf{z}(t, \tilde{\mathsf{y}}_n, \tilde{\mathsf{z}}_n) - \mathsf{z}(t, \tilde{\mathsf{y}}^*, \tilde{\mathsf{z}}^*)|$ satisfies:

$$\upsilon_n(t) \leq |\tilde{\mathsf{z}}_n - \tilde{\mathsf{z}}^*| + \int_0^t |\underline{\mathsf{a}}(\tilde{\mathsf{y}}_n, \mathsf{z}(s, \tilde{\mathsf{y}}_n, \tilde{\mathsf{z}}_n)) - \underline{\mathsf{a}}(\tilde{\mathsf{y}}^*, \mathsf{z}(s, \tilde{\mathsf{y}}^*, \tilde{\mathsf{z}}^*))| \,\mathrm{d}s,$$

and the integrand here can be estimated by $\underline{\Lambda}(|\tilde{\mathsf{y}}_n - \tilde{\mathsf{y}}^*| + \upsilon_n(s))$, where $\underline{\Lambda}$ is a Lipschitz constant of $\underline{\mathsf{a}}$, we have:

$$\upsilon_n(t) \leq |\tilde{\mathsf{z}}_n - \tilde{\mathsf{z}}^*| + \underline{\Lambda}(t|\tilde{\mathsf{y}}_n - \tilde{\mathsf{y}}^*| + \int_0^t \upsilon_n(s)\,\mathrm{d}s).$$

By Gronwall's Lemma, this gives:

$$\upsilon_n(t) \leq (|\tilde{\mathsf{z}}_n - \tilde{\mathsf{z}}^*| + \underline{\Lambda}t|\tilde{\mathsf{y}}_n - \tilde{\mathsf{y}}^*|)\mathrm{e}^{\underline{\Lambda}t}, \qquad \text{for } t > 0.$$

By Assumption B, there exists $t^* = t^*(\tilde{\mathsf{y}}^*)$ such that $|\mathsf{z}(t^*, \tilde{\mathsf{y}}^*, \tilde{\mathsf{z}}^*) - \varphi(\tilde{\mathsf{y}}^*)| < \delta(\epsilon)$ for $\delta(\epsilon)$ defined in $\mathbf{B}'$. Hence, by the estimate for $\upsilon_n(t^*)$, and continuity of φ, there exists $m > t^*$ such that $|\mathsf{z}(t^*, \tilde{\mathsf{y}}_m, \tilde{\mathsf{z}}_m) - \varphi(\tilde{\mathsf{y}}_m)| < \delta(\epsilon)$. Therefore, since (44.7) is autonomous, $\mathsf{z}(m, \tilde{\mathsf{y}}_m, \tilde{\mathsf{z}}_m) = \mathsf{z}(m - t^*, \tilde{\mathsf{y}}_m, \mathsf{z}(t^*, \tilde{\mathsf{y}}_m, \tilde{\mathsf{z}}_m))$ satisfies $|\mathsf{z}(m, \tilde{\mathsf{y}}_m, \tilde{\mathsf{z}}_m) - \varphi(\tilde{\mathsf{y}}_m)| < \epsilon$, contradicting (44.12). □

Let A_n and B be the generators of the semigroups related to (44.6) and (44.7), respectively:

$$\mathrm{e}^{tA_n} f(\tilde{\mathsf{y}}, \tilde{\mathsf{z}}) = f(\mathsf{x}_n(t)), \qquad \mathrm{e}^{tB} f(\tilde{\mathsf{y}}, \tilde{\mathsf{z}}) = f(\tilde{\mathsf{y}}, \mathsf{z}(t, \tilde{\mathsf{y}}, \tilde{\mathsf{z}})).$$

Then, Proposition 44.1 yields:

$$\begin{aligned} A_n f(\mathsf{x}) &= \sum_{j=1}^{K} a_j(\mathsf{x}) \frac{\partial f}{\partial x_j}(\mathsf{x}) + \kappa_n \sum_{j=K+1}^{M} a_j(\mathsf{x}) \frac{\partial f}{\partial x_j}(\mathsf{x}), \\ &=: Af(\mathsf{x}) + \kappa_n Bf(\mathsf{x}) \qquad f \in C^1,\ \mathsf{x} \in J. \end{aligned} \tag{44.13}$$

Lemma 44.3 says that $\lim_{t\to\infty} \mathrm{e}^{tB} = P$, where $Pf(\tilde{\mathsf{y}}, \tilde{\mathsf{z}}) = f(\tilde{\mathsf{y}}, \varphi(\tilde{\mathsf{y}}))$; P is a projection onto the subspace C_{y} of $f \in C$ which do not depend on z. Therefore, we are in the setup of Theorem 42.2 with Q replaced by B, ϵ_n replaced by κ_n^{-1} and $\mathbb{X}_0 = C_{\mathsf{y}}$. To see this, we check that for $f \in C^1 \cap C_{\mathsf{y}}$ we have $A_n f = Af$, and for $g \in C^1$, $\lim_{n\to\infty} \frac{1}{\kappa_n} A_n g = Bg$, proving that assumptions (a) and (b) of this theorem are satisfied. Introducing $S(t)f(\tilde{\mathsf{y}}, \tilde{\mathsf{z}}) = f(\mathsf{y}(t), \tilde{\mathsf{z}})$, where $\mathsf{y}(\cdot)$ solves (44.9), we see that these operators leave J_{y} invariant and form a strongly continuous semigroup there. Since the generator of this semigroup coincides with PA on $C^1 \cap C_{\mathsf{y}}$, and the latter set is its core, the generator equals $\overline{PA}$, and Kurtz's Theorem gives:

$$\lim_{n\to\infty} \mathrm{e}^{tA_n} f = \mathrm{e}^{t\overline{PA}} Pf, \qquad f \in C, t > 0. \tag{44.14}$$

Untangled, this formula reads:

$$\lim_{n\to\infty} f(\mathsf{y}_n(t), \mathsf{z}_n(t)) = f(\mathsf{y}(t), \varphi(\mathsf{y}(t))), \qquad f \in C.$$

Since for f we could in particular take functions assigning to $\mathsf{x} \in J$ its coordinates, this means $\lim_{n\to\infty} \binom{\mathsf{y}_n(t)}{\mathsf{z}_n(t)} = \binom{\mathsf{y}(t)}{\varphi(\mathsf{y}(t))}$, uniformly in initial data $\tilde{\mathsf{y}}$ and $\tilde{\mathsf{z}}$ and almost uniformly in $t \in (0, \infty)$. Additionally, $\mathsf{y}_n(t), n \geq 1$ converge almost uniformly in $t \in [0, \infty)$ since the corresponding coordinate functions belong to the regularity space. This is a **uniform version of the Tikhonov Theorem**; note that Example 26.6 is a particular case of this result.

As the following example shows, the classical assumptions of Tikhonov do not guarantee that the convergence is uniform in initial data. The solution to the Riccati Equation:

$$\frac{\mathrm{d}\mathsf{z}_n}{\mathrm{d}t} = \kappa_n(\mathsf{z}_n^2 - \mathsf{z}_n - 2), \quad \mathsf{z}_n(0) = \tilde{\mathsf{z}},$$

is $\mathsf{z}_n(t, \tilde{\mathsf{z}}) = 2 + \frac{3}{\frac{\tilde{\mathsf{z}}+1}{\tilde{\mathsf{z}}-2}\mathrm{e}^{-3\kappa_n t} - 1}$, for $\tilde{\mathsf{z}} < 2$, and $\mathsf{z}_n(t, 2) = 2$, and there are two singular points here: -1 and 2. As a result, even though:

$$\lim_{n\to\infty} \mathsf{z}_n(t, \tilde{\mathsf{z}}) = -1,$$

for $\tilde{\mathsf{z}} < 2$ and $t > 0$, the convergence is uniform in $\tilde{\mathsf{z}}$ in no interval containing both singular points, for $\sup_{\tilde{\mathsf{z}}\in[-1,2)} |\mathsf{z}_n(t, \tilde{\mathsf{z}}) + 1| = 3, t > 0$.

☞ Chapter's summary

As an application of Kurtz's Theorem we prove a version of Tikhonov's Theorem. The latter describes convergence of solutions to a system of ordinary differential equations in the case where the movement along some chosen axes is much faster than along the remaining ones. In passing (Proposition 44.1) we characterize generators of semigroups related to ordinary differential equations in rectangles.

45

Fast Motion and Frequent Jumps Theorems for Piecewise Deterministic Processes

Piecewise deterministic processes (or PDPs for short), main actors in this chapter, were introduced and studied by M. H. A. Davis [105, 104, 106], who prophesied in 1984: "These (processes) were only isolated rather recently but seen general enough to include as special cases practically all the non-diffusion continuous-time processes of applied probability" ([104], p. i). This was certainly an overstatement, as it is hard to overestimate the role of point processes in nowadays modeling [98]. Nevertheless, the number of models of applied probability involving PDPs is at least surprising.

In between jumps, a typical trajectory of such a process (surprise, surprise!) is an integral curve of one of many given vector fields. These trajectories may lie on manifolds of Euclidean spaces of various degrees, and so the state-space of the process may look like the one in Figure 45.1. A stochastic component

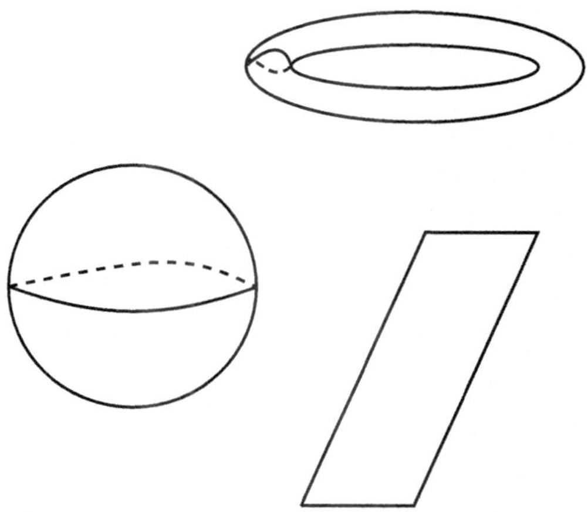

Figure 45.1 A state-space of a piecewise deterministic process

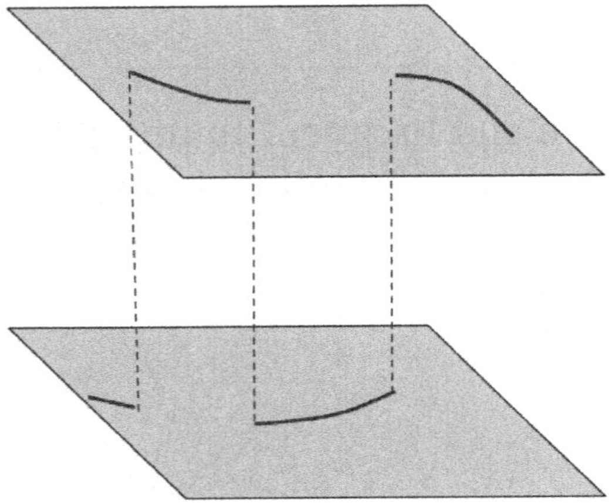

Figure 45.2 Trajectory of the process

in these processes is restricted to jumps between the manifolds. These occur at random times, and intensity of jumps varies from region to region. The latter characteristic distinguishes PDPs from **random evolutions** of Kac, Griego, and Hersh [168, 169], where the jumps occur at a fixed rate.

We will restrict our attention to PDPs with state-space equal to $N+1$ copies of the M-dimensional rectangle J described in the previous chapter, where N is a natural number: the case $N = 1$ is depicted in our Figure 45.2. Following [61], we will study limit behavior of these processes in two cases: that of frequent jumps and that of fast motion along chosen axes. In the first of these cases, the limit process is a deterministic motion along "averaged" curves on another copy of J formed of "lumped" copies $J \times \{i\}$, $i \in \mathcal{N} := \{1, \dots, N\}$ (see Figure 45.3). The other situation of interest is the one in which as in the Tikhonov Theorem, the motion along some axes is much faster than along the remaining ones. Then the limit process is again piecewise deterministic, but its state-space becomes more "flat," as depicted in Figure 45.4. Both theorems are motivated by biological models of the following chapters, and both will be seen to be consequences of Kurtz's Singular Perturbation Theorem.

Turning to the formal description of the first theorem, let:

$$\mathsf{a} = (a_{ij})_{j \in \mathcal{M},\, i \in \mathcal{N}}$$

be a matrix of Lipschitz continuous functions $a_{ij} : J \to \mathbb{R}$. By $\mathsf{a}_i = (a_{ij})_{j \in \mathcal{M}}$, $i \in \mathcal{N}$ we denote the M-dimensional column vectors of a. Suppose

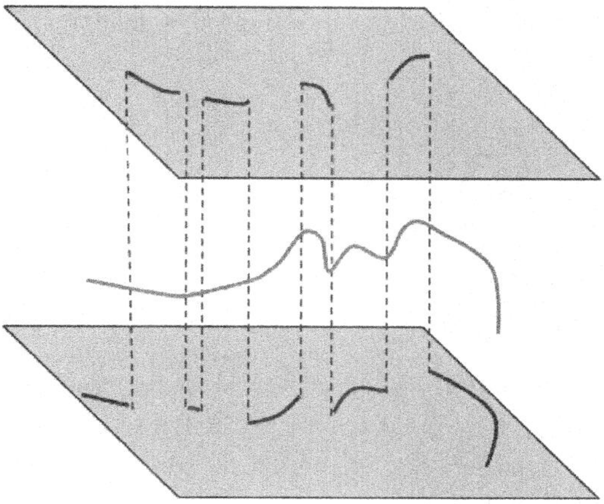

Figure 45.3 Frequent jumps approximation

also that each a_i satisfies the Müller condition (44.2). Then, for each $i \in \mathcal{N}$, the trajectories of the system:

$$\frac{d\mathsf{x}_i(t)}{dt} = \mathsf{a}_i(\mathsf{x}_i(t)), \qquad i \in \mathcal{N}, t \geq 0 \tag{45.1}$$

starting in J remain there forever; our processes move along these trajectories between jumps. Hence, the deterministic part of our processes is determined by the generators A_i of the related Feller semigroups $\{S_i(t),\ t \geq 0\}$ in $C(J)$, where $S_i(t)f(\mathsf{x}) = f(\mathsf{x}_i(t)),\ t \geq 0,\ \mathsf{x} \in J,\ i \in \mathcal{N}$, and $\mathsf{x}_i(t)$ is the solution of (45.1) starting from x, described in Proposition 44.1. Since the state-space of the process is composed of $N + 1$ copies of J, the Banach space $\mathbb{X}$ of continuous functions on the state-space is isometrically isomorphic to the Cartesian product of $N + 1$ copies of $C(J)$ equipped with the norm $\|(f_i)_{i\in\mathcal{N}}\| = \sup_{i\in\mathcal{N}} \|f_i\|$.

The random part of the processes, that is, the jumps, are determined by a sequence of $(N + 1) \times (N + 1)$ intensity matrices $\mathcal{Q}_n = (q_{ij,n})_{i,j\in\mathcal{N}}, n \geq 1$ with C-valued entries; the fact that jumps' intensities depend on the state of the processes is reflected in the fact that the entries depend on x. Inasmuch as a continuous function $h \in C = C(J)$ may be identified with the related multiplication operator $C \ni f \to hf \in C$, the $\mathcal{Q}_n$ may be identified with the bounded multiplication operators in $\mathbb{B}$ given by:

$$\mathcal{Q}_n\ (f_i)_{i\in\mathcal{N}} = \left(\sum_{j=0}^{N} q_{ij,n} f_j\right)_{i\in\mathcal{N}}. \tag{45.2}$$

Hence, our PDPs are related to the semigroups generated by the operators $\mathcal{A}_n$ in $\mathbb{B}$ given by:

$$\mathcal{A}_n\,(f_i)_{i\in\mathcal{N}} = (A_i f_i)_{i\in\mathcal{N}} + \kappa_n\,\mathcal{Q}_n(f_i)_{i=0}^N, \tag{45.3}$$

where $f_i \in \mathcal{D}(A_i)$ and the domain of $\mathcal{A}_n$, denoted $\mathcal{D}$, is the Cartesian product of $\mathcal{D}(A_i)$, $i \in \mathcal{N}$. The sequence $(\kappa_n)_{n\geq 0}$ appearing here is composed of non-negative numbers with $\kappa_0 = 0$, $\kappa_n > 0, n \geq 1$, and $\lim_{n\to\infty}\kappa_n = \infty$; the latter assumption being the reflection of the fact that we want to describe the frequent jumps limit.

Of course, we still need to show that each $\mathcal{A}_n$ generates a Feller semigroups. To this end, note that $\mathcal{A}_0$ is the generator of the Feller semigroup $\{e^{t\mathcal{A}_0}, t \geq 0\}$ given by $e^{t\mathcal{A}_0}(f_i)_{i\in\mathcal{N}} = (e^{tA_i} f_i)_{i\in\mathcal{N}}$; this is the semigroup related to motions along integral curves of (45.1) on each copy of J separately, with no communication between the copies. Since $\mathcal{Q}_n$ is bounded, the Phillips Perturbation Theorem implies that $\mathcal{A}_n = \mathcal{A}_0 + \kappa_n\mathcal{Q}_n$ generates a strongly continuous semigroup. However, $\mathcal{Q}_n$ generates a Feller semigroups too. Hence, by the Trotter Product Formula, so does $\mathcal{A}_n$. Moreover, the same formula makes it clear that our PDPs are "mixtures" of motions along trajectories of (45.1) and Markov chains governed by $\kappa_n\mathcal{Q}_n$. In these Markov chains, a point $(p, i) \in J \times \{i\}$ jumps between its own copies (p, j), $j \in \mathcal{N}$ with intensity matrix $\kappa_n(q_{ij,n}(p))_{i,j\in\mathcal{N}}$.

We assume that the strong (entry-wise) limit $\lim_{n\to\infty}\mathcal{Q}_n =: \mathcal{Q}$ exists and that for $\mathsf{x} \in J$ there exists a unique (up to a constant) left row-eigenvector of $\mathcal{Q}(\mathsf{x})$ corresponding to 0 eigenvalue. (Since $\mathcal{Q}(\mathbf{x})$, $\mathbf{x} \in J$ are intensity matrices, the existence of a unique eigenvector corresponding to zero eigenvalue $\nu_0 = 0$ implies that algebraic multiplicity of ν_0 is 1; see [262], p. 696.) Then there exists the unique, non-negative left eigenvector with the sum of entries equal to 1, denoted $\mathbf{l}_0(\mathsf{x})$, and we assume that:

$$\lim_{t\to\infty} e^{t\mathcal{Q}}\mathbf{f} = \mathcal{P}_0\mathbf{f}, \qquad \mathbf{f} \in \mathbb{B}, \tag{45.4}$$

where:

$$\mathcal{P}_0\mathbf{f} = [\mathbf{l}_0\mathbf{f}] \in \mathbb{B}, \qquad \mathbf{f} \in \mathbb{B}, \qquad [f] = (f)_{i\in\mathcal{N}} \in \mathbb{B},\ f \in C,$$

and $\mathbf{l}_0\mathbf{f}(\mathsf{x})$ is the scalar product of $\mathbf{l}_0(\mathsf{x})$ and $\mathbf{f}(\mathsf{x})$, $\mathsf{x} \in J$. (We stress that some coordinates of this vector may vanish.) Under this assumption we will show that if the entries $q_{ij} \in C$ of $\mathcal{Q} = (q_{ij})_{i,j\in\mathcal{N}}$ are Lipschitz continuous, then so is $\mathsf{x} \mapsto \mathbf{l}_0(\mathsf{x})$ and the limit semigroup is related to the following system of ODEs:

$$\frac{\mathrm{d}\,\mathsf{x}(t)}{\mathrm{d}t} = \mathsf{a}\mathbf{l}_0(\mathsf{x}(t)), \tag{45.5}$$

where $\mathsf{a}\mathbf{l}_0$ is the product of the matrix a and the vector $\mathbf{l}_0$, now treated as a column. Clearly, $\mathsf{a}\mathbf{l}_0$ is Lipschitz continuous on J and since a_i's satisfy the Müller condition (44.2) so do the entries of $\mathsf{a}\mathbf{l}_0$, being convex combinations of a_i's. Thus, the trajectories of (45.5) starting in J remain there forever. By Proposition 44.1, the generator B of the related semigroup $\{S(t), t \geq 0\}$ in C is given by $Bf =< \mathsf{a}\mathbf{l}_0, \nabla f >, f \in C^1$, and C^1 is a core for B. By (45.3), we obtain:

$$Bf =< \mathsf{a}\mathbf{l}_0, \nabla f >= \mathbf{l}_0 \mathcal{A}_0[f] \quad \text{and} \quad [Bf] = \mathcal{P}_0 \mathcal{A}_0[f], \quad f \in C^1. \tag{45.6}$$

Hence, the claim made earlier may be expressed in the following frequent jumps theorem, promised previously.

Theorem 45.1 *Suppose that $\mathcal{Q}_n$ converges strongly to the $\mathcal{Q}$ with Lipschitz continuous entries and* (45.4) *holds. Then, writing $\{\mathrm{e}^{t\mathcal{A}_n}, t \geq 0\}$ for the semigroups generated by $\mathcal{A}_n = \mathcal{A}_0 + \kappa_n \mathcal{Q}_n$, we have:*

$$\lim_{n\to\infty} \mathrm{e}^{t\mathcal{A}_n}\mathbf{f} = \mathcal{S}(t)\mathcal{P}_0\mathbf{f}, \qquad \mathbf{f} \in \mathbb{B}. \tag{45.7}$$

Here $\{\mathcal{S}(t), t \geq 0\}$ given by $\mathcal{S}(t)[f] = [S(t)f]$ is an isometrically isomorphic copy of $\{S(t), t \geq 0\}$ in:

$$\mathbb{B}_0 = \{[f] \in \mathbb{B}; f \in C\}.$$

As always, the limit here is almost uniform in $t \in [0, \infty)$ for $\mathbf{f}$ in the regularity space $\mathbb{B}_0$ and almost uniform in $t \in (0, \infty)$ for the remaining $\mathbf{f}$. What is interesting here is that the limit does not depend on the rate with which $\mathcal{Q}_n$ tends to $\mathcal{Q}$. If $\mathcal{Q}_n(\mathsf{x})$ are not intensity matrices, the result is no longer valid, as the following example shows. Suppose $\mathcal{A}_0$ and $\mathcal{Q}$ are as in the theorem, and $\kappa_n = n$, so that the limit $\lim_{n\to\infty} \mathrm{e}^{t(\mathcal{A}_0+n\mathcal{Q})}\mathbf{f} = \mathcal{S}(t)\mathcal{P}_0\mathbf{f}$ exists. Then, for $\mathcal{Q}_n = \mathcal{Q} - \frac{1}{n}I$ where I is the identity in $\mathbb{B}$, we have $\lim_{n\to\infty} \mathrm{e}^{t(\mathcal{A}_0+n\mathcal{Q}_n)}\mathbf{f} = \lim_{n\to\infty} \mathrm{e}^{-t}\mathrm{e}^{t(\mathcal{A}_0+n\mathcal{Q})}\mathbf{f} = \mathrm{e}^{-t}\mathcal{S}(t)\mathcal{P}_0\mathbf{f}$. At the same time for $\mathcal{Q}_n = \mathcal{Q} - \frac{1}{\sqrt{n}}I$ we have $\lim_{n\to\infty} \mathrm{e}^{t(\mathcal{A}_0+n\mathcal{Q}_n)}\mathbf{f} = \lim_{n\to\infty} \mathrm{e}^{-\sqrt{n}t}\mathrm{e}^{t(\mathcal{A}_0+n\mathcal{Q})}\mathbf{f} = 0, t > 0$.

Postponing temporarily the proof of Theorem 45.1, let us turn to the second main subject of this chapter. Here, we are interested in the situation where the motion along the last $M - K > 0$ axes in our PDP process is relatively faster than that along the remaining ones, where $0 \leq K < M$. More specifically, as in (44.6), systems (45.1) are replaced by the singularly perturbed ones:

$$\frac{\mathrm{d}\mathsf{y}_{i,n}(t)}{\mathrm{d}t} = \overline{\mathsf{a}}_i(\mathsf{y}_{i,n}(t), \mathsf{z}_{i,n}(t)), \qquad \frac{\mathrm{d}\mathsf{z}_{i,n}(t)}{\mathrm{d}t} = \kappa_n\, \underline{\mathsf{a}}_i(\mathsf{y}_{i,n}(t), \mathsf{z}_{i,n}(t)),\ t \geq 0, \tag{45.8}$$

$\mathsf{y}_{i,n} = (x_{i,j,n})^{\mathsf{T}}_{j=1,\dots,K}$, $\mathsf{z}_{i,n} = (x_{i,j,n})^{\mathsf{T}}_{j=K+1,\dots,M}$, $\overline{\mathsf{a}}_i(\cdot) := (a_{ij}(\cdot))^{\mathsf{T}}_{j=1,\dots,K}$ and $\underline{\mathsf{a}}_i(\cdot) := (a_{ij}(\cdot))^{\mathsf{T}}_{j=K+1,\dots,M}$, and κ_n is a sequence of non-negative numbers

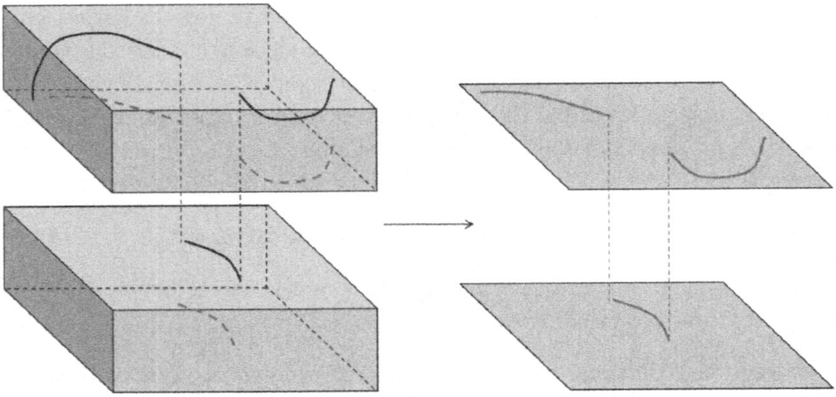

Figure 45.4 Fast deterministic motion approximation

such that $\kappa_0 = 0$ and $\kappa_n \to \infty$. We assume that $a_{ij} : J \mapsto \mathbb{R}$ are Lipschitz continuous in J, the Müller condition (44.2) is satisfied, and the systems are of uniformly Tikhonov type (see Chapter 44).

Hence, the generators of the approximating semigroups are of the form:

$$\mathcal{A}_n + \mathcal{Q}_n,$$

where $\mathcal{Q}_n$ are intensity matrices converging strongly to an intensity matrix $\mathcal{Q}$, and $\mathcal{A}_n(f_i)_{i\in\mathcal{N}} = (A_{n,i}f_i)_{i\in\mathcal{N}}$, for $f_i \in \mathcal{D}(A_{n,i})$ where $A_{n,i}$ are given by (44.13) with a_j replaced by a_{ij}.

Our fast motion theorem says that as $n \to \infty$, the piecewise deterministic process related to $\mathcal{A}_n + \mathcal{Q}_n$ in $\mathbb{B}$, moving along integral curves of (45.8) with jumps given by $\mathcal{Q}_n$, becomes another piecewise deterministic process, with the first coordinates $\mathsf{y}_i(t)$ of the trajectories between jumps being the integral curves of the reduced and modified systems:

$$\frac{\mathrm{d}\mathsf{y}_i(t)}{\mathrm{d}t} = \overline{\mathsf{a}}_i(\mathsf{y}_i(t), \varphi_i(\mathsf{y}_i(t))), \qquad t \geq 0,\ i \in \mathcal{N} \tag{45.9}$$

in J_{y}, and the second coordinates $\mathsf{z}_i(t)$ changing according to the formula $\mathsf{z}_i(t) = \varphi_i(\mathsf{y}_i(t))$, where $\varphi_i(\mathsf{y})$ is the unique vector satisfying:

$$\underline{\mathsf{a}}_i(\mathsf{y}, \varphi_i(\mathsf{y})) = 0.$$

Since z_i is determined by y_i, the latter process may be identified with that moving along the integral curves of (45.9). As a result, the state-space of the limit piecewise deterministic process collapses to $N + 1$ copies of J_{y}, as shown in Figure 45.4, and its jump intensities are governed by an appropriately modified matrix $\mathcal{PQ}$ (see below). If $K = 0$, all equations in (45.8) are singularly perturbed and the y coordinate does not appear there. By Assumption A, there are unique $\varphi_i \in J$ solving $\mathsf{a}_i(\varphi_i) = 0$, $i \in \mathcal{N}$, and the limit process is a Markov chain with states φ_i and intensity matrix $\mathcal{PQ} = (q_{ij}(\tilde{\varphi}_i))_{i,j\in\mathcal{N}}$.

Formally, let $\mathbb{B}_{\mathsf{y}} = (C_{\mathsf{y}})^{N+1}$ be the $(N+1)$-fold Cartesian product of C_{y} and let $\mathcal{P} : \mathbb{B} \mapsto \mathbb{B}_{\mathsf{y}}$ be defined by $\mathcal{P}(f_i)_{i\in\mathcal{N}} = (P_i f_i)_{i\in\mathcal{N}}$, where $P_i f(\mathsf{y}) = f(\mathsf{y}, \varphi_i(\mathsf{y}))$. Also, let $\overline{P_i A_i}$ be the generator of the semigroup of motion along curves of (45.9) (see Chapter 44) and let $\mathcal{A}_0$ be the generator of the Cartesian product of the semigroups generated by the operators $\overline{P_i A_i}$: $\mathrm{e}^{t\mathcal{A}_0}(f_i)_{i\in\mathcal{N}} = (\mathrm{e}^{t\overline{P_i A_i}} f_i)_{i\in\mathcal{N}}$.

Theorem 45.2 *We have:*

$$\lim_{n\to\infty} \mathrm{e}^{t(\mathcal{A}_n+\mathcal{Q}_n)}\mathbf{f} = \mathrm{e}^{t(\mathcal{A}_0+\mathcal{PQ})}\mathcal{P}\mathbf{f}, \qquad \mathbf{f} \in \mathbb{B},\ t > 0. \tag{45.10}$$

As before, for $\mathbf{f} \in \mathbb{B}_{\mathsf{y}}$ the limit is almost uniform in $t \in [0, \infty)$, and in general it is almost uniform in $t \in (0, \infty)$.

The proof of our theorem is immediate once we are armed with the Tikhonov-type theorem of Chapter 44, which in our case yields, by definition of $\mathcal{A}_0$, $\lim_{n\to\infty} \mathrm{e}^{tA_n} = \mathrm{e}^{tA_0}\mathcal{P}$. Hence, Theorem 45.2 follows because convergence is preserved under bounded perturbations and (45.10) is a special case of (29.3).

We are thus left with showing Theorem 45.1. We start by showing that under our assumptions the map $\mathsf{x} \mapsto \mathbf{l}_0(\mathsf{x})$ assigning to x a left eigenvector of $\mathcal{Q}(\mathsf{x})$ may be chosen to be Lipschitz-continuous. We argue as in Theorem 8 in [236], p. 130. Fix $\mathsf{x}_0 \in J$. Since we assume that $\mathbf{l}_0(\mathsf{x}_0)$ is unique (up to a constant), one of the principal minors of $\mathcal{Q}(\mathsf{x}_0)$, say the jth, has nonzero determinant (see Lemma 9 in [236]). Therefore the jth entry in $\mathbf{l}_0(\mathsf{x}_0)$ is nonzero, for otherwise the entire $\mathbf{l}_0(\mathsf{x}_0)$ would need to be zero, and without loss of generality we may assume it equals 1. By continuity, we may assume the same is true for x in a neighborhood of x_0. Then, denoting by $\mathbf{l}_{0,*}(\mathsf{x})$ the vector $\mathbf{l}_0(\mathsf{x})$ with the jth entry removed, we obtain:

$$\mathbf{l}_{i,*}(\mathsf{x})\mathcal{M}_j(\mathsf{x}) = -\mathbf{c}_j(\mathsf{x})$$

or:

$$\mathbf{l}_{i,*}(\mathsf{x}) = -\mathbf{c}_j(\mathsf{x})[\mathcal{M}_j(\mathsf{x})]^{-1},$$

where $\mathcal{M}_j(\mathsf{x})$ and $\mathbf{c}_j(\mathsf{x})$ are the jth principal minor and the jth row of $\mathcal{Q}(\mathsf{x})$ (with the jth entry removed), respectively. With this definition, $\mathsf{x} \mapsto \mathbf{l}_0(\mathsf{x})$ is locally Lipschitz continuous, since the entries of $\mathcal{Q}$ are Lipschitz continuous. Then the classic result says that this map is Lipschitz continuous on the whole of the compact set J (see, e.g., [339], p. 107 or Exercise 45.1), proving our claim.

Next, we fix x, write a nonzero left eigenvector as $\mathbf{l}_0(\mathsf{x}) = (l_i)_{i\in\mathcal{N}}$, and claim that $(|l_i|)_{i\in\mathcal{N}}$ is a left eigenvector of $\mathcal{Q}(\mathsf{x})$, as well. To show this, we argue as on p. 310 in [305]: for simplicity we write q_{ij} for $q_{ij}(\mathsf{x})$. By assumption for all $j \in \mathcal{N}$, $l_j q_j = \sum_{i\neq j} l_i q_{ij}$, where $q_j = -q_{jj}$. Hence, $|l_j|q_j \le \sum_{i\neq j} |l_i|q_{ij}$. If contrary to our claim one of these inequalities is strict, then:

$$\sum_{j\in\mathcal{N}} |l_j|q_j < \sum_{j\in\mathcal{N}} \sum_{i\neq j} |l_i|q_{ij} = \sum_{i\in\mathcal{N}} |l_i| \sum_{j\neq i} q_{ij} = \sum_{i\in\mathcal{N}} |l_i|q_i,$$

a contradiction.

Finally, by continuity and since $\mathbf{l}_0(\mathsf{x})$ is nonzero, the map assigning the sum of entries of $\mathbf{l}_0(\mathsf{x})$ to an $\mathsf{x} \in J$ is bounded away from 0. Dividing $\mathbf{l}_0(\mathsf{x})$ by the sum of its entries if necessary, we obtain that $\mathbf{l}_0(\mathsf{x})$ is the unique left eigenvector with non-negative entries summing to 1. Since taking the absolute value of coordinates and dividing by the sum of coordinates do not destroy Lipschitz continuity, we conclude that the map assigning to x the unique left probability eigenvector of $\mathcal{Q}(\mathsf{x})$ corresponding to 0 eigenvalue, is Lipschitz continuous. Without this information, we could not claim that the Cauchy problem related to the system (45.5) is well posed.

Proof of Theorem 45.1. Note that we are in the setup of Theorem 42.2 with $\mathbb{X}, A_n, Q$ and P replaced by $\mathbb{B}, \mathcal{A}_n, \mathcal{Q}$ and $\mathcal{P}_0$, respectively. Condition (a) in Theorem 42.2 is clearly satisfied for $D = \{[f], f \in C^1\}$, since $[f] \in \mathcal{D}(\mathcal{A}_0)$ and $\mathcal{Q}_n[f] = 0$. Moreover, for $\mathbf{f} \in D(\mathcal{A}_0)$, we have $\lim_{n\to\infty} \frac{1}{\kappa_n}\mathcal{A}_n\mathbf{f} = \mathcal{Q}\mathbf{f}$, proving (b) with $D' = D(\mathcal{A}_0)$. The set $D(\mathcal{A}_0)$, being dense in $\mathbb{B}$, is a core for $\mathcal{Q}$ since $\mathcal{Q}$ is bounded. Using (45.6) we check that condition (c) is also satisfied. This shows (45.7). □

The following proposition allows checking assumption (45.4) in practice.

Proposition 45.3 *Assume that:*

(DE) for each $\mathsf{x} \in J$, $\mathcal{Q}(\mathsf{x})$ has $N+1$ real distinct eigenvalues $\nu_i(\mathsf{x})$, $i \in \mathcal{N}$, or

(SG) for each $\mathsf{x} \in J$, $\mathcal{Q}(\mathsf{x})$ is irreducible and its spectral gap is positive.

Then (45.4) *holds.*

Matrices:

$$Q = \begin{pmatrix} -1 & 1 & 0 \\ 0 & -1 & 1 \\ 0 & 2 & -2 \end{pmatrix} \text{ and } Q = \begin{pmatrix} -2 & 1 & 1 \\ 1 & -2 & 1 \\ 1 & 1 & -2 \end{pmatrix}$$

show that it may happen that (DE) is satisfied while (SQ) is not, and vice versa.

For the proof of the proposition, assume first (DE) holds. Since $\mathcal{Q}(\mathsf{x})$ is an intensity matrix, one of the eigenvalues is zero and the others are negative. Let $\nu_i(\mathsf{x})$ be the eigenvalue of $\mathcal{Q}(\mathsf{x})$ with the ith smallest absolute value. Then, ν_i are continuous functions of x ([236], Thm. 6, p. 130). A straightforward modification of the proof of Theorem 8, p. 130 in [236] shows that there are continuous functions $\mathbf{l}_i$ such that $\mathbf{l}_i(\mathsf{x})$ is a left eigenvector corresponding to $\nu_i(\mathsf{x})$. Also, as we have already seen, $\mathbf{l}_0$ may be chosen to be a probability vector (but does not need to be strictly positive), and $\mathsf{x} \mapsto \mathbf{l}_0(\mathsf{x})$ is Lipschitz continuous.

Let $\mathcal{L}(\mathsf{x})$ be the $(N+1)\times(N+1)$ matrix with the ith row equal to $\mathbf{l}_i(\mathsf{x})$. Since the rows form a basis of $\mathbb{R}^{N+1}$, $\mathcal{L}(\mathsf{x})$ is invertible. The columns of its inverse $\mathcal{R}(\mathsf{x})$ are the right eigenvectors $\mathbf{r}_i(\mathsf{x})$ of $\mathcal{Q}(\mathsf{x})$. This implies that $\mathbf{r}_i$ are continuous, because taking an inverse of an invertible matrix is a continuous map. Because of the choice of $\mathbf{l}_0$, entries of $\mathbf{r}_0$ equal 1.

As it was the case with $\mathcal{Q}$, matrices $\mathcal{R}$ and $\mathcal{L}$ may be identified with bounded multiplication operators in $\mathbb{B}$. Clearly, $\mathcal{L}$ is the inverse of $\mathcal{R}$.

For $\mathbf{f} = (f_i)_{n\ge 1} \in \mathbb{B}$ and $g \in C$ let $g\mathbf{f} \in \mathbb{B}$ denote $(gf_i)_{n\ge 1}$.

Lemma 45.4 *$\mathbb{B}$ may be represented as the direct sum $\mathbb{B} = \bigoplus_{i=0}^N \mathbb{B}_i$, where:*

$$\mathbb{B}_i := \mathrm{Ker}\,(\mathcal{Q} - \nu_i) = \{\mathbf{f} \in \mathbb{B}; \mathbf{f} = h\mathbf{r}_i, h \in C\}, \qquad i \in \mathcal{N}, \tag{45.11}$$

and ν_i is seen as a bounded operator in $\mathbb{B}$ mapping $\mathbf{f}$ to $\nu_i\mathbf{f}$.

Proof The key to the proof is the fact that $\mathcal{Q}$ commutes with the operators $g : \mathbf{f} \mapsto g\mathbf{f}$; the argument is quite the same as in the elementary case where $\mathcal{Q}$ does not depend on x.

For $\mathbf{f} \in \mathbb{B}$ there exist a unique $\mathbf{h} = (h_i)_{n\ge 1} \in \mathbb{B}$ such that $\mathbf{f} = \sum_{i=0}^N h_i\mathbf{r}_i$. To see this, let $\mathbf{h} = \mathcal{L}\mathbf{f}$. Then, $\mathbf{f} = \mathcal{R}\mathbf{h} = \sum_{i=0}^N h_i\mathbf{r}_i$. On the other hand, if $\mathbf{f} = \sum_{i=0}^N h_i'\mathbf{r}_i$ for $\mathbf{h}' = (h_i')_{n\ge 1}$, then $0 = \sum_{i=0}^N (h_i - h_i')\mathbf{r}_i = \mathcal{R}(\mathbf{h} - \mathbf{h}')$ implying $\mathbf{h} = \mathbf{h}'$ since $\mathcal{R}$ is invertible.

Since $\mathbb{B}_i$s are clearly closed, it remains to show (45.11). The vectors of the form $h\mathbf{r}_i$, $h \in C$ belong to $\mathrm{Ker}\,(\mathcal{Q} - \nu_i)$ because $\mathcal{Q}$ commutes with the operator $h : \mathbf{f} \mapsto h\mathbf{f}$. Conversely, suppose $0 \ne \mathbf{f} = \sum_{j=0}^N h_j\mathbf{r}_j \in \mathbb{B}_i$ and for some $j_0 \ne i$ and $\mathsf{x}_0 \in J$, $h_{j_0}(\mathsf{x}_0) \ne 0$. Then, $\nu_i\mathbf{f} = \mathcal{Q}\mathbf{f} = \mathcal{Q}(\sum_{j=0}^N h_j\mathbf{r}_j) = \sum_{j=0}^N h_j\nu_j\mathbf{r}_j$ implying $\sum_{j=0}^N h_j(\nu_j - \nu_i)\mathbf{r}_j = 0$. Therefore, for all j, $h_j(\nu_j - \nu_i) = 0$, and $\nu_{j_0}(\mathsf{x}_0) = \nu_i(\mathsf{x}_0)$. This contradicts (DE). □

Multiplying the column vector $\mathbf{r}_i$ by the row vector $\mathbf{l}_i$ we obtain a matrix $\mathcal{P}_i = \mathbf{r}_i\mathbf{l}_i$; this definition agrees with the previously introduced definition of $\mathcal{P}_0$. Seen as operators in $\mathbb{B}$, $\mathcal{P}_i$ are idempotent since $\mathbf{l}_i\mathbf{r}_i = \mathcal{I}$. Since $\mathbf{r}_i\mathbf{l}_i(\sum_{j=0}^N h_j\mathbf{r}^j) = h_i\mathbf{r}_i$, $\mathcal{P}_i$ is the projection on $\mathbb{B}_i$ with the kernel $\bigoplus_{j\ne i}^N \mathbb{B}_j$ (comp. Thms. 3.2.11 and 3.2.14 in [261]) and we have $\mathcal{I} = \sum_{i=0}^N \mathcal{P}_i$. Moreover,

$\mathcal{Q}\mathbf{f} = \mathcal{Q}(\sum_{i=0}^N h_i\mathbf{r}_i) = \sum_{i=0}^N h_i\mathcal{Q}\mathbf{r}_i = \sum_{i=0}^N \nu_i h_i\mathbf{r}_i = \sum_{i=1}^N \nu_i\mathcal{P}_i\mathbf{f}$, $\mathbf{f} \in \mathbb{B}$, that is, $\mathcal{Q} = \sum_{i=1}^N \nu_i\mathcal{P}_i$. This gives immediately, $\mathcal{Q}^n = \sum_{i=1}^N \nu_i^n\mathcal{P}_i$, $n \geq 1$ and:

$$\mathrm{e}^{t\mathcal{Q}} = \mathcal{P}_0 + \sum_{i=1}^N \mathrm{e}^{t\nu_i}\mathcal{P}_i. \tag{45.12}$$

Since ν_i, $i \in \{1, \dots, N\}$ are negative, we see that (DE) implies (45.4).

Next assume (SG) holds. More specifically, we assume that for all $\mathsf{x} \in J$, $\mathcal{Q}(\mathsf{x})$ is nonzero, and for some $k \geq 1$, the matrix $(\gamma^{-1}\mathcal{Q} + I)^k$, where $\gamma = \gamma(\mathsf{x}) = \max_i(-q_{ii}(\mathsf{x}))$, has all entries positive. By the Perron Theorem [236, 305], there exists a unique, strictly positive, left, row-eigenvector $\mathbf{l}_0 = (l_i)_{i=0,\dots,N}$ of $\mathcal{K} = \gamma^{-1}\mathcal{Q} + I$ corresponding to the eigenvalue 1. Clearly, $\mathbf{l}_0$ is an eigenvector of $\mathcal{Q}$ corresponding to eigenvalue 0, and we have already seen that $\mathsf{x} \mapsto \mathbf{l}_0(\mathsf{x})$ is Lipschitz continuous. The spectral gap is defined as the smallest nonzero eigenvalue $\lambda = \lambda(x)$ of $I - \frac{1}{2}(\mathcal{K} + \mathcal{K}^*)$ where the entries k_{ij} and k_{ij}^* of $\mathcal{K}$ and $\mathcal{K}^*$ are related by $k_{ij}^* = \frac{k_{ji}l_j}{l_i}$. We assume that $\lambda(x) > 0$, which implies $\lambda_0 := \inf_{\mathsf{x}\in J}\lambda(\mathsf{x}) > 0$. Then, see [305] p. 328, $|(\mathrm{e}^{t\gamma^{-1}\mathcal{Q}})_{ij} - l_j| \leq \sqrt{\frac{1}{l_i}}\mathrm{e}^{-\lambda_0 t}$. Since $l_i > 0$ is a continuous function of x, it follows that there exists a constant $K_{\mathcal{Q}}$ such that:

$$\|\mathrm{e}^{t\gamma^{-1}\mathcal{Q}}\mathbf{f} - \mathcal{P}_0\mathbf{f}\| \leq K_{\mathcal{Q}}\mathrm{e}^{-\lambda_0 t}\|\mathbf{f}\|.$$

Using positivity of γ, we see that this implies (45.4). □

Exercise 45.1 Assume (S, d) is a compact metric space and $f : S \to \mathbb{R}$ is locally Lipschitz continuous, that is, for each $p \in S$ there exists $\epsilon > 0$ and M_p such that $|f(q) - f(q')| \leq M_p d(p, q)$, as long as $d(p, q), d(p, q') < \epsilon$. Show that f is (globally) Lipschitz: there exists M such that $|f(p) - f(q)| \leq Md(p, q)$ for $p, q \in S$. *Hint:* assume that the hypothesis is not true, choose p_n, q_n such that $|f(p_n) - f(q_n)| > nd(p_n, q_n)$ and pass to convergent subsequences.

☞ Chapter's summary

Two convergence theorems for piecewise deterministic processes are proved. The first of them deals with the case in which, much as in the Tikhonov's Theorem, motions along some chosen axes are "very fast." The other is devoted to the case in which jumps are very frequent. In the first theorem, the state-space of the limit process is composed of less-dimensional rectangles. In the second, the rectangles do not lose dimensions but get lumped together into a single rectangle.

46

Models of Gene Regulation and Gene Expression

Motivations of Theorem 45.1 include Kepler–Elston's model of stochastic **gene regulation** ([204], section titled "Regulated systems II: mutual repressors"). In the model, two proteins responsible for a gene's expression are considered. They are products of the same gene and at the same time act as mutual repressors. If none of them is bound to a regulatory sequence, called *operator*, in the gene, both are being produced at a common, constant rate. When one of them is bound to the operator, production of the other is repressed and its existing molecules gradually degrade. Proteins leave the operator at random times, thus triggering production of the protein previously repressed (see Figure 46.1).

These assumptions are formalized as follows. Let $\gamma \in \{0, 1, 2\}$ denote the state of the operator, 0 referring to the unoccupied state, 1 to being occupied by a dimer of the first protein, and 2 to being occupied by a dimer of the second protein. Then, in the intervals where $\gamma = i$ is constant, the vector $\mathsf{x}_i = (x_{ij})_{j=1,2}$ of normalized levels of proteins satisfies:

$$\frac{\mathrm{d}\mathsf{x}_i(t)}{\mathrm{d}t} = (F_j(i) - x_{ij}(t))_{j=1,2}, \tag{46.1}$$

where $F_1(0) = F_1(1) = F_2(0) = F_2(2) = 1$ and $F_1(2) = F_2(1) = 0$, $x_{ij} \in [0, 1]$.

Hence, we are in the setup of Theorem 45.1 with $N = 2$, $M = 2$,

$$\mathsf{v} = (0, 0), \mathsf{w} = (1, 1) \quad \text{and} \quad \mathsf{a}_i(x_j)_{j=1,2} = (F_j(i) - x_j)_{j=1,2}.$$

Clearly a_i, $i \in \mathcal{N}$ are Lipschitz-continuous and the Müller condition (44.2) is satisfied.

In [204], γ jumps from 1 or 2 to 0 with the same constant intensity α_0, and since the proteins bind to the operator as dimers, the intensities α_1 and α_2 of jumps from 0 to 1 and 2, respectively, are proportional to the squares

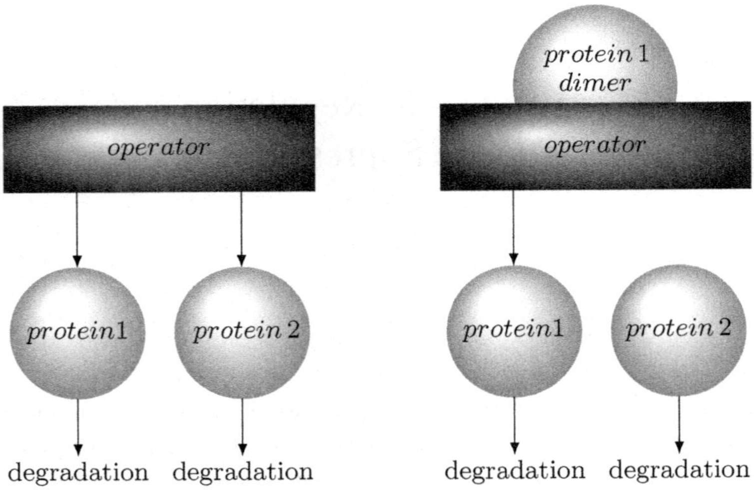

Figure 46.1 Gene expression with two mutually repressing proteins. When a dimer of one protein binds to the operator, production of the other is repressed, and vice versa.

of the corresponding protein levels. Here, we replace these particular intensities by arbitrary non-negative Lipschitz-continuous functions $\alpha_0, \alpha_1, \alpha_2$ on $J = [0, 1]^2$. This leads to:

$$\mathcal{Q} = \begin{pmatrix} -\alpha_1 - \alpha_2 & \alpha_1 & \alpha_2 \\ \alpha_0 & -\alpha_0 & 0 \\ \alpha_0 & 0 & -\alpha_0 \end{pmatrix};$$

we assume merely that $\alpha_0 > 0$ and $\sigma := \alpha_0 + \alpha_1 + \alpha_2 > \alpha_0$. Then, $\mathcal{Q}(\mathsf{x})$ has three distinct eigenvalues, $\nu_0(\mathsf{x}) = 0$, $\nu_1(\mathsf{x}) = -\alpha_0(\mathsf{x})$, $\nu_2(\mathsf{x}) = -\sigma(\mathsf{x})$, and we check that $\mathbf{l}_0 = (\frac{\alpha_0}{\sigma}, \frac{\alpha_1}{\sigma}, \frac{\alpha_2}{\sigma})$.

By Proposition 45.3 and Theorem 45.1 with $\mathcal{Q}_n = \mathcal{Q}$ and arbitrary κ_n, the semigroups of the corresponding piecewise deterministic processes converge to the semigroup describing deterministic motion along the interval curves of:

$$\frac{\mathrm{d}\mathsf{x}(t)}{\mathrm{d}t} = \left(\frac{\alpha_0 + \alpha_j}{\sigma} - x_j(t)\right)_{j=1,2}, \tag{46.2}$$

because $\mathsf{a}(\mathsf{x})\mathbf{l}_0(\mathsf{x}) = (\frac{\alpha_0+\alpha_j}{\sigma} - x_j)_{j=1,2}$. Intuitively, as noted in [204], when $\kappa_n \to \infty$, γ reaches its (conditional) statistical equilibrium with probabilities of being at 0, 1, and 2 equal to $\frac{\alpha_0}{\sigma}$, $\frac{\alpha_1}{\sigma}$, and $\frac{\alpha_2}{\sigma}$, respectively. Therefore $F_j(i)$ in

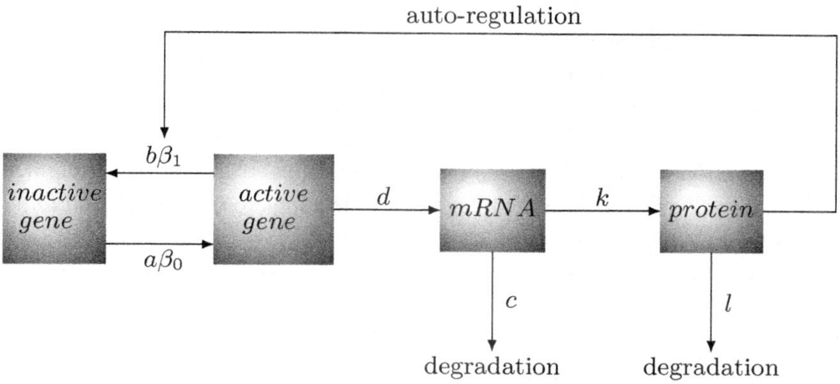

Figure 46.2 Simplified diagram of auto-regulated gene expression (redrawn with insignificant changes from [69])

(46.1) may be replaced by conditional expected values $\mathbb{E}F_j(\gamma)$, $j = 1, 2$; this leads to (46.2).

Another motivation for Theorem 45.1 is the model of stochastic **gene expression** due to Lipniacki et al. [69, 172, 242, 241], see also [50, 60]. The model describes stochastic mechanisms governing production and degradation of proteins in a cell, with intermediary role of mRNA in the process. It is assumed that the cell contains several copies of a gene, activated and inactivated independently. When at least one copy of the gene is activated, transcription of mRNA is initiated that stimulates production of proteins. Increasing level of proteins may in turn stimulate deactivation of gene copies and degradation of proteins and mRNA (so-called auto-regulation; see Figure 46.2).

Formally this process can be described as follows. Let $\gamma \in \mathcal{N} = \{0, 1, \ldots, N\}$ denote the number of active copies of the gene, x_1 the level of mRNA and x_2 the level of proteins. For constant $\gamma = i$, $\mathsf{x}_i = (x_{ij})_{j=1,2}$ satisfies:

$$\frac{\mathrm{d}\mathsf{x}_i(t)}{\mathrm{d}t} = \begin{pmatrix} -cx_{i1}(t) + di \\ kx_{i1}(t) - lx_{i2}(t) \end{pmatrix}, \tag{46.3}$$

where c and d denote the mRNA degradation and production rates per active gene copy, while k and l are the protein's translation and degradation rates, respectively.

Taking $M = 2$, $\mathsf{v} = (0, 0)$,

$$\mathsf{w} = (Ndc^{-1}, Nkd(lc)^{-1}) \quad \text{and} \quad \mathsf{a}_i(\mathsf{x}) = (-cx_1 + di, kx_1 - lx_2),$$

we see that a_i are Lipschitz-continuous and the Müller condition is satisfied.

In [242], intensity of activation of a single gene's copy is assumed to be constant, while deactivation rate is proportional to the level of proteins. Here we use more general rates of the form $a\beta_0$, $b\beta_1$, where β_0 and β_1 are arbitrary, non-negative, Lipschitz continuous functions, and a and b are real positive parameters. Because gene copies are activated and deactivated independently, conditionally on $\gamma(t) = i$, intensities of activation of another gene copy and of deactivation of an active gene copy are $(N - i)a\beta_0$ and $ib\beta_1$, respectively.

We consider the case in which activation/deactivation rates become infinite by replacing a and b by a_n and b_n, and letting a_n, b_n tend to infinity in such a way that $r := \lim_{n\to\infty} \frac{a_n}{b_n} \geq 0$ exists. Jump intensities of the approximating processes are then described by $b_n \mathcal{Q}_n$ where $\mathcal{Q}_n = (q_{ij,n})_{i,j\in\mathcal{N}}$ and:

$$q_{ij,n} = \begin{cases} i\beta_1, & \text{if } j = i-1, \quad 1 \leq i \leq N, \\ -(N-i)\frac{a_n}{b_n}\beta_0 - i\beta_1, & \text{if } j = i, \quad 0 \leq i \leq N, \\ (N-i)\frac{a_n}{b_n}\beta_0, & \text{if } j = i+1, \quad 0 \leq i \leq N-1, \\ 0, & \text{otherwise.} \end{cases} \tag{46.4}$$

We will show later that (Lemma 46.1), assuming that $r\beta_0 + \beta_1$ is strictly positive, for $\mathsf{x} \in J = [0, Ndc^{-1}] \times [0, Nkd(lc)^{-1}]$ the limit matrix $\mathcal{Q}(\mathsf{x}) = \lim_{n\to\infty} \mathcal{Q}_n(\mathsf{x})$ has $N+1$ distinct eigenvalues:

$$\nu_i(\mathsf{x}) = -i(r\beta_0(\mathsf{x}) + \beta_1(\mathsf{x})), \; i \in \mathcal{N},$$

and $\mathbf{l}_0 = \left(\frac{1}{(r\beta_0+\beta_1)^N}\binom{N}{i}(r\beta_0)^i\beta_1^{N-i}\right)_{i\in\mathcal{N}}$.

By Proposition 45.3 and Theorem 45.1 with $\kappa_n = b_n$, the semigroups of the corresponding piecewise deterministic processes converge to the semigroup of the ODE:

$$\frac{\mathrm{d}\mathsf{x}(t)}{\mathrm{d}t} = \begin{pmatrix} -cx_1(t) + \frac{Nrd\beta_0}{r\beta_0+\beta_1} \\ kx_1(t) - lx_2(t) \end{pmatrix}, \tag{46.5}$$

for $\mathsf{a}(\mathsf{x})\mathbf{l}_0(\mathsf{x}) = \left(\begin{smallmatrix} -cx_1 + \frac{Nrd\beta_0(\mathsf{x})}{r\beta_0(\mathsf{x})+\beta_1(\mathsf{x})} \\ kx_1 - lx_2 \end{smallmatrix}\right)$. This proves the hypothesis made in [242].

Let us also consider the case of large protein's translation and degradation rates k and l (this case is not discussed in the published version of [242], but the related hypothesis was stated in the preprint preceding publication). To this end we replace k and l in (46.3) by κ_n and $\eta\kappa_n$, where $(\kappa_n)_{n\geq 1}$ satisfies the usual assumptions and η is a positive constant. Replacing x_i in (46.3) by $\binom{y_{i,n}}{z_{i,n}} \in J$, we rewrite (46.3) to the form compatible with (45.8):

$$\frac{\mathrm{d}y_{i,n}(t)}{\mathrm{d}t} = -cy_{i,n}(t) + di, \qquad \frac{\mathrm{d}z_{i,n}(t)}{\mathrm{d}t} = \kappa_n(y_{i,n}(t) - \eta z_{i,n}(t)), \quad i \in \mathcal{N}; \tag{46.6}$$

so that now $y_{i,n}$ is the level of proteins and $z_{i,n}$ is the level of mRNA (we changed the font of y and z to stress that they are scalars here). Then we are in the setup of Theorem 45.2 with $K = 1$, and $\overline{\mathsf{a}_i}(y, z) = -cy + di$, $\underline{\mathsf{a}_i}(y, z) = y - \eta z$. Note also that k/l does not change with n, and neither does $J = [0, Ndc^{-1}] \times [0, Nd(\eta c)^{-1}]$. Jumps of the approximating processes are governed by $\mathcal{Q}_n = b\mathcal{Q} = (q_{ij})$, with $q_{ij} \in C$ defined by (46.4) where a_n, b_n are replaced by fixed parameters a, b.

The systems (46.6) are of uniformly Tikhonov-type. Indeed, for $\tilde{\mathsf{y}} \in J_{\mathsf{y}} = [0, Ndc^{-1}]$, $\tilde{\mathsf{z}} = \varphi(\tilde{\mathsf{y}}) = \frac{\tilde{\mathsf{y}}}{\eta} \in J_{\mathsf{z}} = [0, Nd(\eta c)^{-1}]$ is the unique solution to $\underline{\mathsf{a}_i}(\tilde{\mathsf{y}}, \tilde{\mathsf{z}}) = 0$, and φ is Lipschitz continuous (in this example, $\underline{\mathsf{a}_i}$ does not depend on i and neither do φ_i and $\tilde{\mathsf{z}}_i$). Also, $\mathsf{z}_i(t) - \frac{\mathsf{y}}{\eta} = (\mathsf{z}_i(0) - \frac{\mathsf{y}}{\eta})\mathrm{e}^{-\eta t}$ for $i \in \mathcal{N}$, where $\mathsf{z}_i(t)$ are the solutions of the boundary layer systems corresponding to (46.6) and Assumption B is satisfied for all i, too.

By Theorem 45.2, if the proteins' translation and degradation rates are large, but stay in a fixed proportion η, in the limit process the level of mRNA is governed by:

$$\frac{\mathrm{d}\mathsf{y}_i(t)}{\mathrm{d}t} = -c\mathsf{y}_i(t) + di, \quad i \in \mathcal{N},$$

with the level of proteins given by $\mathsf{z}_i(t) = \frac{\mathsf{y}_i(t)}{\eta}$ and jumps driven by $\mathcal{L}\mathcal{Q} = (Lq_{ij})$, where $Lq_{ij}(\mathsf{y}) = q_{ij}(\mathsf{y}, \frac{\mathsf{y}}{\eta})$. Note that for $N = 1$ and rescaled coefficients, this result has been obtained in Example 26.6.

Lemma 46.1 *Suppose $r\beta_0 + \beta_1$ is strictly positive on J. Then, for $\mathsf{x} \in J$, the limit matrix $\mathcal{Q}(\mathsf{x}) = \lim_{n\to\infty} \mathcal{Q}_n(\mathsf{x})$ has $N + 1$ distinct eigenvalues $\nu_i(\mathsf{x}) = -i(r\beta_0(\mathsf{x}) + \beta_1(\mathsf{x}))$, $i \in \mathcal{N}$.*

Proof Let $\mathcal{Q}(N) = (q_{ij})_{i,j\in\mathcal{N}}$ denote the limit $(N + 1) \times (N + 1)$ matrix $\mathcal{Q}$ in the Lipniacki model with $r = 1$ (i.e., q_{ij}'s are given by (46.4) with $a_n/b_n = 1$) and assume that $\beta_0 + \beta_1 = 1$. We will prove that the characteristic polynomial of $\mathcal{Q}(N)$ is:

$$\det(\lambda - \mathcal{Q}(N)) = \lambda(\lambda + 1)(\lambda + 2)\ldots(\lambda + N), \qquad N \geq 1. \tag{46.7}$$

To this end, we perform the following operations on the matrix $\lambda - \mathcal{Q}(N)$: starting from the right, to each column we add the sum of columns preceding it, obtaining the matrix with the same determinant:

$$\begin{pmatrix} N\beta_0+\lambda & \lambda & \lambda & \dots & \dots & \lambda & \lambda \\ -\beta_1 & (N-1)\beta_0+\lambda & \lambda & \dots & \dots & \dots & \lambda \\ 0 & -2\beta_1 & (N-2)\beta_0+\lambda & \dots & \dots & \dots & \lambda \\ \dots & 0 & -3\beta_1 & \dots & \dots & \dots & \dots \\ \vdots & \vdots & \ddots & \ddots & \vdots & \vdots & \vdots \\ 0 & \dots & \dots & \dots & 2\beta_0+\lambda & \lambda & \lambda \\ 0 & \dots & \dots & 0 & -(N-1)\beta_1 & \beta_0+\lambda & \lambda \\ 0 & \dots & \dots & \dots & 0 & -N\beta_1 & \lambda \end{pmatrix}.$$

Next, starting from the top down, from each row we subtract the one lying below it, leaving the last row unchanged. Applying the Laplace expansion along the last column of the resulting matrix and using $\beta_0 + \beta_1 = 1$ we obtain $\det(\lambda - \mathcal{Q}(N)) = \lambda \det(\lambda + 1 - \mathcal{Q}(N-1))$. Since the case $N = 1$ is easily checked, (46.7) follows by induction.

Dropping the simplifying assumption ($r = 1$, $\beta_0 + \beta_1 = 1$) we replace β_0 by $r\beta_0$ and divide the matrix by $r\beta_0 + \beta_1 > 0$, to see that the eigenvalues of the limit matrix are $-i(r\beta_0 + \beta_1)$, $i \in \mathcal{N}$. □

A deeper insight into the way biological systems, like those described by (46.1) and (46.3) work, may be gained by reading expertly and reader-friendly written monograph by Uri Alon [3].

☞ Chapter's summary

We discuss two motivating examples of the main theorems of the previous chapter: models of gene regulation and gene expression. As an application of the theorems we conclude that under certain conditions, these models may be reduced to simpler ones.

47

Oligopolies, Manufacturing Systems, and Climate Changes

Piecewise deterministic Markov models such as those in Theorem 45.1 are present not only in mathematical biology, but also in other applied sciences. In this chapter we exhibit three examples of this type, taken from [61].

First, we consider a **manufacturing system** ([149, 293, 310, 352]) where items are produced at the rate $u \geq 0$ and sold at the demand rate $z \geq 0$. By $\mathsf{x}(t) \in \mathbb{R}$ we denote the *surplus*, that is, the difference between cumulative production and cumulative demand at time t and by $a > 0$ we denote the spoilage rate of a finished product. The rate of production $u = u(\gamma)$ depends on a number γ of active parallel production machines. We consider a so-called *failure-prone system*, where each machine is subject to breakdown and repair, so it can either be operational or out of order. We assume that the system consists of two identical machines, that is, $\gamma \in \mathcal{N} = \{0, 1, 2\}$. Then, in the intervals where $\gamma = i$ is constant, the surplus is governed by the ODE:

$$\frac{\mathrm{d}\mathsf{x}_i(t)}{\mathrm{d}t} = -a\mathsf{x}_i(t) + u(i) - z. \tag{47.1}$$

Suppose that $u(0) = 0$ and $u(1) \leq u(2)$. In the notation of Theorem 45.1, we have $M = 1$, $J = [\frac{-z}{a}, \frac{u(2)-z}{a}]$, and $\mathsf{a}_i(\mathsf{x}) = -a\mathsf{x} + u(i) - z$; the latter functions are Lipschitz continuous and satisfy (44.2).

Let $\mu > 0$ be the intensity of machine failures and let $\xi(\mathsf{x}) > 0$ be the intensity of repairs. We suppose that ξ depends on x, because if the surplus is low, machines are repaired quicker than if it is high. There is a single repair crew; if two machines are out of order, the repair crew is working on the one that failed first. If machines break down independently, this leads to the intensity matrix:

$$\mathcal{Q}(\mathsf{x}) = \begin{pmatrix} -\xi(\mathsf{x}) & \xi(\mathsf{x}) & 0 \\ \mu & -\mu - \xi(\mathsf{x}) & \xi(\mathsf{x}) \\ 0 & 2\mu & -2\mu \end{pmatrix}.$$

Eigenvalues of $\mathcal{Q}(\mathsf{x})$ equal 0 and $\frac{1}{2}((-1)^k\sqrt{\mu(\mu+4\xi(\mathsf{x}))}-(3\mu+2\xi(\mathsf{x})))$, $k=1,2$, and:

$$\mathbf{l}_0(\mathsf{x})=\left(\frac{2\mu^2}{\sigma(\mathsf{x})},\frac{2\mu\xi(\mathsf{x})}{\sigma(\mathsf{x})},\frac{\xi^2(\mathsf{x})}{\sigma(\mathsf{x})}\right),$$

where $\sigma(\mathsf{x})=2\mu^2+2\mu\xi(\mathsf{x})+\xi^2(\mathsf{x})$. Taking $\mathcal{Q}_n=\mathcal{Q}$ and assuming ξ to be Lipschitz continuous we obtain, by Proposition 45.3 and Theorem 45.1, that if the breakdowns and repairs occur in a much faster scale than production, the surplus satisfies the ODE:

$$\frac{\mathrm{d}\mathsf{x}(t)}{\mathrm{d}t}=-a\mathsf{x}(t)+\frac{2\mu\xi(\mathsf{x})}{\sigma(\mathsf{x})}u(1)+\frac{\xi^2(\mathsf{x})}{\sigma(\mathsf{x})}u(2)-z.$$

Our next example is that of **oligopoly** (see, e.g., [174]). We consider M players in a market with random switches of economic situation $\gamma(t)$, described by two market modes: recession $\gamma(t)=0$ and prosperity $\gamma(t)=1$. Let the stochastic investment rate $u_{ij}\in\mathbb{R}$ of the jth firm ($j\in\mathcal{M}=\{1,\dots,M\}$) in the ith market mode ($i\in\mathcal{N}=\{0,1\}$) depend on i and j; we assume that $u_{0j}\le u_{1j}$ for $j\in\mathcal{M}$. In the models of optimal control or in stochastic games, u_{ij}'s usually depend on t and are treated as control tools [174]. Here we disregard the elements of control and games, leaving the process uncontrolled. If $\gamma=i$ is constant, the market production capacity $\mathsf{x}_i=(x_{ij})^{\mathsf{T}}_{j\in\mathcal{M}}$ satisfies:

$$\frac{\mathrm{d}\mathsf{x}_i(t)}{\mathrm{d}t}=\mathsf{u}_i-\mathsf{x}_i(t),\quad i\in\mathcal{N},\tag{47.2}$$

where $\mathsf{u}_i=(u_{ij})_{j\in\mathcal{M}}$, and the capacity deterioration rate is set to 1 for simplicity. In the notation of Theorem 45.1, for $i\in\mathcal{N}$ we have $\mathsf{a}_i(\mathsf{x})=\mathsf{u}_i-\mathsf{x}$, $\mathsf{v}=(u_{0j})_{j\in\mathcal{M}}$, $\mathsf{w}=(u_{1j})_{j\in\mathcal{M}}$ and $J=[u_{01},u_{11}]\times\cdots\times[u_{0M},u_{1M}]\subset\mathbb{R}^M$. Clearly a_i are Lipschitz continuous and the Müller condition is satisfied.

Intensity matrix $\mathcal{Q}$ of market switches is determined by the following normalized jump intensities: $q_{00}=-1$, $q_{11}=-s_M$, $s_M(\mathsf{x})=x_1+\cdots+x_M$, $\mathsf{x}=(x_j)_{j\in\mathcal{M}}$ (if the total supply is too large, a transition from prosperity to recession is highly probable). The negative eigenvalue of $\mathcal{Q}(\mathsf{x})$ equals $-1-s_M$. Furthermore, $\mathbf{l}_0=(\frac{s_M}{1+s_M},\frac{1}{1+s_M})$. By Proposition 45.3 and Theorem 45.1 with $\mathcal{Q}_n=\mathcal{Q}$ and arbitrary κ_n, if market switches are frequent, the semigroups of the related piecewise deterministic processes converge to the semigroup describing deterministic motion along the integral curves of:

$$\frac{\mathrm{d}\mathsf{x}(t)}{\mathrm{d}t}=\frac{s_M}{1+s_M}\mathsf{u}_0+\frac{1}{1+s_M}\mathsf{u}_1-\mathsf{x}(t).\tag{47.3}$$

Consider also the case in which market switches occur rarely when compared to the time scale of the economic activity, comp. [174]. To this end, we replace

(47.2) by the singularly perturbed equations of type (45.8):

$$\frac{d\mathsf{z}_{i,n}(t)}{dt} = \kappa_n(\mathsf{u}_i - \mathsf{z}_{i,n}(t)), \quad i \in \mathcal{N}, \tag{47.4}$$

so that the marked production capacity is now denoted by z and not by x. Here all coordinates are singularly perturbed, $J_\mathsf{z} = J$, $J_\mathsf{y} = \{0\}$ and $\underline{\mathsf{a}}_i(\mathsf{z}) = \mathsf{u}_i - \mathsf{z}$ are Lipschitz-continuous and satisfy the Müller condition. Thus we are in the setup of Theorem 45.2 with $K = 0$. Intensities of jumps between integral curves of the full system (47.4) are given by $\mathcal{Q}$ described earlier.

The systems (47.4) are of uniformly Tikhonov type because $\varphi_i = \mathsf{u}_i$ is the unique root of $\underline{\mathsf{a}}_i$ and as in the previous example the solutions of the related boundary layer system converge to u_i, regardless of the initial condition. Hence, by Theorem 45.2, in the limit as $n \to \infty$, we obtain the Markov chain jumping between the points u_i with intensity matrix $\mathcal{PQ} = (P_i q_{ij})$, where $P_0 q_{00} = -\sum_{m=1}^{M} u_m^0$ and $P_1 q_{11} = -1$.

In our final example, we consider a piecewise deterministic process inspired by the stochastic game of **interactions between economy and climate** described in [174] and [173]. It is a model of economic activity in a changing climate, based on the observation that the natural environment evolves slowly when compared to the economic growth.

The model assumes that in the global economy there are M players, called nations. The world climate may be in one of several climate modes, $i \in \mathcal{N}$. For the ith climate mode let $\mathsf{z}_i \in \mathbb{R}^M$ denote the M-dimensional vector of the accumulated capital of M nations. In accordance with a widely accepted view (see, e.g., [174, 173] and references given there), we assume that climate changes are to some extent stimulated by the human economic activity, since the latter leads to accumulation of greenhouse gases (GHG, e.g., CO_2) in the atmosphere. Let $\mathsf{y}_i \in \mathbb{R}$ be the average concentration of GHG in the ith climate mode. The levels of y_i and z_i satisfy:

$$\frac{d\mathsf{y}_i(t)}{dt} = -\beta \mathsf{y}_i(t) + \sigma_i \mathsf{z}_i(t), \qquad \frac{d\mathsf{z}_i(t)}{dt} = \kappa(\mathsf{u}_i - \alpha \mathsf{z}_i(t)), \tag{47.5}$$

where $\alpha = \operatorname{diag}\{\alpha_1, \ldots, \alpha_M\}$ is the $M \times M$ diagonal matrix of capital depreciation rates $\alpha_j > 0$, $\mathsf{u}_i = (u_{ij})_{j \in \mathcal{M}}$ is the vector of investment rates, $\beta > 0$ is the GHG natural elimination rate, $\sigma_i = (\sigma_{ij})_{j \in \mathcal{M}}$ is the vector of GHG production rates $\sigma_{ij} \geq 0$ in the national economies, $\sigma_i \mathsf{z}_i$ is the scalar product of σ_i and z_i, and κ is a positive parameter.

Suppose that there are three possible climate modes, so that $\mathcal{N} = \{0, 1, 2\}$. The climate may either remain unchanged ($i = 0$) or experience a sudden change, resulting from catastrophes caused by the increasing concentration of

Figure 47.1 "Indian Summer" or "Golden Polish Autumn"

GHG. For instance, the climate may change due to a slowdown in the ocean currents (the so-called *thermohaline circulation*, $i = 1$) or due to the Antarctic and Greenland ice sheet collapse ($i = 2$). We assume that intensities of switches in the climate modes are given by the matrix $\mathcal{Q}(\mathsf{y}) = (q_{ij}(\mathsf{y}))_{i,j\in\mathcal{N}}$, where q_{ij} are climate transition rates continuous in y. We assume $i = 2$ is the cemetery state, that is,

$$\mathcal{Q}(\mathsf{y}) = \begin{pmatrix} -q_{01}(\mathsf{y}) - q_{02}(\mathsf{y}) & q_{01}(\mathsf{y}) & q_{01}(\mathsf{y}) \\ 0 & -q_{12}(\mathsf{y}) & q_{12}(\mathsf{y}) \\ 0 & 0 & 0 \end{pmatrix}.$$

Let us note now that the economic activity takes place in a faster time scale than the climate changes. To incorporate this information in our model we replace the parameter κ in (47.5) by a sequence $(\kappa_n)_{n\geq 1}$, satisfying the usual assumptions. Then we are in the setup of Theorem 45.2 with Lipschitz-continuous functions $\overline{\mathsf{a}}_i$ and $\underline{\mathsf{a}}_i$ given by:

$$\overline{\mathsf{a}}_i(\mathsf{y},\mathsf{z}) = -\beta\mathsf{y} + \sigma_i\mathsf{z} \quad \text{and} \quad \underline{\mathsf{a}}_i(\mathsf{y},\mathsf{z}) = \mathsf{u}_i - \alpha\mathsf{z},$$

$N = 2$, $K = 1$ and $M + 1$ instead of M (the model contains M production variables and one variable describing the level of GHG). To show this we introduce:

$$\alpha_{\max} = \max_{i\in\mathcal{N}} \alpha_i, \quad \sigma_{\max} = \max_{i\in\mathcal{N}} \sum_{j=1}^{M} \sigma_{ij}, \quad u_{\max} = \max_{ij} u_{ij}$$

$$\alpha_{\min} = \min_{i\in\mathcal{N}} \alpha_i, \quad \sigma_{\min} = \min_{i\in\mathcal{N}} \sum_{j=1}^{M} \sigma_{ij}, \quad u_{\min} = \min_{ij} u_{ij}.$$

It follows that if the jth coordinate of z equals $v = u_{\min}\alpha_{\max}^{-1}$ (or $w = u_{\max}\alpha_{\min}^{-1}$, respectively), then the jth coordinate of $\underline{\mathsf{a}}_i(\mathsf{y}, \mathsf{z}) = \mathsf{u}_i - \alpha\mathsf{z}$, is non-negative (or, nonpositive, respectively). Since $\overline{\mathsf{a}}_i(\mathsf{y}, \mathsf{z}) \geq 0$ for $\mathsf{y} = v_0 := v\beta^{-1}\sigma_{\min}$ and $\overline{\mathsf{a}}_i(\mathsf{y}, \mathsf{z}) \leq 0$ for $\mathsf{y} = w_0 := w\beta^{-1}\sigma_{\max}$, provided $\mathsf{z} \in J_{\mathsf{z}} = [v, w] \times \cdots \times [v, w]$, this shows that the system (47.5) satisfies the Müller condition in $J = J_{\mathsf{y}} \times J_{\mathsf{z}}$, where $J_{\mathsf{y}} = [v_0, w_0]$. Moreover, the equilibrium point $\varphi_i(\mathsf{y})$ does not depend of y and equals $\alpha^{-1}\mathsf{u}_i \in J_{\mathsf{z}}$. We check that Assumption B is satisfied, similarly as in the previous examples, using explicit form of the solution to the boundary layer system of (47.5). Thus (47.5) are of uniformly Tikhonov-type.

By Theorem 45.2 the original process moving along integral curves of (47.5) with jumps determined by $\mathcal{Q}$, converges as $n \to \infty$ to the limit process, moving along integral curves of:

$$\frac{\mathrm{d}\mathsf{y}_i(t)}{\mathrm{d}t} = -\beta\mathsf{y}_i(t) + \sigma_i\alpha^{-1}\mathsf{u}_i,$$

with $\mathsf{z}_i(t) = \alpha^{-1}\mathsf{u}_i$ and jumps described by $\mathcal{P}\mathcal{Q} = \mathcal{Q}$.

☞ Chapter's summary

We continue with applications of theorems of Chapter 45. We present models of manufacturing systems, oligopolies, and interactions between economy and climate.

48

Convex Combinations of Feller Generators

Let $\mathcal{S}$ be a locally compact Hausdorff space, and let $\alpha_i, i = 0, \ldots, N$, where N is an integer, be non-negative continuous functions on $\mathcal{S}$ with $\sum_{i=0}^N \alpha_i = 1$. Furthermore, let $A_i, i = 0, \ldots, N$ and A be generators of Feller semigroups $\{e^{tA_i}, t \ge 0\}$ and $\{e^{tA}, t \ge 0\}$ in $C_0(\mathcal{S})$ with related Feller processes $X_i = \{X_i(t), t \ge 0\}$ and $X = \{X(t), t \ge 0\}$. Finally, suppose that $\mathcal{D} := \bigcap_{i=0}^N \mathcal{D}(A_i)$ is a core for A and that $Af = \sum_{i=0}^N \alpha_i A_i f$ for $f \in \mathcal{D}$. Then, it is natural to expect that X may be described as follows: conditional on being at $p \in \mathcal{S}$, X behaves like X_i with probability $\alpha_i(p)$.

Unfortunately, none of the previously introduced formulae for the semigroup $\{e^{tA}, t \ge 0\}$ support such an interpretation. If we assume $N = 1$ and α_0 and α_1 to be constants, we end up in the setup of the Trotter Product Formula to obtain:

$$e^{tA} = \lim_{n\to\infty} \left[e^{\frac{\alpha_0 t}{n} A_0} e^{\frac{\alpha_1 t}{n} A_1} \right]^n ,$$

so that the process $\{X(t), t \ge 0\}$ is a "mixture" of $\{X_0(\alpha_0 t), t \ge 0\}$ and $\{X_1(\alpha_1 t), t \ge 0\}$. However, this is still far from the interpretation we desire to have. In the case in which A_0 or A_1 are bounded, $\{e^{tA}, t \ge 0\}$ may be expressed explicitly as the Dyson–Phillips perturbation series, but neither this formula allows for the desired interpretation.

On the other hand, we could use the intuitions developed in the previous chapters and approximate X by a sequence $\mathcal{X}_n, n \ge 0$ of Feller processes in an extended space $\mathcal{S}^{N+1}$ composed of $N + 1$ copies $\mathcal{S} \times \{i\}, i = 0, 1, \ldots, N$ of $\mathcal{S}$. These processes are "mixtures" of two components. The first of these starting on $\mathcal{S} \times \{i\}$, behaves like X_i and never leaves $\mathcal{S} \times \{i\}$. The second is composed of independent Markov chains indexed by $p \in \mathcal{S}$: conditional on starting from a $(p, i), i = 0, \ldots, N$ it is a Markov chain with states $(p, j), j = 0, 1, \ldots, N$ and certain intensity matrix depending on n and $\alpha_j(p), j = 0, \ldots, N$. This matrix is chosen so that $(\alpha_i(p))_{i=0,\ldots,N}$ is its unique invariant probability measure. Put

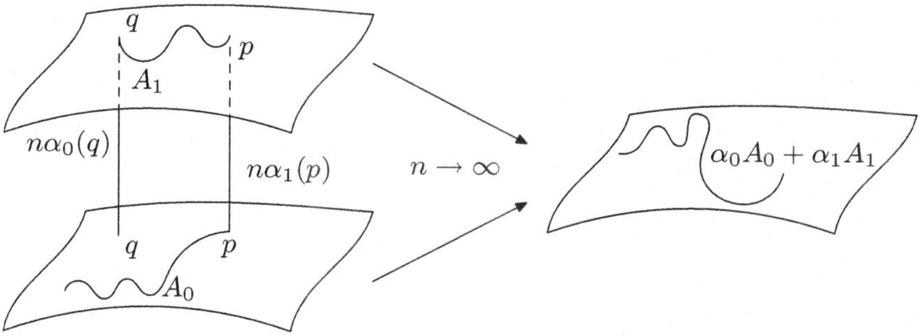

Figure 48.1 State-space merging for $N = 1$ (redrawn with insignificant changes from [51])

otherwise, $\mathcal{X}_n$'s are somewhat similar to piecewise deterministic processes of Chapter 45 in that their behavior may be described as follows: starting at the ith copy of $\mathcal{S}$, they stay there, behaving like X_i until random time τ when they jump to jth copy of $\mathcal{S}$ to behave like X_j there. At the time τ, conditional on being at (p, i) they jump to (p, j); the distribution of τ and the probability of choosing j are determined by the intensity matrix changing with p.

Now, increasing jump intensities as $n \to \infty$, corresponds to convergence of the Markov chains described above to their equilibria. In the limit, points of $\mathcal{S}^{N+1}$ with same first coordinates get glued together (the chain is at the equilibrium) to form another copy of $\mathcal{S}$, and the limit process defined there may indeed be thought of as that behaving with probability $\alpha_i(p)$ like X_i – conditional on being at p (see Figure 48.1).

Thus, in a sense we reverse the procedure known from Theorem 45.1. There, we show that the limit of piece-wise deterministic processes with frequent jumps is a deterministic motion along "averaged" curves of the deterministic motions involved. Here, given a convex combination of generators of Feller semigroup, we approximate the related process by an appropriate averaging procedure; this idea is developed in [51, 60, 72].

Specifically, the processes $\{\mathcal{X}_n(t), n \geq 0\}$ are related to the Feller semigroups $\{\mathcal{T}_n(t), t \geq 0\}$ generated by the operators:

$$\mathcal{A}_n(f_i)_{i=0,\dots,N} = (A_i f_i)_{i=0,\dots,N} + n\mathcal{Q}(f_i)_{i=0,\dots,N}, \quad f_i \in \mathcal{D}(A_i), \tag{48.1}$$

where $C_0(\mathcal{S}^{N+1})$ is identified via isometric isomorphism with the Cartesian product of $N+1$ copies of $C_0(\mathcal{S})$, and:

$$\mathcal{Q} = \left(q_{ij}\right)_{i,j=0,\dots,N} \tag{48.2}$$

is an operator matrix whose entries are multiplication operators related to functions q_{ij}, and q_{ij} have the following properties.[1] (a) For all $p \in S$, the scalar matrix $\left(q_{i,j}(p)\right)_{i,j=0,\dots,N}$ is an intensity matrix (describing the chain referred to earlier) with unique stationary measure $\alpha_i(p)$, and (b) defining $\mathcal{P}$ as the operator mapping $(f_i)_{i=0,\dots,N} \in [C_0(S)]^{N+1}$ to the vector with all coordinates equal $\sum_{i=0}^N \alpha_i f_i$, we have:

$$\lim_{t\to\infty} \mathrm{e}^{t\mathcal{Q}} = \mathcal{P} \qquad \text{(strongly).} \tag{48.3}$$

For example, the matrix:

$$\mathcal{Q} = \begin{pmatrix} \alpha_0 - 1 & \alpha_1 & \alpha_2 & \dots & \alpha_N \\ \alpha_0 & \alpha_1 - 1 & \alpha_2 & \dots & \alpha_N \\ \alpha_0 & \alpha_1 & \alpha_2 - 1 & \dots & \alpha_N \\ \vdots & \vdots & \ddots & \ddots & \ddots \\ \alpha_0 & \alpha_1 & \alpha_2 & \dots & \alpha_N - 1 \end{pmatrix}, \tag{48.4}$$

possesses these properties. For, $(\alpha_i(p))_{i\in\mathcal{N}}$ is the unique stationary measure for $Q(p)$, and we have $\mathcal{Q} = \mathcal{P} - \mathcal{I}$, where $\mathcal{I}$ is the identity on $\mathbb{Y}$, and $\mathcal{P}$ is idempotent. Hence (see Exercise 1.4 (a)),

$$\mathrm{e}^{t\mathcal{Q}} = \mathrm{e}^{-t} \sum_{n=0}^{\infty} \frac{t^n \mathcal{P}^n}{n!} = \mathrm{e}^{-t}(\mathcal{I} + (\mathrm{e}^t - 1)\mathcal{P}),$$

implying (48.3).

Elements of $C_0(\mathcal{S})$, the space of continuous functions vanishing at infinity, defined on $\mathcal{S}$, a locally compact Hausdorff space, will be denoted by f, g, h, and so on. The Cartesian product of $N+1$ copies of this space will be denoted by $\mathbb{B}$, and its elements by $\mathbf{f}, \mathbf{g}, \mathbf{h}$, $(f_0, f_1, \dots, f_N)$, $(f_i)_{i\in\mathcal{N}}$, etc. We will also write $[f]$ for the element $(f, f, \dots, f)$, where $f \in \mathbb{X}$. Operators in $C_0(\mathcal{S})$ will be denoted A, B, etc. and these in $\mathbb{B}$ by $\mathcal{A}, \mathcal{B}$, etc. with possible subscripts; the only exception to this rule are multiplication operators α_i. Finally, we put:

$$\operatorname{conv} \mathbf{f} = \operatorname{conv}(f_i)_{i=0,\dots,N} = \sum_{i=0}^{N} \alpha_i f_i.$$

Theorem 48.1 *Let A_i, $i = 0, \dots, N$ and A be generators of Feller semigroups $\{\mathrm{e}^{tA_i}, t \ge 0\}$ and $\{\mathrm{e}^{tA}, t \ge 0\}$ in $C_0(\mathcal{S})$. Also suppose that $Af = \sum_{i=0}^N \alpha_i A_i f = \operatorname{conv}(A_i f)_{i=0,\dots,N}$ for $f \in \mathcal{D} := \bigcap_{i=0}^N D(A_i)$ and that this last set is a core for A. Finally, let $\mathcal{A}_n$, $n \ge 0$ with $D(\mathcal{A}_n) = D(A_0) \times D(A_1) \times \dots \times D(A_N) \subset \mathbb{Y}$ be*

[1] In what follows, for notational convenience, we will not distinguish between a bounded, continuous function h on $\mathcal{S}$ and the related multiplicative operator $C_0(\mathcal{S}) \ni f \to hf \in C_0(\mathcal{S})$.

given by (48.1) and suppose that condition (48.3) *is fulfilled. Then, $\mathcal{A}_n$'s generate contraction semigroups in $\mathbb{B}$ and:*

$$\lim_{n\to\infty} \mathrm{e}^{t\mathcal{A}_n}\mathbf{f} = [\mathrm{e}^{tA}\mathrm{conv}\,\mathbf{f}], \qquad \mathbf{f} \in \mathbb{B}. \tag{48.5}$$

We are actually able to prove a more abstract version of this result, where α_i are bounded, commuting operators in a Banach space $\mathbb{X}$, such that $\sum_{i=0}^{N} \alpha_i = I_{\mathbb{X}}$. Similarly as in the case in which $\mathbb{X} = C_0(\mathcal{S})$, elements of $\mathbb{X}$ will be denoted x, y, z and so on with possible subscripts, and elements of the Cartesian product $\mathbb{Y}$ of $N+1$ copies of $\mathbb{X}$ will be denoted by $\mathbf{x}, \mathbf{y}, \mathbf{z}$, $(x_0, x_1, \ldots, x_N)$, $(x_i)_{i\in\mathcal{N}}$, and so on. We will also write $[x]$ for the element $(x, x, \ldots, x)$, where $x \in \mathbb{X}$, and $\mathrm{conv}(x_i)_{i=0,\ldots,N} = \sum_{i=0}^{N} \alpha_i x_i$.

Theorem 48.2 *Let A_i, $i \in \mathcal{N}$ and A be the generators of semigroups $\{\mathrm{e}^{tA_i}, t \geq 0\}$ and $\{\mathrm{e}^{tA}, t \geq 0\}$ in a Banach space $\mathbb{X}$. Suppose that $Ax = \mathrm{conv}(A_i x)_{i=0,\ldots,N}$ for $x \in \mathcal{D} := \bigcap_{i=0}^{N} \mathcal{D}(A_i)$ and that this last set is a core for A. Finally, let $\mathcal{A}_n$, $n \geq 0$ with $D(\mathcal{A}_n) = D(A_0) \times \mathcal{D}(A_1) \times \cdots \times D(A_N) \subset \mathbb{Y}$ be given by:*

$$\mathcal{A}_n\,(x_i)_{i\in\mathcal{N}} = (A_i x_i)_{i\in\mathcal{N}} + n\mathcal{Q}\,(x_i)_{n\geq 1} \qquad x_i \in D(A_i),\, i = 0, \ldots, N \tag{48.6}$$

where $\mathcal{Q}$ is a bounded linear operator in $\mathbb{Y}$ such that:

$$\lim_{t\to\infty} \mathrm{e}^{t\mathcal{Q}}\mathbf{x} = [\mathrm{conv}\,\mathbf{x}], \tag{48.7}$$

and suppose that $\mathcal{A}_n$'s generate equibounded semigroups $\{\mathcal{T}_n(t), t \geq 0\}$ in $\mathbb{Y}$. Then,

$$\lim_{n\to\infty} \mathrm{e}^{t\mathcal{A}_n}\mathbf{x} = [\mathrm{e}^{tA}\mathrm{conv}\,\mathbf{x}], \qquad \mathbf{x} \in \mathbb{X}. \tag{48.8}$$

Note that our theorem involves "double convex combination" in that in the limit we have the semigroup generated by a convex combination of generators, acting on a convex combination of initial conditions. Moreover, Theorem 48.1 is a special case of Theorem 48.2. Indeed, under its assumptions, $\mathcal{A}_0$ is the generator of the semigroup $\{\mathcal{T}_0(t), t \geq 0\}$ given by:

$$\mathcal{T}_0(t)\,(f_i)_{i\in\mathcal{N}} = \left(\mathrm{e}^{tA_i} f_i\right)_{i\in\mathcal{N}}.$$

Because $\mathcal{Q}$ is a bounded, the Phillips Perturbation Theorem implies that $\mathcal{A}_n$ generates a strongly continuous semigroup. However, $\mathcal{Q}$ is also the generator of a contraction semigroup. Hence, by the Trotter Product Formula, so is $\mathcal{A}_n = \mathcal{A}_0 + n\mathcal{Q}$, implying that all the assumptions of Theorem 48.2 are satisfied (assumption (48.3) is a special case of (48.7) – see (48.9)).

As is often the case, it is much easier to prove our theorem than it is to discover it; with the experience of the previous chapters, the reader will not be surprised to learn that the theorem is a direct consequence of that of Kurtz. To show this, first we note that in notations introduced earlier,

$$\mathcal{P}\mathbf{x} = [\text{conv}\,\mathbf{x}]. \tag{48.9}$$

Next, let $\mathcal{A} = \mathcal{A}_0$. For $x \in \mathbb{X}$, $\mathcal{P}[x] = [x]$ and so, by (42.2), $\mathcal{Q}[x] = 0$. Taking $x \in \mathcal{D}$ we see that $\mathcal{A}_n[x] = \mathcal{A}[x]$, showing that $[x]$ is a member of the domain of the extended limit $\mathcal{A}_{\text{ex}}$ of $\mathcal{A}_n$ and $\mathcal{A}_{\text{ex}}[x] = \mathcal{A}[x]$. Similarly, for any $\mathbf{x}$ in the common domain of $\mathcal{A}_n$, $\lim_{n\to\infty} \frac{1}{n}\mathcal{A}_n\mathbf{x} = \mathcal{Q}\mathbf{x}$. Finally, let $\mathbb{X}_0 = \mathit{Range}\,\mathcal{P}$ be the subspace of vectors of the form $[x], x \in \mathbb{X}$. Then, $\mathcal{P}\mathcal{A}[x] = [Ax], x \in \mathcal{D}$ is closable and generates the strongly continuous semigroup in $\mathbb{X}_0$ given by $\mathrm{e}^{\overline{\mathcal{P}\mathcal{A}}t}[x] = [\mathrm{e}^{tA}x]$. Hence, all the assumptions of Kurtz's Singular Perturbation Theorem are satisfied (with $\epsilon_n = \frac{1}{n}$, and Q, A_n and A replaced by their calligraphic versions), and we are done.

Turning to examples, let us start with the case of Poisson processes with changing colors, inspired by the "single gene, no feedback" section of [204] (see in particular equations (4), (5), and (21) there).

Example 48.3 Let $\alpha_i = (\alpha_i(p))_{p\ge1}, i = 0, \dots, N$ be convergent sequences with $\sum_{i=0}^N \alpha_i = 1$, and $c_i, i = 0, \dots, N$ be positive constants. (In other words we are in $\mathcal{S} = \mathbb{N}$, and $C_0(S)$ is c_0.) The state-space of the sequence $\{\mathcal{X}_n(t), t \ge 0\}, n \ge 1$ of processes to be described is $\mathbb{N} \times \{0, \dots, N\}$ with the first coordinate denoting the state of a regular Poisson process, and the second its changing color. The process $\mathcal{X}_n(t), t \ge 0$, starting at a point (p, i) waits there for an exponential time with parameter $c_i + n(1 - \alpha_i(p))$ to jump, with probability $\frac{c_i}{c_i+n(1-\alpha_i(p))}$ to $(p + 1, i)$ or, with probability $\frac{n\alpha_j(p)}{c_i+n(1-\alpha_i(p))}$, to $(p, j), j \ne i$. Equivalently, $\mathcal{X}_n(t), t \ge 0$ is a "mixture" of two processes: the first of them, at the i-th copy of $\mathbb{N}$, is a Poisson process with intensity c_i; the other is composed of independent p-indexed color change Markov chains with values in $\{(p, 0), \dots, (p, N)\}$ and intensity matrix nQ where Q is given by (48.4). This mixture is a Feller process and its generator (after suitable identification of isomorphic objects) is:

$$\mathcal{A}_n\,(x_i)_{i\in\mathcal{N}} = (c_iPx_i)_{i\in\mathcal{N}} + nQ\,(x_i)_{i\in\mathcal{N}},$$

where $(x_i)_{i\in\mathcal{N}}$ belongs to the space $\mathbb{B}$, the Cartesian product of $N + 1$ copies of $c_0 = c_0(\mathbb{N})$, and P is the operator in c_0 given by $P(\xi_p)_{p\ge1} = (\xi_{p+1} - \xi_p)_{p\ge1}$.

For large n, the probability of Poisson-type transition $(p, i) \to (p + 1, i)$ is small (provided $\alpha_i(p) \ne 1$), while transitions $(p, i) \to (p, j)$ occur relatively often. As a result, before Poisson-type transition occurs, the color change process reaches its equilibrium, so that we are at (p, j) with probability $\alpha_j(p)$. Hence, in the limit, the intensity of Poisson-type transition becomes state dependent and equals $\sum_{i=0}^{N} c_i\alpha_i(p)$. The same is true in the case in which $\alpha_i(p) = 1$, when the process, after waiting for an exponential time with parameter c_i jumps right away to $(p+1, i)$. In other words, processes $\{\mathcal{X}_n(t), t \geq 0\}, n \geq 1$ approximate the process in $\mathbb{N}$ with generator:

$$A\left(\xi_p\right)_{p\geq 1} = \left(\sum_{i=0}^{N} c_i\alpha_i(p)\xi_{p+1} - \xi_p\right)_{p\geq 1},$$

that is, a pure birth process with intensities averaged over equilibrium states. Theorem 48.2 makes these intuitions precise.

Example 48.4 A classical situation in which a convex combination (with constant coefficients) of Feller generators is also a generator, is the one where these generators are related to Lévy processes in $\mathbb{R}^n$ – this follows directly by the Lévy–Khintchine Formula, see, for example, [146, 193]. It is precisely in this context that Hille and Phillips write of a cone of generators [180]. By Hunt's Theorem [178, 183], this statement is also true for generators of Lévy processes with values in a Lie group.

For a still more concrete example, let a and b be two positive constants. The operator $Af = af'' + bf'$ with maximal domain in $C_0(\mathbb{R})$ is known to generate a Feller semigroup in this space (see, e.g., [247], Theorem 1, p. 38); the constant a is the diffusion coefficient, and b is the drift coefficient of the related Feller process. Writing $Af = \frac{a}{a+b}[(a+b)f''] + \frac{b}{a+b}[(a+b)f']$, by Theorem 48.1, we have the following alternative description of this process: at each instant of time, with probability $\frac{b}{a+b}$ the process moves to the right with speed $a+b$ or, with probability $\frac{a}{a+b}$, it behaves like the (rescaled) Brownian motion.

Example 48.5 Assume that A, A_0, and A_1 are generators of Feller semigroups in $C_0(\mathcal{S})$, a_0 and a_1 are positive constants, $Af = a_0A_0f + a_1A_1f$ for $f \in \mathcal{D}(A) \cap \mathcal{D}(B)$ and the last set is a core for A. Let $\{X_0(t), t \geq 0\}$, $\{X_1(t), t \geq 0\}$ and $\{X(t), t \geq 0\}$ be the corresponding Feller processes. Writing $A = \frac{a_0}{a_0+a_1}((a_0+a_1)A_0) + \frac{a_1}{a_0+a_1}((a_0+a_1)A_1)$ we see that $\{X(t), t \geq 0\}$ may be interpreted as a process evolving, at each instant of time, as $\{X_0((a_0+a_1)t), t \geq 0\}$ with probability $\frac{a_0}{a_0+a_1}$ or as $\{X_1((a_0+a_1)t), t \geq 0\}$ with probability $\frac{a_1}{a_0+a_1}$.

☞ Chapter's summary

We provide an approximation of the semigroup generated by a convex combination of Feller generators that supports the intuition that the related process chooses its dynamics randomly from the set of generators forming the convex combination. The approximation involves state-space lumping/merging as in Figure 48.1 (for a systematic theory of state-space merging/lumping, see the monographs [220, 219, 221]). The number of examples of convex combinations of Feller generators is at least surprising, see, for example, articles on fast-switching dynamical systems [307] and fast-switching diffusions [185, 350, 351]. A number of examples will be presented in the following chapters.

49

The Dorroh Theorem and the Volkonskii Formula

Following [51], let us study in more detail a very special, but curious case where in the convex combination of the previous chapter, all the operators but one are zero. To simplify notation, we will write A instead of A_0 (the notation A for convex combination of generators will not be used here). In what follows, the reader will hopefully forgive me this change in notation, persuaded that it was worthwhile dropping unnecessary subscripts.

So, let A be the generator of a Feller semigroup $\{e^{tA}, t \ge 0\}$ in $C_0(\mathcal{S})$ and let α be a continuous function on $\mathcal{S}$ with $0 < \alpha_0 \le \alpha(p) \le 1$ for some α_0 and all $p \in \mathcal{S}$. Then, αA with domain $D(A)$ is the generator of another Feller semigroup, say $\{e^{t\alpha A}, t \ge 0\}$, in $C_0(\mathcal{S})$. Indeed, by Dorroh's Theorem [116] (see also the interesting extension in [170]; for further results and further literature see [128]) αA generates a strongly continuous semigroup of contractions in $C_0(\mathcal{S})$, and since αA satisfies the positive maximum principle, this semigroup is composed of positive contractions.

Let us write αA as $\alpha A + (1 - \alpha)B$ where B is the zero operator (*i.e.*, $Bx = 0$, $x \in C_0(\mathcal{S})$) and recall that $e^{tB} = I_{C_0(\mathcal{S})}, t \ge 0$. In particular, B generates the "no movement" process. Theorem 48.1 shows now that a Feller process $\{X_{\alpha A}(t), t \ge 0\}$ related to αA may be described as follows: conditional on being at $p \in \mathcal{S}$, with probability $\alpha(p)$ the process behaves like X_A and with probability $1 - \alpha(p)$ it stays at p. As a result, the process "slows down" as at each point it hesitates whether to move or to stay, and the smaller α is, the slower the process moves.

This agrees with the explicit formula for $\{X_{\alpha A}(t), t \ge 0\}$, known as **Volkonskii's Formula** (see [298], pp. 277–278) involving **change of time** (see [296, 298]), the procedure of changing the speed at which the original process runs through its path. To explain this in more detail, let us assume that $\mathcal{S}$ is, additionally, metrizable and separable, so that there exists a Markov process $\{X_A(t), t \ge 0\}$ related to A with paths that are right-continuous and have left limits (see, e.g., [296], p. 91). Consider the process $\Lambda(t) = \int_0^t \frac{\mathrm{d}s}{\alpha(X_A(s))}$; since it is increasing, we may define $\Theta(t) = \Lambda^{-1}(t)$. Then, $\Theta(t), t \geq 0$ is a time change needed here (see Definition 1.2, p. 180 of [296], see also Proposition 1.1, p. 179 in the same monograph), and Volkonskii's Formula says that the time-changed process:

$$\tilde{X}(t) = X_A(\Theta(t))$$

is a version of the process related to αA. Since $t \leq \Lambda(t) \leq \frac{1}{\alpha_0} t$, we have:

$$\alpha_0 t \leq \Theta(t) \leq t;$$

hence, $\{\tilde{X}(t), t \geq 0\}$ is indeed a "slowed-down" process $\{X_A(t), t \geq 0\}$.

The following direct calculation of the resolvent of αA gives an independent proof of the fact that this operator is the generator of a Feller semigroup, and of Volkonskii's Formula. We recall that by the Feynman–Kac Formula, for any positive c, $A - c\frac{\beta}{\alpha}$ generates the contraction semigroup $\mathrm{e}^{t(A - c\frac{\beta}{\alpha})} f(p) = E_p \, \mathrm{e}^{-c\int_0^t \frac{\beta}{\alpha}(X_A(s))\,\mathrm{d}s} f(X_A(t))$, $p \in \mathcal{S}, f \in C_0(\mathcal{S})$, where E_p stands for expected value conditional on $X_A(0) = p$, and $\beta = 1 - \alpha$. Writing:

$$(\lambda - \alpha A)^{-1} = \left(\lambda - A + \frac{\lambda}{\alpha} - \lambda\right)^{-1} \alpha^{-1} = \left[\lambda - \left(A - \lambda\frac{\beta}{\alpha}\right)\right]^{-1} \alpha^{-1},$$

and taking $c = \lambda$ in the Feynman–Kac Formula, we see that:

$$\begin{aligned}(\lambda - \alpha A)^{-1} f(p) &= \int_0^\infty \mathrm{e}^{-\lambda t} E_p \, \mathrm{e}^{-\lambda \int_0^t \frac{\beta}{\alpha}(X_A(s))\,\mathrm{d}s} \frac{f}{\alpha}(X_A(t))\,\mathrm{d}t \\ &= E_p \int_0^\infty \mathrm{e}^{-\lambda\Lambda(t)} \frac{f}{\alpha}(X_A(t))\,\mathrm{d}t \\ &= \int_0^\infty \mathrm{e}^{-\lambda t} E_p \, f(X_A(\Theta(t)))\,\mathrm{d}t,\end{aligned}$$

for $p \in \mathcal{S}, \lambda > 0, f \in C_0(\mathcal{S})$, with the last equality following by a change of variables – see [296], p. 8 for an appropriate version. The Laplace transform being injective on bounded measurable functions, we have $\mathrm{e}^{t\alpha A} f(p) = E_p f(X_A(\Theta(t)))$ for almost all $t > 0$. Since $t \mapsto E_p f(X_A(\Theta(t)))$ is right-continuous, this equality is true for all $t \geq 0$, completing the proof.

As a corollary to this example and the Trotter Product Formula we see that the process related to a convex combination of two generators may be equivalently seen as a mixture of two time-changed processes.

Dorroh's Theorem [116] cited earlier is a prototype of a whole class of **multiplicative perturbation theorems** – see [128], pp. 201–204 and notes on p. 237. The fact that αA is a generator may also be shown in a straightforward way by going along typical lines for proving such theorems, that is, by replacing multiplicative perturbation by an additive one and using theorems on additive perturbations. Here are the details: we define the operator D as $Df = \alpha Af - Af$ with domain equal to $D(A)$, and note that for every $r \in \mathbb{R}$ and $f \in D(A)$, $[A + rD]f = [(1-r) + r\alpha]Af$. This implies that, as A satisfies the positive maximum principle, so do operators $A + rD$, $r \in [0,1]$. Hence, they are dissipative. Moreover, $\|Df\| \le \sup_{p\in\mathcal{S}} |\alpha(p) - 1| \|Ax\| \le (1-\alpha_0)\|Af\|$ for all $f \in D(A)$, with $0 \le 1 - \alpha_0 < 1$. Therefore, conditions of Theorem 3.2 of [284] (or Theorem 4.11 of [16]) are satisfied, and we conclude that $\alpha A = A + D$ generates a strongly continuous contraction semigroup $\{T_{\alpha A}(t), t \ge 0\}$ in $C_0(\mathcal{S})$. Again, since αA satisfies the positive maximum principle, $\{T_{\alpha A}(t), t \ge 0\}$ is composed of positive contractions.

In the Dorroh Theorem, the assumption $\sup_{p\in\mathcal{S}} \alpha(p) \le 1$ may be relaxed to $\sup_{p\in\mathcal{S}} \alpha(p) = \alpha_1 < \infty$. Indeed, $\alpha_1 A$ is the generator of a Feller semigroup and we may write αA as $\frac{\alpha}{\alpha_1}(\alpha_1 A)$ so that the role of A will be played by $\alpha_1 A$. On the other hand, the assumption $\alpha(p) \ge \alpha_0 > 0$ is essential. This is seen from the following example. Let $\mathcal{S} = [0,\infty]$ and let A be the generator of the Brownian motion with reflecting boundary at $p = 0$, that is, let $A = \frac{\mathrm{d}^2}{\mathrm{d}p^2}$ with domain $\mathcal{D}(A)$ composed of twice differentiable functions f such that there exist finite limits $\lim_{p\to 0} f''(p)$ and $\lim_{p\to\infty} f''(p)$, and $f'(0+) = 0$. Also, let $\alpha(p) = p^2 \wedge 1$. Then αA is the generalized Laplacian of W. Feller with canonical scale $s(p) = p - 1$ and canonical measure $m(p) = 2 - \frac{2}{p}$, $p > 0$ – see [132, 247, 353]. We have $\int_0^1 m(p)\,\mathrm{d}s(p) = -\infty$. In other words, in Feller's boundary classification [132, 193, 247, 353], 0 is an inaccessible boundary for αA. Hence, by Theorem 1, p. 38 of [247], αA with maximal domain in $C_0[0,\infty]$ generates a Feller semigroup. Therefore, αA restricted to the domain of A is not the generator of a semigroup.

Contrary to what the above example may seem to suggest, when α is not strictly positive, in general there may be many extensions of αA that generate Feller semigroups in $C_0(\mathcal{S})$. To see this, let us consider $\mathbb{X} = C[0,\infty]$ and $A = \frac{\mathrm{d}}{\mathrm{d}p}$ with maximal domain; A generates the semigroup $\{T_A(t), t \ge 0\}$ of translations $T_A(t)f(p) = f(p+t)$ in $C[0,\infty]$. We note that the function $\alpha(p) = 2\sqrt{|1-p|} \wedge 2$ is not Lipschitz continuous at $p = 1$. As a result, the solutions to the Cauchy problem related to $p'(t) = \alpha(p)$ are not unique. In particular, we may define two semiflows $\{\pi_t^i, t \ge 0\}$, $i = 1, 2$ in $[0,\infty)$ such that $\frac{\mathrm{d}\pi_t^i}{\mathrm{d}t} = \alpha(\pi_t^i)$. The first of these is determined by:

$$\pi_t^1(p) = \begin{cases} p + 2t\sqrt{1-p} - t^2, & 0 \le t \le \sqrt{1-p},\ p \in [0,1), \\ p + t^2 + 2t\sqrt{p-1}, & 0 \le t \le 1 - \sqrt{p-1},\ p \in [1,2), \\ p + 2t, & t \ge 0,\ p \ge 2, \end{cases}$$

and the second is the same as the first except that for $p = 1$ we put $\pi_t^2(p) = 1, t \geq 0$. Since, as $t \to 0+$, $\pi_t^i(p)$ converges to p uniformly in $p \in [0, \infty)$, the corresponding semigroups $\{S_i(t), t \geq 0\}$ in $C[0, \infty]$ given by $S_i(t)f(p) = f(\pi_t^i(p))$ are strongly continuous. Moreover, for $f \in \mathcal{D}(A)$ we have, by the Lagrange Theorem, $\frac{1}{t}(S_i(t)f(p) - f(p)) = Af(\xi_i)\frac{\pi_t^i(p)-\pi_0^i(p)}{t} = Af(\xi_i)\frac{\mathrm{d}\pi_t^i(p)}{\mathrm{d}t}_{|t=\tau_i} = Af(\xi_i)\alpha(\pi_{\tau_i}^i(p))$, where $\xi_i = \xi_i(t, p) \in (p, \pi_t^i(p))$ and $\tau_i = \tau_i(t, p) \in (0, t)$. Hence, by uniform convergence of π_t^i and uniform continuity of Af and α, $Af(\xi_i)$ converges uniformly to $Af(p)$ and $\alpha(\pi_{\tau_i}^i(p))$ converges to $\alpha(p)$, as $t \to 0+$. This shows that generators of these semigroups (which are different, semigroups being different) are extensions of αA.

In the same way as the semiflows and the corresponding semigroups presented here differ in their behavior at $p = 1$, in general the various semigroups related to extensions of αA will differ in their behavior at the "boundary" where $\alpha = 0$.

The literature devoted to the problem of when $\alpha\frac{\mathrm{d}}{\mathrm{d}p}$, where α is assumed continuous, generates a semigroup in a space of continuous functions is abundant and includes [30, 31, 109, 110]; I would like to thank Charles Batty for these references.

☞ Chapter's summary

The curious case in which all but one generators in a convex combination are equal zero is studied: the related stochastic process randomly slows down. In particular, we express this process with the help of Volkonskii's change of time formula. Our path leads through the archetypical example of multiplicative perturbations, that is, the Dorroh Theorem, and we complete the chapter with a short discussion of when the operator of first derivative composed with bounded multiplication operator is a generator.

50

Convex Combinations in Biological Models

$\sum_{i=0}^{n} \binom{n}{i} p^i (1-p)^{n-i}$  $_i$

We have noted previously that the main theorem of Chapter 48 is in a sense a converse to the frequent jumps theorem of Chapter 45. In other words, we have already encountered examples of convex combinations of Feller generators. For instance, in the model of gene regulation, the generators of semigroups describing motions along the integral curves of (46.1) are for $f \in C^1$ given by:

$$A_i f(x_1, x_2) = (F_1(i) - x_1)\frac{\partial f}{\partial x_1}(x_1, x_2) + (F_2(i) - x_2)\frac{\partial f}{\partial x_2}(x_1, x_2),$$

$i = 0, 1, 2$, while the limit equation (46.2) is related to the generator:

$$Af(x_1, x_2) = \left(\frac{\alpha_0 + \alpha_1}{\sigma} - x_1\right)\frac{\partial f}{\partial x_1}(x_1, x_2) + \left(\frac{\alpha_0 + \alpha_2}{\sigma} - x_2\right)\frac{\partial f}{\partial x_2}(x_1, x_2).$$

Hence, recalling the definition of F_i's we obtain:

$$A = \sigma^{-1}\sum_{i=0}^{2} \alpha_i A_i,$$

which is a convex combination since $\sum_{i=0}^{2} \alpha_i = \sigma$.

Similarly, in the model of gene expression, the generators of deterministic motions (46.3) in the approximating processes are:

$$A_i f(x_1, x_2) = (-cx_1 + di)\frac{\partial f}{\partial x_1}(x_1, x_2) + (kx_1 - lx_2)\frac{\partial f}{\partial x_2}(x_1, x_2),$$

$i \in \{0, \dots, N\}$, while in the limit the generator is (see (46.5)):

$$Af(x_1, x_2) = (-cx_1 + Nd\alpha(x_1, x_2))\frac{\partial f}{\partial x_1}(x_1, x_2) + (kx_1 - lx_2)\frac{\partial f}{\partial x_2}(x_1, x_2),$$

where $\alpha(x_1, x_2) = \frac{r\beta_0(x_1,x_2)}{r\beta_0(x_1,x_2)+\beta_1(x_1,x_2)}$. Noting that:

$$\sum_{i=0}^{N} i\binom{N}{i}\alpha^i(1-\alpha)^{N-i} = N\alpha,$$

we obtain:

$$A = \sum_{i=0}^{N} \binom{N}{i}\alpha^i(1-\alpha)^{N-i}A_i,$$

a binomial-type convex combination. This result agrees with our intuition: since the gene copies are switched off and on independently, and in the case of frequent jumps, given the state (x_1, x_2), the probability that a single gene copy is active is $\alpha(x_1, x_2)$, then $\binom{N}{i}\alpha^i(x_1, x_2)(1-\alpha_i(x_1, x_2))^{N-i}$ is the probability that there are exactly i active gene copies.

Finally, let us consider the model of dynamics of a *solea solea* or *Engraulis encrasicholus* population with both age and vertical structures, presented in [11] (see also [22, 23, 306]). In the model, the fish habitat is divided into N spatial patches and the fish densities or age profiles n_i in the i-th patch satisfy the following system of equations:

$$\begin{aligned}\frac{\partial n_i(t,a)}{\partial t} + \frac{\partial n_i(t,a)}{\partial a} &= -\mu_i(a)n_i(t,a) + \epsilon^{-1}\sum_{j=1}^{N} k_{ij}(a)n_j(t,a),\\ n_i(t,0) &= \int_0^\infty b_i(a)n_i(t,a)\,\mathrm{d}a, \quad i = 1,\dots,N,\\ n_i(0,a) &= \phi_i(a), \quad i = 1,\dots,N, \end{aligned} \tag{50.1}$$

where "t" stands for time, "a" stands for age, ϕ_i is the initial population distribution in patch i, and μ_i and b_i are age-specific and patch-specific mortality and birth rates.

In the absence of the terms $\epsilon^{-1}\sum_{j=1}^{N} k_{ij}(a)n_j(t,a)$ each of the patches could have been treated separately, and the population densities there would satisfy the McKendrick Equation [268, 302] (called also Sharpe–Lotka–McKendrick Equation [341, 342], or Lotka–McKendrick Equation [184]). In this equation, the term $\frac{\partial n_i(t,a)}{\partial a}$ describes aging process (without the other factors, the operator $-\frac{\partial}{\partial a}$ would result in shifting the age profile to the right), μ_i describes deaths (the term $-\mu_i n_i$ takes away a portion of the age profile shifted by $-\frac{\partial}{\partial a}$ – see (50.4), further on), and the boundary condition $n_i(t,0) = \int_0^\infty b_i(a)n_i(t,a)\,\mathrm{d}a$ describes the process of births ($b_i(a)n_i(t,a)$ may be thought of as the number of individuals born to all fish at age a living in patch i at time t).

The matrix $k(a) = (k_{ij}(a))$ is composed of intensities of movements between patches that occur on a daily basis: the sum of entries in each column of the matrix is zero. The factor ϵ^{-1} (with $\epsilon \ll 1$) corresponds to the fact that the age-related processes and vertical migrations (between the patches) occur at different time scales, a day being the fast time scale as compared to the fish lifetime.

The main question addressed in [11] is that of whether in approaching modeling such populations one may disregard the vertical migration to work with a model that has been aggregated, or averaged, over the whole water column. To this end, the authors assume that the matrix k is irreducible, and hence possesses the unique normalized right eigenvector $\mathbf{v}(a) = (v_i(a))_{i=1,\ldots,N}$ corresponding to the simple dominant eigenvalue 0 [155]. Moreover, they consider the case in which the population distribution over the patches does not change in time and agrees with the stable population distribution dictated by the matrix k:

$$\frac{n_i(t,a)}{n(t,a)} = v_i(a), \quad i = 1,\ldots,N, a \geq 0,$$

where $n = \sum_{i=1}^{N} n_i$. This is tacitly assuming that migrations governed by k hold so fast, as compared to aging processes, that the population distribution over the patches reaches the (age-specific) equilibrium much before the aging process intervenes. As in our theorems, this corresponds to letting in (50.1) $\epsilon \to 0$. In such a simplified, aggregated model, the population density satisfies the McKendrick Equation with averaged birth and mortality rates:

$$\frac{\partial n(t,a)}{\partial t} + \frac{\partial n(t,a)}{\partial a} = -\mu_{\mathrm{a}}(a)n(t,a),$$
$$n(t,0) = \int_0^\infty b_{\mathrm{a}}(a)n(t,a)\,\mathrm{d}a, \qquad n(0,a) = \phi_{\mathrm{a}}(a), \qquad (50.2)$$

where "a" stands for "aggregated," $\mu_{\mathrm{a}} = \sum_{i=1}^{N} v_i\mu_i$ and $b_{\mathrm{a}} = \sum_{i=1}^{N} v_i b_i$. Here, the weights v_i mark the underlying, hidden spatial structure.

Equation (50.1) is related to a semigroup $\{\mathcal{T}_\epsilon(t), t \geq 0\}$ of operators in $\mathbb{Y}$, the Cartesian product of N copies of L^1 (with norm $\|(\phi_i)_{i=1,\ldots,N}\|_{\mathbb{Y}} = \sum_{i=1}^{N} \|\phi_i\|_{L^1}$), where L^1 be the space of Lebesgue integrable functions on $[0,\infty)$. The generator of this semigroup is $\mathcal{A}_\epsilon = \mathcal{A} + \epsilon^{-1}\mathcal{K}$ where:

$$\mathcal{A}(\phi_i)_{i=1,\ldots,N} = (A_i\phi_i)_{i=1,\ldots,N}, \phi_i \in \mathcal{D}(A_i),$$

$$\mathcal{K}(\phi_i)_{i=1,\ldots,N} = \left(\sum_{j=1}^{N} k_{ij}\phi_j\right)_{i=1,\ldots,N}, \phi_i \in L^1,$$

and A_i is the operator in L^1 given by:

$$A_i\phi(a) = -\phi'(a) - \mu_i(a)\phi(a)$$

with domain $\mathcal{D}(A_i)$ composed of absolutely continuous functions ϕ such that $\phi' \in L^1$ and $\phi(0) = \int_0^\infty b_i(a)\phi(a)\,\mathrm{d}a$, $i = 1, \ldots, N$.

On the other hand, (50.2) is related to the semigroup $\{S(t), t \geq 0\}$ in L^1 generated by the operator A_a given by:

$$A_\mathrm{a}\phi(a) = -\phi'(a) - \mu_\mathrm{a}(a)\phi(a)$$

with domain $\mathcal{D}(A_\mathrm{a})$ composed of absolutely continuous functions ϕ such that $\phi' \in L^1$ and $\phi(0) = \int_0^\infty b_\mathrm{a}(a)\phi(a)\,\mathrm{d}a$.

To see that we are actually dealing with a semigroup generated by a convex combination of generators, we need to pass to the space $C = C[0, \infty]$ of continuous functions on $[0, \infty)$ with limits at infinity. To this end, we assume that mortality and birth rates μ_i, b_i, $i = 1, \ldots, N$ belong to C. Then, as we shall see shortly, the predual $A_{i,*}$ to A_i in C is given by:

$$A_{i,*}f(a) = f'(a) - \mu_i(a)f(a) + b_i(a)f(0),$$

and a similar statement is true for $A_{\mathrm{a},*}$. Notably, all of these operators share the same domain C^1 composed of differentiable functions $f \in C$ with $f' \in C$. Moreover, $A_{\mathrm{a},*}$ is a convex combination of $A_{i,*}$ with (age dependent) $\alpha_i = v_i$:

$$A_{\mathrm{a},*}f = \sum_{i=1}^N \alpha_i A_{i,*}f, \qquad f \in C^1.$$

Therefore, we are left with showing the claim on the predual of A_i. To simplify notation, we drop the subscript to consider the operator A in L^1 given by:

$$A\phi(x) = -\phi'(x) - \mu(x)\phi(x) \quad \text{(almost surely)},$$

with domain $\mathcal{D}(A)$ composed of absolutely continuous functions ϕ such that $\phi' \in L^1$ and $\phi(0) = \int_0^\infty b(x)\phi(x)\,\mathrm{d}x$. Here, $b \in C$ and $\mu \in C$ are non-negative. We will show that A generates a strongly continuous semigroup $\{T_t, t \geq 0\}$ in L^1 such that its predual in C is generated by:

$$A_*f(x) = f'(x) - \mu f(x) + Bf(x) \tag{50.3}$$

where $Bf(x) = f(0)b(x)$, with domain C^1 composed of differentiable functions $f \in C$ with $f' \in C$. We note that B is bounded with $\|B\| = b_{\max}$, where $b_{\max} = \max_{x\geq 0} b(x)$. Clearly,

$$A_0 f(x) = f'(x) - \mu(x)f(x)$$

(with domain equal to C^1) generates a strongly continuous contraction semigroup $\{P(t), t \geq 0\}$ in C, given by:

$$P(t)f(x) = f(x+t)e^{-\int_x^{x+t} \mu(y)\,dy}, \qquad f \in C, t \geq 0.$$

By the Phillips Perturbation Theorem, it follows that A_* generates the semigroup $\{T(t), t \geq 0\}$ given by $T(t) = \sum_{n=0}^{\infty} T_n(t)$ where:

$$T_0(t) = P(t), \text{ and } T_{n+1}(t) = \int_0^t P(t-s)BT_n(s)\,ds, n \geq 0,$$

and we have $\|T_n(t)\| \leq \frac{t^n b_{\max}^n}{n!}, n \geq 0, \|T(t)\| \leq e^{b_{\max} t}$. It is easy to see that $P^*(t)$ the dual to $P(t)$ leaves L^1 invariant, and is given by:

$$P^*(t)\phi(x) = \begin{cases} e^{-\int_0^t \mu(x-s)\,ds}\phi(x-t), & x \geq t, \\ 0, & x < t. \end{cases} \tag{50.4}$$

Moreover, $\{P^*(t), t \geq 0\}$ is a strongly continuous contraction semigroup in L^1. Since:

$$P(t)Bf = f(0)P(t)b, \qquad t \geq 0, f \in C, \tag{50.5}$$

introducing $k_f(s) = e^{-\int_0^s \mu(u)\,du} f(s)$, we see that:

$$P(t)BP(s)f = k_f(s)P(t)b.$$

Hence,

$$T_1(t) = \int_0^t k_f(t-s)P(s)b\,ds = k_f(t) * P(t)b,$$

and so, by induction and (50.5),

$$T_{n+1}(t) = [k_b(t)]^{n*} * k_f(t) * P(t)b = [k_b(t)]^{n*} * T_1(t),$$

where "$n*$" denotes the n-th convolution power.

We check directly that the dual $T_1^*(t)$ to $T_1(t)$ leaves L^1 invariant (even though B^* does not have this property); $T_1^*(t)$ as restricted to L^1 is given by the formula:

$$T_1^*(t)\phi(x) = \begin{cases} e^{-\int_0^x \mu(y)\,dy} \int_0^\infty b(s+t-x)e^{-\int_s^{s+t-x} \mu(u)\,du}\phi(s)\,ds, & t \geq x, \\ 0, & t < x. \end{cases}$$

Hence, so do $T_n^*(t), t \geq 0, n \geq 2$ and we have:

$$T_{n+1}^*(t) = [k_b(t)]^{n*} * T_1^*(t), n \geq 0, t \geq 0. \tag{50.6}$$

Clearly, $\|T_n^*(t)\| \leq \frac{t^n b_{\max}^n}{n!}, n \geq 0$, and so $T^*(t) := \sum_{n=0}^{\infty} T_n^*(t)$, with the series converging almost uniformly, leaves L^1 invariant as well. We note that, by (50.6),

$$T^*(t) = P^*(t) + T_1^*(t) + k(t) * T_1^*(t)$$

where $k(t) = \sum_{n=1}^{\infty}[k_b(t)]^{*n}$ (comp. [184], Chapter 1). The last series converges almost uniformly in $t \geq 0$, being dominated by $\sum_{n=1}^{\infty} \frac{b_{\max}^n t^{n-1}}{n!}$; in particular k is continuous. $\{T^*(t), t \geq 0\}$ is a semigroup, since so is $\{T(t), t \geq 0\}$, and its strong continuity follows by strong continuity of $\{P^*(t), t \geq 0\}$ and $\{T_1^*(t), t \geq 0\}$.

Finally, for $\lambda > b_{\max}$, the solution to the resolvent equation for A:

$$(\lambda + \mu(x))\phi(x) + \phi'(x) = \psi(x), \qquad \psi \in L^1,$$

is:

$$(\lambda - A)^{-1}\psi(x) = \phi(x) = \frac{(b, H_\lambda^*\psi)}{1 - (e_\lambda, b)} e_\lambda + H_\lambda^*\psi(x)$$

where:

$$e_\lambda(x) = \mathrm{e}^{-\lambda x}\mathrm{e}^{-\int_0^x \mu(y)\,\mathrm{d}y}, \quad H_\lambda^*\psi(x) = \int_0^x \mathrm{e}^{-\lambda y}\mathrm{e}^{-\int_{x-y}^x \mu(z)\,\mathrm{d}z}\psi(x-y)\,\mathrm{d}y,$$

and for $f \in C$ and $\phi \in L^1$ we write $(f, \phi) := \int_0^\infty f(x)\phi(x)\,\mathrm{d}x$. On the other hand,

$$\int_0^\infty \mathrm{e}^{-\lambda t}P^*(t)\,\mathrm{d}t = H_\lambda^*, \qquad \int_0^\infty \mathrm{e}^{-\lambda t}T_1^*(t)\phi\,\mathrm{d}t = (b, H_\lambda^*\phi)e_\lambda,$$

and $\int_0^\infty \mathrm{e}^{-\lambda t}k_b(t)\,\mathrm{d}t = (e_\lambda, b)$. Therefore,

$$\begin{aligned}\int_0^\infty \mathrm{e}^{-\lambda t}T^*(t)\psi\,\mathrm{d}t &= H_\lambda^*\psi + \left(1 + \int_0^\infty \mathrm{e}^{-\lambda t}k(t)\,\mathrm{d}t\right)\int_0^\infty \mathrm{e}^{-\lambda t}T_1^*(t)\psi\,\mathrm{d}t \\ &= H_\lambda^*\psi + \sum_{n=0}^{\infty}(e_\lambda, b)^n\,(b, H_\lambda^*\psi)\,e_\lambda = (\lambda - A)^{-1}\psi.\end{aligned}$$

This implies that A is the generator of $\{T^*(t), t \geq 0\}$, and completes our proof.

It should be noted that, even though the setup of L^1 is natural for population dynamic models, including the McKendrick model, the perspective of the space of measures leads to their significant unification [301, 302]. Hence, the approach via $C[0, \infty]$ (as the predual to the space of measures) presented earlier may have some merit in itself.

Exercise 50.1 Describe the convex combinations of Feller generators involved in the models of a manufacturing system and of an oligopoly of Chapter 47.

Exercise 50.2 Let us consider the problem of modeling a McKendrick-type population with very large death rate. (I thank Mustapha Mokhtar-Kharoubi for suggesting this problem.) For the limit to be nontrivial, we need to balance the death rate with the growth rate. To this end, we take $\mathbb{X} = C_0[0, \infty)$, the space of continuous functions on $\mathbb{R}^+$ vanishing at infinity.[1] We consider the sequence of operators given by (cf. (50.3))

$$A_n f = f' - n\mu f + nf(0)b = \left(\frac{\mathrm{d}}{\mathrm{d}x} - nM + nB\right) f, \qquad (50.7)$$

on a common domain D composed of continuously differentiable members of $\mathbb{X}$ such that $f' \in \mathbb{X}$ and $\mu f \in \mathbb{X}$. Here, μ and b are positive functions on $\mathbb{R}^+$ such that $b \in \mathbb{X}$ and $q := \frac{b}{\mu} \in \mathbb{X}$. We assume $q(0) = 1$ and $\|q\| \le 1$, so that the birth rate does not exceed the death rate, and for newborns they are equal. Also, M and B are operators described in Exercise 1.5, while $\frac{\mathrm{d}}{\mathrm{d}x}$ is defined for continuously differentiable f such that $f' \in \mathbb{X}$; thus, $D = D(M) \cap D(\frac{\mathrm{d}}{\mathrm{d}x})$.

(a) The formula:

$$T_n(t)f(x) = f(x+t)\mathrm{e}^{-n\int_x^{x+t} \mu(y)\,\mathrm{d}y}$$

defines a strongly continuous semigroup in $\mathbb{X}$. Show that the domain of its generator G_n contains D and:

$$G_n f = f' - n\mu f, \qquad f \in D.$$

Moreover, check that $C^1_{00}[0, \infty)$, the space of continuously differentiable functions with compact supports is contained in D, and is left invariant by the semigroup. By Exercise 15.2, conclude that $C^1_{00}[0, \infty)$, is a core for G_n, and so is D.

(b) By (a) and the Phillips Perturbation Theorem, A_n is closable and its closure $A_n = G_n + nB$ is a generator. Use (1.8) to show that $\mathrm{e}^{t(M+B)}, t \ge 0$ are contractions. Then, use the Lie–Trotter Formula to show that $\mathrm{e}^{tA_n}, t \ge 0$ are contractions, as well.

(c) Assume $q \in D$. Then, $A_n q = q'$, and for $f \in D$, $\lim_{n\to\infty} \frac{1}{n} A_n f = -\mu f + Bf$. By (1.8), $\lim_{t\to\infty} \mathrm{e}^{t(M+B)} f = f(0)q := Pf$. Use Kurtz's Theorem to conclude that in this case:

$$\lim_{n\to\infty} \mathrm{e}^{tA_n} f = \mathrm{e}^{-at} f(0)q, \qquad (50.8)$$

where $a := -q'(0) \le 0$ since, by assumption, $q(0) = 1$ and $\|q\| \le 1$. In particular, the regularity space is spanned by q.

[1] We work in $C_0[0, \infty)$ rather than in $L^1(\mathbb{R}^+)$, because the limit involves a point mass concentrated at the origin, the phenomenon $L^1(\mathbb{R}^+)$ is not likely to capture. Also, we prefer $C_0[0, \infty)$ over $C[0, \infty]$, because we want to use the handy formula (1.8).

Since the dual to the operator appearing on the right-hand side of (50.8) maps a measure on $\mathbb{R}^+$ to a scalar multiple of the Dirac measure at 0, the formula may be interpreted by saying that under the stated conditions, the limit population is composed solely of newborn individuals, and a necessary and sufficient condition for persistence of the population is $q'(0) = 0$.

☞ Chapter's summary

We give a few examples of convex combinations of Feller generators in biological models. These include models of gene expression and gene regulation, and a model of fish dynamics. The latter model uses the important McKendrick semigroup as a building block. For an example of a convex combination of Feller generators in actuarial sciences, see [72, Example 7].

51

Recombination

In this chapter we present another example of irregular convergence of semigroups. This example involves both Kurtz's Lemma and a convex combination of generators, but not in a context of Feller operators, but Markov operators.

To start with biological background, recall that recombination, one of the basic factors influencing genetic variability, is a process by which genetic material is transferred between chromosomes. Its models in population genetics have a long history, reflecting its importance. Recently there have been at least three monographs that thoroughly explain recombination, mostly in the context of the coalescent in reverse time. These are Durrett [119], Hein et al. [175], and Wakeley [335]. Recombination is providing additional genetic variability to a population, which supplements mutation and can counterbalance the effects of genetic drift and selective sweeps that, themselves, strive to reduce the variability. In this way, it contributes to the evolutionary process (see, for example, [156]).

In **cross-over**, a particular type of recombination taking place in the course of meiotic division, homologous (alike) chromosomes exchange parts of their genetic material as depicted in Figure 51.1. For any two loci (i.e., genes or other sequences with given locations along the chromosome), the probability that there is a cross-over (or rather an odd number of cross-overs) between them, is called the recombination fraction, r. If the **alleles** (variants of sequences) at these two loci were (A_1, B_1) on chromosome 1 and (A_2, B_2) on chromosome 2, respectively, and recombination occurs between them, then they are replaced by (A_1, B_2) and (A_2, B_1), respectively.

Figure 51.1 Cross-over

In this chapter we study semigroups related to a continuous-time Moran-type model, discussed in [65], of a population under mutations, genetic drift and recombination, where there are only two recombining loci. (The general case of $n \geq 2$ loci is dealt with in the next chapter.) To this end, we consider a population of $M = 2N$ individuals. Each individual is represented as a pair of integer-valued random variables; the i–th individual being a pair (X_i, Y_i), $i = 1, \dots, 2N$. Here, X_i and Y_i describe allelic types at the first and second locus of the ith individual; if alleles are numbered, X_i and Y_i may be thought as integer-valued random variables (changing in time). We assume that these pairs are exchangeable (i.e., that permutations of the pairs do not change the distribution of the entire population), that each pair evolves in time as a pair of independent, nonexplosive Markov chains, and that each individual (pair) evolves independently from the other ones, but with the same transition probabilities; this models mutations at two sites of a chromosome in each individual. More specifically, we assume that the process of mutations on the first site in each individual is described by means of a strongly continuous semigroup $\{S_X(t), t \geq 0\}$ of Markov operators in l^1. As in Chapter 27, this means that if $x \in l^1$ is the distribution of the type of an allele at time 0 then $S_X(t)x$ is the distribution of allele types at time t. Analogously, we assume that the process of mutations on the second site is governed by a semigroup $\{S_Y(t), t \geq 0\}$.

Because we assume that new mutations occur at the first site independently of new mutations at the second site, the dynamics of the joint two-dimensional distributions in an individual is described by means of the tensor product semigroup $\{S_{XY}(t), t \geq 0\}$, $S_{XY}(t) = S_X(t) \otimes S_Y(t)$ (see, e.g., [269], or the next chapter). In other words, if m is a matrix of joint distribution of allelic types at time 0, then at time t the joint distribution is $S_X(t)mS_Y^*(t)$, where $S_X(t)$ and $S_Y(t)$ are identified with matrices and $*$ denotes the transpose. The natural Banach space where S_{XY} is defined is the space $\mathcal{M}$ of infinite matrices with absolutely summable entries. We note that $\mathcal{M}$ is isometrically isomorphic to l^1 and we may meaningfully speak of distributions, Markov operators, and so on in this space.

We also incorporate recombination and genetic drift in the model by assuming that each individual's life-length is an exponential random variable with parameter λ and that at the moment of individual's death, the pair (X_i, Y_i) by which it is represented is replaced by another pair in the following manner. With probability $1 - r$, where $r \in [0, 1]$ is a given parameter, one of the pairs (X_j, Y_j), $j = 1, \dots, 2N$ is drawn at random to replace the deceased one; with probability r the X – variable is drawn at random first and the Y-variable next, independently from the result of the first draw. As a consequence, a new pair becomes one of the already existing pairs (including the one just deceased) (X_j, Y_j), each of them with probability $(1-r)\frac{1}{2N} + r\frac{1}{(2N)^2}$, or a "mixed one" (X_j, Y_k), $j \neq k$ each with probability $r\frac{1}{(2N)^2}$. The same procedure may be described in the following way. Three numbers i, j, and k are chosen with replacement from $1, 2, \ldots, 2N$; then X_i is replaced by X_j, and Y_i is replaced by Y_j with probability $1-r$ or by Y_k with probability r. By symmetry of the procedure just described it is clear that after an individual's death the members of the population remain exchangeable (details may be found in [65]).

The first vital question is: can we argue as in Chapter 27 to find a formula or an equation describing evolution of joint distributions of a pair of individuals? It may seem surprising initially, but the answer is in the negative, and the guilty party is recombination. To see this, note that if we trace the genealogy of a chosen pair of individuals back in time and if we are lucky enough, then the first event we encounter is the one where a "dying" individual was replaced by a different, but entire individual. In other words, we find a common ancestor of the considered pair. However, we could also encounter a situation in which the dying individual inherited parts of his genetic makeup from two different individuals. In this case, the distribution of allelic types of our pair turns out to depend on the distribution of allelic types of three individuals, with the third individual "hitchhiking" to the considered genealogy (see, e.g., [315] and the papers cited there). This does not allow finding a closed formula, similar to (27.1). In fact, in populations with recombination, a distribution of any sample of individuals depends on the distributions of larger samples. Hence, in studying such populations we cannot simplify the analysis by considering samples, but need to take into account the entire population.

We could circumvent the difficulty by taking a slightly more modest approach [65]. Instead of describing the entire distribution of a randomly chosen pair of individuals, say the ith and the jth individual ($i \neq j, i, j = 1, \ldots, 2N$), that is, instead of describing the distribution of the four-dimensional random vector $(X_i(t), Y_i(t), X_j(t), Y_j(t))$, let us try to describe the pair of two distributions, composed of the distribution of the two-dimensional vector $(X_i(t), Y_i(t))$ and of the distribution of the vector $(X_i(t), Y_j(t))$ (see Figure 51.2). (By exchangeability, none of these distributions depends on the choice of i and j.)

Let $E(t)$ denote the distribution of $(X_i(t), Y_i(t))$, and let $F(t)$ denote the distribution of $(X_i(t), Y_j(t))$. As we shall see, these two objects are nicely coupled, and the pair forms a closed system in the sense that $(E(t), F(t)), t \geq 0$ depends merely on $(E, F) := (E(0), F(0))$ and not on the distributions of larger vectors. In fact, there is a semigroup in the Cartesian product $\mathcal{M}^2 = \mathcal{M} \times \mathcal{M}$ of two copies of $\mathcal{M}$, governing the dynamics of the pair. (We note that with the norm $\left\| \binom{E}{F} \right\| = \|E\|_{\mathcal{M}} + \|F\|_{\mathcal{M}}$, the space $\mathcal{M}^2$ is isometrically isomorphic to l^1.

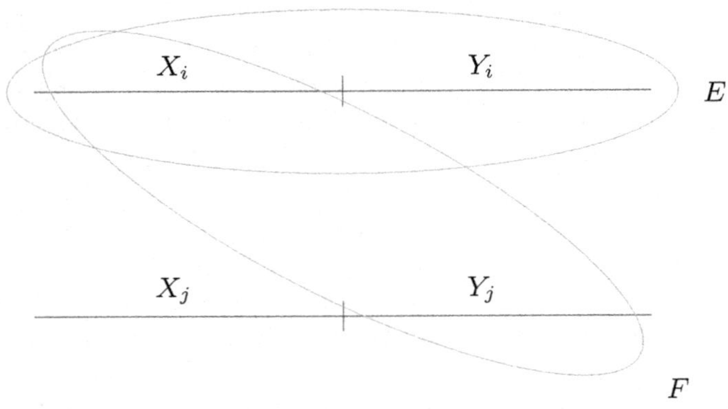

Figure 51.2 A closed system of two distributions, E and F

Hence, as it was the case with $\mathcal{M}$, it is meaningful and quite intuitive to speak of distributions and Markov operators in this space.)

To see this let us consider first the case in which in the interval $[0, t]$ there were no death/birth events. In other words, in this time interval each site in each individual was undergoing an independent process of mutations. In this case, the distribution of $(X_i(t), Y_i(t))$ is $S_{XY}(t)E$. Analogously, the distribution of $(X_i(t), Y_j(t))$ is $S_{XY}(t)F$ and so the dynamic is governed by the Cartesian product semigroup:

$$\mathcal{S}(t)\begin{pmatrix} E \\ F \end{pmatrix} = \begin{pmatrix} S_{XY}(t)E \\ S_{XY}(t)F \end{pmatrix}. \tag{51.1}$$

Moreover, at random times, the action of this semigroup is perturbed by death/birth events. To deal with these, we denote by $\binom{E(t)}{F(t)}$ and $\binom{\tilde{E}(t)}{\tilde{F}(t)}$ the considered pair of distributions right before and right after a birth/death event that took place at time t, respectively, and claim that:

$$\begin{pmatrix} \tilde{E}(t) \\ \tilde{F}(t) \end{pmatrix} = \begin{pmatrix} 1 - r\frac{2N-1}{(2N)^2} & r\frac{2N-1}{(2N)^2} \\ \frac{2}{(2N)^2} & 1 - \frac{2}{(2N)^2} \end{pmatrix} \begin{pmatrix} E(t) \\ F(t) \end{pmatrix}. \tag{51.2}$$

To prove this claim, assume without loss of generality that the randomly chosen individuals were the individuals number 1 and 2, and the birth/death event involved individuals number i, j, and k. If $i \neq 1$, the distribution $E(t)$ does not change, because $(X_1(t), Y_1(t))$ is left intact. Neither does $E(t)$ change if $i = 1$ but there is no recombination event; in this case $(X_1(t), Y_1(t))$ is replaced by $(X_j(t), Y_j(t))$, which is a pair of variables with the same distribution as that of $(X_1(t), Y_1(t))$. In the case of recombination, $(X_1(t), Y_1(t))$ is replaced by $(X_j(t), Y_k(t))$. The distribution of this vector is either $E(t)$ (if $j = k$), or $F(t)$

(in the other case). Since the probability of recombination event with $j \neq k$ is $r\frac{2N-1}{(2N)^2}$,

$$\tilde{E}(t) = \left(1 - r\frac{2N-1}{(2N)^2}\right)E(t) + r\frac{2N-1}{(2N)^2}F(t),$$

proving the first part of formula (51.2). The second part is proven similarly.

The matrix on the right-hand side of (51.2), denoted Θ in what follows, describes a change in $\binom{E}{F}$ caused by a single birth/death event. Noting that the time to the first death/birth event is an exponential random variable with parameter $2\lambda N$ (as a minimum of $2N$ independent, exponential random variables with parameter λ), we see that all changes caused by birth/death events up to time $t \geq 0$ are described by the operator (compare (9.1)):

$$e^{-2\lambda Nt}e^{2\lambda Nt\Theta} = e^{-2\lambda Nt}\sum_{n=0}^{\infty}\frac{(2\lambda Nt)^n}{n!}\Theta^n.$$

As a result, the dynamics of the $\binom{E}{F}$ pair is described by the sum of two operators:

$$\mathcal{G} = \mathcal{A} + \lambda\mathcal{Q} \tag{51.3}$$

where $\mathcal{A}$ is the generator of the semigroup $\{\mathcal{S}(t), t \geq 0\}$ responsible for mutations, and $\mathcal{Q} = 2N(\Theta - I_{\mathcal{M}^2})$ so that $\lambda\mathcal{Q}$ is the generator of the semigroup responsible for birth/death events ($I_{\mathcal{M}^2}$ denotes the identity operator in $\mathcal{M}^2$). Since the latter operator is bounded, $\mathcal{G}$ is the generator of a strongly continuous semigroup, and since both $\mathcal{A}$ and $\lambda\mathcal{Q}$ generate Markov semigroups, so does $\mathcal{G}$.

Formula (51.3) may be rewritten as follows. By (51.1), the domain $D(\mathcal{A})$ of the generator of $\{\mathcal{S}(t), t \geq 0\}$ is $D(G) \times D(G)$ where G is the generator of $\{S_{XY}(t), t \geq 0\}$, and $\mathcal{A}\binom{E}{F} = \binom{GE}{GF}$, for $\binom{E}{F} \in D(\mathcal{A})$. Hence, for such $\binom{E}{F}$, the vector $\binom{E(t)}{F(t)} = e^{t\mathcal{G}}\binom{E}{F}$ satisfies:

$$\begin{aligned}\frac{dE(t)x}{dt} &= GE(t) - r\lambda\frac{2N-1}{2N}E(t) + r\lambda\frac{2N-1}{2N}F(t),\\ \frac{dF(t)x}{dt} &= GE(t) + \frac{\lambda}{N}E(t) - \frac{\lambda}{N}F(t), \qquad t \geq 0.\end{aligned} \tag{51.4}$$

We are interested in the limit of our semigroups as $\lambda \to \infty$; this is the case where the birth/death events are relatively faster than mutations. (Comp. [65] where the limit as $t \to \infty$ is considered, and applications to linkage disequilibrium and microsatellites are discussed.) For notational convenience and to stress dependence on λ, we will write $\mathcal{G}_\lambda$ for the generator defined in (51.3). In this notation, we are interested in the limit:

$$\lim_{\lambda\to\infty} \mathrm{e}^{t\mathcal{G}_\lambda}.$$

To solve this problem, we note first that $\mathcal{Q} = \begin{pmatrix} -a & a \\ b & -b \end{pmatrix}$ where $a = r\frac{2N-1}{2N}$ and $b = \frac{1}{N}$. Thus, by Exercise 13.3,

$$\mathrm{e}^{t\mathcal{Q}} = \frac{1}{a+b}\begin{pmatrix} b + a\mathrm{e}^{-(a+b)t} & a - a\mathrm{e}^{-(a+b)t} \\ b - b\mathrm{e}^{-(a+b)t} & a + b\mathrm{e}^{-(a+b)t} \end{pmatrix}.$$

It follows that:

$$\lim_{t\to\infty} \mathrm{e}^{t\mathcal{Q}} = \begin{pmatrix} \frac{b}{a+b} & \frac{a}{a+b} \\ \frac{b}{a+b} & \frac{a}{a+b} \end{pmatrix} = \begin{pmatrix} \frac{2}{r(2N-1)+2} & \frac{r(2N-1)}{r(2N-1)+2} \\ \frac{2}{r(2N-1)+2} & \frac{r\lambda(2N-1)}{r(2N-1)+2} \end{pmatrix} =: \mathcal{P}.$$

Hence, choosing a sequence $(\lambda_n)_{n\ge1}$ with $\lim_{n\to\infty}\lambda_n = \infty$ and defining $A_n = \mathcal{G}_{\lambda_n}$ we are in the setup of Kurtz's Theorem 42.2 with $\epsilon_n = \lambda_n^{-1}, \mathbb{X} = \mathcal{M}^2$ and $\mathbb{X}' :=$ *Range* $\mathcal{P}$ composed of $\binom{E}{F}$ such that $E = F$. Next, taking

$$A = \mathcal{A} \text{ and } D = \{\begin{pmatrix} E \\ E \end{pmatrix}; E \in D(G)\} \subset D(\mathcal{A}) \cap \mathbb{X}'$$

we see that $\lim_{n\to\infty} A_n\binom{E}{E} = \mathcal{A}\binom{E}{E}$ (because $\mathcal{Q}\binom{E}{E} = 0$), establishing condition (a) in Kurtz's Theorem. Moreover, the operator $\mathcal{P}\mathcal{A}$ with domain D generates the strongly continuous semigroup:

$$\mathrm{e}^{t\mathcal{P}\mathcal{A}}\begin{pmatrix} E \\ E \end{pmatrix} = \begin{pmatrix} S_{XY}(t)E \\ S_{XY}(t)E \end{pmatrix}$$

in $\mathbb{X}'$, proving condition (c) in the theorem. Finally, for $\binom{E}{F} \in D(\mathcal{A})$, $\lim_{n\to\infty}\lambda_n^{-1}A_n\binom{E}{F} = \mathcal{Q}\binom{E}{F}$, showing that (b) in the theorem is satisfied with $D' = D(\mathcal{A})$. Kurtz's Theorem in this case yields (see the definition of a and b):

$$\lim_{\lambda\to\infty} \mathrm{e}^{t\mathcal{G}_\lambda}\begin{pmatrix} E \\ F \end{pmatrix} = \mathrm{e}^{t\mathcal{P}\mathcal{A}}\mathcal{P}\begin{pmatrix} E \\ F \end{pmatrix} = \begin{pmatrix} S_{XY}(t)[\frac{b}{a+b}E + \frac{a}{a+b}F] \\ S_{XY}(t)[\frac{b}{a+b}E + \frac{a}{a+b}F] \end{pmatrix}. \tag{51.5}$$

From the mathematical viewpoint, what we have obtained in the limit is again a convex combination of generators and initial conditions, much as in the cases described in the previous chapters, except that in a different type of space. In fact, we could generalize the proof presented here to deal with the case where each component in the Cartesian product semigroup (51.1) is different; the limit then would involve a semigroup generated by a convex combination of the corresponding generators.

Biologically, however, such a generalization does not make much sense: as long as we restrict ourselves to distributions, the process of mutations is the same regardless of whether we look at (X_i, Y_i) or at (X_i, Y_j). What is noteworthy here is that in the limit we obtain a convex combination of initial conditions E and F. Moreover, it is actually quite important that the limit involves merely the mutation semigroup $\{S_{XY}(t), t \geq 0\}$, one of the building blocks of the model. For formula (51.5) states that if death/birth events occur on a faster time-scale than mutations, distributions E and F influence each other via formula (51.2) so often, that in the limit they are "mixed together" (a similar conclusion may be drawn for large times t; see [65]). As a result, $E(t)$ may be approximated by the value of $S_{XY}(t)$ on a convex combination of $E(0)$ and $F(0)$ (and the same is true for $F(t)$):

$$E(t) \sim S_{XY}(t)\left[\frac{2}{r(2N-1)+2}E(0) + \frac{r(2N-1)}{r(2N-1)+2}F(0)\right]. \tag{51.6}$$

In other words, *a model with recombination reduces to a model without recombination*, with the former influencing the latter merely via coefficients of the convex combination shown here. The reader should notice the role of the recombination coefficient, r, or rather of the related compound $s = r(2N-1)$ in (51.6). In particular, for $s \ll 1$ the influence of the $F(0)$ term can be neglected and for $s \gg 1$ the latter term becomes dominant. Hence, interestingly, in sufficiently large populations, even if the recombination coefficient is quite small, the "pure" distribution of pairs (X_i, Y_i) does not have as much influence as the mixed one of (X_i, Y_j).

Exercise 51.1 Show that the death/birth events do not destroy exchangeability of the pairs of random variables considered in the model of this chapter.

Exercise 51.2 Show the other part of formula (51.2), the one involving $\tilde{F}(t)$.

☞ Chapter's summary

If birth/death events occur relatively more often than mutations, distributions of pairs (X_i, Y_j) and (X_i, Y_i) (see Figure 51.2) in a model with recombinations are well-approximated by those in a model without recombination. Formally, formula (51.5) holds. In this case, it is not the convex combination of generators, but rather the convex combination of initial conditions obtained in the limit, that is of importance.

52

Recombination (Continued)

In this chapter, following [68], we generalize the findings of Chapter 51 to an arbitrary (but finite) number of recombination sites.

We start with the case in which there are two such sites. As before, we consider a population of $2N$ individuals, but now each individual is represented as a triple of random variables describing three loci on a chromosome in linear order; the i-th individual being a triple (X_i, Y_i, Z_i), $i = 1, \dots, 2N$. The two recombination sites lie between X_i and Y_i, and between Y_i and Z_i (see Figure 52.1).

A distribution of such a triple may be identified with a member of the space $\mathcal{M}_3$ of absolutely summable three-dimensional matrices $\mathrm{m} = (\mu_{i,j,k})_{i,j,k\in\mathbb{N}}$. More generally, the distribution of an n-tuple of random variables may be identified with a member of the space $\mathcal{M}_n$ of absolutely summable n-dimensional matrices $\mathrm{m} = (\mu_{i_1,\dots,i_n})_{i_1,\dots,i_n\in\mathbb{N}}$; the norm in this space is:

$$\|m\|_{\mathcal{M}_n} = \sum_{i_1,\dots,i_n\in\mathbb{N}} |\mu_{i_1,\dots,i_n}|.$$

Note that $\mathcal{M}_n$ is isometrically isomorphic to l^1 and hence it is meaningful to speak about distributions and Markov operators in $\mathcal{M}_n$.

On the other hand, it will be convenient to view $\mathcal{M}_n$ as a tensor product of n copies of $l^1 : \mathcal{M}_n = (l^1)^{n\otimes}$; the tensor product:

$$x_1 \otimes x_2 \otimes \cdots \otimes x_n$$

of $x_j = (\xi_{i,j})_{i\in\mathbb{N}} \in l^1$, $j = 1, \dots, n$, is:

$$x_1 \otimes x_2 \otimes \cdots \otimes x_n = (\xi_{i_1,1}\xi_{i_2,2}\cdots\xi_{i_n,n})_{i_1,\dots,i_n\in\mathbb{N}} \in \mathcal{M}_n.$$

More specifically, $\mathcal{M}_n$ is isometrically isomorphic to the completion of the algebraic tensor product with respect to the projective norm [108, 304].

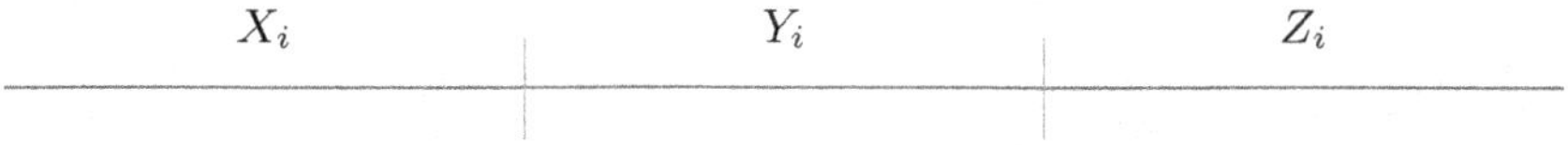

Figure 52.1 An individual (a chromosome) represented as a triple of random variables; two recombination sites are marked with grey vertical lines

Coming back to our model: we assume that the triples describing individuals are exchangeable, and that each of them evolves in time as a triple of independent Markov chains, independent of the other triples, but with the same transition probabilities; this models mutation at three loci of a chromosome in each individual. More specifically, the process of mutation at the first locus in each individual is described by means of a strongly continuous semigroup $\{S_X(t), t \geq 0\}$ of Markov operators in l^1. As before, this means that if $x \in l^1$ is the distribution of allele types at time 0 then $S_X(t)x$ is the distribution of allele types at time t. The process of mutation on the second locus is governed by a semigroup $\{S_Y(t), t \geq 0\}$, and the process of mutation on the third locus is governed by a semigroup $\{S_Z(t), t \geq 0\}$. Then it is the tensor product semigroup $\{S(t), t \geq 0\}$,

$$S(t) = S_X(t) \otimes S_Y(t) \otimes S_Z(t), \tag{52.1}$$

that describes evolution of distributions at three loci.

To explain this in more detail, assume temporarily that at time 0 the distribution of the allelic types is concentrated at the point $(i, j, k) \in \mathbb{N}^3$. By assumption of independence, the probability that at time t the allelic types will be (i', j', k') is the product of three transition probabilities: from i to i', from j to j' and from k to k' in the three Markov chains described by the semigroups $\{S_X(t), t \geq 0\}$, $\{S_Y(t), t \geq 0\}$, and $\{S_Z(t), t \geq 0\}$, respectively. This is the same as saying that the joint distribution of the allelic types at time t is $S_X(t)e_i \otimes S_Y(t)e_j \otimes S_Z(t)e_k$, where $e_\alpha = (\delta_{\alpha,\beta})_{\beta\in\mathbb{N}} \in l^1$ and $\delta_{\alpha,\beta}$ is the Dirac delta. In general, if the initial distribution is $\mathrm{m} = (\mu_{i,j,k})_{i,j,k\in\mathbb{N}}$, by the total probability formula it follows that the joint distribution of the allelic types at time t is:

$$\sum_{i,j,k\in\mathbb{N}} \mu_{i,j,k} S_X(t)e_i \otimes S_Y(t)e_j \otimes S_Z(t)e_k.$$

This formula serves as the definition of $S_X(t) \otimes S_Y(t) \otimes S_Z(t)$ for all $\mathrm{m} \in \mathcal{M}_n$ (m does not need to be a distribution), and so-defined $\{S(t), t \geq 0\}$ is a strongly continuous semigroup of Markov operators.

More generally, if $\{S_i(t), t \geq 0\}$, $i = 1, \ldots, n$ are strongly continuous semigroups of Markov operators in l^1, then an analogous formula defines the tensor product semigroup $\{S_1(t) \otimes \cdots \otimes S_n(t), t \geq 0\}$; this is a strongly continuous

Markov semigroup in $\mathcal{M}_n$ describing evolution of n independent Markov chains.

We also incorporate recombination and genetic drift in the model by assuming that each individual's life-length is an exponential random variable with parameter λ and that at the moment of an individual's death, the triple by which it is represented is replaced by another triple in the following manner. With probability $1 - r$, where $r = r_1 + r_2$ with $r_1, r_2 \in [0, 1]$ such that $r \in (0, 1]$, are given parameters, one of the triples (X_j, Y_j, Z_j), $j = 1, \ldots, 2N$ is drawn at random to replace the deceased one. With probability r_1 the $X-$variable is drawn at random first and the par (Y, Z) next, independently of the result of the first draw (this models recombination after the first locus). With probability r_2, the pair (X, Y) is drawn first and the variable Z next, independently of the first draw (this models recombination after the second locus). As a consequence, a new triple becomes one of the already existing triples (including the one just deceased) (X_j, Y_j, Z_j), each of them with probability $(1 - r)\frac{1}{2N} + r\frac{1}{(2N)^2}$, or one of the two types of "mixed ones": either (X_j, Y_k, Z_k), $j \neq k$ each with probability $r_1\frac{1}{(2N)^2}$ or (X_j, Y_j, Z_k), $j \neq k$ each with probability $r_2\frac{1}{(2N)^2}$.

As before, by symmetry of the procedure just described it is clear that after an individual's death the members of the population remain exchangeable. This fact may also be deduced from Exercise 51.1 if we note that either the pair (X, Y) or the pair (Y, Z) can be treated as a single compound locus.

We proceed to finding a closed system of distributions hidden in the model. To recall, in the case of a single recombination site, we considered a pair of two-dimensional distributions. Here, we will need a five-tuple of three-dimensional distributions. To describe them, note that by exchangeability of the triples the distribution of (X_a, Y_a, Z_b) where $a \neq b$ does not depend on the choice of a and b; we will denote it by D_{112}. The same is true of the distributions of (X_a, Y_a, Z_a), (X_a, Y_b, Z_a), (X_a, Y_b, Z_b) and (X_a, Y_b, Z_c) where a, b, and c are distinct numbers; we denote these distributions by $D_{111}, D_{121}, D_{122}$, and D_{123}, respectively.

Vector D with coordinates D_{111}, D_{112}, $D_{121}, D_{122}, D_{123}$ (note the lexicographic order) is a member of the Cartesian product $\mathcal{M}_3^5$ of five copies of $\mathcal{M}_3$. More generally, the Cartesian product $\mathcal{M}_n^m$ of m copies of $\mathcal{M}_n$ provides a way of gathering information on distributions of m n-tuples of variables. The norm in this space is $\|(m_i)_{i=1,\ldots,m}\| = \sum_{i=1}^m \|m_i\|_{\mathcal{M}_n}$. Again, $\mathcal{M}_n^m$ is isomorphically isomorphic to l^1 and we may speak of distributions and Markov operators in this space. In particular, the Cartesian product of m Markov semigroups in $\mathcal{M}_n$ is a Markov semigroup in $\mathcal{M}_n^m$. It is the product, say:

$$\{\mathcal{S}(t), t \geq 0\}, \tag{52.2}$$

of five copies of the mutation semigroup $S(t)$ defined in (52.1) that describes evolution of the vector D in the absence of death/birth events.

To describe the semigroup governing these events, we proceed as in the previous chapter to find a formula that allows deriving from D, the vector of distributions right before such an event, the vector $\tilde{D}$ of the corresponding distributions right after this event. As in (51.2) we will find a transition matrix Θ such that:

$$\tilde{D} = \Theta D. \tag{52.3}$$

In fact, we will write Θ as a convex combination:

$$\Theta = (1-r)\Theta_0 + r_1\Theta_1 + r_2\Theta_2 \tag{52.4}$$

of three transition matrices, corresponding to the cases of no recombination, and of recombination after the first and after the second locus, respectively.

To this end, let (X_a, Y_a, Z_a) and $(\tilde{X}_a, \tilde{Y}_a, \tilde{Z}_a)$, $a = 1, \ldots, 2N$ be the triples representing individuals in the population immediately before and immediately after an individual's death. We note that if there was no recombination, none of D_{112}, D_{121}, and D_{122} has changed unless $i = 1, j = 2$ or $i = 2, j = 1$. In this last case $\tilde{D}_{112} = \tilde{D}_{121} = \tilde{D}_{122} = D_{111}$. Hence,

$$\Theta_0 = \begin{pmatrix} 1 & 0 & 0 & 0 & 0 \\ \frac{2}{(2N)^2} & 1-\frac{2}{(2N)^2} & 0 & 0 & 0 \\ \frac{2}{(2N)^2} & 0 & 1-\frac{2}{(2N)^2} & 0 & 0 \\ \frac{2}{(2N)^2} & 0 & 0 & 1-\frac{2}{(2N)^2} & 0 \\ 0 & \frac{2}{(2N)^2} & \frac{2}{(2N)^2} & \frac{2}{(2N)^2} & 1-\frac{6}{(2N)^2} \end{pmatrix}, \tag{52.5}$$

where the form of the last row is justified as follows. If $i = 1, j = 2$, then $(\tilde{X}_1, \tilde{Y}_2, \tilde{Z}_3) = (X_2, Y_2, Z_3)$ and so $\tilde{D}_{123} = D_{112}$; similarly we show that this equality is true when $i = 2$ and $j = 1$. Analogously, $\tilde{D}_{123} = D_{122}$ if either $i = 2, j = 3$ or $i = 3, j = 2$, and $\tilde{D}_{123} = D_{121}$ if either $i = 3, j = 1$ or $i = 1, j = 3$. In the remaining cases $\tilde{D}_{123} = D_{123}$.

To find the three rows in the middle of Θ_1 we consider recombination between the first two loci, by listing the possible cases in Table 52.1.

This gives Θ_1 in the form:

$$\begin{pmatrix} 1-\frac{2N-1}{(2N)^2} & 0 & 0 & \frac{2N-1}{(2N)^2} & 0 \\ \frac{2N+1}{(2N)^3} & \frac{2N-2}{2N}+\frac{2N-1}{(2N)^3}+\frac{2N-1}{(2N)^2} & \frac{2N-1}{(2N)^3} & \frac{2N-1}{(2N)^3} & \frac{(2N-1)(2N-2)}{(2N)^3} \\ \frac{2N+1}{(2N)^3} & \frac{2N-1}{(2N)^3} & \frac{2N-2}{2N}+\frac{2N-1}{(2N)^3}+\frac{2N-1}{(2N)^2} & \frac{2N-1}{(2N)^3} & \frac{(2N-1)(2N-2)}{(2N)^3} \\ \frac{2}{(2N)^2} & 0 & 0 & 1-\frac{2}{(2N)^2} & 0 \\ 0 & \frac{2}{(2N)^2} & \frac{2}{(2N)^2} & \frac{2}{(2N)^2} & 1-\frac{6}{(2N)^2} \end{pmatrix}.$$

Table 52.1 **Calculation of Θ_1.** *If $i = 1$ and recombination took place after the first locus, $(\tilde{X}_1, \tilde{Y}_1, \tilde{Z}_1) = (X_j, Y_k, Z_k)$ and so $(\tilde{X}_1, \tilde{Y}_2, \tilde{Z}_1) = (X_j, Y_2, Z_k)$. Considering all possible cases for j and k we obtain four entries in the middle row of Table 52.1. The remaining entries in the table are obtained similarly.*

	$i \neq 1, 2$	$i = 1$, $j \neq 2$, $k \neq 2$	$i = 1$, $j = 2$, $k \neq 2$	$i = 1$, $j = 2$, $k = 2$	$i = 1$, $j \neq 2$, $k = 2$	$i = 2$, $k \neq 1$	$i = 2$, $k = 1$
$\tilde{D}_{112}$	D_{112}	$D_{112}, j = k$ $D_{123}, j \neq k$	D_{121}	D_{111}	D_{122}	D_{112}	D_{111}
$\tilde{D}_{121}$	D_{121}	$D_{121}, j = k$ $D_{123}, j \neq k$	D_{112}	D_{111}	D_{122}	D_{121}	D_{111}
$\tilde{D}_{122}$	D_{122}	D_{122}	D_{111}	D_{111}	D_{122}	D_{122}	D_{111}

Obtaining the first row here is straightforward, and the last row is obtained by noting that: (a) for $i = 1$, $\tilde{D}_{123} = D_{112}$ provided $j = 2$, $\tilde{D}_{123} = D_{121}$ provided $j = 3$, and $\tilde{D}_{123} = D_{123}$ in the remaining cases, (b) for $i = 2$, $\tilde{D}_{123} = D_{112}$ provided $k = 1$, $\tilde{D}_{123} = D_{122}$ provided $k = 3$, and $\tilde{D}_{123} = D_{123}$ in the remaining cases, (c) for $i = 3$, $\tilde{D}_{123} = D_{121}$ provided $k = 1$, $\tilde{D}_{123} = D_{122}$ provided $k = 2$, and $\tilde{D}_{123} = D_{123}$ in the remaining cases, and (d) for $i \geq 4$, $\tilde{D}_{123} = D_{123}$.

To cover the case of recombination after the second locus we note that our model is symmetric with respect to numbering loci. More specifically, if the loci were numbered from the last one to the first, the distributions D_{111}, $D_{112}, D_{121}, D_{122}, D_{123}$ would have become $D_{111}, D_{122}, D_{121}, D_{112}, D_{123}$, which amounts to transposition of D_{112} and D_{122}. Since such a numbering transposes recombination loci, the matrix Θ_2 may be obtained from Θ_1 by interchanging columns 2 and 4 and, next, interchanging rows 2 and 4. This completes the proof of (52.3).

It follows that the dynamics of D is governed by the generator:

$$\mathcal{G}_\lambda = \mathcal{A} + \lambda \mathcal{Q} \tag{52.6}$$

where $\mathcal{A}$ is the generator of the semigroup (52.2) responsible for mutations, and $\mathcal{Q} = 2N(\Theta - I_{\mathcal{M}_3^5})$.

Lemma 52.1 *The transition matrix Θ is ergodic.*

Proof For a transition matrix $\Phi = (\phi_{ij})$ where i, j are, say, in the set $\{1, \ldots, \kappa\}$, where κ is a natural number, we define the Dobrushin's ergodicity coefficient α as:

$$\alpha = \alpha(\Phi) = \min_{1\le i,j\le \kappa} \sum_{k=1}^{\kappa} \min(\phi_{ik}, \phi_{jk}) = 1 - \frac{1}{2} \max_{1\le i,j\le \kappa} \sum_{k=1}^{\kappa} |\phi_{i,k} - \phi_{j,k}|, \tag{52.7}$$

see [186] Sections 1.11.3–1.11.5. This coefficient provides an efficient way of studying asymptotic behavior of Φ in that if $\alpha(\Phi) > 0$ then the matrix Φ is ergodic and there exists a probability vector $(\pi_i)_{i=1,\dots,\kappa}$ such that:

$$\|\Phi^n - \Pi\| \le (1 - \alpha(\Phi))^n \tag{52.8}$$

where Π is a $\kappa \times \kappa$ matrix with all rows equal to $(\pi_i)_{i=1,\dots,\kappa}$, and $\|\cdot\|$ denotes the maximum of all absolute values of entries of a matrix ([186] Section 4.1.3).

To estimate the Dobrushin's coefficient of Θ we note that $\beta = 2 - 2\alpha$ (which is the maximum appearing in (52.7)) is a convex function of Φ, and so α is concave. Therefore, $\alpha(\Theta) \ge (1-r)\alpha(\Theta_0) + r_1\alpha(\Theta_1) + r_2\alpha(\Theta_2)$. Since for Θ_0 this maximum is attained for $i = 1$ and $j = 5$, and equals 2, $\alpha(\Theta_0) = 0$. Similarly, the maximum for Θ_1 is attained simultaneously for $(i, j) = (1, 2), (1, 3), (1, 5), (2, 4), (3, 4)$, and $(4, 5)$ (provided $2N \ge 3$), and equals $2 - \frac{1}{N^2}$. Hence, $\alpha(\Theta_1) = \frac{2}{(2N)^2}$. Finally, since interchanging rows and columns do not influence the value of α, $\alpha(\Theta_2) = \alpha(\Theta_1)$. Hence, $\alpha(\Theta) \ge r\alpha(\Theta_1) \ge \frac{2r}{(2N)^2} > 0$. □

The ergodicity of the transition matrix Θ implies that:

$$\lim_{n\to\infty} \Theta^n D = \mathcal{P}, \qquad D \in \mathcal{M}_3^5, \tag{52.9}$$

where $\mathcal{P}$ maps D into the vector with all coordinates equal $\pi_1 D_{111} + \pi_2 D_{112} + \pi_3 D_{121} + \pi_4 D_{122} + \pi_5 D_{123}$. It follows that:

$$\lim_{t\to\infty} \mathrm{e}^{tQ} D = \mathcal{P}D, \qquad D \in \mathcal{M}_3^5. \tag{52.10}$$

This shows that the key structure and properties of the last chapter's model carry over to the case of two recombination loci, and we have the following analogue of (51.5):

$$\lim_{\lambda\to\infty} \mathrm{e}^{t\mathcal{G}_\lambda} D = \mathrm{e}^{t\mathcal{P}\mathcal{A}} \mathcal{P}D = \mathcal{S}(t)\mathcal{P}D. \tag{52.11}$$

Again, all coordinates in the limit object are the same; if individual's life-length is short as compared to the time of observation, recombination events mix coordinates of D making them identical, and we have:

$$D_{111}(t) \sim S(t)(\pi_1 D_{111}(0) + \pi_2 D_{112}(0) + \pi_3 D_{121}(0) + \pi_4 D_{122}(0) + \pi_5 D_{123}(0)).$$

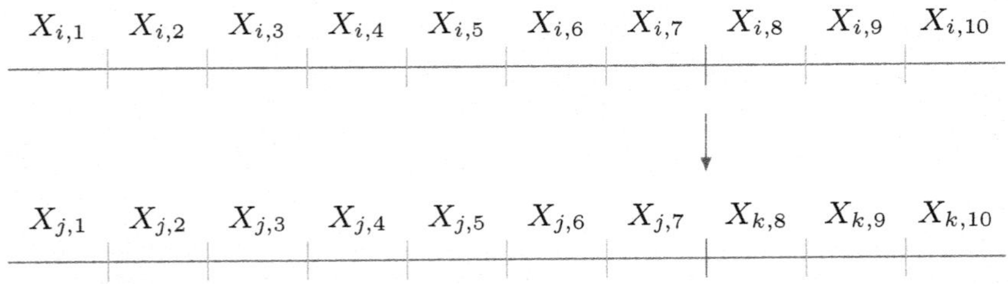

Figure 52.2 A recombination event at the seventh recombination site for $n = 10$

In other words, as in the previous chapter, recombination influences the model merely through the stationary distribution of Θ, and this is regardless of the way mutations are modeled.

Finally, we turn to the general case. Here, because of the complexity of the model, explicit calculation of the matrix Θ is impossible. Nevertheless, we can prove that such a matrix exists and show that it is ergodic.

Here are the details. In the general model there are n loci, and hence $n - 1$ possible recombination sites. The population is composed of $2N$ individuals represented by n-tuples of random variables $(X_{a,b})_{b=1,\ldots,n}$, $a = 1, \ldots, 2N$, a being the individual number and b being the locus number. In between death/recombination events each locus in each individual evolves independently with mutation on the b-th locus described by a Markov semigroup $\{S_b(t), t \geq 0\}$. The time to such an event is exponential with parameter $2\lambda N$. At the moment of death three numbers i, j, and k are chosen with replacement from $1, \ldots, 2N$, and then:

- with probability $1 - r$ (probability of no recombination), the entire n-tuple $(X_{i,l})_{l=1,\ldots,n}$ is replaced by the n-tuple $(X_{j,l})_{l=1,\ldots,n}$
- for $m = 1, \ldots, n - 1$, with the probability r_m (recombination after locus m), the $X_{i,l}$, $l = 1, \ldots, m$ are replaced by the $X_{j,l}$, $l = 1, \ldots, m$, and the $X_{i,l}$, $l = m + 1, \ldots, n$ are replaced by $X_{k,l}$, $l = m + 1, \ldots, n$ (see Figure 52.2).

Above, the positive numbers r_m, $m = 1, .., n - 1$ are such that $\sum_{m=1}^{n-1} r_m = r \in (0, 1]$. Arguing as before we see that such a procedure does not lead out of the class of exchangeable n-tuples.

The distributions forming a closed system are labelled as $D_{i_1,\ldots,i_n}$ where the multi-indexes $(i_1, \ldots, i_n)$ satisfy the following properties:

1. i_1 is 1,
2. $i_\alpha \leq \max(i_1, \ldots, i_{\alpha-1}) + 1, \alpha \geq 2;$

such multi-indexes will be called regular. There are ϖ_n regular multi-indexes, where ϖ_n is the Bell number, the number of ways to partition a set of n elements into subsets [165]. For every partition we have a natural order of its elements (subsets) where the first subset is the one containing the element 1 and the kth is the one containing the smallest number not included in the previous $k-1$ subsets (provided such number exists). To such naturally ordered partition we assign the regular multi-index by labeling elements of the kth subset with label k, and this map is injective. On the other hand, given a regular multi-index, we obtain a partition by collecting all numbers with the same index into one subset. Such assignment of a partition is injective, since the multi-index agrees with the labeling obtained from the natural order.

We arrange all the distributions $D_{i_1,\dots,i_n}$ in the lexical order, thus forming vector D. Similarly, we form the vector $\tilde{D}$ of the distributions $\tilde{D}_{i_1,\dots,i_n}$, and consider the way a coalescence/recombination event influences it. Suppose the recombination occurred after the sth locus; we are interested in $\tilde{D}_{i_1,\dots,i_n}$ and we know that the ith individual died to be replaced partly by the jth and partly by the kth individual. Then, $\tilde{D}_{i_1,\dots,i_n}$ equals $D_{j_1,\dots,j_n}$ where the multi-index $(j_1,\dots,j_n)$ is formed as follows. First, all occurrences of i up to and including the sth place in $(i_1,\dots,i_n)$ are replaced by j, and all the remaining occurrences are replaced by k. Then, the newly formed multi-index is transformed into a regular multi-index as follows. First, we change all occurrences of i_1 to 1, if the first condition of regularity is not yet met. Next, we look for the first place, say i_α, where the second requirement is not met. If there is no such place, we are done. Otherwise, we replace i_α and all its occurrences by the smallest integer larger than all i_β preceding i_α, and we continue this procedure until the multi-index is regular.

As a result, $\tilde{D}_{i_1,\dots,i_n}$ is a convex combination of all possible $D_{j_1,\dots,j_n}$'s; each choice of i, j, and k leading from $\tilde{D}_{i_1,\dots,i_n}$ to $D_{j_1,\dots,j_n}$ adds the term $\frac{1}{(2N)^3}D_{j_1,\dots,j_n}$ to this combination (all choices of i, j, and k are equally likely). All coefficients of this combination are, themselves, linear combinations of 1, $(2N)^{-1}$, $(2N)^{-2}$ and $(2N)^{-3}$.

Hence, there exists a $\varpi_n \times \varpi_n$ matrix Θ such that:

$$\tilde{D} = \Theta D.$$

Since ϖ_n is a fast growing sequence ([165], p. 693, e.g., $\varpi_4 = 15$, $\varpi_5 = 52$, and $\varpi_9 = 21147$), finding an explicit form of Θ by hand is not advisable. However, we have the following fundamental lemma, generalizing Lemma 52.1.

Lemma 52.2 *The transition matrix Θ is ergodic.*

Proof The index $(1, \dots, 1)$ is an aperiodic state for Θ, since the one-step transition probability from this state to itself is positive. Hence, Θ being finite, it suffices to show that all other states communicate with $(1, \dots, 1)$. Let $(i_1, \dots, i_n)$ be an arbitrary regular multi-index. Consider a recombination event: let s be the recombination site number, i be the number of the individual to be replaced, j be the number of the individual supplying the loci with numbers 1 through s, and k be the number of the individual supplying the loci with numbers $s+1$ through n. Taking $s = 1, i = j = 1$, and $k = i_2$ we jump from $(1, \dots, 1)$ to $(i_1, i_2, \dots, i_2)$, i_1 being equal to 1 by assumption. After arriving at $(i_1, i_2, \dots, i_l, \dots, i_l)$, $l \ge 2$ we choose $s = l, i = j = i_l, k = i_{l+1}$ to jump to $(i_1, i_2, \dots, i_l, i_{l+1}, \dots, i_{l+1})$. Hence, after at most $n-1$ jumps, we arrive at $(i_1, \dots, i_n)$.

Conversely, starting from $(i_1, \dots, i_n)$, we choose $s = n-1, i = j = i_n, k = i_{n-1}$ to jump to $(i_1, \dots, i_{n-2}, i_{n-1}, i_{n-1})$. After arriving at:

$$(i_1, \dots, i_{n-l-1}, i_{n-l}, \dots, i_{n-l}), \quad 1 \le l \le n-2$$

we choose $s = n-l-1, i = j = i_{n-l}$ and $k = i_{n-l-1}$, to jump to $(i_1, \dots, i_{n-l-2}, i_{n-l-1}, \dots, i_{n-l-1})$. Hence, after at most $n-1$ jumps we arrive at $(i_1, \dots, i_1) = (1, \dots, 1)$, proving our claim. □

To conclude: the findings of the previous chapter are valid in the case of any finite number of loci, as well. Frequent recombinations mix distributions of the considered system to such an extend that these distributions become identical. Moreover, $D_{11\dots1}(t)$ may be approximated by the value of the mutation operator $S(t)$ on a convex combination of all coordinates of the initial condition with coefficients in the combination coming from the stationary state of the matrix Θ. For more details, see [68].

Exercise 52.1 Show that (52.9) implies (52.10).

Exercise 52.2 Discuss the details of construction of the mutation semigroup in the general case of n loci.

☞ Chapter's summary

Formula (52.11) extends (51.5): this result says that the findings of the previous chapter are still valid if the number of recombination sites is finite. If mutations are relatively rare as compared to birth/death events, a model with recombinations reduces to the model without recombinations provided initial condition is appropriately modified. This modified initial condition contains information on recombination coefficient and 'mixed' distributions.

53

Averaging Principle of Freidlin and Wentzell: Khasminskii's Example

The famous observation of Freidlin and Wentzel [151], [153] says that $\mathbb{R}^d$-valued stochastic processes may converge to stochastic processes on graphs. How closely this observation is related to phenomena this book is devoted to can be judged by reading the following first sentences from their original paper:

> Let $X^\epsilon(t)$, $\epsilon > 0$ be a family of Markov processes on a space M. It is possible that as $\epsilon \to 0$ the process $X^\epsilon(t)$ moves faster and faster in some directions, whereas the motion in other direction does not accelerate. This is the situation where one can expect that the so-called averaging principle works: We can identify the points in the state-space M in the "fast" directions, obtaining a new space $Y(M)$ (Y is the mapping effecting the identification). The "fast" motion across $Y(M)$ is not a Markov process in general, but in its "fast" time it is nearly one because the characteristics of the "fast" motion depend on the "slow" variables and vary slowly compared to the "fast" motion itself. The slow process $Y(X^\epsilon(t))$ also is not a Markov one, but the averaging principle means that it converges in some sense to a Markov process $Y(t)$ on $Y(M)$ as $\epsilon \to 0$, and the characteristics of the process $Y(X^\epsilon(t))$ are obtained by averaging the characteristics of the process $Y(X^\epsilon(t))$ over the "fast" directions with respect to the stationary distribution of the fast Markov process.

Our goal is to present a semigroup-theoretical approach to two typical instances of such convergence. In this chapter, based on [59], we consider the following example of Khasminskii [206] (for the other instance, see Chapter 58): Imagine a Brownian motion in an ellipsoidal annulus (see Figure 53.1) with reflection at the boundary (i.e., with the Neumann boundary conditions). If we assume that diffusion is fast along the ellipses, then it is natural to expect that the limit process resembles a diffusion on an interval: points on each ellipse are lumped together so that the ellipse shrinks to a single point of the limit interval.

We will deduce this hypothesis from the abstract result explained below later (Proposition 53.1). Let $\mathbb{X}$ and $\mathbb{Y}$ be two Banach spaces, and let $\mathbb{X}\hat{\otimes}_\varepsilon\mathbb{Y}$ be their **injective tensor product** [108, 304], that is, the completion of the set of linear

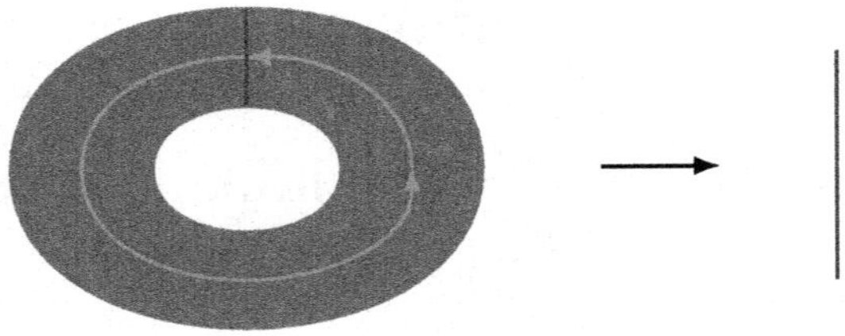

Figure 53.1 State-space collapse in Khasminskii's Example. In the terminology of the quote, "fast directions" are ellipses, and since each ellipse is characterized by its intersection point with an interval, $Y(M)$ may be identified with the interval (redrawn from [59] with permission of Elsevier)

combinations:

$$u = \sum_{i=1}^{n} x_i \otimes y_i, \tag{53.1}$$

of simple tensors in the **injective norm**:

$$\|u\|_\varepsilon = \sup \left| \sum_{i=1}^{n} F(x_i)G(y_i) \right|$$

where the supremum is taken over all representations (53.1) and all bounded linear functionals $F \in \mathbb{X}^*$ and $G \in \mathbb{X}^*$ of norm at most 1. We note that:

$$\|x \otimes y\|_\varepsilon = \|x\|_\mathbb{X} \, \|y\|_\mathbb{Y}, \qquad x \in \mathbb{X}, y \in \mathbb{Y}. \tag{53.2}$$

Next, let A and Q be the generators of equibounded semigroups in $\mathbb{X}$ and $\mathbb{Y}$, respectively, and assume that there is a $y_* \in \mathbb{Y}$ of norm 1, and a linear functional $\psi \in \mathbb{Y}^*$ such that $\psi y_* \neq 0$ and:

$$\lim_{t\to\infty} \mathrm{e}^{tQ} y = (\psi y) y_*, \, y \in \mathbb{Y}. \tag{53.3}$$

Then $y_* \in \ker Q$ and $\psi y_* = 1$. Also, given $\epsilon > 0$, let:

$$S_\epsilon(t) = \mathrm{e}^{tA} \otimes_\varepsilon \mathrm{e}^{\epsilon^{-1} tQ} \tag{53.4}$$

be the injective tensor product semigroup in $\mathbb{X}\hat{\otimes}_\varepsilon\mathbb{Y}$ determined by:

$$S_\epsilon(t)u = \sum_{i=1}^{n} (\mathrm{e}^{tA} x_i) \otimes (\mathrm{e}^{\epsilon^{-1} tQ} y_i)$$

for u of the form (53.1). The operators $S_\epsilon(t), t \geq 0, \epsilon > 0$ are bounded in $\mathbb{X}\hat{\otimes}_\varepsilon\mathbb{Y}$ with $\|S_\epsilon(t)\|_{\mathcal{L}(\mathbb{X}\hat{\otimes}_\varepsilon\mathbb{Y})} = \|\mathrm{e}^{tA}\|_{\mathcal{L}(\mathbb{X})} \|\mathrm{e}^{\epsilon^{-1} tQ}\|_{\mathcal{L}(\mathbb{Y})}$, so that by assumption they are equibounded. A simple calculation based on (53.2) shows that $\lim_{t\to 0} S_\epsilon(t)u = u$, for tensors of the form (53.1), and since the set of such tensors is dense in

$\mathbb{X}\hat{\otimes}_\varepsilon\mathbb{Y}$ and the operators are equibounded, $\{S_\epsilon(t), t \ge 0\}$ are strongly continuous semigroups.

Let A_ϵ be the generator of $\{S_\epsilon(t), t \ge 0\}$, and let D be the set of tensors u that can be represented in the form (53.1) with all $x_i \in D(A)$, $y_i \in D(Q)$. Using (53.2) again, we show that $u \in D$ belongs to $D(A_\epsilon)$ and:

$$A_\epsilon u = \sum_{i=1}^{n}(Ax_i)\otimes y_i + \frac{1}{\epsilon}\sum_{i=1}^{n} x_i \otimes (Qy_i). \tag{53.5}$$

Since $D(A)$ is dense in $\mathbb{X}$ and invariant for $\{\mathrm{e}^{tA}, t \ge 0\}$, and a similar statement is true for $D(Q)$, D is dense in $\mathbb{X}\hat{\otimes}_\varepsilon\mathbb{Y}$ and invariant for $\{S_\epsilon(t), t \ge 0\}$. It follows, see Exercise 15.2, that D is a core for A_ϵ.

Let $\mathfrak{X} \subset \mathbb{X}\hat{\otimes}_\varepsilon\mathbb{Y}$ be the subspace of simple tensors of the form $x \otimes y_*$, where $x \in \mathbb{X}$. Clearly, $\mathfrak{X}$ is isometrically isomorphic to $\mathbb{X}$, with natural isomorphism $x \otimes y_* \mapsto x$ (recall $\|y_*\|_{\mathbb{Y}} = 1$). Let $\mathfrak{A}$ be the copy of A in $\mathfrak{X}$, that is, let $\mathfrak{A}(x \otimes y_*) = (Ax)\otimes y_*$ for $x \in D(A)$. The operator $\mathfrak{A}$ generates the semigroup in $\mathfrak{X}$ given by $\mathrm{e}^{t\mathfrak{A}}(x\otimes y_*) = (\mathrm{e}^{tA}x)\otimes y_*$. Also, let $(\mathrm{e}^{t\mathfrak{Q}})_{t\ge 0}$ be the tensor product semigroup:

$$\mathrm{e}^{t\mathfrak{Q}} = \mathrm{e}^{t0} \otimes_\varepsilon \mathrm{e}^{tQ} = I_{\mathbb{X}} \otimes_\varepsilon \mathrm{e}^{tQ}, \qquad t \ge 0.$$

For u of the form (53.1),

$$\lim_{t\to\infty} \mathrm{e}^{t\mathfrak{Q}}u = \sum_{i=1}^{n} x_i \otimes (\psi y_i)y_* = \left(\sum_{i=1}^{n}(\psi y_i)x_i\right)\otimes y_*.$$

Hence, the set of such tensors being dense in $\mathbb{X}\hat{\otimes}_\varepsilon\mathbb{Y}$, $P := \lim_{t\to\infty}\mathrm{e}^{t\mathfrak{Q}}$ exists in the strong topology. Clearly, $\overline{\mathit{Range}\,P} = \mathit{Range}\,P = \mathfrak{X}$.

Proposition 53.1 *We have:*

$$\lim_{\epsilon\to 0} \mathrm{e}^{tA_\epsilon}u = \mathrm{e}^{t\mathfrak{A}}Pu, \qquad u \in \mathbb{X}\hat{\otimes}_\varepsilon\mathbb{Y}, t > 0,$$

with the limit almost uniform in $t \in (0,\infty)$. For $u \in \mathfrak{X}$, the relation is true for $t = 0$, as well, and the limit is almost uniform in $t \in [0,\infty)$.

Proof Let $x \in D(A)$, that is, $x \otimes y_* \in D(\mathfrak{A})$. By $Qy_* = 0$ and equation (53.5), we have $A_\epsilon(x\otimes y_*) = (Ax)\otimes y_*$, so that $x \otimes y_*$ belongs to $D(A_{\mathrm{ex}})$, where A_{ex} is the extended limit of A_ϵ as $\epsilon \to 0$, and $A_{\mathrm{ex}}(x\otimes y_*) = (Ax)\otimes y_*$. Similarly, for u of the form (53.1) with all $y_i \in D(Q)$ (such u form a core for $\mathfrak{Q}$), $\lim_{\epsilon\to 0}\epsilon A_\epsilon u = \sum_{i=1}^{n} x_i \otimes (Qy_i) = \mathfrak{Q}u$, and, as we have seen, the strong limit $\lim_{t\to\infty}\mathrm{e}^{t\mathfrak{Q}}$ exists. Finally, $\overline{P\mathfrak{A}} = P\mathfrak{A} = \mathfrak{A}$ is a generator. Therefore, all assumptions of Kurtz's Theorem are satisfied and the proposition follows. □

Remark 53.2 It is clear from the proof that the same theorem is valid for the projective tensor norm and, in fact, for any tensor norm (see, e.g., [108] Chapter II or [304] Chapter 6 for the definition). We have chosen the injective tensor norm merely to fix our attention (in Khasminskii's Example this is exactly the norm that is needed).

Coming back to our main subject: in proving the hypothesis of Khasminskii it is natural to apply a stretching transformation first so that ellipses become circles; one term of the Laplace operator generating the related semigroup is then scaled by a factor, and so for simplicity we assume that we are in an annulus (a ring), where the radiuses of points lie between, say, 1 and 2, and the (pre-)generator is the half of the two-dimensional Laplacian with Neumann boundary conditions.

The next step is to pass to polar coordinates: then the (pre-)generator has the form:

$$B_\epsilon u(r,\alpha) = \frac{1}{2}\left[\frac{\partial^2 u(r,\alpha)}{\partial r^2} + \frac{1}{r}\frac{\partial u(r,\alpha)}{\partial r} + \frac{1}{\epsilon r^2}\frac{\partial^2 u(r,\alpha)}{\partial \alpha^2}\right], \tag{53.6}$$

where $\epsilon \ll 1$ indicates that the angular motion is much faster than the radial.

As we will explain next, this operator "almost" comes from a tensor product semigroup of the form (53.4) (see (53.12)), and this will allow using Proposition 53.1. To this end, we recall, see Exercise 20.4, that the operator $Af = h^2 f'' + hf'$, where $h(r) = r$, $r \in [1,2]$, with domain composed of twice continuously differentiable functions satisfying $f'(1) = f'(2) = 0$, is the generator of a conservative Feller semigroup in $C[1,2]$ ($C[1,2]$ is the space of continuous functions on $[1,2]$). Similarly, see Exercise 1.12, $Qf = f''$ with domain equal to the set of twice continuously differentiable functions on $[0,2\pi]$ and such that $f', f'' \in C_{\mathrm{p}}[0,2\pi]$, generates a conservative Feller semigroup in $C_{\mathrm{p}}[0,2\pi]$, where $C_{\mathrm{p}}[0,2\pi]$ is the space of continuous, "periodic" functions on $[0,2\pi]$, that is, functions f such that $f(0) = f(2\pi)$. Moreover, see Exercise 32.4, the strong limit $\lim_{t\to\infty} \mathrm{e}^{tQ} f = \frac{1}{2\pi}\int_0^{2\pi} f$ exists for $f \in C_{\mathrm{p}}[0,2\pi]$, that is, assumption (53.3) is satisfied with $y_* \equiv 1$ and $\psi f = \frac{1}{2\pi}\int_0^{2\pi} f$.

Therefore, as we have seen A_ϵ defined in (53.4) satisfy the assumptions of the theorem of Kurtz. By Exercise 44.2, if we can find a nice operator B such that $B_\epsilon = BA_\epsilon$ where B_ϵ is defined by (53.6) (on a yet unspecified but sufficiently large domain; see (53.12) further on), we can conclude that the limit of $\lim_{\epsilon\to 0} \mathrm{e}^{B_\epsilon}$ exists. This is precisely our goal.

Before we accomplish this, we need to say a few words about the space $C[1,2]\hat{\otimes}_\varepsilon C_{\mathrm{p}}[0,2\pi]$ and operators A_ϵ in this space. Since $C_{\mathrm{p}}[0,2\pi]$ may be identified with the space of continuous functions on a circle, the injective tensor product of $C[1,2]$ and $C_p[0,2\pi]$ is $C_p(R)$, the space of continuous functions on a surface of a cylinder, that is, the space of continuous functions on the rectangle $R = [1,2] \times [0,2\pi]$, that are "periodic" with respect to α, that is, $f(r,0) = f(r,2\pi)$, $r \in [0,2\pi]$. (A fundamental and basic example says that the injective tensor product of two spaces of continuous functions on compact sets K_1 and K_2, respectively, is the space of jointly continuous functions on the Cartesian product $K_1 \times K_2$, see, for example, [304], p. 50.) This means that simple tensors, that is, functions of the form:

$$(f \otimes g)(r,\alpha) = f(r)g(\alpha), \ f \in C[1,2], g \in C_p[0,2\pi], r \in [1,2], \alpha \in [0,2\pi],$$

form a linearly dense set in $C_p(R)$; this may be deduced from the Stone–Weierstrass Theorem without appealing to [304], p. 50.

For tensors of the form $u = \sum_{i=1}^n f_i \otimes g_i$ where $f_i \in D(A)$ and $g_i \in D(Q)$, the generator of $(e^{tA_\epsilon})_{t\ge 0}$ defined in (53.4) is given by:

$$A_\epsilon u(r,\alpha) = r^2 \frac{\partial^2 u(r,\alpha)}{\partial r^2} + r\frac{\partial u(r,\alpha)}{\partial r} + \frac{1}{\epsilon}\frac{\partial^2 u(r,\alpha)}{\partial \alpha^2}. \tag{53.7}$$

Lemma 53.3 *This formula extends to u in the set:*

$$D' = \left\{ u \in C_p^2(R) \mid \frac{\partial u(r,\alpha)}{\partial r}_{|r=1} = \frac{\partial u(r,\alpha)}{\partial r}_{|r=2} = 0 \right\},$$

where $C_p^2(R)$ is the subspace of functions with continuous partial derivatives up to the second order belonging to $C_p(R)$.

Proof For the sake of this argument, let $A_\epsilon^\sharp$ denote the operator given by (53.7) with domain D'. This operator is then densely defined (since $D \subset D'$) and satisfies the maximum principle:

$$u \in D(A_\epsilon^\sharp) \text{ and } u(r_0,\alpha_0) = \sup_{(r,\alpha)\in R} u(r,\alpha) \quad \Longrightarrow \quad A_\epsilon^\sharp u(r_0,\alpha_0) \le 0.$$

Moreover, the range of $\lambda I_{C_p(R)} - A_\epsilon^\sharp$ is for $\lambda > 0$ dense in $C_p(R)$ (since so is $(\lambda I_{C_p(R)} - A)_{|D}$, D being a core for A_ϵ). Hence, by the result already cited in Chapter 20, $A_\epsilon^\sharp$ is closable and its closure generates a Feller semigroup (see [49, 132, 193]). Since $(A_\epsilon)_{|D} \subset A_\epsilon^\sharp$, we have $A_\epsilon = \overline{(A_\epsilon)_{|D}} \subset \overline{A_\epsilon^\sharp}$. Thus we must have $D' \subset D(A_\epsilon)$, for otherwise we would have two generators with one being a proper extension of the other, which we know is impossible (see, e.g., the proof of Corollary 8.4). □

Having explained the background, we note the following explicit formula for $(e^{tQ})_{t\ge 0}$:

$$e^{tQ} f(\alpha) = E\, f(w_\alpha(2t)), \qquad f \in C_p[0,2\pi], \alpha \in [0,2\pi], \tag{53.8}$$

where $w_\alpha(t) = (\alpha + w(t)) \bmod 2\pi$, $w(t), t \geq 0$ is a standard Brownian motion, and E stands for expected value. It follows that:

$$e^{t\mathfrak{Q}} u(r, \alpha) = E\, u(r, w_\alpha(2t)) \text{ and } Pu(r, \alpha) = \frac{1}{2\pi} \int_0^{2\pi} u(r, \alpha)\, d\alpha; \qquad (53.9)$$

Pu is a projection of u on the space $\mathfrak{X}$ of functions in $C_p(R)$ that do not depend on α. Therefore, Proposition 53.1 yields:

$$\lim_{t\to\infty} e^{tA_\epsilon} u = e^{t\mathfrak{A}} Pu, \qquad u \in C_p(R), \qquad (53.10)$$

where $\mathfrak{A}$ is defined on $u = f \otimes 1_{[0,2\pi]}$ with $f \in D(A)$ by:

$$\mathfrak{A}u = (Af) \otimes 1_{[0,2\pi]}, \text{ i.e., } \mathfrak{A}u(r, \alpha) = r^2 \frac{\partial^2 u(r, \alpha)}{\partial r^2} + r \frac{\partial u(r, \alpha)}{\partial r}. \qquad (53.11)$$

Now, defining B_ϵ by (53.6) on D', we see that:

$$A_\epsilon u = 2h^2 B_\epsilon u \quad \text{and} \quad B_\epsilon u = \frac{1}{2h^2} A_\epsilon u, \qquad u \in D', \qquad (53.12)$$

where h is identified with a function on R (i.e., $h(r, \alpha) = r$). This key relation proves that we are in the setup of Exercise 44.2. Here are the details:

The multiplication operator B mapping u to $\frac{1}{2h^2} u$ is an isomorphism of $C_p(R)$. It follows that:

$$\overline{B_\epsilon} = BA_\epsilon \qquad (\text{with } D(\overline{B_\epsilon}) = D(A_\epsilon)),$$

because A_ϵ is closed and D' is its core. By Dorroh's Theorem $\overline{B_\epsilon}$ is a generator of a Feller semigroup, and to apply Exercise 42.3 it suffices to check that the strong limit:

$$P_B = \lim_{t\to\infty} e^{tB\mathfrak{Q}} \qquad (53.13)$$

exists, and that $P_B B\mathfrak{A}$ is a generator.

Starting with the first claim, we note that $B\mathfrak{Q}u(r, \alpha) = \frac{1}{r^2} \frac{\partial^2 u(r,\alpha)}{\partial \alpha^2}$, and:

$$e^{tB\mathfrak{Q}} u(r, \alpha) = E\, u(r, w_\alpha(2r^{-2}t)).$$

The latter relationship may either be checked directly or deduced from the first part of (53.9) and the Volkonskii Formula. It follows immediately that the limit (53.13) exists with $P_B = P$.

Also, by (53.11), the operator $\mathfrak{A}_r$ defined by:

$$\mathfrak{A}_r u := P_B B\mathfrak{A}u = \left(\frac{1}{2h^2} Af\right) \otimes 1_{[0,2\pi]} =: (A_r f) \otimes 1_{[0,2\pi]},$$

for $u = f \otimes 1_{[0,2\pi]}$, $f \in D(A)$, is a generator in $\mathfrak{X}$ as so is A_r in $C[1, 2]$ (the latter statement is true, again, by Dorroh's Theorem). Exercise 42.3 now shows that:

$$\lim_{t\to\infty} \mathrm{e}^{tB_\epsilon} u = \mathrm{e}^{t\mathfrak{A}_\mathrm{r}} Pu, \qquad u \in C_\mathrm{p}(R), \tag{53.14}$$

which was our goal.

To explain: The operator $\mathfrak{A}_\mathrm{r}$ is the radial part of B_ϵ, and a copy of A_r in $\mathfrak{X}$. The latter operator is given by:

$$A_\mathrm{r} f(r) = \frac{1}{2}\left[\frac{\mathrm{d}^2 f(r)}{\mathrm{d}r^2} + \frac{1}{r}\frac{\mathrm{d}f(r)}{\mathrm{d}r}\right].$$

Except for reflecting barriers' conditions ($f'(1) = f'(2) = 0$), A_r is formally the same as the generator of the two-dimensional Bessel process, that is, of the process modeled as a distance of a two-dimensional standard Brownian motion from the origin ([188, p. 60], [296, p. 252]). This shows that (53.14) is a semigroup-theoretical formulation of the Khasminskii example: as the diffusion's angular speed increases to infinity, the semigroups describing Brownian motions on an annulus converge to a system involving diffusion only in the radial direction, that is, the two-dimensional Bessel process with reflecting barriers at $r = 1$ and $r = 2$.

Exercise 53.1 Since we have already bumped into projective and injective **tensor norms**, it is a high time to consider an example showing basic differences between these norms. To this end, let $\mathbb{X} = c_0$ be the space of sequences $x = (\xi_i)_{i\ge1}$ converging to zero, with the usual supremum norm $\|x\|_{c_0} = \sup_{i\ge1}|\xi_i|$, and let $\mathbb{Y} = l^1$ be the space of absolutely summable sequences $y = (\eta_j)_{j\ge1}$ with the norm $\|y\|_{l^1} = \sum_{j\ge1}|\eta_j|$. For $x \in c_0$ and $y \in l^1$ we define an infinite matrix $x \otimes y$:

$$x \otimes y = \left(\xi_i \eta_j\right)_{i,j\ge1}, \tag{53.15}$$

termed the tensor product of x and y.

(a) Let $\mathfrak{M}$ be the space of matrices $(u_{i,j})_{i,j\ge1}$ for which:

$$\|u\|_\pi := \sum_{j\ge1} \sup_{i\ge1} |u_{i,j}| \tag{53.16}$$

is finite. Prove that $\|\cdot\|_\pi$ is a complete norm in $\mathfrak{M}$, and that linear combinations:

$$u = \sum_{i=1}^n x_i \otimes y_i \tag{53.17}$$

(where $n \in N$, $x_i \in \mathbb{X}$, $y_i \in \mathbb{Y}$ may vary) form a dense set in $\mathfrak{M}_0$, the (closed) subspace of $\mathfrak{M}$ of matrices $(u_{i,j})_{i,j\ge1}$ for which "rows" $(u_{i,j})_{i\ge1}$,

$j \geq 1$ belong to c_0. Check moreover that:

$$\|x \otimes y\|_\pi = \|x\|_{c_0} \, \|y\|_{l^1}. \tag{53.18}$$

Matrices of the form $x \otimes y$ are called **simple tensors**, their linear combinations are termed tensors, and the space $\mathbb{X} \otimes \mathbb{Y}$ of these combinations is the tensor product of $\mathbb{X}$ and $\mathbb{Y}$. The norm $\| \cdot \|_\pi$ is called the **projective norm**, and $\mathfrak{M}_0$, the completion of $\mathbb{X} \otimes \mathbb{Y}$ in this norm, is termed the **projective tensor product**, and denoted $\mathbb{X}\hat{\otimes}_\pi\mathbb{Y} = c_0\hat{\otimes}_\pi l^1$.

(b) In $\mathbb{X} \otimes \mathbb{Y}$ we introduce the **injective norm**:

$$\|u\|_\varepsilon = \sup_{i\geq 1} \sum_{j\geq 1} |u_{i,j}|.$$

Check that:

$$\|x \otimes y\|_\varepsilon = \|x\|_{c_0} \, \|y\|_{l^1}, \tag{53.19}$$

while:

$$\|u\|_\varepsilon \leq \|u\|_\pi, \qquad u \in \mathbb{X} \otimes \mathbb{Y}. \tag{53.20}$$

Prove also that the completion of $\mathbb{X} \otimes \mathbb{Y}$ in $\| \cdot \|_\varepsilon$ is the space of matrixes u for which $Iu = (\sum_{j\geq 1} |u_{i,j}|)_{i\geq 1}$ belongs to c_0. The latter space is termed the **injective tensor product**, and denoted $c_0\hat{\otimes}_\varepsilon l^1$. As suggested by (53.20), as sets:

$$c_0\hat{\otimes}_\pi l^1 \supset c_0\hat{\otimes}_\varepsilon l^1.$$

Find a sample matrix $u \in c_0\hat{\otimes}_\varepsilon l^1 \setminus c_0\hat{\otimes}_\pi l^1$.

Exercise 53.2 Prove the properties of y_* stated in the text, that is, that $y^* \in \ker Q$ and $\psi y_* = 1$. *Hint:* show first that $e^{tQ} y_* = y_*, t \geq 0$.

☞ Chapter's summary

As an example of the averaging principle of Freidlin and Wentzell, we consider the Khasminskii's approximation of diffusions on ellipsoidal annulus with fast motion in angular direction, by a (radial) diffusion on an interval. The latter diffusion is the two-dimensional Bessel process. This result is derived from an abstract theorem on convergence of semigroups in tensor product spaces.

54

Comparing Semigroups

Kurtz's Theorem 42.2 seems to provide an efficient way of dealing with most irregular convergence problems that do not fit into the scheme of uniformly holomorphic semigroups of Chapter 31. However, in some examples, the structure of Kurtz's Theorem is not so obvious at first glance. Furthermore, the theorem in question has inherited from its predecessor, the Trotter–Kato Theorem, both its elegance and the main deficiency – no information on the rate of convergence, the subject neglected throughout this book, as well. Hence, it may be profitable to, at least roughly, look at alternative approaches to the question of irregular convergence.

Let us start with the method of **comparing semigroups** [50, 52]. To explain this, suppose that we already succeeded in characterizing the regularity space $\mathbb{X}_0$ of a sequence $(\mathrm{e}^{tA_n})_{t\geq 0}$ of equibounded semigroups and that we want to establish convergence outside of $\mathbb{X}_0$. This convergence often amounts to null convergence $\lim_{n\to\infty} \mathrm{e}^{tA_n}x = 0$ for $t > 0$ and x in a subspace, say $\mathbb{X}_1$, in a sense complementary to $\mathbb{X}_0$ in $\mathbb{X}$. (For example, in Kurtz's Theorem $\mathbb{X}_0 = \mathit{Range}P$ and $\mathbb{X}_1 = \mathit{Ker}\, P$.) Of necessity, these x satisfy $\lim_{n\to\infty}(\lambda - A_n)^{-1}x = 0$, $\lambda > 0$, i.e., they are values of A_{ex} on 0. To prove this null convergence we look for a sequence of semigroups $(\mathrm{e}^{tB_n})_{t\geq 0}$ which are close to $(\mathrm{e}^{tA_n})_{t\geq 0}$ and are known to converge to zero on X_1. Typically, $D(A_n) \subset D(B_n)$ and e^{tB_n} leaves $D(A_n)$ invariant, so that we may write:

$$\mathrm{e}^{tA_n}x - \mathrm{e}^{tB_n}x = -\int_0^t \frac{\mathrm{d}}{\mathrm{d}s}\left(\mathrm{e}^{(t-s)A_n}\mathrm{e}^{sB_n}x\right)\mathrm{d}s$$
$$= \int_0^t \mathrm{e}^{(t-s)A_n}(A_n - B_n)\mathrm{e}^{sB_n}x\,\mathrm{d}s, \quad x \in D(A_n). \tag{54.1}$$

If $A_n - B_n$ is a bounded operator, this equality extends to the whole of $\mathbb{X}$ and equiboundedness reduces the problem to proving that:

$$\lim_{n\to\infty} \int_0^t \|(A_n - B_n)\mathrm{e}^{sB_n}x\|\,\mathrm{d}s = 0, \qquad x \in \mathbb{X}_1. \tag{54.2}$$

To illustrate, we consider the semigroups of Chapter 12 (see equation (12.5)) related to the abstract telegraph equation. We will show that:

$$\lim_{\epsilon\to 0} \mathrm{e}^{tA_\epsilon}\begin{pmatrix}0\\ y\end{pmatrix} = \begin{pmatrix}0\\ 0\end{pmatrix}, \qquad y \in \mathbb{X}, t > 0. \tag{54.3}$$

To construct the semigroups $(\mathrm{e}^{tB_\epsilon})_{t\ge 0}$ we will compare $(\mathrm{e}^{tA_\epsilon})_{t\ge 0}$ to, we recall that by the Kisyński Theorem, the operator $B\binom{x}{y} = \binom{y}{Ax}$ with domain $D(B) = D(A) \times \mathbb{X}_{\mathrm{Kis}}$ generates the group $\{G(t), t \in \mathbb{R}\}$ of operators in $\mathbb{B} = \mathbb{X}_{\mathrm{Kis}} \times \mathbb{X}$, given by $G(t)\binom{x}{y} = \begin{pmatrix} C(t)x + \int_0^t C(s)y\,\mathrm{d}s \\ \frac{dC(t)x}{\mathrm{d}t} + C(t)y\end{pmatrix}$. Hence, the operators $B_\epsilon\binom{x}{y} = \begin{pmatrix}-\epsilon^{-2}x + \epsilon^{-1}y \\ \epsilon^{-1}Ax - \epsilon^{-2}y\end{pmatrix}$, $0 < \epsilon < 1$ with common domain $D(B)$, generate the groups $\mathrm{e}^{tB_\epsilon} = \mathrm{e}^{-t\epsilon^{-2}}G(t\epsilon^{-1})$. In particular, if $M \ge 1$ and $\omega \in \mathbb{R}$ are chosen so that $\|G(t)\| \le M\mathrm{e}^{\omega t}, t \ge 0$ then $\|\mathrm{e}^{tB_\epsilon}\| \le M\mathrm{e}^{\frac{t\omega}{\epsilon} - \frac{t}{\epsilon^2}} \underset{\epsilon\to 0}{\longrightarrow} 0$, for $t > 0$. On the other hand, for $y \in \mathbb{X}_{\mathrm{Kis}}$ and $t \ge 0$, calculating as in (54.1),

$$\begin{aligned} \mathrm{e}^{tB_\epsilon}\begin{pmatrix}0\\ y\end{pmatrix} - \mathrm{e}^{tA_\epsilon}\begin{pmatrix}0\\ y\end{pmatrix} &= \int_0^t \mathrm{e}^{(t-s)A_\epsilon}(B_\epsilon - A_\epsilon)\mathrm{e}^{sB_\epsilon}\begin{pmatrix}0\\ y\end{pmatrix}\mathrm{d}s \\ &= -\int_0^t \mathrm{e}^{(t-s)A_\epsilon}\begin{pmatrix}\frac{1}{\epsilon^2}\mathrm{e}^{-\frac{s}{\epsilon^2}}\int_0^{\frac{s}{\epsilon}} C(u)y\,\mathrm{d}u \\ 0\end{pmatrix}\mathrm{d}s. \end{aligned} \tag{54.4}$$

Since $\mathbb{X}_{\mathrm{Kis}}$ is dense in $\mathbb{X}$ (in the norm in $\mathbb{X}$), the right-most and left-most expressions here are equal for all $y \in \mathbb{X}$. Therefore, the operators $\mathrm{e}^{tA_\epsilon}, t \ge 0, 0 < \epsilon < 1$ being equibounded, we need to show that $\int_0^t \frac{1}{\epsilon^2}\mathrm{e}^{-\frac{s}{\epsilon^2}}\|\int_0^{\frac{s}{\epsilon}} C(u)y\,\mathrm{d}u\|\,\mathrm{d}s = \int_0^{\frac{t}{\epsilon}} \frac{1}{\epsilon}\mathrm{e}^{-\frac{s}{\epsilon}}\|\int_0^s C(u)y\,\mathrm{d}u\|\,\mathrm{d}s$ converges to zero, as $\epsilon \to 0$. Since $\int_0^s C(u)y\,du$ tends to 0 as $s \to 0$, the result follows because the measures on $[0, \infty)$ with densities $s \mapsto \frac{1}{\epsilon}\mathrm{e}^{-\frac{s}{\epsilon}}$ converge weakly, as $\epsilon \to 0$, to the Dirac measure at 0.

It is instructive to compare our semigroups also with the semigroups generated by $B_\epsilon\binom{x}{y} = \binom{0}{\epsilon^{-2}y}$. Then $\mathrm{e}^{tB_\epsilon}\binom{x}{y} = \binom{x}{\mathrm{e}^{-t\epsilon^{-2}}y}$ and the calculation (54.4) gives:

$$\mathrm{e}^{tB_\epsilon}\begin{pmatrix}0\\ y\end{pmatrix} - \mathrm{e}^{tA_\epsilon}\begin{pmatrix}0\\ y\end{pmatrix} = -\int_0^t \mathrm{e}^{(t-s)A_\epsilon}\begin{pmatrix}\frac{1}{\epsilon}\mathrm{e}^{-\frac{s}{\epsilon^2}}y \\ 0\end{pmatrix}\mathrm{d}s, \tag{54.5}$$

proving that the norm of the difference on the left-hand side does not exceed $\epsilon K\|y\|_{\mathbb{X}_{\text{Kis}}} \int_0^{t\epsilon^{-2}} \mathrm{e}^{-u}\,\mathrm{d}u \le \epsilon K\|y\|_{\mathbb{X}_{\text{Kis}}}$, where K is the constant such that $\|\mathrm{e}^{tA_\epsilon}\| \le K$ for all $t \ge 0$ and $\epsilon > 0$. This combined with the density argument, gives a slightly shorter proof of (54.3), and has an additional advantage: since the rate of convergence of e^{tB_ϵ} is clearly seen, we obtain the rate of null convergence of e^{tA_ϵ} for $y \in \mathbb{X}_{\text{Kis}}$ (see the next chapter for more on this approximation).

Of course, the applicability of the method of comparing semigroups is not restricted to irregular convergence. If we can guess the limit semigroup, we can replace B_n by the generator of the limit semigroup in (54.1) and proceed with natural estimates, and often a careful analysis gives rates of convergence for initial data in a dense set. See, for example, [163] p. 48, and the proof of Chernoff's Product Formula given at p. 51, which in fact provides such rates. Moreover, each approximating semigroup may, as in Chapter 24, be defined in a different Banach space.

We conclude this chapter with the example of this type I have learned from A. Bátkai (see also [29]). As in Chapter 24, we assume that approximating equibounded semigroups $(\mathrm{e}^{tA_n})_{t\ge 0}$ are defined in spaces $\mathbb{X}_n$ and the limit semigroup $(\mathrm{e}^{tA})_{t\ge 0}$ is defined in a space $\mathbb{X}$, where $\mathbb{X}_n$ converge to $\mathbb{X}$ in the sense that $\lim_{n\to\infty} \|P_n x - x\|_n = 0$, where P_n are certain equibounded operators $P_n : \mathbb{X} \to \mathbb{X}_n$. Furthermore, we assume that there exists a dense subspace $\mathbb{Y} \subset D(A)$ of $\mathbb{X}$, left invariant by $(\mathrm{e}^{tA})_{t\ge 0}$, equipped with a norm $\|\cdot\|_{\mathbb{Y}}$ stronger than that in $\mathbb{X}$, such that $(\mathrm{e}^{tA})_{t\ge 0}$ restricted to $\mathbb{Y}$ is a strongly continuous semigroup of equibounded operators there.

Now, assume that $P_n\mathbb{Y} \subset D(A_n)$, $n \ge 1$ and that there exist constants $C > 0$ and $p \in \mathbb{N}$ such that for all $x \in \mathbb{Y}$, $\|A_nP_nx - P_nAx\| \le \frac{C}{n^p}\|x\|_{\mathbb{Y}}$. Then calculating as in (54.1), we obtain:

$$\begin{aligned}\mathrm{e}^{tA_n}P_nx - P_n\mathrm{e}^{tA}x &= -\int_0^t \frac{\mathrm{d}}{\mathrm{d}s}\left(\mathrm{e}^{(t-s)A_n}P_n\mathrm{e}^{sA}x\right)\,\mathrm{d}s \\ &= \int_0^t \mathrm{e}^{(t-s)A_n}(A_nP_n - P_nA)\mathrm{e}^{sA}x\,\mathrm{d}s, \qquad x \in \mathbb{Y}. \qquad (54.6)\end{aligned}$$

If we assume, without loss of generality, that all the semigroups involved here have the same bound M, this implies that:

$$\|\mathrm{e}^{tA_n}P_nx - P_n\mathrm{e}^{tA}x\|_{\mathbb{X}} \le \int_0^t M^2\frac{C}{n^p}\,\mathrm{d}s\|x\|_{\mathbb{Y}} = tM^2\frac{C}{n^p}\|x\|_{\mathbb{Y}}.$$

The moral is that for $x \in \mathbb{Y}$ the semigroups converge with the rate inherited from convergence of generators and that, since $\mathbb{Y}$ is dense, the same is true for all $x \in \mathbb{X}$, obviously except for the rate.

As an example, let us consider the following model of the **spread of an infectious disease** [231] (see [260] for a recent generalization of this model, where further references may also be found). In the model, a population consists of n individuals, some of which are ill. Each of them may recover independently of the others, and the time to recovery is exponential with parameter r. A healthy member of the population may become infected by contacting those already ill, and the chance of being infected in a short interval is proportional to the number of infected individuals he can contact. We assume furthermore that each individual becomes infected independently of the others, and independently of the recoveries. In other words, denoting by $X_n(t), t \geq 0$ the number of infected individuals at time t, we obtain a continuous-time Markov chain with jump intensities:

$$k \to k-1 \text{ with intensity } rk, \qquad k \to k+1 \text{ with intensity } \lambda k(n-k),$$

where λ is a constant describing infectiousness of the disease. The generator A_n of the process $\frac{X(t)}{n}, t \geq 0$ with λ replaced by λn^{-1}, in the space $C(S_n)$ of continuous functions (i.e., sequences) on the set $S_n = \{0, \frac{1}{n}, \frac{2}{n}, \ldots, \frac{n-1}{n}, 1\}$ is given by:

$$A_n f\left(\frac{k}{n}\right) = \frac{\lambda}{n} k(n-k) f\left(\frac{k+1}{n}\right) + rkf\left(\frac{k-1}{n}\right) - \left[\frac{\lambda}{n} k(n-k) + rk\right] f\left(\frac{k}{n}\right).$$

We will show that for large n, the process described here is well-approximated by the deterministic process of moving along integral curves of the equation:

$$\frac{\mathrm{d}x}{\mathrm{d}t} = \lambda x(1-x) - rx, \quad x(0) = x_0 \in [0, 1]. \tag{54.7}$$

To this end, let A be the generator of the semigroup in $C[0, 1]$ given by $T(t)f(x_0) = f(x(t))$ where $x(t)$ is the solution to (54.7) starting at x_0. For $f \in C^2[0, 1]$, that is, in the set of twice continuously differentiable functions (with one-sided derivatives at the ends of the interval) the generator is given by:

$$Af(x) = (\lambda x(1-x) - rx)f'(x).$$

Also, $C^2[0, 1]$ is invariant for the semigroup, since for f in this set, $f(x(t))$ is twice continuously differentiable in initial condition (see Chapter 1 in [327], see also Exercise 54.4). Hence, we may proceed with $\mathbb{X} = C[0, 1]$ and $\mathbb{Y} = C^2[0, 1]$.

The operators $P_n : \mathbb{X} \to \mathbb{X}_n$ are given here by $P_n f(\frac{k}{n}) = f(\frac{k}{n})$, and the difference $A_n P_n f(\frac{k}{n}) - P_n A f(\frac{k}{n})$ equals, by the Taylor Formula,

$$\begin{aligned} & n\lambda\frac{k}{n}\left(1-\frac{k}{n}\right)\left[f\left(\frac{k+1}{n}\right)-f\left(\frac{k}{n}\right)-\frac{1}{n}f'\left(\frac{k}{n}\right)\right] \\ & + nr\frac{k}{n}\left[f\left(\frac{k-1}{n}\right)-f\left(\frac{k}{n}\right)+\frac{1}{n}f'\left(\frac{k}{n}\right)\right] \\ & = \frac{1}{2n}\left\{\lambda\frac{k}{n}\left(1-\frac{k}{n}\right)f''\left(\frac{k}{n}+\theta_1\frac{1}{n}\right)+r\frac{k}{n}f''\left(\frac{k}{n}-\theta_2\frac{1}{n}\right)\right\} \end{aligned}$$

for certain $\theta_1, \theta_2 \in (0, 1)$. Hence, the norm of the difference does not exceed:

$$\frac{1}{2n}\frac{\lambda+4r}{4}\sup_{x\in[0,1]}|f''(x)| \le \frac{1}{2n}\frac{\lambda+4r}{4}\|f\|_{C^2[0,1]}.$$

This proves our claim and gives the rate of convergence of the semigroups for $f \in C^2[0, 1]$, since all the semigroups involved here are contractions.

Exercise 54.1 Do Exercise 42.2 by the method of comparing semigroups.

Exercise 54.2 Use Kurtz's Lemma to prove (54.3).

Exercise 54.3 For a discrete-time analogue of (54.1) consider a semigroup of operators, say $(\mathrm{e}^{tA})_{t\ge 0}$, and the question of convergence of the families $P_n^{[t/h_n]}, t \ge 0, n \ge 1$ to this semigroup, where P_n are bounded linear operators and $h_n, n \ge 1$ is a positive null sequence (i.e., $\lim_{n\to\infty} h_n = 0$ and $h_n > 0$). Since for this convergence it is necessary and sufficient to have $\lim_{n\to\infty} \|P_n^{[t/h_n]} - \mathrm{e}^{[t/h_n]h_n A}\| = 0$, define $R_n = \mathrm{e}^{h_n A}$, and show that:

$$P_n^{[t/h_n]} - R_n^{[t/h_n]} = \sum_{k=1}^{[t/h_n]} \left(P_n^{[t/h_n]-k} R_n^k - P_n^{[t/h_n]-k+1} R_n^{k-1}\right).$$

In the particular case where all the operators considered here are equipowerbounded, this implies: $\|P_n^{[t/h_n]} - R_n^{[t/h_n]}\| \le [t/h_n] M \|P_n - R_n\|$, for some constant M. Use this to reestablish the Bernstein-type approximation (18.8) of a semigroup. Discuss the rate of convergence. *Hint:* write $P_n - R_n = h_n(\frac{P_n - I_{\mathbb{X}}}{h_n} - \frac{R_n - I_{\mathbb{X}}}{h_n})$.

Exercise 54.4 Show that the solution to (54.7) is:

$$x(t) = \begin{cases} \frac{aC\mathrm{e}^{\mu t}}{1+C\mathrm{e}^{\mu t}}, & t \ge 0, \ \text{provided } x_0 \ne a, \\ a, & t \ge 0, \ \text{provided } x_0 = a, \end{cases}$$

where $a = 1 - \frac{r}{\lambda}$ and $C = \frac{x_0}{a - x_0}$.

☞ Chapter's summary

A way to prove irregular convergence is often to compare the semigroups in question with other semigroups which are close to the original ones and, on the other hand, are known or can be proved to converge. We explain the method in more detail, taking as a case study the semigroups related to the abstract telegraph equation. A question of convergence rate is also discussed and the model of infectious disease is studied.

55

Relations to Asymptotic Analysis

As we have seen in the previous chapter, the method of comparing semigroups hinges on an intelligent guess of the semigroups to be compared with the semigroups whose convergence we are to show. A hint for such a guess may be obtained from the very form of the semigroups involved, or by physical or biological considerations underlying the model in question. However, a systematic way of *deriving* the form of the searched-for semigroups is provided by **asymptotic analysis**, an alternative, ingenuous method of proving convergence theorems; see [263]. Additionally, in many cases the semigroups derived in this way seem to be in a sense optimal [17].

In this chapter, we discuss this method briefly in the context of the **Chapman–Enskog expansion method**. Asymptotic analysis begins with decomposition of the underlying Banach space into a so-called **hydrodynamic space** and a complementary space called **kinetic space**. As we shall see, the first of them roughly corresponds to the regularity space of the semigroups under study, and the second is the space of null convergence. Then, the semigroups are decomposed into parts corresponding to these two subspaces.

Once this is achieved, these parts are expanded into (truncated) formal series in ϵ^{-p}, where ϵ is the small parameter involved and $p > 0$ is a certain power, characterizing the problem in question; in the Chapman–Enskog method it is merely the kinetic part that is being expanded, and the hydrodynamic part is left intact. Then, comparing like terms in the expansion allows calculating the **bulk part** of the approximating semigroups, which corresponds to the limit semigroup in the regularity space. This part, however, is not able to describe transient phenomena occurring because of passing from the whole space to its subspace. Asymptotic analysis takes care of this by "blowing up" the time near $t = 0$, that is, by introducing fast time $\tau = \frac{t}{\epsilon}$, and redoing analysis described above in this new time. This allows calculating the **initial layer**, which sometimes gives the form of the semigroups we should compare our semigroups with. More complicated situations, however, require calculating **boundary layers** and **corner layers**, which are handled similarly, see [263] or [22].

For example, let us consider the question of convergence of the semigroups of (12.5); a thorough analysis of a more general telegraph system is given in [17]. The related Cauchy problem is:

$$\frac{dx}{dt} = \epsilon^{-1} y,$$
$$\frac{dy}{dt} = \epsilon^{-1} A x - \epsilon^{-2} y, \qquad x(0) = \overset{o}{x}, y(0) = \overset{o}{y}. \tag{55.1}$$

We note the somewhat unusual way of denoting initial values used here; this convention is forced by the fact that we will need many indexes for x and y in a minute.

In this example, the space is already decomposed into the hydrodynamical and kinetic parts, played by $\mathbb{X}_{\text{Kis}} \times \{0\}$ and $\{0\} \times \mathbb{X}$, respectively. To start asymptotic analysis, first we write x and y in the form: $x = \overline{x}$ and $y = \overline{y}_0 + \epsilon \overline{y}_1$ and insert these to (55.1); the line over x and y is reserved for denoting the bulk part of x and y. This renders:

$$\frac{d\overline{x}}{dt} = \frac{1}{\epsilon}(\overline{y}_0 + \epsilon \overline{y}_1),$$
$$\frac{d\overline{y}_0}{dt} + \epsilon \frac{d\overline{y}_1}{dt} = \epsilon^{-1} A\overline{x} - \epsilon^{-2}(\overline{y}_0 + \epsilon \overline{y}_1), \qquad x(0) = \overset{o}{x}, y(0) = \overset{o}{y}. \tag{55.2}$$

Comparing like terms related to ϵ^{-2}, ϵ^{-1} and ϵ^{0}, we obtain the relations:

$$\overline{y}_0 = 0, \qquad \frac{d\overline{x}}{dt} = \overline{y}_1, \text{ and } A\overline{x} = \overline{y}_1;$$

the remaining terms are disregarded since they disappear as $\epsilon \to 0$, anyway. This implies that the bulk part in the hydrodynamical space should satisfy $\frac{d\overline{x}}{dt} = A\overline{x}$, $x(0) = \overset{o}{x}$, in agreement with our findings from Chapter 12 in which we proved that the limit semigroup is generated by the part of A in the Kisyński space $\mathbb{X}_{\text{Kis}}$.

So, we hope that $x - \overline{x}$ and $y - \epsilon \overline{y}_1$ converge to 0 as $\epsilon \to 0$. However, since $\overline{y}_1(0) = A\overset{o}{x}$ and $y(0) = \overset{o}{y}$, it is clear that this cannot be true at $t = 0$: as $\epsilon \to 0$ there are transient phenomena hidden in the initial layer. To find the formula for the latter, we consider the fast time $\tau = t\epsilon^{-2}$. In terms of τ, the system (55.1) reads:

$$\frac{dx}{d\tau} = \epsilon y,$$
$$\frac{dy}{d\tau} = \epsilon A x - y, \qquad x(0) = \overset{o}{x}, y(0) = \overset{o}{y}. \tag{55.3}$$

Inserting the initial layer expansion $x = \tilde{x}$ and $y = \tilde{y}_0 + \epsilon\tilde{y}_1$ into these equations, and comparing like terms as before, we obtain:

$$\frac{\mathrm{d}\tilde{x}}{\mathrm{d}\tau} = 0 \text{ and } \frac{\mathrm{d}\tilde{y}_0}{\mathrm{d}\tau} = -\tilde{y}_0.$$

The first of these relations tells us that $\tilde{x}$ is constant. On the other hand, we know that for large times, the influence of $\tilde{x}$ should be negligible. Hence, we take $\tilde{x} = 0$. The other formula combined with $\tilde{y}(0) = \overset{o}{y}$ forces $\tilde{y}(\tau) = \mathrm{e}^{-\tau}\overset{o}{y}$. Coming back to the original time, we see that the initial layer has the form $\tilde{y}(t) = \mathrm{e}^{-t\epsilon^{-2}}\overset{o}{y}$. We note that this was exactly the analysis presented here that hinted the usage of the semigroups e^{tB_ϵ} in (54.5).

Having completed these formal considerations, we still need to prove that the approximation we have found works well. To this end we introduce $u = x - \overline{x}$ and $v = y - \epsilon\frac{\mathrm{d}\overline{x}}{\mathrm{d}t} - \tilde{y}$, where $\overline{x}(t) = \mathrm{e}^{tA_p}\overset{o}{x}$ and $\tilde{y}(t) = \mathrm{e}^{-t\epsilon^{-2}}\overset{o}{y}$, and check that, for $\overset{o}{x} \in D(A_p^2)$ and $\overset{o}{y} \in \mathbb{X}$, u and v are differentiable with:

$$\begin{aligned}
\frac{\mathrm{d}u}{\mathrm{d}t} &= \frac{\mathrm{d}x}{\mathrm{d}t} - \frac{\mathrm{d}\overline{x}}{\mathrm{d}t} = \epsilon^{-1}y - \epsilon^{-1}(y - v + \tilde{y}) = \epsilon^{-1}v + \epsilon^{-1}\tilde{y},\\
\frac{\mathrm{d}v}{\mathrm{d}t} &= \frac{\mathrm{d}y}{\mathrm{d}t} - \epsilon\frac{\mathrm{d}^2\overline{x}}{\mathrm{d}t^2} - \frac{\mathrm{d}\tilde{y}}{\mathrm{d}t} = \epsilon^{-1}Ax - \epsilon^{-2}y - \epsilon\frac{\mathrm{d}^2\overline{x}}{\mathrm{d}t^2} + \epsilon^{-2}\tilde{y}\\
&= \epsilon^{-1}Au - \epsilon^{-2}v - \epsilon\frac{\mathrm{d}^2\overline{x}}{\mathrm{d}t^2}.
\end{aligned}$$

Hence,

$$\frac{\mathrm{d}}{\mathrm{d}t}\begin{pmatrix}u\\ v\end{pmatrix} = A_\epsilon\begin{pmatrix}u\\ v\end{pmatrix} + f(t), \qquad \text{where } f(t) = \begin{pmatrix}\epsilon^{-1}\mathrm{e}^{-t\epsilon^{-2}}\overset{o}{y}\\ -\epsilon\mathrm{e}^{tA}A^2\overset{o}{x}\end{pmatrix}, \tag{55.4}$$

with the initial condition $u(0) = 0$ and $v(0) = -\epsilon A\overset{o}{x}$. Thus, by the variation of constants formula:

$$\begin{pmatrix}u(t)\\ v(t)\end{pmatrix} = -\epsilon\mathrm{e}^{tA_\epsilon}\begin{pmatrix}0\\ A\overset{o}{x}\end{pmatrix} + \int_0^t \mathrm{e}^{(t-s)A_\epsilon}f(s)\,\mathrm{d}s. \tag{55.5}$$

In order to prove that the right-hand side converges to 0 we consider the case $\overset{o}{y} = 0$ first. Then, $\|f(t)\| \le \epsilon\, const.\|A^2\overset{o}{x}\|, t \ge 0$ and the result easily follows, since the initial condition clearly converges to 0, as well (and the semigroups generated by A_ϵ are equicontinuous). By the way, this combined with density argument reestablishes the main theorem of Chapter 12 and gives the rates of convergence for $\overset{o}{x} \in D(A_p^2)$.

By linearity, we are left with checking the case $\overset{o}{x} = 0$. However, for such initial data:

$$\begin{pmatrix}u(t)\\ v(t)\end{pmatrix} = \mathrm{e}^{tA_\epsilon}\begin{pmatrix}0\\ \overset{o}{y}\end{pmatrix} - \mathrm{e}^{tB_\epsilon}\begin{pmatrix}0\\ \overset{o}{y}\end{pmatrix},$$

(55.5) reduces to (54.5), and the simple estimate given in Chapter 54 completes the proof.

Incidentally, this example shows nicely how these two approaches: asymptotic analysis and semigroup-theoretical methods, intertwine and complement each other. For more on this subject, see the already cited [17]. But again, the big moral is that using asymptotic analysis may help us guess the semigroups we want to compare our semigroups with.

Exercise 55.1 Use asymptotic analysis to prove the main result of Chapter 48.

☞ Chapter's summary

A brief discussion of asymptotic analysis is given. A magnificent tool in itself (see, e.g., [25, 263]), the asymptotic analysis may also be considered as a way to provide intelligent hints of the form of the limit semigroup and of semigroups we need to compare the original semigroups with.

56

Greiner's Theorem

In Chapter 50, we have encountered a curious phenomenon: the boundary condition in the limit semigroup related to fish dynamics (see (50.2)) is a convex combination of boundary conditions involved in the approximation (see (50.1)). This effect is very similar to that observed in Chapter 48, where convex combinations of Feller generators resulting from "averaging" the stochastic processes involved were studied. As we have seen in Chapter 50, these two effects are in a sense dual: under additional regularity conditions on the model's parameters, the predual of the McKendrick semigroup may be constructed in a space of continuous functions, and the perturbation of a boundary condition becomes a perturbation of the generator.

Taking all this into account we arrive at the following tempting hypothesis (similar to that discussed in Chapter 50). Let α be a bounded linear operator in a Banach space $\mathbb{X}$, and let A be a closed operator in this space. Suppose that for Φ_1, Φ_2 mapping $D(A)$ to another Banach space $\mathbb{Y}$, $A_{|\ker \Phi_1}$ and $A_{|\ker \Phi_1}$ are generators of strongly continuous semigroups in $\mathbb{X}$. Assume finally that $A_{|\ker \Phi_a}$ where $\Phi_a = \Phi_1\alpha + \Phi_2\beta$ and $\beta = I_{\mathbb{X}} - \alpha$, is a generator also. Then, it is tempting to think of the later semigroup as describing dynamics, which is a result of averaging the semigroups generated by $A_{|\ker \Phi_1}$ and $A_{|\ker \Phi_1}$. In Chapter 57, following [18], we use the abstract setup of Greiner [167] to provide an approximation of the semigroup generated by $A_{|\ker \Phi_a}$ by semigroups built from those generated by $A_{|\ker \Phi_1}$ and $A_{|\ker \Phi_1}$, that supports this intuition.

Note that the problem posed here is in a sense converse to the result of [11, 21, 22, 23, 25] (where convergence of solutions to (50.1) as $\epsilon \to 0$ was established using asymptotic analysis, even in a much more general case, that is, without the assumption of irreducibility of migration matrix): there, a complex model is reduced to a simpler one involving convex combination of the boundary conditions while here, given a generator with convex combination

Figure 56.1 Ancient bounds of a town in a slightly perturbed condition

of boundary conditions, we want to construct an appropriate approximating sequence of semigroups.

What we want to do in this chapter is to prepare the reader for the next one, by presenting Greiner's ideas in a nutshell.

Let $\mathbb{X}$ and $\mathbb{Y}$ be two Banach spaces, $A : D(A) \to \mathbb{X}$ be a closed operator in $\mathbb{X}$, and $L : D(A) \to \mathbb{Y}$ be a linear operator that is continuous with respect to the graph norm in $D(A)$. Moreover, assume L to be surjective, and suppose that A_0, defined as the restriction of A to $\ker L$, generates a semigroup of operators in $\mathbb{X}$. The main question studied in [167] was whether, given $F \in \mathcal{L}(\mathbb{X}, \mathbb{Y})$, the operator A_F defined as the restriction of A to $\ker(L - F)$ is the generator as well. While in general (see [167, Example 1.5]) the answer is in the negative, Greiner's first fundamental theorem [167, Thm. 2.1] establishes that A_F is the generator for any F provided there is a constant $\gamma > 0$ such that for λ larger than some λ_0:

$$\|Lx\| \geq \lambda\gamma \|x\|, \qquad \text{for all } x \in \ker(\lambda - A). \tag{56.1}$$

To recall the basic structure related to assumption (56.1) (see [167, Lemma 1.2]), we fix $\lambda > \lambda_0$ in the resolvent set of A_0, and consider an $x \in D(A)$. Since A_0 is a generator, we can choose $x_1 \in D(A_0)$ so that $(\lambda - A_0)x_1 = (\lambda - A)x$.

Writing $x_2 = x - x_1$, we have:

$$x = x_1 + x_2 \tag{56.2}$$

with $x_1 \in \ker L = D(A_0)$ and $x_2 \in \ker(\lambda - A)$. Such a decomposition of x is unique, because otherwise we could find a nonzero $y \in D(A_0) \cap \ker(\lambda - A)$, implying that y is an eigenvector of A_0 related to λ, a contradiction.

Since L is assumed to be surjective, condition (56.1) shows, by the open mapping theorem, that L restricted to $\ker(\lambda - A)$ is an isomorphism of $\ker(\lambda - A)$ (with the graph norm) and $\mathbb{Y}$. Following Greiner, by L_λ we denote the inverse of $L_{|\ker(\lambda - A)}$.

Before going further, let's see some examples.

Example 56.1

(a) Let A be the second derivative in $C[0, \infty]$ with $D(A) = C^2[0, \infty]$, and let $Lf = f''(0)$. Clearly, A_0 is the generator of the Brownian motion stopped at $x = 0$. For fixed $\lambda > 0$, $\ker(\lambda - A)$ is spanned by $e_{\sqrt{\lambda}}$, where $e_\lambda(x) = \mathrm{e}^{-\lambda x}$, $x \geq 0$. Since $Le_{\sqrt{\lambda}} = -\lambda$ while $\|e_{\sqrt{\lambda}}\| = 1$, condition (56.1) holds with $\gamma = 1$ (in fact, we have equality there). Moreover, for $f \in D(A)$ and $\lambda > 0$, we take $C = f''(0)\lambda^{-1}$ to see that:

$$f = f_1 + f_2 := (f - Ce_{\sqrt{\lambda}}) + Ce_{\sqrt{\lambda}},$$

with $f_1 \in \ker L = D(A_0)$ and $f_2 \in \ker(\lambda - A)$. Hence, the isomorphism L_λ maps $C \in \mathbb{R}$ to $Ce_{\sqrt{\lambda}} \in \ker(\lambda - A)$.

(b) In the motivating example of fish population dynamics,

$$A : W^{1,1}(\mathbb{R}^+) \to L^1(\mathbb{R}^+) \qquad A\phi = -\phi',$$

and $L : W^{1,1}(\mathbb{R}^+) \to \mathbb{R}$ is given by $L\phi = \phi(0)$. Then, A_0 generates the semigroup of translations to the right: $\mathrm{e}^{tA_0}\phi(x)$ equals $\phi(x - t)$ for $x > t$ and equals 0 for $x \leq t$. In this case, $\ker(\lambda - A)$, $\lambda > 0$ is spanned by e_λ defined as in point (b). Since $Le_\lambda = 1$ and $\|e_\lambda\| = \frac{1}{\lambda}$, condition (56.1) is satisfied with $\gamma = 1$ (again, we have equality there). A $\phi \in D(A)$ may be represented as:

$$\phi = \phi_1 + \phi_2 := (\phi - \phi(0)e_\lambda) + \phi(0)e_\lambda,$$

and the isomorphism L_λ maps $C \in \mathbb{R}$ to $Ce_\lambda \in \ker(\lambda - A)$.

(c) Let us look at Rotenberg's model, presented in Exercise 4.7, from the perspective of Greiner's Theorem. To recall, we are working in the Banach space $\mathbb{X} = L^1(\Omega)$ of absolutely integrable functions ϕ on $\Omega = (0, 1) \times (a, b)$. Here, $\mathbb{Y}$ is chosen to be the space $L^1(a, b)$ of absolutely integrable functions φ

on (a, b). For the domain of the operator A we choose ϕ of the form:

$$\phi(x, v) = \varphi(v) + \int_0^x \psi(y, v)\, \mathrm{d}y,$$

where $\varphi \in \mathbb{Y}$ and, for almost all v, $\psi(\cdot, v)$ is absolutely integrable over $(0, 1)$, and define $(A\phi)(x, v) = -v\psi(x, v)$. Moreover, we let $L\phi = \varphi$. Again, A_0 is a generator with:

$$\mathrm{e}^{tA_0}\phi(x, v) = \begin{cases} \phi(x - vt), & x > vt, \\ 0, & x \le vt. \end{cases}$$

The kernel of $\lambda - A$ is composed of functions of the form:

$$\phi(x, v) = \varphi(v)\mathrm{e}^{-\lambda v^{-1}x},$$

where $\varphi \in \mathbb{Y}$. For such ϕ,

$$\begin{aligned} \lambda\|\phi\|_{\mathbb{X}} &= \int_a^b |\varphi(v)|\lambda \int_0^1 \mathrm{e}^{-\lambda v^{-1}x}\, \mathrm{d}x\, \mathrm{d}v = \int_a^b v(1 - \mathrm{e}^{-\lambda v^{-1}})|\varphi(v)|\, \mathrm{d}v \\ &\le b\|\varphi\|_{\mathbb{Y}}, \end{aligned}$$

proving that condition (56.1) is fulfilled with $\gamma = b^{-1}$; recall that $L\phi = \varphi$. We leave it to the reader to work out the decomposition (56.2) in this case. □

Since in all applications to be presented later on, the semigroups generated by A_0 are contractions, in the statement of the Greiner Theorem we restrict ourselves to such semigroups.

Theorem 56.2 (Greiner) *Suppose* (56.1) *holds and* A_0 *is the generator of a semigroup of contractions. Then, for any* F, *so is* $A_F - \|F\|\gamma^{-1}$.

Proof For simplicity of notation let $\gamma_F = \|F\|\gamma^{-1}$. By (56.1), the operator L_λ, as a map from $\mathbb{Y}$ to $\mathbb{X}$ has a norm not exceeding $(\lambda\gamma)^{-1}$. Therefore, for $\lambda > \gamma_F$, $\|L_\lambda F\|_{\mathcal{L}(\mathbb{X})} \le \frac{\|F\|}{\lambda\gamma} < 1$, and so $I - L_\lambda F$ is invertible with:

$$\|(I - L_\lambda F)^{-1}\| \le \sum_{n=0}^{\infty} \|L_\lambda F\|^n_{\mathcal{L}(\mathbb{X})} = \frac{1}{1 - \frac{\|F\|}{\lambda\gamma}} = \frac{\lambda}{\lambda - \gamma_F}. \tag{56.3}$$

Next, fix $\lambda > \gamma_F$ and $y \in \mathbb{X}$, and suppose $x \in D(A_F)$ solves:

$$\lambda x - A_F x = y. \tag{56.4}$$

Let $x_\lambda := x - L_\lambda F x$. Then, $x_\lambda \in D(A_0)$ and:

$$\lambda x - A_0 x = y. \tag{56.5}$$

For, L being the left inverse of L_λ, $Lx_\lambda = Lx - LL_\lambda Fx = Lx - Fx = 0$, and:

$$\lambda x_\lambda - A_0 x_\lambda = \lambda x - \lambda L_\lambda Fx - Ax + AL_\lambda Fx = \lambda x - Ax = y,$$

since $L_\lambda Fx$ belongs to $\ker(\lambda - A)$ and A_F is a restriction of A.

Conversely, if for some $x \in D(A)$, $x_\lambda := x - L_\lambda Fx$ belongs to $D(A_0)$ and (56.5) holds, then $Lx = Lx_\lambda + LL_\lambda Fx = LL_\lambda Fx = Fx$, i.e., $x \in D(A_F)$, and:

$$\lambda x - A_F x = \lambda x_\lambda + \lambda L_\lambda Fx - Ax_\lambda - AL_\lambda Fx = \lambda x_\lambda - Ax_\lambda = y;$$

this shows that such x solves (56.5).

It follows that for $\lambda > \gamma_F$, the resolvent equation for A_F has a unique solution and:

$$(\lambda - A_F)^{-1} = (I - L_\lambda F)^{-1}(\lambda - A_0)^{-1}, \tag{56.6}$$

implying, by (56.3),

$$(\lambda - \gamma_F)\|(\lambda - A_F)^{-1}\| \le (\lambda - \gamma_F)\frac{\lambda}{\lambda - \gamma_F}\frac{1}{\lambda} = 1,$$

since $\lambda(\lambda - A_0)^{-1}$ are contractions. By the Hille–Yosida Theorem, the proof will be complete, once we show that $D(A_F)$ is dense in $\mathbb{X}$.

To this end, note that $\lim_{\lambda\to\infty} \|L_\lambda F\| = 0$, and so $(I - L_\lambda F)^{-1}$ converges, as $\lambda \to \infty$, to I in the operator topology. Since $\lim_{\lambda\to\infty} \lambda(\lambda - A_0)^{-1}x = x, x \in \mathbb{X}$ by assumption, the formula for $(\lambda - A_F)^{-1}$ implies that $\lim_{\lambda\to\infty} \lambda(\lambda - A_F)^{-1}x = x, x \in \mathbb{X}$. This shows that A_F is densely defined. □

Example 56.3

(a) Let $F \in C[0, \infty]$. By the Riesz Representation Theorem, there is a (signed) Borel measure μ such that $Ff = \int_{[0,\infty]} f \,\mathrm{d}\mu$. By Greiner's Theorem, operator A from Example 56.1 (a) with domain equal to the kernel of $L - F$ is a generator. In particular, we obtain a generation theorem for the elementary return Brownian motion (see (3.7)); note, however, that the information on the growth of this semigroup gained from Greiner's Theorem is far from being satisfactory. Neither does the theorem tell us anything about preservation of positivity, and so on.

(b) Turning to Example 56.1 (b), we see that Greiner's Theorem shows that for any $b \in L^\infty(0, \infty)$, operator A with domain composed of absolutely continuous functions ϕ with derivative in $L^1(\mathbb{R}^+)$ and $\phi(0) = \int_0^\infty b(x)\phi(x)\,\mathrm{d}x$ is the generator of a semigroup in $L^1(\mathbb{R}^+)$, such that $\|\mathrm{e}^{tA_b}\| \le \mathrm{e}^{t\|b\|_{L^\infty(0,\infty)}}$. This is the McKendrick semigroup: for other proofs of existence of this semigroup see, for example, [54, 128, 129, 184].

(c) In Rotenberg's model (Example 56.1(c) and Exercise 4.7), we would like to define $F : \mathbb{X} \to \mathbb{Y}$ by:

$$(F\phi)(v) = pv^{-1} \int_a^b w\phi(1, w)K(w, v)\,\mathrm{d}w, \qquad v \in (a, b). \tag{56.7}$$

Unfortunately, this is not a bounded linear operator, and the Greiner Theorem is not applicable for the Rotenberg model. However, since (56.1) holds, the theorem clearly applies to any bounded linear operator $F : \mathbb{X} \to \mathbb{Y}$. □

Before completing this chapter, we note that formula (56.6) is worth commenting. One of its immediate consequences is that the resolvent of A_F depends continuously on F (say, in the strong topology). This yields the following corollary on continuous dependence of semigroups on boundary conditions (cf. Chapter 3):

Corollary 56.4 *Suppose $F_n \in \mathcal{L}(\mathbb{X}, \mathbb{Y})$ are bounded linear operators converging strongly to $F \in \mathcal{L}(\mathbb{X}, \mathbb{Y})$. Assume also that $\lim_{n\to\infty} \|F_n\| = \|F\|$. Then:*

$$\lim_{n\to\infty} \mathrm{e}^{A_{F_n} t} = \mathrm{e}^{A_F t}.$$

Proof By Greiner's Theorem, $B_n = A_{F_n} - \gamma_{F_n}$, where $\gamma_{F_n} = \|F_n\|\gamma^{-1}$, are generators of contraction semigroups. Since:

$$(\lambda - B_n)^{-1} = \left(\lambda + \gamma_{F_n} - A_{F_n}\right)^{-1}, \qquad \lambda > 0,$$

formula (56.6) yields:

$$(\lambda - B_n)^{-1} = (I - L_{\lambda+\gamma_{F_n}} F_n)^{-1}(\lambda + \gamma_{F_n} - A_0)^{-1}.$$

If $F \neq 0$, assuming without loss of generality:

$$\frac{1}{2}\|F\| \le \|F_n\| \le \frac{3}{2}\|F\|, \qquad n \ge 1, \tag{56.8}$$

we see that:

$$(I - L_{\lambda+\gamma_{F_n}} F_n)^{-1} = \sum_{k=0}^{\infty} \left(L_{\lambda+\gamma_{F_n}} F_n\right)^k$$

where:

$$\|L_{\lambda+\gamma_{F_n}} F_n\| \le \frac{\|F_n\|}{\left(\lambda + \frac{\|F_n\|}{\gamma}\right)\gamma} \le \frac{\frac{3}{2}\|F\|}{\lambda\gamma + \frac{1}{2}\|F\|} \le \frac{3}{4}$$

provided $\lambda > \frac{3}{2\gamma}\|F\|$. If $F = 0$, we fix $\lambda > 0$ and instead of (56.8) assume without of generality that $\|F_n\| \le \frac{3}{4}\lambda\gamma$, $n \ge 1$ to obtain $\|L_{\lambda+\gamma_{F_n}} F_n\| \le \frac{3}{4}$, as before.

Moreover, for each k, $\lim_{n\to\infty}(L_{\lambda+\gamma_{F_n}}F_n)^k = (L_{\lambda+\gamma_F}F)^k$. Hence, for such λ,

$$\lim_{n\to\infty}(I - L_{\lambda+\gamma_{F_n}}F_n)^{-1} = \sum_{k=0}^{\infty}\left(L_{\lambda+\gamma_F}F\right)^k = (I - L_{\lambda+\gamma_F}F)^{-1},$$

by the Lebesgue Dominated Convergence Theorem, and, therefore,

$$\lim_{n\to\infty}(\lambda - B_n)^{-1} = (I - L_{\lambda+\gamma_F}F)^{-1}(\lambda + \gamma_F - A_0)^{-1} = (\lambda + \gamma_F - A_F)^{-1}.$$

By the Trotter–Kato Theorem, it follows that $\lim_{n\to\infty}\mathrm{e}^{tB_n} = \mathrm{e}^{tA_F-\gamma_F t}$. The thesis is now immediate by $\mathrm{e}^{tA_{F_n}} = \mathrm{e}^{tB_n+\gamma_{F_n}t}$. □

Exercise 56.1 Find decomposition (56.2) in Rotenberg's model.

Exercise 56.2 In Example 56.1 (c) introduce new L by:

$$(L\phi)(v) = \phi(0, v) - q\phi(1, v),$$

where $q \in (0, 1]$. Check that so modified A_0 generates a contraction semigroup, and that (56.1) is satisfied with $\gamma = b^{-1}$. *Hint:* for $\phi \in L^1(\Omega)$, let:

$$\tilde{\phi}(x - k, v) = q^k\phi(x, v), x \in [0, 1), k = 0, 1, \ldots.$$

Then, $\mathrm{e}^{tA_0}\phi(x, v) = \tilde{\phi}(x - vt, v)$. This semigroup describes the Rotenberg model in which all cells pass their maturation speed to their daughters.

☞ Chapter's summary

We prove the theorem of Greiner, which allows perturbing boundary conditions of a generator. The setup of the theorem will be used in the next chapter in dealing with generators related to convex combinations of boundary conditions.

57

Fish Population Dynamics and Convex Combination of Boundary Conditions

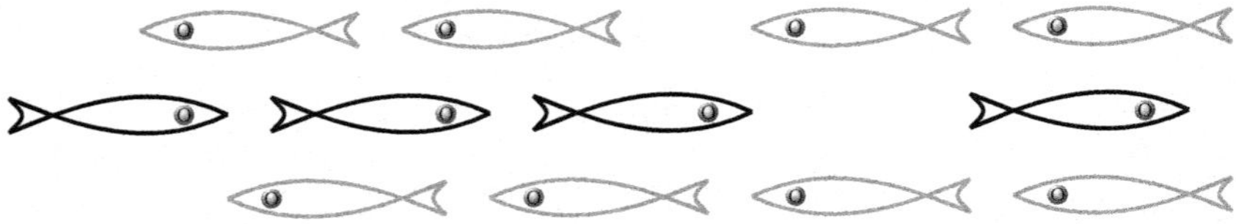

We come back to the question mentioned at the beginning of Chapter 56. Can the semigroup related to a convex combination of boundary conditions be thought of as an average of the semigroups related to the boundary conditions involved in the convex combination?

To answer this question in the affirmative, we follow [18] and throughout this chapter assume that A and L are as in the setup of the Greiner Theorem; in particular, A_0 is the generator of a contraction semigroup and (56.1) holds. Also, given a bounded linear operator $\alpha \in \mathcal{L}(\mathbb{X})$, and two operators $F_1, F_2 \in \mathcal{L}(\mathbb{X}, \mathbb{Y})$, we define:

$$F_{\mathrm{a}} = F_1\alpha + F_2\beta$$

where $\beta = I_{\mathbb{X}} - \alpha$ ("a" for "average"). By Greiner's Theorem, $A_i := A_{F_i}$ and $A_{\mathrm{a}} := A_{F_{\mathrm{a}}}$, are generators with $D(A_i) = \ker \Phi_i$, where $\Phi_i = L - F_i, i = 1, 2$ and $D(A_{\mathrm{a}}) = \ker \Phi_{\mathrm{a}}$, where $\Phi_{\mathrm{a}} = L - F_{\mathrm{a}}$.

Our main goal is to approximate $(\mathrm{e}^{tA_{\mathrm{a}}})_{t\geq 0}$ by means of semigroups built from $(\mathrm{e}^{tA_1})_{t\geq 0}$ and $(\mathrm{e}^{tA_2})_{t\geq 0}$. To this end, we introduce operators $\mathcal{A}_\kappa, \kappa > 0$, in:

$$\mathbb{B} := \mathbb{X} \times \mathbb{X}$$

given by:

$$D(\mathcal{A}_\kappa) = D(A_1) \times D(A_2) = \ker \Phi_1 \times \ker \Phi_2,$$

$$\mathcal{A}_\kappa = \begin{pmatrix} A_1 & 0 \\ 0 & A_2 \end{pmatrix} + \kappa \begin{pmatrix} -\beta & \alpha \\ \beta & -\alpha \end{pmatrix} =: \mathcal{A}_0 + \kappa \mathcal{Q}.$$

By Greiner's Theorem,

$$\|e^{t\mathcal{A}_0}\| \le e^{\omega t}, \qquad t \ge 0, \tag{57.1}$$

for some $\omega \in \mathbb{R}$. We assume that:

$$\mathcal{P} := \mathcal{Q} + I_{\mathbb{X}\times\mathbb{X}} = \begin{pmatrix} \alpha & \alpha \\ \beta & \beta \end{pmatrix} \tag{57.2}$$

is a contraction in $\mathbb{B}$. We note that $\mathcal{P}$ is idempotent, hence:

$$\begin{aligned} e^{\kappa t\mathcal{Q}} &= e^{-\kappa t}e^{\kappa t\mathcal{P}} = e^{-\kappa t}\left[I_{\mathbb{X}\times\mathbb{X}} + (e^{t\kappa} - 1)\mathcal{P}\right] \\ &= e^{-\kappa t}I_{\mathbb{X}\times\mathbb{X}} + (1 - e^{-\kappa t})\mathcal{P}. \end{aligned} \tag{57.3}$$

It follows that $\|e^{t\mathcal{Q}}\| \le 1$ and for the semigroups generated by $\mathcal{A}_\kappa$ (which exist by the Phillips Perturbation Theorem) we have, by the Trotter Product Formula,

$$\|e^{t\mathcal{A}_\kappa}x\| \le \lim_{n\to\infty}\left\|\left[e^{\frac{t}{n}\mathcal{A}_0}e^{\frac{\kappa t}{n}\mathcal{Q}}\right]^n x\right\| \le e^{\omega t}\|x\|, \qquad x \in \mathbb{X}\times\mathbb{X},$$

so that

$$\|e^{t\mathcal{A}_\kappa}\| \le e^{\omega t}, \qquad \kappa > 0, t \ge 0. \tag{57.4}$$

Operator $\mathcal{P}$ is a projection on the subspace $\mathbb{B}' \subset \mathbb{B}$ of vectors of the form $\binom{\alpha x}{\beta x}$; the latter space is isomorphic to $\mathbb{X}$ with isomorphism $\mathcal{I} : \mathbb{X} \to \mathbb{B}'$ given by:

$$\mathcal{I}x = \begin{pmatrix} \alpha x \\ \beta x \end{pmatrix}.$$

Theorem 57.1 *In this setup, assume that α leaves $D(A)$ invariant. Then:*

$$\lim_{\kappa\to+\infty} e^{t\mathcal{A}_\kappa}\begin{pmatrix} x_1 \\ x_2 \end{pmatrix} = \mathcal{I}e^{tA_a}\mathcal{I}^{-1}\mathcal{P}\begin{pmatrix} x_1 \\ x_2 \end{pmatrix} = \begin{pmatrix} \alpha e^{tA_a}(x_1 + x_2) \\ \beta e^{tA_a}(x_1 + x_2) \end{pmatrix}, \tag{57.5}$$

for $t > 0, x_1, x_2 \in \mathbb{X}$. For $\binom{x_1}{x_2} \in \mathbb{B}'$ the same is true for $t = 0$ as well, and the limit is almost uniform in $t \in [0, \infty)$; for other $\binom{x_1}{x_2}$ the limit is almost uniform in $t \in (0, \infty)$.

Intuitively, this result may be explained as follows. The components of the semigroup $(e^{t\mathcal{A}_0})_{t\ge 0}$ are uncoupled, while in $(e^{t\mathcal{A}_\kappa})_{t\ge 0}$ the coupling is realized by the operator $\mathcal{Q}$, which may be thought of as describing a Markov chain switching one dynamics into the other (the jumps' intensities are state-dependent; see examples given later). As $\kappa \to \infty$, the Markov chain reaches its statistical equilibrium, so that with "probability" α it chooses the first dynamics, and with "probability" β, it chooses the second dynamics. This results in a convex combination of boundary conditions in the limit semigroup, thus confirming our intuitions on $(e^{tA_a})_{t\geq 0}$. (Again, compare Chapter 48.)

Theorem 57.2 *Under conditions of the previous theorem, let:*

$$\mathcal{B} = \begin{pmatrix} B_1 & 0 \\ 0 & B_2 \end{pmatrix},$$

where B_1 and B_2 are bounded linear operators. Then,

$$\lim_{\kappa\to+\infty} e^{t(\mathcal{A}_\kappa+\mathcal{B})}\begin{pmatrix} x_1 \\ x_2 \end{pmatrix} = \mathcal{I}e^{t(A_a+B_1\alpha+B_2\beta)}\mathcal{I}^{-1}\mathcal{P}\begin{pmatrix} x_1 \\ x_2 \end{pmatrix}, \qquad (57.6)$$

for $t > 0, x_1, x_2 \in \mathbb{X}$. For $\binom{x_1}{x_2} \in \mathbb{B}'$ the same is true for $t = 0$ as well, and the limit is almost uniform in $t \in [0, \infty)$; for other $\binom{x_1}{x_2}$ the limit is almost uniform in $t \in (0, \infty)$.

Remark 57.3 For Theorems 57.1 and 57.2, in addition to (56.1) and (57.1), we assume that $\mathcal{P}$, defined in (57.2), is a contraction in $\mathbb{B}$ and α leaves $D(A)$ invariant. While the nature of the first and the last conditions is transparent, the other two require a comment. As already mentioned, together they imply stability condition (57.4) (which, as we know, is a common assumption in convergence theorems), and in fact our theorems remain true if we simply assume (57.4). However, for the sake of applications it is more convenient to assume the two conditions discussed above. Out of these two, the one requiring $\mathcal{P}$ to be a contraction seems to be most restrictive, apparently excluding spaces with supremum norm. On the other hand, this assumption is often satisfied in L^1-type spaces. (Similarly, the "dual" theorem in Chapter 48 is designed for spaces of continuous functions.) In particular, if $\mathbb{X}$ is an AL-space, that is, a Banach lattice such that:

$$\|x+y\| = \|x\| + \|y\|, \qquad x, y \geq 0,$$

and $\mathbb{B}$ is equipped with the order "$\binom{x}{y} \geq 0$ iff $x \geq 0$ and $y \geq 0$" and the norm $\|\binom{x}{y}\| = \|x\| + \|y\|$, then $\mathcal{P}$ is a contraction provided α and β are positive operators. For, in such a case,

$$\left\|\mathcal{P}\begin{pmatrix} x \\ y \end{pmatrix}\right\| = \left\|\begin{pmatrix} \alpha(x+y) \\ \beta(x+y) \end{pmatrix}\right\| = \|\alpha(x+y)\| + \|\beta(x+y)\| = \|x+y\| = \left\|\begin{pmatrix} x \\ y \end{pmatrix}\right\|,$$

for $x, y \geq 0$, and $\mathcal{P}$ is positive. Hence, by Lemma 13.1,

$$\|\mathcal{P}\| = \sup_{\|\binom{x}{y}\|=1, \binom{x}{y}\geq 0} \left\|\mathcal{P}\begin{pmatrix} x \\ y \end{pmatrix}\right\| = 1,$$

as desired. □

For the proof of the theorems we need a set of lemmas.

Lemma 57.4 *Let $\mathfrak{A}_\kappa$ denote the extension of operator $\mathcal{A}_\kappa$ to the maximal domain $D(A) \times D(A)$ and let $\Phi : \mathbb{X} \times \mathbb{X} \to \mathbb{Y} \times \mathbb{Y}$ be given by $\Phi\binom{x_1}{x_2} = \binom{\Phi_1 x_1}{\Phi_2 x_2}$. For $x \in D(A_a)$, we define:*

$$x_{1,\kappa} = \alpha x + L_\kappa \tilde{x},$$
$$x_{2,\kappa} = \beta x - L_\kappa \tilde{x},$$

where $\tilde{x} := F_1 \alpha x - L\alpha x = L\beta x - F_2 \beta x \in \mathbb{Y}$, by $D(A_a) = \ker \Phi_a$. Then,

$$\mathfrak{A}_\kappa \begin{pmatrix} x_{1,\kappa} \\ x_{2,\kappa} \end{pmatrix} = \begin{pmatrix} A\alpha x \\ A\beta x \end{pmatrix}, \lim_{\kappa \to +\infty} \begin{pmatrix} x_{1,\kappa} \\ x_{2,\kappa} \end{pmatrix} = \begin{pmatrix} \alpha x \\ \beta x \end{pmatrix} \text{ and } \lim_{\kappa \to +\infty} \Phi \begin{pmatrix} x_{1,\kappa} \\ x_{2,\kappa} \end{pmatrix} = 0. \tag{57.7}$$

Proof Using $AL_\kappa \tilde{x} = \kappa L_\kappa \tilde{x}$, we get:

$$\mathcal{A}_0 \begin{pmatrix} x_{1,\kappa} \\ x_{2,\kappa} \end{pmatrix} = \begin{pmatrix} A\alpha x + \kappa L_\kappa \tilde{x} \\ A\beta x - \kappa L_\kappa \tilde{x} \end{pmatrix}$$

while:

$$\mathcal{Q} \begin{pmatrix} x_{1,\kappa} \\ x_{2,\kappa} \end{pmatrix} = \begin{pmatrix} -\beta L_\kappa \tilde{x} - \alpha L_\kappa \tilde{x} \\ \beta L_\kappa \tilde{x} + \alpha L_\kappa \tilde{x} \end{pmatrix} = \begin{pmatrix} -L_\kappa \tilde{x} \\ L_\kappa \tilde{x} \end{pmatrix};$$

this shows the first part. Next, by (56.1), we have $\|L_\kappa \tilde{x}\| \le \frac{1}{\kappa \gamma} \|\tilde{x}\|$. It follows that $\lim_{\kappa \to +\infty} \binom{x_{1,\kappa}}{x_{2,\kappa}} = \binom{\alpha x}{\beta x}$ and:

$$\Phi \begin{pmatrix} x_{1,\kappa} \\ x_{2,\kappa} \end{pmatrix} = \begin{pmatrix} L\alpha x + \tilde{x} - F_1 \alpha x - F_1 L_\kappa \tilde{x} \\ L\beta x - \tilde{x} + F_2 \beta x + F_2 L_\kappa \tilde{x} \end{pmatrix} \underset{\kappa \to \infty}{\longrightarrow} \begin{pmatrix} L\alpha x + \tilde{x} - F_1 \alpha x \\ L\beta x - \tilde{x} + F_2 \beta x \end{pmatrix} = \begin{pmatrix} 0 \\ 0 \end{pmatrix}. \tag{57.8}$$

This completes the proof. □

Unfortunately, in general $\binom{x_{1,\kappa}}{x_{2,\kappa}}$ does not belong to $D(\mathcal{A}_\kappa)$. Our main goal is to modify this vector appropriately without altering its crucial properties (see Lemma 57.7). To this end, we need the following two lemmas.

Lemma 57.5 *Let $\lambda > \max(\lambda_0, \omega)$. Then $\ker(\lambda - \mathfrak{A}_\kappa)$ is composed of vectors of the form:*

$$\begin{pmatrix} x_1 \\ x_2 \end{pmatrix} = \begin{pmatrix} \kappa(\lambda + \kappa - A_0)^{-1} \alpha h_\lambda - h_{\lambda+\kappa} \\ h_\lambda + h_{\lambda+\kappa} - \kappa(\lambda + \kappa - A_0)^{-1} \alpha h_\lambda \end{pmatrix}, \tag{57.9}$$

where $h_\lambda \in \ker(\lambda - A)$ and $h_{\lambda+\kappa} \in \ker(\lambda + \kappa - A)$ may be chosen arbitrarily.

Proof The operator $\mathcal{I} = \begin{pmatrix} \alpha & -I_{\mathbb{X}} \\ \beta & I_{\mathbb{X}} \end{pmatrix}$ is an isomorphism of $\mathbb{B}$ with the inverse $\begin{pmatrix} I_{\mathbb{X}} & I_{\mathbb{X}} \\ -\beta & \alpha \end{pmatrix}$. A vector $x = \binom{x_1}{x_2}$ satisfies $\lambda x - \mathfrak{A}_\kappa x = 0$ iff for $y = \mathcal{I}^{-1}x$ we have:

$$\lambda y - \mathcal{I}^{-1}\mathfrak{A}_\kappa \mathcal{I} y = 0. \tag{57.10}$$

On the other hand,

$$\mathcal{I}^{-1}\mathfrak{A}_\kappa\mathcal{I} = \begin{pmatrix} A\alpha + A\beta & 0 \\ -\beta A\alpha + \alpha A\beta & \beta A + \alpha A - \kappa \end{pmatrix} = \begin{pmatrix} A & 0 \\ \alpha A - A\alpha & A - \kappa \end{pmatrix}.$$

Hence, (57.10) is satisfied for $y = \binom{y_1}{y_2}$ iff:

$$\lambda y_1 - Ay_1 = 0 \quad \text{and} \quad (\lambda + \kappa)y_2 - Ay_2 = (\alpha A - A\alpha)y_1.$$

The first condition here means that y_1 is a member, say h_λ of $\ker(\lambda - A)$. Consequently, the other condition may be written in the form:

$$(\lambda + \kappa)y_2 - Ay_2 = (\lambda + \kappa - A)\alpha h_\lambda - \kappa\alpha h_\lambda$$

or $(\lambda + \kappa - A)(y_2 - \alpha h_\lambda) = -\kappa\alpha h_\lambda$. Therefore, by (56.2), $y_2 - \alpha h_\lambda$ differs from $-\kappa(\lambda + \kappa - A_0)^{-1}\alpha h_\lambda$ by an element of $\ker(\lambda + \kappa - A)$, say $h_{\lambda+\kappa}$. This is equivalent to saying that:

$$y_2 = \alpha h_\lambda + h_{\lambda+\kappa} - \kappa(\lambda + \kappa - A_0)^{-1}\alpha h_\lambda.$$

To complete the proof it suffices to calculate $\binom{x_1}{x_2} = \mathcal{I}\binom{y_1}{y_2}$. □

Lemma 57.6 *There is λ_1 such that for $\lambda > \lambda_1$ and all $\kappa > 0$*

$$\|\Phi_1 x_1\| + \|\Phi_2 x_2\| \geq \|x_1\| + \|x_2\| \text{ for all } \begin{pmatrix} x_1 \\ x_2 \end{pmatrix} \in \ker(\lambda - \mathfrak{A}_\kappa). \tag{57.11}$$

Proof Fix $\lambda > \max(\lambda_0, \omega)$ and $\kappa > 0$. By Lemma 57.5, a vector $\binom{x_1}{x_2} \in \ker(\lambda - \mathfrak{A}_\kappa)$ is of the form (57.9). Using (57.1) and then (56.1),

$$\begin{aligned} \|x_1\| + \|x_2\| &\leq \frac{2\kappa\|\alpha\|}{\lambda\gamma(\lambda+\kappa-\omega)}\|Lh_\lambda\| + \frac{2}{\gamma(\lambda+\kappa)}\|Lh_{\lambda+\kappa}\| + \frac{1}{\gamma\lambda}\|Lh_\lambda\| \\ &\leq \frac{K}{\lambda}\left(\|Lh_\lambda\| + \|Lh_{\lambda+\kappa}\|\right) \end{aligned}$$

for an appropriately chosen constant $K > 0$. Since F_1 and F_2 are bounded operators,

$$\|F_1 x_1\| + \|F_2 x_2\| \leq \frac{K}{\lambda}\left(\|Lh_\lambda\| + \|Lh_{\lambda+\kappa}\|\right),$$

with possibly different K.

On the other hand, the map $\mathbb{Y} \times \mathbb{Y} \ni (y_1, y_2) \mapsto \|y_2\| + \|y_1 + y_2\|$ is a norm in $\mathbb{Y} \times \mathbb{Y}$ and $\|y_2\| + \|y_1 + y_2\| \leq 2(\|y_1\| + \|y_2\|)$. Hence, by the open mapping theorem, $\|y_1\| + \|y_2\| \le \mu(\|y_2\| + \|y_1 + y_2\|)$ for some constant $\mu > 0$ and all $y_1, y_2 \in \mathbb{Y}$. It follows that (recall $(\lambda + \kappa - A_0)^{-1}\alpha h_\lambda \in D(A_0) = \ker L$),

$$\begin{aligned}
\|\Phi_1 x_1\| + \|\Phi_2 x_2\| &= \|Lh_{\lambda+\kappa} + F_1 x_1\| + \|Lh_\lambda + Lh_{\lambda+\kappa} - F_2 x_2\| \\
&\ge \|Lh_{\lambda+\kappa}\| + \|Lh_\lambda + Lh_{\lambda+\kappa}\| - \|F_1 x_1\| - \|F_2 x_2\| \\
&\ge \left(\frac{1}{\mu} - \frac{K}{\lambda}\right)(\|Lh_\lambda\| + \|Lh_{\lambda+\kappa}\|) \\
&\ge \frac{K}{\lambda}(\|Lh_\lambda\| + \|Lh_{\lambda+\kappa}\|) \\
&\ge \|x_1\| + \|x_2\|,
\end{aligned}$$

provided $\lambda \ge 2K\mu$. □

The map Φ from Lemma 57.4 is surjective. Arguing as in the proof of the decomposition (56.2), we see that any vector $v \in \mathbb{X} \times \mathbb{X}$ may be uniquely written as $v = v_1 + v_2$ where $v_1 \in \ker \Phi = D(\mathcal{A}_\kappa)$, and $v_2 \in \ker(\lambda - \mathfrak{A}_\kappa)$, where $\lambda > \max(\lambda_1, \omega)$ and $\kappa > 0$ are fixed (λ_1 is defined in Lemma 57.6). Hence, Φ restricted to $\ker(\lambda - \mathfrak{A}_\kappa)$ is surjective as well, and inequality (57.11) shows that it is invertible, with inverse, say $J_{\lambda,\kappa}$, of norm at most 1.

Lemma 57.7 *For $x \in D(A_a)$, there are $\binom{y_{1,\kappa}}{y_{2,\kappa}} \in D(A_1) \times D(A_2) = D(\mathcal{A}_\kappa)$ such that:*

$$\lim_{\kappa\to+\infty} \begin{pmatrix} y_{1,\kappa} \\ y_{2,\kappa} \end{pmatrix} = \begin{pmatrix} \alpha x \\ \beta x \end{pmatrix} \quad \textit{and} \quad \lim_{\kappa\to+\infty} \mathcal{A}_\kappa \begin{pmatrix} y_{1,\kappa} \\ y_{2,\kappa} \end{pmatrix} = \begin{pmatrix} A\alpha x \\ A\beta x \end{pmatrix}.$$

Proof Take $\lambda > \lambda_1$ and define (see Lemma 57.4):

$$\begin{pmatrix} y_{1,\kappa} \\ y_{2,\kappa} \end{pmatrix} := \begin{pmatrix} x_{1,\kappa} \\ x_{2,\kappa} \end{pmatrix} - J_{\lambda,\kappa}\Phi \begin{pmatrix} x_{1,\kappa} \\ x_{2,\kappa} \end{pmatrix}.$$

Then, $\Phi\binom{y_{1,\kappa}}{y_{2,\kappa}} = \Phi\binom{x_{1,\kappa}}{x_{2,\kappa}} - \Phi\binom{x_{1,\kappa}}{x_{2,\kappa}} = 0$, proving that $\binom{y_{1,\kappa}}{y_{2,\kappa}} \in D(\mathcal{A}_\kappa)$. Also, by Lemma 57.4,

$$\lim_{\kappa\to+\infty} \begin{pmatrix} y_{1,\kappa} \\ y_{2,\kappa} \end{pmatrix} = \begin{pmatrix} \alpha x \\ \beta x \end{pmatrix} - \lim_{\kappa\to+\infty} J_{\lambda,\kappa}\Phi \begin{pmatrix} x_{1,\kappa} \\ x_{2,\kappa} \end{pmatrix} = \begin{pmatrix} \alpha x \\ \beta x \end{pmatrix},$$

by (56.1) and since $\left\| J_{\lambda,\kappa}\Phi\binom{x_{1,\kappa}}{x_{2,\kappa}} \right\| \le \left\| \Phi\binom{x_{1,\kappa}}{x_{2,\kappa}} \right\| \underset{\kappa\to\infty}{\longrightarrow} 0$ by (57.8). Similarly,

$$\begin{aligned}
\lim_{\kappa\to+\infty} \mathcal{A}_\kappa \begin{pmatrix} y_{1,\kappa} \\ y_{2,\kappa} \end{pmatrix} &= \begin{pmatrix} A\alpha x \\ A\beta x \end{pmatrix} + \lim_{\kappa\to+\infty} \mathfrak{A}_\kappa J_{\lambda,\kappa}\Phi \begin{pmatrix} x_{1,\kappa} \\ x_{2,\kappa} \end{pmatrix} \\
&= \begin{pmatrix} A\alpha x \\ A\beta x \end{pmatrix} + \lambda \lim_{\kappa\to+\infty} J_{\lambda,\kappa}\Phi \begin{pmatrix} x_{1,\kappa} \\ x_{2,\kappa} \end{pmatrix} \\
&= \begin{pmatrix} A\alpha x \\ A\beta x \end{pmatrix}.
\end{aligned}$$

□

Lemma 57.7 combined with Kurtz's Theorem is all we need to prove Theorem 57.1. To see this we start by noting that, although the semigroups generated by $\mathcal{A}_\kappa, \kappa \ge 0$ in $\mathbb{B}$ are not equibounded, those generated by $\mathcal{A}_\kappa - \omega I, \kappa \ge 0$, are (by (57.4)). It is easy to see that Kurtz's Theorem holds in such a case as well. Next, we take $Q = \mathcal{Q}$, and then (42.1) follows by (57.3) with $P = \mathcal{P}$. Therefore, $\mathbb{B}' = Range\mathcal{P}$ is the space of vectors of the form $\binom{\alpha x}{\beta x}$, $x \in \mathbb{X}$. Let $\mathcal{A}$ be the operator in $\mathbb{B}$ given by:

$$\mathcal{A}\begin{pmatrix}\alpha x\\ \beta x\end{pmatrix} = \begin{pmatrix}A\alpha x\\ A\beta x\end{pmatrix}, \qquad D(\mathcal{A}) = \left\{\begin{pmatrix}\alpha x\\ \beta x\end{pmatrix}; x \in D(A)\right\}.$$

This operator is well defined since we assumed that α leaves $D(A)$ invariant. By Lemma 57.7, condition (a) in Kurtz's Theorem is satisfied with $D = \mathcal{I}D(A_{\rm a})$: more precisely, for $x \in D$, $(x, \mathcal{A}x)$ is a member of the extended limit of $\mathcal{A}_\kappa$. Next, for $x_1 \in D(A_1), x_2 \in D(A_2)$, we have $\lim_{\kappa\to+\infty} \kappa^{-1}\mathcal{A}_\kappa\binom{x_1}{x_2} = \mathcal{Q}\binom{x_1}{x_2}$ so that (b) in Kurtz's Theorem is satisfied (for $\mathcal{A}\, D' = D(A_1) \times D(A_2)$. The latter set is a core for $\mathcal{Q}$, $\mathcal{Q}$ being bounded. Finally, $D \cap \mathbb{B}' = D = \mathcal{I}D(A_{\rm a})$ and:

$$\mathcal{P}\mathcal{A}\mathcal{I}x = \mathcal{P}\mathcal{A}\begin{pmatrix}\alpha x\\ \beta x\end{pmatrix} = \begin{pmatrix}\alpha A x\\ \beta A x\end{pmatrix} = \mathcal{I}Ax = \mathcal{I}A_{\rm a}x, \qquad x \in D(A_{\rm a}).$$

This shows that $\mathcal{P}\mathcal{A}$ is an isomorphic copy in $\mathbb{B}'$ of the generator $A_{\rm a}$ in $\mathbb{X}$, and hence is a generator, as well. Therefore, Kurtz's Theorem is applicable, and (57.5) is a particular case of (42.5).

Finally, Theorem 57.2 is a direct consequence of Theorem 57.1 and the principle established in Chapter 29 (see in particular formula (29.3)).

Example 57.8 Let us come back to the motivating example of the fish population dynamics, where:

$$A : W^{1,1}(\mathbb{R}^+) \to L^1(\mathbb{R}^+) \qquad A\phi = -\phi',$$

and $L : W^{1,1}(\mathbb{R}^+) \to \mathbb{R}$ is given by $L\phi = \phi(0)$. We have seen in Example 56.3 (a) that Greiner's Theorem applies here.

For $b_i \in L^\infty(\mathbb{R}^+)$, $i = 1, 2$, the functionals $F_i\phi = \int_0^\infty b_i(a)\phi(a)\,da$ are linear and bounded. Hence, A_{F_i} generates a semigroup of operators and so does $A_{F_i} + B_i$, where given $\mu_i \in L^\infty(\mathbb{R}^+)$, B_i is a (bounded) multiplication operator $\phi \mapsto -\mu_i\phi$. It is well known (see, e.g., [54, 128, 184]) that there is ω such that $\|e^{t(A_{F_i}+B_i)}\| \le e^{\omega t}$, $i = 1, 2$, implying (57.1).

Let $\alpha \in W^{1,\infty}(\mathbb{R}^+)$ satisfy $0 \le \alpha \le 1$. Then the related multiplication operator (denoted in what follows by the same letter) is bounded in $L^1(\mathbb{R}^+)$, and leaves $D(A) = W^{1,1}(\mathbb{R}^+)$ invariant. Moreover, the related operator $\mathcal{P}$ (see (57.2)) in $L^1(\mathbb{R}^+) \times L^1(\mathbb{R}^+)$, equipped with the norm $\left\|\binom{\phi_1}{\phi_2}\right\| = \|\phi_1\|_{L^1(\mathbb{R}^+)} +$

$\|\phi_2\|_{L^1(\mathbb{R}^+)}$, is a contraction (see Remark 57.3). Hence, all assumptions of Theorem 57.2 are satisfied. This establishes that the general model (50.1) (with $N = 2$ and normalized matrix k) may be approximated by the averaged one (50.2). For the general case, see [21, 22, 23, 25].

Example 57.9 Let $\mathbb{X} = L^1(\mathbb{R})$. For our purposes, it will be convenient to identify $\phi \in L^1(\mathbb{R})$ with the pair $(\phi_j)_{j\in\mathbb{J}}$ of functions on $\mathbb{R}^+$ defined by $\phi_j(a) = \phi(ja)$, $a \geq 0$, where $\mathbb{J} = \{-1, 1\}$. Certainly $\phi_j \in L^1(\mathbb{R}^+)$, that is, we identify $L^1(\mathbb{R})$ with $L^1(\mathbb{R}^+) \times L^1(\mathbb{R}^+)$. With this identification in mind, and given constants $k_j \geq 0, \sigma_j > 0, j \in \mathbb{J}$, we define an operator A in $L^1(\mathbb{R}^+)$ by:

$$A\left(\phi_j\right)_{j\in\mathbb{J}} = \left(\sigma_j^2 \phi_j''\right)_{j\in\mathbb{J}}, \qquad D(A) = W^{2,1}(\mathbb{R}^+) \times W^{2,1}(\mathbb{R}^+),$$

and $L : D(A) \to \mathbb{R}^2$, by

$$L\left(\phi_j\right)_{j\in\mathbb{J}} = \left(\phi_j'(0) - k_j\phi_j(0)\right)_{j\in\mathbb{J}}.$$

The semigroup generated by A_0, the restriction of A to $\ker L$, is composed of Markov operators in $L^1(\mathbb{R})$ and describes dynamics of distributions of two independent elastic Brownian motions on two half-axes, with two different elasticity coefficients (k_j) in each half-axis, and no communication between the half-axes.

The subspace $\ker(\lambda - A)$ is composed of vectors of the form $(\phi_j)_{j\in\mathbb{J}} = (C_i e_{\lambda_j})_{j\in\mathbb{J}}$, where C_j are arbitrary constants, $\lambda_j = \frac{\sqrt{\lambda}}{\sigma_j}$, and e_λ was defined in the previous example. For such vectors and L^1-type norm in $\mathbb{R}^2$,

$$\begin{aligned} \|L\left(\phi_j\right)_{j\in\mathbb{J}}\| &= \sum_{j\in\mathbb{J}} |\phi_j'(0) - k_j\phi_j(0)| = \sum_{j\in\mathbb{J}} \left(\frac{\sqrt{\lambda}}{\sigma_i} + k_j\right)|C_j| \\ &\geq \frac{\sqrt{\lambda}}{\sigma} \sum_{j\in\mathbb{J}} |C_j|, \\ \lambda\|\left(\phi_j\right)_{j\in\mathbb{J}}\| &= \sqrt{\lambda}\sum_{j\in\mathbb{J}} \sigma_j|C_j| \leq \sigma\sqrt{\lambda}\sum_{j\in\mathbb{J}} |C_j|, \qquad \lambda > 0, \end{aligned}$$

where $\sigma = \max\{\sigma_1, \sigma_{-1}\}$. Hence, condition (56.1) is satisfied with $\gamma = \sigma^{-2}$.

Therefore, given $F_{j,i} \in L^\infty(\mathbb{R})$, $j \in \mathbb{J}$, $i = 1, 2$, A restricted to $(\phi_j)_{j\in\mathbb{J}} = \phi \in L^1(\mathbb{R})$ satisfying:

$$\phi_j'(0) - k_j\phi_j(0) = \int_{\mathbb{R}} F_{j,i}\phi, \qquad j \in \mathbb{J}, \tag{57.12}$$

is the generator of a semicontractive semigroup in $L^1(\mathbb{R})$ for both i, and (57.1) is satisfied. Nonlocal boundary conditions (57.12) do not seem to have any clear probabilistic interpretation, comp. [145, p. 470]. Nevertheless, they may model two populations occupying (one-dimensional) domains with a joint boundary, in which the flux of individuals across the common boundary is proportional not only to the difference of the densities at the boundary but also to the (weighted) difference of total numbers of individuals in each domain.

For sufficiently regular $0 \le \alpha \le 1$, the related multiplication operator in $L^1(\mathbb{R})$ leaves $D(A)$ invariant, and the related operator $\mathcal{P}$ is a contraction (see Remark 57.3). Therefore Theorem 57.1 ensures that in the limit (57.5) we obtain (an isomorphic copy) of the semigroup generated by A restricted to $(\phi_j)_{j\in\mathbb{J}} = \phi \in L^1(\mathbb{R})$, which satisfy:

$$\phi_j'(0) - k_j\phi_j(0) = \int_{\mathbb{R}} (\alpha F_{j,1} + \beta F_{j,2})\phi, \qquad j \in \mathbb{J};$$

the right-hand side here is a convex combination of the right-hand sides in (57.12).

☞ Chapter's summary

Let α be an operator in a Banach space $\mathbb{X}$, and let $\Phi_1, \Phi_2 \in \mathcal{L}(\mathbb{X}, \mathbb{Y})$ be two operators to another Banach space $\mathbb{Y}$. Furthermore, let $(\mathrm{e}^{tA_{\mathrm{a}}})_{t\ge 0}$ be a semigroup in $\mathbb{X}$ generated by A_{a} with domain equal to $\ker \Phi_{\mathrm{a}}$, where $\Phi_{\mathrm{a}} = \Phi_1\alpha + \Phi_2\beta$, $\alpha \in \mathcal{L}(\mathbb{X})$ is a bounded operator and $\beta = I_{\mathbb{X}} - \alpha$. We show that $(\mathrm{e}^{tA_{\mathrm{a}}})_{t\ge 0}$ can be approximated by a family of semigroups in $\mathbb{X} \times \mathbb{X}$ obtained by "averaging" two semigroups with the generators being the realizations of the same formal expression but restricted to different domains, equal to $\ker \Phi_1$ and $\ker \Phi_2$, respectively. This result is motivated by a model of dynamics of *solea solea* from Chapter 50.

58

Averaging Principle of Freidlin and Wentzell: Emergence of Transmission Conditions

Let us start with the example that was one of the motivations for the averaging principle of Freidlin and Wentzell [151], [153] (another motivation for this principle was discussed in Chapter 53). Imagine N diffusions (with different diffusion and drift coefficients) on N copies of an interval with Neumann boundary conditions at the ends (see Figure 58.1). Suppose also that, as in Chapter 48, these diffusions are coupled by Markov chains: while at the ith copy, the process behaves according to the rules governed by operator:

$$A_i = a_i \frac{d^2}{dx^2} + b_i \frac{d}{dx}$$

but after a random time depending on its position, may jump to another copy of the interval to behave according to the rules described by the operator defined there. In distinction to the situation of Chapter 48 we assume, however, that on the left part of the interval no communication is possible: the intensities of jumps are zero here.

Freidlin and Wentzell's result says that ([151, Theorem 5.1]) as the intensities of jumps (in the right part of the interval) tend to infinity, the corresponding processes converge weakly to a diffusion on a graph formed by identifying corresponding points of all the right-parts of the intervals (see Figure 58.2). The generator of the limit process is a convex combination of the generators

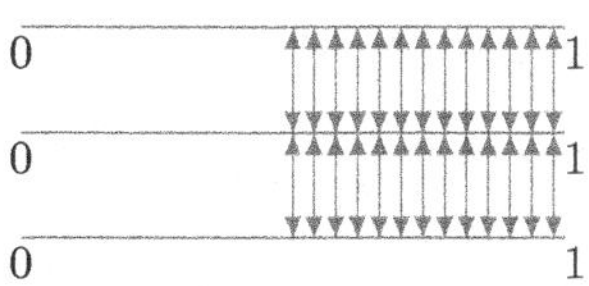

$A_1 = a_1 \frac{d^2}{dx^2} + b_1 \frac{d}{dx}$ with reflection at endpoints

$A_2 = a_2 \frac{d^2}{dx^2} + b_2 \frac{d}{dx}$ with reflection at endpoints

$A_3 = a_3 \frac{d^2}{dx^2} + b_3 \frac{d}{dx}$ with reflection at endpoints

Figure 58.1 Three diffusions on three copies of the unit interval coupled by Markov chains in the right part of the interval copies (redrawn from [56])

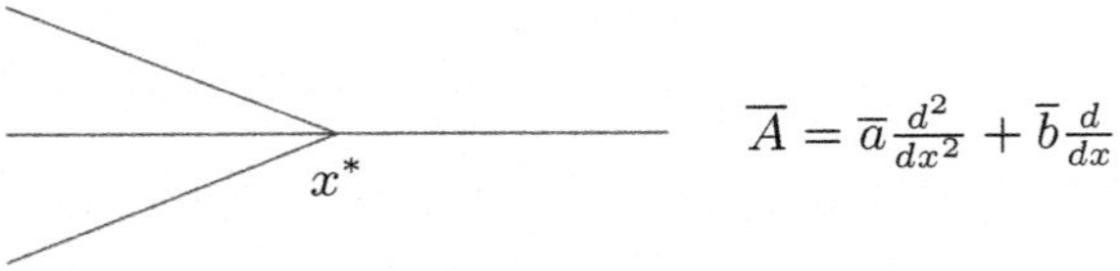

Figure 58.2 The limit process on a graph is generated by an averaged operator $\overline{A}$ (a convex combination of the involved operators A_i) with Neumann boundary conditions at the graph's ends, and balance, transmission conditions at point x^* where the segments meet, that is, at the new vertex of the graph (redrawn from [56])

of the involved diffusions – a phenomenon thoroughly studied in Chapter 48. However, a new phenomenon is observed here as well: at the junction x^* where the intervals meet, transmission conditions need to be introduced. They are of the form:

$$\sum_{i=1}^{N} a_i(x^*)\pi_i(x^*)f'_{i,-}(x^*) = \left(\sum_{i=1}^{N} a_i(x^*)\pi_i(x^*)\right) f'_+(x^*) \tag{58.1}$$

where $\pi_i(x^*), i = 1, \dots, N$ are probabilities of the equilibrium state of the Markov chain at x^*, $f'_{i,-}(x^*)$ is the left-hand derivative of f at x^* calculated on the ith interval, and $f'_+(x^*)$ is the right-hand derivative of f at x^* calculated on the edge formed by amalgamating right parts of the original intervals.

In this chapter, following [56], we drop secondary features of the example to focus on the reason for emergence of this transmission condition and its unique form. More specifically, we consider the case of two intervals (i.e., we take $N = 2$), and assume that diffusion coefficients are constant throughout these intervals and there is no drift at all, so that $a_i(x) = a_i > 0$ and $b_i(x) = 0$ for $x \in [0, 1]$, and some constants $a_i > 0$. (The effect of convex combination has been studied in Chapter 48 for processes much more general than diffusion processes, so we would gain no generality by introducing variable coefficients here.) Moreover, we assume that for some $x^* \in (0, 1)$, the Kolmogorov matrix does not depend on $x \in [x^*, 1]$ and equals:

$$\begin{pmatrix} -\beta & \beta \\ \alpha & -\alpha \end{pmatrix} \tag{58.2}$$

where $\alpha, \beta > 0$ are given constants (so that the stationary distribution is $(\frac{\alpha}{\alpha+\beta}, \frac{\beta}{\alpha+\beta})$); for $x \in [0, x^*)$, the Kolmogorov matrix is 0.

Since the operator related to such a choice of Kolmogorov matrices (see later, equation (58.3)) does not leave the space of continuous functions invariant, we will work in the Hilbert space $\mathbb{H}$ of pairs $f = (f_1, f_2)$ of square integrable functions on the unit interval, equipped with the scalar product:

$$f \circ g = \int_0^1 (\alpha f_1 \overline{g_1} + \beta f_2 \overline{g_2}).$$

For $\kappa > 0$, we define a sesquilinear form:

$$\mathfrak{a}_\kappa = \mathfrak{a} + \kappa \mathfrak{q}$$

where for $f = (f_1, f_2)$ and $g = (g_1, g_2)$,

$$\mathfrak{a}(f, g) = \int_0^1 (\alpha a_1 f_1' \overline{g_1'} + \beta a_2 f_2' \overline{g_2'})$$

while:

$$\mathfrak{q}(f, g) = \alpha\beta \int_{x^*}^1 (f_1 - f_2)\overline{(g_1 - g_2)}.$$

Here, the domain V of the form $\mathfrak{a}$ is the space of pairs $f = (f_1, f_2)$ where $f_i \in H^1[0, 1]$, that is, f_i are absolutely continuous with square integrable derivatives $f_i' \in L^2[0, 1]$, while the domain of $\mathfrak{q}$ is the whole of $\mathbb{H}$.

It is clear that both $\mathfrak{a}$ and $\mathfrak{q}$ are symmetric, and:

$$\mathfrak{a}[f] \geq 0, \text{ for } f \in V, \quad \text{and} \quad \mathfrak{q}[f] \geq 0 \text{ for } f \in \mathbb{H}.$$

A direct calculation shows that the operator related to $\mathfrak{a}$ (in the sense of Chapter 5) is:

$$A(f_1, f_2) = (a_1 f_1'', a_2 f_2'')$$

with domain composed of pairs (f_1, f_2) where $f_i \in H^2[0, 1]$ (i.e., f_i are continuously differentiable functions $f_i \in L^2[0, 1]$ with $f_i' \in H^1[0, 1]$) satisfying $f_i'(0) = f_i'(1) = 0$. Moreover, the operator related to $\mathfrak{q}$ is given by:

$$Q(f_1, f_2) = (\beta(f_2 - f_1)1_{[x^*,1]}, \alpha(f_1 - f_2)1_{[x^*,1]}), \tag{58.3}$$

where $1_{[x^*,1]}$ is the indicator function of the interval $[x^*, 1]$. It follows that both A and Q generate semigroups of self-adjoint contraction operators, and the same is true for the operators $A_\kappa = A + \kappa Q$ related to forms $\mathfrak{a}_\kappa$.

The problem is that of finding the limit:

$$\lim_{\kappa \to +\infty} e^{tA_\kappa}.$$

To this end, we introduce $\mathbb{H}_0$ as the subspace of $\mathbb{H}$ composed of (f_1, f_2) such that $(f_1 - f_2)1_{[x^*,1]} = 0$; such pairs may be identified with square integrable functions on a Y-shaped graph obtained by removing the middle segment in the left-hand part in Figure 58.2.

Let $V_0 = \mathbb{H}_0 \cap V$ and let:

$$\mathfrak{b}(f,g) = \int_0^{x^*} (\alpha a_1 f_1' \overline{g_1'} + \beta a_2 f_2' \overline{g_2'}) + \int_{x^*}^1 (\alpha a_1 + \beta a_2) f_1' \overline{g_1'};$$

by definition of $\mathbb{H}_0$, f_1 and g_1 in the second integral may be replaced by f_2 and g_2, respectively, without altering $\mathfrak{b}$. Again,

$$\mathfrak{b}[f] = \int_0^{x^*} (\alpha a_1 |f_1'|^2 + \beta a_2 |f_2'|^2) + \int_{x^*}^1 (\alpha a_1 + \beta a_2)|f_1'|^2 \geq 0,$$

and $\mathfrak{b}$ is symmetric. Integration by parts shows that the related operator B in $\mathbb{H}_0$ is given by:

$$B(f_1, f_2) = \left(\chi_1 f_1'', \chi_2 f_2''\right),$$

where $\chi_i = a_i 1_{[0,x^*)} + (\frac{\alpha}{\alpha+\beta} a_1 + \frac{\beta}{\alpha+\beta} a_2) 1_{[x^*,1]}$. Its domain is composed of pairs (f_1, f_2) such that:

(A) f_i are continuous on $[0,1]$, and continuously differentiable on $[0,x^*]$ and $[x^*,1]$ separately (one-sided derivatives at x^* may differ),
(B) $(f_i)'_{|[0,x^*]} \in H^1[0,x^*]$ and $(f_i)'_{|[x^*,1]} \in H^1[x^*,1]$,
(C) we have $f_i'(1) = f_i'(0) = 0$ and:

$$(\alpha a_1 + \beta a_2) f_{1,+}'(x^*) = \alpha a_1 f_{1,-}'(x^*) + \beta a_2 f_{2,-}(x^*), \tag{58.4}$$

where $+$ and $-$ denote the right-sided and left-sided derivatives, respectively. (Again, $f_{1,+}'(x^*)$ on the left-hand side may be replaced by $f_{2,+}'(x^*)$.)

Certainly, (58.4) is a counterpart of (58.1): these conditions inform of a flux balance (cf. eq. (4.2) and its discussion); we stress that along with this transmission condition, the continuity condition at x^* is tacitly assumed (as implied by (A)).

Theorem 58.1 *We have:*

$$\lim_{\kappa\to+\infty} e^{A_\kappa t} f = e^{tB} P f, \qquad t > 0, f \in \mathbb{H} \tag{58.5}$$

strongly and almost uniformly in $t > 0$, where projection $P \in \mathcal{L}(\mathbb{H})$ is given by $P(f_1, f_2) = (g_1, g_2)$,

$$g_i = f_i 1_{[0,x^*)} + \left(\frac{\alpha}{\alpha+\beta} f_1 + \frac{\beta}{\alpha+\beta} f_2\right) 1_{[x^*,1]}.$$

To prove this theorem, we need a "degenerate convergence version" of Theorem 5.5. In this version, the limit "upper bound" form needs not be densely defined, and the limit semigroup exists on a subspace of the original space. This result is due to B. Simon [313] (while Theorem 5.5 and its fuller version presented in [201] is due to T. Kato); it has recently been generalized by Ch. Batty and A. F. M. ter Elst to the case of series of sectorial forms [32]. We adopt the argument from the latter paper to prove B. Simon's Theorem.

Theorem 58.2 *Let $0 \le \mathfrak{c}_1 \le \mathfrak{c}_2 \le \dots$ be positive symmetric closed sesquilinear forms in a Hilbert space $\mathbb{H}$. Define the form $\mathfrak{c}$ by:*

$$D(\mathfrak{c}) = \left\{ x \in \bigcap_{n\ge 1} D(\mathfrak{c}_n) : \sup_{n\ge 1} \mathfrak{c}_n[x] < \infty \right\},$$

and $\mathfrak{c}(x, y) = \lim_{n\to\infty} \mathfrak{c}_n(x, y)$. Then, $\mathfrak{c}$ is closed, positive, and symmetric. Moreover, denoting by C the closed operator related to $\mathfrak{c}$, defined in $\mathbb{H}_0 = \overline{D(\mathfrak{c})}$, and by C_n the closed operators related to $\mathfrak{c}_n$ defined in $\mathbb{H}$, we have:

$$\lim_{n\to\infty} (\lambda - C_n)^{-1} y = (\lambda - C)^{-1} y, \qquad y \in \mathbb{H}_0,$$
$$\lim_{n\to\infty} (\lambda - C_n)^{-1} y = 0, \qquad y \in \mathbb{H}_0^{\perp}.$$

Proof In the example of Freidlin and Wentzel's averaging principle the limit form $\mathfrak{c}$ will be closed "by inspection." Hence, instead of proving that $\mathfrak{c}$ is closed in general we refer the reader to [102, 201, 297] or [313], and note that B. Simon's argument, based on a canonical decomposition of a form into closable and singular parts, is of special beauty.

For definiteness, we note that the main assumption means that for all $n \ge 1$, the domain $D(\mathfrak{c}_n)$ of $\mathfrak{c}_n$ contains $D(\mathfrak{c}_{n+1})$, and:

$$\mathfrak{c}_n[x] \le \mathfrak{c}_{n+1}[x], \qquad x \in D(\mathfrak{c}_{n+1}).$$

(a) Since the forms $\mathfrak{c}_n$ increase with n, the resolvents $(\lambda - C_n)^{-1}$ decrease (see calculation (5.12)). Moreover, they are bounded from below by 0 (compare (5.11)). It follows (see, e.g., [49, p. 108], [201, p. 254] or [264, Chapter 4]) that the strong limit:

$$R_\lambda = \lim_{n\to\infty} (\lambda - C_n)^{-1}$$

exists and is a bounded, self-adjoint operator.

(b) For $y \in \mathbb{H}$ and $\lambda > 0$, let $x_n = (\lambda - C_n)^{-1} y$. Then, for all $z \in \mathbb{H}$,

$$\mathfrak{c}_n(x_n, z) + \lambda(x_n, z) = (y, z); \tag{58.6}$$

(note that in this proof, in distinction to the convention used in this chapter, $(\cdot, \cdot)$ denotes again the scalar product (in $\mathbb{H}$) and not a pair). Upon taking $z = x_n$, it follows that:

$$\mathfrak{c}_n[x_n] + \lambda \|x_n\|^2 \le \|y\| \, \|x_n\|$$

and then:

$$\|x_n\| \le \lambda^{-1}\|y\| \quad \text{and} \quad \mathfrak{c}_n[x_n] \le \lambda^{-1}\|y\|^2. \tag{58.7}$$

(c) We know that $(x_n)_{n\ge1}$ converges strongly to an $x \in \mathbb{H}$. We claim that it converges weakly to x in all the Hilbert spaces $D(\mathfrak{c}_m)$, $m \ge 1$. This is because, as seen from the following estimate, this sequence is bounded in $D(\mathfrak{c}_m)$ (if we disregard the first $m-1$ elements, which need not belong to this space): for $n \ge m$, we have:

$$\|x_n\|_{D(\mathfrak{c}_m)} = a_m[x_n] + \|x_n\| \le a_n[x_n] + \|x_n\| \le \lambda^{-1}[\|y\| + \|y\|^2]. \tag{58.8}$$

For if the sequence does not converge weakly to x, we can choose its subsequence that converges weakly to an $x' \in D(\mathfrak{c}_m)$ different from x (recall, see, e.g., [224, Lemma 5.69] or [254, p. 149], that from a bounded sequence in a Hilbert space one may choose a weakly convergent one). By Mazur's Theorem (see, e.g., [235, p. 102]), there is a sequence of convex combinations of elements of this subsequence that converges to x' in the strong topology of $D(\mathfrak{c}_m)$. Hence, all but finite number of these convex combinations lie in a ball centered at x' with radius $\epsilon := \frac{1}{3}\|x - x'\|$ (since the norm in $D(\mathfrak{c}_m)$ is stronger than that in $\mathbb{H}$). This, however, is impossible, since without loss of generality we may assume that all elements of the original sequence belong to the ball centered in x and radius ϵ. This contradiction proves the claim. Therefore, $x \in \bigcap_{m\ge1} D(\mathfrak{c}_m)$.

(d) With self-explanatory notations, for $z \in D(\mathfrak{c}_m)$,

$$|(x, z)_{D(\mathfrak{c}_m)}| = \lim_{n\to\infty} |(x_n, z)_{D(\mathfrak{c}_m)}| \le \limsup_{n\to\infty} \|x_n\|_{D(\mathfrak{c}_m)}\|z\|_{D(\mathfrak{c}_m)}.$$

Hence (see (58.8)),

$$\begin{aligned}\mathfrak{c}_m[x] \le \|x\|_{D(\mathfrak{c}_m)} &= \sup_{\|z\|_{D(\mathfrak{c}_m)}=1} |(x, z)_{D(\mathfrak{c}_m)}| \le \limsup_{n\to\infty} \|x_n\|_{D(\mathfrak{c}_m)} \\ &\le \lambda^{-1}[\|y\| + \|y\|^2]. \end{aligned} \tag{58.9}$$

It follows that $x \in D(\mathfrak{c})$.

(e) Our next aim is to show that for $z \in D(\mathfrak{c})$,

$$\mathfrak{c}(x, z) + \lambda(x, z) = (y, z). \tag{58.10}$$

We will deduce this from (58.6): clearly, it suffices to show that we have $\lim_{n\to\infty} \mathfrak{c}_n(x_n, z) = \mathfrak{c}(x, z)$. Since $\lim_{n\to\infty} \mathfrak{c}_n(x, z) = \mathfrak{c}(x, z)$, we are left with showing that $\lim_{n\to\infty} \mathfrak{c}_n(x - x_n, z) = 0$. To this end, given $\epsilon > 0$, we choose n_0 so that $(\mathfrak{c}_n - \mathfrak{c}_{n_0})[z] \le \epsilon^2$, for $n \ge n_0$. Then, for sufficiently large n, we have

$|\mathfrak{c}_{n_0}(x - x_n, z)| \leq \epsilon$ (by (b)), and:

$$\begin{aligned}
|\mathfrak{c}_n(x - x_n, z)| &\leq |\mathfrak{c}_{n_0}(x - x_n, z)| + |(\mathfrak{c}_n - \mathfrak{c}_{n_0})(x - x_n, z)| \\
&\leq \epsilon + \sqrt{(\mathfrak{c}_n - \mathfrak{c}_{n_0})[x - x_n]}\sqrt{(\mathfrak{c}_n - \mathfrak{c}_{n_0})[z]} \text{ (since } \mathfrak{c}_n - \mathfrak{c}_{n_0} \geq 0) \\
&\leq \epsilon(1 + \sqrt{(\mathfrak{c}_n - \mathfrak{c}_{n_0})[x - x_n]}) \\
&\leq \epsilon(1 + \sqrt{\mathfrak{c}_n[x - x_n]}) \\
&\leq \epsilon\left(1 + \sqrt{2\mathfrak{c}_n[x_n] + 2\mathfrak{c}_n[x]}\right) \\
&\leq \epsilon\left(1 + \sqrt{\frac{2}{\lambda}\|y\|^2 + \frac{2}{\lambda}(\|y\|^2 + \|y\|)}\right) \text{ (by (58.7) and (58.9)).}
\end{aligned}$$

This completes the proof of (58.10).

(f) For $y \perp \overline{D(\mathfrak{c})}$ (closure in the strong topology in $\mathbb{H}$), equation (58.10) says that $x \in D(C)$, and $Cx = \lambda x$, implying $x = 0$ (because C is the generator of a contraction semigroup). This means that for such y, $\lim_{n\to\infty}(\lambda - C_n)^{-1}y = 0$. On the other hand, if $y \in \overline{D(\mathfrak{c})}$, then (58.10) shows that $x \in D(C)$ and $\lambda x - Cx = y$., that is, $\lim_{n\to\infty}(\lambda - C_n)^{-1}y = (\lambda - C)^{-1}y$. □

Proof of Theorem 58.1 Since $\mathfrak{q} \geq 0$, the forms $\mathfrak{a}_\kappa$ increase with κ, so that we are in the setup of Simon's Theorem. It is clear that:

$$\sup_{\kappa>0} \mathfrak{a}_\kappa[f] < \infty$$

iff $\mathfrak{q}[f] = 0$, that is, iff $f \in \mathbb{H}_0$. Therefore, the limit form coincides with $\mathfrak{a}$ restricted to V_0 and this equals $\mathfrak{b}$ (see Exercise 58.2). It follows that:

$$\lim_{\kappa\to+\infty}(\lambda - A_\kappa)^{-1}f = (\lambda - B)^{-1}f, \qquad f \in \mathbb{H}_0,$$

that is, that for $D(B)$ is a subset of the domain of the extended limit A_{ex} of A_κ, $\kappa \to \infty$ and $A_{\mathrm{ex}}g = Bg$, $g \in D(B)$. Hence, condition (a) of the Kurtz Singular Perturbation Theorem is satisfied with $D = D(B)$.

We check the remaining conditions. For all $f \in D(A)$,

$$\lim_{\kappa\to+\infty}\kappa^{-1}A_\kappa f = Qf$$

while $D(A)$ is dense in $\mathbb{H}$ and Q is bounded, proving condition (b). Next, Q and P are related by the following formula:

$$Q = (\alpha + \beta)(P - I_{\mathbb{H}}).$$

It follows that $\lim_{t\to\infty} \mathrm{e}^{tQ} = P$ (see (57.3)). Finally, for $f \in D(B)$ we have:

$$PAf = Bf.$$

In particular, PA is is closed, proving assumption (c) in Kurtz's Theorem and completing the proof. □

Exercise 58.1 Prove that operators A, Q, and B correspond to forms $\mathfrak{a}$, $\mathfrak{q}$, and $\mathfrak{b}$, respectively.

Exercise 58.2 Check that on V_0, the forms $\mathfrak{a}$, $\mathfrak{b}$, and $\mathfrak{a}_\kappa$ coincide.

Exercise 58.3 Prove that $(\lambda - B)^{-1} f \circ f \le (\lambda - A_\kappa)^{-1} f \circ f$, $f \in \mathbb{H}_0$.

Exercise 58.4 Prove an analogue of (33.6) (i.e., of the main result in Chapters 33 and 34) in $L^2(\Omega)$, by completing the following steps (we keep notations from the chapters just mentioned).

1. For each $\kappa > 0$ define an operator $\mathcal{A}_\kappa$ in $L^2(S)$ with domain $D(\mathcal{A}_\kappa)$ composed of all f with the property that f restricted to Ω_i is a member of $H^2(\Omega_i)$, $i = 1, 2, 3$ and the following transmission conditions hold:

$$\begin{aligned} f'(0) &= 0, & \kappa f'(r_3^-) &= k_{32}[f(r_3^+) - f(r_3^-)], \\ \kappa f'(r_3^+) &= k_{23}[f(r_3^+) - f(r_3^-)], & \kappa f'(r_2^-) &= k_{21}[f(r_2^+) - f(r_2^-)], \\ \kappa f'(r_2^+) &= k_{12}[f(r_2^+) - f(r_2^-)], & \kappa f'(r_1) &= -k_{10} f(r_1), \end{aligned} \tag{58.11}$$

(of course, at the interval's endpoints, the derivatives are one-sided). The operator is then defined as:

$$\mathcal{A}_\kappa f = \kappa \sigma f''.$$

2. Equip $L^2(\Omega)$ with the scalar product (which is evidently equivalent to the natural one):

$$f \circ g = \sum_{i=1}^{3} w_i \int_{\Omega_i} f \overline{g}$$

where:

$$\begin{aligned} w_1 &= k_{21} k_{32} \sigma_2 \sigma_3, \\ w_2 &= k_{12} k_{32} \sigma_1 \sigma_3, \\ w_3 &= k_{12} k_{23} \sigma_1 \sigma_2. \end{aligned}$$

(This scalar product allows working with symmetric operators.) Also, introduce the sesquilinear form:

$$\begin{aligned} \mathfrak{a}_\kappa(f, g) = \kappa \sum_{i=1}^{3} w_i \sigma_i \int_{I_i} f' \overline{g'} & \\ + \sigma_3 w_3 k_{32} [f(r_3^+) - f(r_3^-)] & \overline{[g(r_3^+) - g(r_3^-)]} \\ + \sigma_1 w_1 k_{12} [f(r_2^+) - f(r_2^-)] & \overline{[g(r_2^+) - g(r_2^-)]} \\ + \sigma_1 w_1 k_{10} f(r_1) \overline{g(r_1)}, & \end{aligned} \tag{58.12}$$

on $V \subset L^2(\Omega)$, where V is composed of g with the property that g restricted to Ω_i belongs to $H^1(\Omega_i)$. Prove that $\mathfrak{a}_\kappa$ is the form related to $\mathcal{A}_\kappa$. *Hint:* use the relations:

$$\sigma_3 w_3 k_{32} = \sigma_2 w_2 k_{23} \qquad \text{and} \qquad \sigma_2 w_2 k_{21} = \sigma_1 w_1 k_{12}$$

following directly from the definition of w_i's.

3. Let $\mathbb{H}_0$ be composed of functions that are constant on each interval Ω_i, and thus may be identified with $\mathbb{R}^3$. Prove that the limit $\lim_{\kappa\to+\infty} \mathfrak{a}_\kappa[f]$ exists and is finite iff $f \in \mathbb{H}_0$. Moreover, writing $f = (f_1, f_2, f_3)$ and $g = (g_1, g_2, g_3)$ for $f, g \in \mathbb{H}_0$ (where $f_i = f_{|\Omega_i}$ is identified with a real number, and similarly for g), the limit form is given by:

$$\mathfrak{q}(f, g) = \sigma_3 w_3 k_{32}[f_2 - f_3]\overline{(g_2 - g_3)} + \sigma_1 w_1 k_{12}[f_1 - f_2]\overline{(g_1 - g_2)} + \sigma_1 w_1 k_{10} f_1 \overline{g_1}. \tag{58.13}$$

Check that the related (bounded) operator Q is given by:

$$\begin{pmatrix} f_1 \\ f_2 \\ f_3 \end{pmatrix} \mapsto Q \begin{pmatrix} f_1 \\ f_2 \\ f_3 \end{pmatrix} \qquad \text{(matrix multiplication)},$$

where Q was defined in Chapter 33 (right after (33.6)).

4. Conclude that, as in (33.6),

$$\lim_{\kappa\to\infty} \mathrm{e}^{t\mathcal{A}_\kappa} = \mathrm{e}^{tQ} P,$$

where P is the orthogonal projection onto $\mathbb{H}_0$:

$$Pf = \left(\frac{1}{|\Omega_1|} \int_{\Omega_1} f, \frac{1}{|\Omega_2|} \int_{\Omega_2} f, \frac{1}{|\Omega_3|} \int_{\Omega_3} f \right).$$

To this end, combine B. Simon's and T. G. Kurtz's theorems.

☞ Chapter's summary

The approximation procedure considered in this chapter is a close relative of those of Chapters 48 and 57. In Chapter 48, the generator of a limit semigroup was a convex combination of generators of original semigroups. In Chapter 57 it was the domain of the generator of the limit semigroup that we focused on: this domain was the kernel of a convex combination of operators describing domains of generators of original semigroups. In other words, the boundary or transmission conditions were changed in the limit, as a result of averaging. Here, in this chapter, the averaging process leads to another phenomenon: it forces *emergence* of transmission conditions at a point where they were not needed before.

59

Averaging Principle Continued: L^1-Setting

The approach of the previous chapter is quite elegant: the quadratic forms contain all the information needed for the limit theorem. The arguable elegance, however, comes perhaps at the cost of blurring the mechanism of emergence of transmission conditions in the limit. To explain: the information about these conditions is compressed, or, so to say, "zipped" in the quadratic form $\mathfrak{b}$ corresponding to the operator B. While it is seen without a shadow of a doubt that the limit form cannot be anything other than $\mathfrak{b}$, from the perspective of forms it is still somewhat difficult to grasp the way transmission conditions come into existence. In other words, with forms, the picture is quite clear, but some part of the mystery is still there – unless you master the connection between the operator and the form. Therefore, in this chapter, we present another approach, where calculations are much more explicit, if a bit complex. As we shall see, convergence of semigroups involved may be deduced from convergence of solutions of a four-dimensional linear system of equations (see (59.9)).

To this end, we follow [56] to work with the semigroups of Markov operators in the space of absolutely integrable functions. These semigroups are responsible for evolution of densities of the Markov processes involved (and are perhaps a bit more natural than those of the previous chapter). More specifically, we work with the space $\mathbb{X} = L^1(\mathbb{R}) \times L^1(\mathbb{R})$ identified with the space of integrable functions $f : \mathbb{R} \times \{1, 2\} \to \mathbb{R}$. In other words, each pair $(f_1, f_2) \in \mathbb{X}$ is identified with such a function, defined by $f(x, i) = f_i(x)$, $i = 1, 2$, $x \in \mathbb{R}$. The norm in $\mathbb{X}$ is given by:

$$\|f\| = \|f_1\|_{L^1(\mathbb{R})} + \|f_2\|_{L^1(\mathbb{R})}.$$

Given two diffusion coefficients $a_1, a_2 > 0$ we define:

$$\mathcal{A}\begin{pmatrix} f_1 \\ f_2 \end{pmatrix} = \begin{pmatrix} a_1 f_1'' \\ a_2 f_2'' \end{pmatrix},$$

for $f_1, f_2 \in W^{2,1}(\mathbb{R})$. Then, $\mathcal{A}$ generates a semigroup of Markov operators in $\mathbb{X}$: the reader has noticed probably that we got rid of reflecting (Neumann) boundary conditions (which have no bearing on the phenomenon under study) and have allowed Brownian particles to diffuse freely on two copies of the real axis. Next, for given intensities $\alpha, \beta > 0$ we define bounded linear operators in $\mathbb{X}$ by:

$$\mathcal{Q}\begin{pmatrix} f_1 \\ f_2 \end{pmatrix} = \begin{pmatrix} (\alpha f_2 - \beta f_1)1_{[0,\infty)} \\ (\beta f_1 - \alpha f_2)1_{[0,\infty)} \end{pmatrix}$$

and $\mathcal{P} = \frac{1}{\alpha+\beta}\mathcal{Q} + I_{\mathbb{X}}$, so that:

$$\mathcal{P}\begin{pmatrix} f_1 \\ f_2 \end{pmatrix} = \begin{pmatrix} f_1 1_{(-\infty,0)} + \alpha'(f_1 + f_2)1_{[0,\infty)} \\ f_2 1_{(-\infty,0)} + \beta'(f_1 + f_2)1_{[0,\infty)} \end{pmatrix}, \tag{59.1}$$

where:

$$\alpha' = \frac{\alpha}{\alpha+\beta} \quad \text{and} \quad \beta' = \frac{\beta}{\alpha+\beta}.$$

It is easy to see that $\mathcal{P}$ is a Markov operator; it follows that $\mathcal{Q}$ generates a semigroup $(\mathrm{e}^{t\mathcal{Q}})_{t\geq 0}$ of such operators. This semigroup describes the process in which states (x, i) where $x < 0$ are absorbing, while $(x, 1)$ and $(x, 2)$ communicate as states of a Markov chain with intensity matrix (58.2). (In particular, it is the point 0 that plays now the role of x^* of the previous chapter.) Therefore, as in Chapter 45, the Phillips Perturbation Theorem combined with the Trotter Product Formula, implies that for each $\kappa > 0$ the operator:

$$\mathcal{A}_\kappa = \mathcal{A} + \kappa\mathcal{Q}$$

generates a semigroup of Markov operators in $\mathbb{X}$. (This semigroup is in a sense dual to that generated by A_κ of the previous chapter.)

Since we want to find the limit:

$$\lim_{\kappa\to+\infty} \mathrm{e}^{t\mathcal{A}_\kappa},$$

we turn to studying the resolvent equation $\lambda f - \mathcal{A}_\kappa f = g$, for $\lambda > 0$ and $g \in \mathbb{X}$. As we shall see, the solution $f \in D(\mathcal{A})$ (which exists and is unique, $\mathcal{A}_\kappa$ generating the contraction semigroup) may be found in a quite explicit way.

To begin with, we note that on the left half-axis, the resolvent equation takes the form:

$$\lambda f_i(x) - a_i f_i''(x) = g_i(x), \qquad x < 0, i = 1, 2.$$

Basic principles of ordinary differential equations tell us that there are solutions to this equation that are integrable on the left half-axis, and they are given by (compare the beginning of Chapter 3):

$$f_i(x) = l_i \mathrm{e}^{\sqrt{\frac{\lambda}{a_i}}x} + \frac{1}{2\sqrt{\lambda a_i}} \int_{-\infty}^{0} \mathrm{e}^{-\sqrt{\frac{\lambda}{a_i}}|x-y|} g_i(y)\,\mathrm{d}y, \qquad x < 0, \tag{59.2}$$

where l_1, l_2 are (yet) unknown constants.

On the right half-axis, the resolvent equation takes the form:

$$\begin{aligned} (\lambda + \beta) f_1 - a_1 f_1'' - \alpha f_2 &= g_1, \\ (\lambda + \alpha) f_2 - a_2 f_2'' - \beta f_1 &= g_2. \end{aligned} \tag{59.3}$$

More precisely, this is (a part of) the resolvent equation for $\mathcal{A}_1$; the general case will be recovered later by replacing each instance of α and β by $\kappa\alpha$ and $\kappa\beta$, respectively. Moreover, g_1, g_2 are now treated as members of $L^1(\mathbb{R}^+)$ and solutions are also sought in this space. The question of existence of these solutions is answered in Lemma 59.1, below, but we need to make some preparatory remarks for this result.

The quadratic equation:

$$(\lambda + \alpha - a_2 t)(\lambda + \beta - a_1 t) = \alpha\beta$$

has precisely two real solutions:

$$t_i = \frac{(\lambda + \alpha)a_1 + (\lambda + \beta)a_2 - (-1)^i\sqrt{\Delta}}{2a_1 a_2}, \qquad i = 1, 2,$$

where:

$$\begin{aligned} \Delta &= [(\lambda + \alpha)a_1 + (\lambda + \beta)a_2]^2 - 4a_1a_2[(\lambda + \alpha)(\lambda + \beta) - \alpha\beta] \\ &> [(\lambda + \alpha)a_1 + (\lambda + \beta)a_2]^2 - 4a_1a_2(\lambda + \alpha)(\lambda + \beta) \\ &= [(\lambda + \alpha)a_1 - (\lambda + \beta)a_2]^2 \geq 0. \end{aligned}$$

Moreover, since $\sqrt{\Delta} < (\lambda + \alpha)a_1 + (\lambda + \beta)a_2$, these solutions are positive; we note that:

$$t_1 - t_2 = \frac{\sqrt{\Delta}}{a_1 a_2}. \tag{59.4}$$

We look for (the first coordinate of) solutions of (59.3) of the form[1]:

$$f_1 = \sum_{i=1}^{2} \left(k_i e_{t_i} + (-1)^i G_{t_i} h_i \right), \tag{59.5}$$

where k_1, k_2 are constants, $e_t(x) = \mathrm{e}^{-\sqrt{t}x}$, $x \geq 0$ and, for $t > 0$ and $h \in L^1(\mathbb{R}^+)$,

$$G_t h(x) = \int_0^\infty \mathrm{e}^{-\sqrt{t}|x-y|} h(y)\,\mathrm{d}y, \qquad x \geq 0, \tag{59.6}$$

[1] Such a form is suggested by straightforward, but lengthy analysis involving standard methods of ODEs with constant coefficients, accompanied with the care to obtain solutions in $L^1(\mathbb{R}^+)$.

and:

$$h_i = \frac{1}{2\sqrt{t_i \Delta}}(\alpha g_2 + (\lambda + \alpha - a_2 t_i) g_1).$$

We note that G_t is a bounded linear operator in $L^1(\mathbb{R}^+)$.

Lemma 59.1 *The pair* (f_1, f_2)*, where* f_1 *is given by* (59.5)*, and:*

$$f_2 = \alpha^{-1}((\lambda + \beta) f_1 - a_1 f_1'' - g_1)$$

solves (59.3) *(for any constants* k_1, k_2*).*

Proof By the very definition of f_2, the first equality in (59.3) is satisfied, and we are left with proving the second one. To this end, we note that $G_t h \in W^{2,1}(\mathbb{R}^+)$, and:

$$(G_t h)'' = t G_t h - 2\sqrt{t} h, \qquad h \in L^1(\mathbb{R}^+). \tag{59.7}$$

It follows that:

$$a_1 f_1'' = a_1 \sum_{i=1}^{2} t_i (k_i e_{t_i} + (-1)^i G_{t_i} h_i) + 2\sqrt{t_1} a_1 h_1 - 2\sqrt{t_2} a_1 h_2.$$

Since:

$$2\sqrt{t_1} a_1 h_1 - 2\sqrt{t_2} a_1 h_2 = \frac{a_1 a_2}{\sqrt{\Delta}}(t_2 - t_1) g_1 = -g_1, \qquad \text{(see (59.4))}$$

we obtain:

$$\alpha f_2 = \sum_{i=1}^{2} (\lambda + \beta - a_1 t_i)(k_i e_{t_i} + (-1)^i G_{t_i} h_i). \tag{59.8}$$

Therefore, invoking (59.7) (again) and the definition of h_i (for the first time), we see that $\alpha((\lambda + \alpha) f_2 - a_2 f_2'')$ equals:

$$\begin{aligned}
&\sum_{i=1}^{2} [(\lambda + \alpha)(\lambda + \beta) - a_1 t_i (\lambda + \alpha) - a_2 t_i (\lambda + \beta) + a_1 a_2 t_i^2](k_i e_{t_i} + (-1)^i G_{t_i} h_i) \\
&\quad + 2a_2 \sqrt{t_2}(\lambda + \beta - a_1 t_2) h_2 - 2a_2 \sqrt{t_1}(\lambda + \beta - a_1 t_1) h_1 \\
&\quad = \alpha\beta \sum_{i=1}^{2} (k_i e_{t_i} + (-1)^i G_{t_i} h_i) + \frac{\alpha a_2}{\sqrt{\Delta}}(\lambda + \beta - a_1 t_2) g_2 \\
&\quad + \frac{(\lambda + \alpha - a_2 t_2)(\lambda + \beta - a_1 t_2)}{\sqrt{\Delta}} g_1 - \frac{\alpha a_2}{\sqrt{\Delta}}(\lambda + \beta - a_1 t_1) g_2 \\
&\quad - \frac{(\lambda + \alpha - a_2 t_1)(\lambda + \beta - a_1 t_1)}{\sqrt{\Delta}} g_1,
\end{aligned}$$

since t_is are roots of the quadratic. For the same reason, the terms involving g_1 cancel out, both being equal to $\alpha\beta g_1$. To summarize, using (59.4) one more time,

$$\alpha((\lambda+\alpha)f_2 - a_2 f_2'') = \alpha\beta f_1 + \frac{\alpha a_1 a_2}{\sqrt{\Delta}}(t_1 - t_2)g_2 = \alpha\beta f_1 + \alpha g_2.$$

This proves the second equality in (59.3). □

Now, the pair $(f_1, f_2) \in \mathbb{X}$ defined by (59.2), (59.5), and (59.8) belongs to $D(\mathcal{A})$ iff:

$$f_i(0-) = f_i(0) \quad \text{and} \quad f_i'(0-) = f_i(0), \quad i = 1, 2.$$

The first of these conditions (compatibility of values) may be written as:

$$l_1 + D_1 = \sum_{i=1}^{2}(k_i + (-1)^i C_i),$$
$$l_2 + D_2 = \sum_{i=1}^{2} E_i(k_i + (-1)^i C_i),$$

where:

$$C_i := (G_{t_i} h_i)(0), \qquad D_i := \frac{1}{2\sqrt{\lambda a_i}} \int_{-\infty}^{0} \mathrm{e}^{\sqrt{\frac{\lambda}{a_i}}y} g_i(y)\,\mathrm{d}y \quad \text{and}$$
$$E_i := \frac{\lambda + \beta - a_1 t_i}{\alpha}.$$

Since:

$$(G_t h)'(0) = \sqrt{t}(G_t h)(0),$$

the other condition (compatibility of derivatives) reads (see also Exercise 59.2):

$$\sqrt{\frac{\lambda}{a_1}} l_1 - \sqrt{\frac{\lambda}{a_1}} D_1 = \sum_{i=1}^{2} \sqrt{t_i}((-1)^i C_i - k_i),$$
$$\sqrt{\frac{\lambda}{a_2}} l_2 - \sqrt{\frac{\lambda}{a_2}} D_2 = \sum_{i=1}^{2} \sqrt{t_i} E_i((-1)^i C_i - k_i).$$

Hence, we have the following linear system of equations for l_1, l_2, k_1, k_2:

$$l_1 - k_1 - k_2 = C_2 - C_1 - D_1,$$
$$l_2 - E_1 k_1 - E_2 k_2 = E_2 C_2 - E_1 C_1 - D_2,$$
$$\sqrt{s_1} l_1 + \sqrt{t_1} k_1 + \sqrt{t_2} k_2 = \sqrt{t_2} C_2 - \sqrt{t_1} C_1 + \sqrt{s_1} D_1,$$
$$\sqrt{s_2} l_2 + \sqrt{t_1} E_1 k_1 + \sqrt{t_2} E_2 k_2 = \sqrt{t_2} E_2 C_2 - \sqrt{t_1} E_1 C_1 + \sqrt{s_2} D_2,$$

where for further simplicity of notation $s_i = \frac{\lambda}{a_i}$. The Jordan–Gauss elimination method (stopped one step before completion) yields now:

$$
\begin{aligned}
l_1 - k_1 - k_2 &= C_2 - C_1 - D_1, \\
l_2 - E_1 k_1 - E_2 k_2 &= E_2 C_2 - E_1 C_1 - D_2, \\
(\sqrt{s_1} + \sqrt{t_1})k_1 + (\sqrt{s_1} + \sqrt{t_2})k_2 &= (\sqrt{t_2} - \sqrt{s_1})C_2 \\
&\quad + (\sqrt{s_1} - \sqrt{t_1})C_1 + 2\sqrt{s_1}D_1, \\
E_1(\sqrt{s_2} + \sqrt{t_1})k_1 + E_2(\sqrt{s_2} + \sqrt{t_2})k_2 &= E_2(\sqrt{t_2} - \sqrt{s_2})C_2 \\
&\quad + E_1(\sqrt{s_2} - \sqrt{t_1})C_1 + 2\sqrt{s_2}D_2.
\end{aligned}
\tag{59.9}
$$

Before continuing, we present a lemma summarizing asymptotic behavior of constants and functions appearing in the definition of f_1 and f_2.

Lemma 59.2 *Let:*

$$a_{\mathrm{a}} = \alpha' a_1 + \beta' a_2.$$

If each occurence of α and β is replaced by $\kappa\alpha$ and $\kappa\beta$, respectively, then:

(i) $\lim_{\kappa\to+\infty} \frac{\sqrt{\Delta}}{\kappa} = (\alpha+\beta)a_{\mathrm{a}}, \quad \lim_{\kappa\to+\infty} \frac{t_1}{\kappa} = \frac{\alpha+\beta}{a_1 a_2} a_{\mathrm{a}}, \quad \lim_{\kappa\to+\infty} t_2 = \frac{\lambda}{a_{\mathrm{a}}},$

(ii) $\lim_{\kappa\to+\infty} E_1 = -\frac{a_1}{a_2}, \quad \lim_{\kappa\to+\infty} E_2 = \frac{\beta}{\alpha},$

(iii) $\lim_{\kappa\to+\infty} h_2 = \frac{1}{2}\frac{\alpha'}{\sqrt{\lambda a_{\mathrm{a}}}}(g_1 + g_2) := h_{2,\infty},$

(iv) $\lim_{\kappa\to+\infty} C_2 = \frac{\alpha'}{2\sqrt{\lambda a_{\mathrm{a}}}} \int_0^\infty \mathrm{e}^{-\sqrt{\frac{\lambda}{a_{\mathrm{a}}}}y}(g_1(y) + g_2(y))\,\mathrm{d}y =: C_{2,\infty},$

(v) $\lim_{\kappa\to+\infty} \sqrt{t_1} h_1 = \frac{\alpha'}{2a_{\mathrm{a}}}(g_2 - \frac{\beta a_2}{\alpha a_1} g_1),$

(vi) $\lim_{\kappa\to+\infty} \sqrt{t_1} C_1 = 0.$

Proof Except for (vi), all claims are immediate by standard calculus, if proven consecutively. To show (vi), we note that for each $t > 0$,

$$L^1(\mathbb{R}^+) \ni h \mapsto F_t h := \int_0^\infty \mathrm{e}^{-\sqrt{t}y} h(y)\,\mathrm{d}y$$

is a bounded linear functional of norm 1. Moreover, for $h \in L^1(\mathbb{R}^+) \cap L^\infty(\mathbb{R}^+)$,

$$|F_t h| \le \frac{1}{\sqrt{t}} \|h\|_{L^\infty(\mathbb{R}^+)},$$

implying $\lim_{t\to\infty} F_t h = 0$. Therefore,

$$\lim_{t\to\infty} F_t h = 0, \qquad h \in L^1(\mathbb{R}^+),$$

because $L^1(\mathbb{R}^+) \cap L^\infty(\mathbb{R}^+)$ is dense in $L^1(\mathbb{R}^+)$ and the functionals are equi-bounded. Since $\sqrt{t_1}C_1 = F_{t_1}\sqrt{t_1}h_1$ and $\lim_{\kappa\to+\infty} t_1 = \infty$, this combined with (v) completes the proof. □

We are finally able to state the first of the two main results of this chapter: it provides information on convergence of resolvents.

Proposition 59.3 *If each occurence of α and β is replaced by $\kappa\alpha$ and $\kappa\beta$, respectively, then:*

(a) For κ large enough, (59.9) *has a unique solution. Moreover, equations* (59.2), (59.5), *and* (59.8) *with k_1, k_2, l_1, l_2 calculated from* (59.9), *give the solution to the resolvent equation for $\mathcal{A}_\kappa$.*

(b) As $\kappa \to \infty$, the solutions to the resolvent equations for $\mathcal{A}_\kappa$ converge to:

$$(f_1, f_2) = R_\lambda(g_1, g_2)$$

defined by:

$$f_i(x) = \begin{cases} l_i e^{\sqrt{\frac{\lambda}{a_i}}x} + \frac{1}{2\sqrt{\lambda a_i}}\int_{-\infty}^0 e^{-\sqrt{\frac{\lambda}{a_i}}|y-x|} g_i(y)\,dy, & x < 0, \\ k e^{-\sqrt{\frac{\lambda}{a_a}}x} + \frac{\alpha'}{2\sqrt{\lambda a_a}}\int_0^\infty e^{-\sqrt{\frac{\lambda}{a_a}}|x-y|}(g_1+g_2)(y)\,dy, & x \ge 0, i = 1, \\ \frac{\beta}{\alpha} f_1(x) & x \ge 0, i = 2, \end{cases} \tag{59.10}$$

where:

$$k = \frac{\left[\frac{\beta}{\alpha}\left(\sqrt{\frac{\lambda}{a_a}} - \sqrt{\frac{\lambda}{a_2}}\right) + \frac{a_1}{a_2}\left(\sqrt{\frac{\lambda}{a_a}} - \sqrt{\frac{\lambda}{a_1}}\right)\right] C_{2,\infty} + 2\sqrt{\frac{\lambda}{a_2}}D_2 + 2\frac{a_1}{a_2}\sqrt{\frac{\lambda}{a_1}}D_1}{\frac{\beta}{\alpha}\left(\sqrt{\frac{\lambda}{a_a}} + \sqrt{\frac{\lambda}{a_2}}\right) + \frac{a_1}{a_2}\left(\sqrt{\frac{\lambda}{a_a}} + \sqrt{\frac{\lambda}{a_1}}\right)},$$

$$l_1 = k + C_{2,\infty} - D_1 \quad \text{and} \quad l_2 = \frac{\beta}{\alpha}k + \frac{\beta}{\alpha}C_{2,\infty} - D_2. \tag{59.11}$$

Proof (a) For the main determinant, say W, of the last two equations in (59.9), we have:

$$\lim_{\kappa\to+\infty} \frac{W}{\sqrt{t_1}} = \frac{\beta}{\alpha}\left(\sqrt{\frac{\lambda}{a_2}} + \sqrt{\frac{\lambda}{a_a}}\right) + \frac{a_1}{a_2}\left(\sqrt{\frac{\lambda}{a_1}} + \sqrt{\frac{\lambda}{a_a}}\right) \neq 0.$$

Since $\lim_{\kappa\to+\infty}\sqrt{t_1} = \infty$, this shows uniqueness of k_1 and k_2 in (59.9), which in turn implies uniqueness of l_1 and l_2 (by the first two equations). The rest is clear.

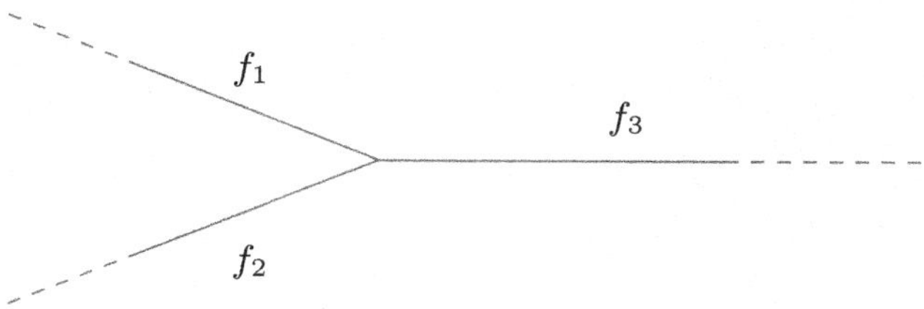

Figure 59.1 The "infinite y"-shaped graph; its three edges, when identified with three half-axes are domains of three functions (redrawn from [56])

(b) Using Cramer's rule for the last two equations in (59.9) and Lemma 59.2, we see that $\lim_{\kappa\to+\infty} k_1 = 0$ and $\lim_{\kappa\to+\infty} k_2 = k$ for k defined in (59.11). Passing to the limit in the first two equations (and using Lemma 59.2) we obtain $\lim_{\kappa\to+\infty} l_1 = k + C_{2,\infty} - D_1$ and $\lim_{\kappa\to+\infty} l_2 = \frac{\beta}{\alpha}k + \frac{\beta}{\alpha}C_{2,\infty} - D_2$. By (59.5) and (59.8), this shows that the solutions of the resolvent equations converge to (f_1, f_2) defined by formula (59.10) because the map $t \mapsto e_t \in L^1(\mathbb{R}^+)$ is norm-continuous (with $\lim_{t\to\infty} e_t = 0$), and the map $t \mapsto G_t \in \mathcal{L}(L^1(\mathbb{R}^+))$ is strongly continuous. □

All that is left to do now is to interpret, or decipher, this result and provide the link with the Freidlin–Wentzell transmission conditions. To this end, let:

$$\mathbb{X}_0 = \{(f_1, f_2) \in \mathbb{X}; \alpha f_2 1_{\mathbb{R}^+} = \beta f_1 1_{\mathbb{R}^+}\}.$$

This subspace of $\mathbb{X}$ is isometrically isomorphic to $L^1(Y)$, the space of integrable functions on an "infinite y"-shaped graph depicted in Figure 59.1. The latter space in turn may be identified with:

$$\mathbb{Y} = L^1(\mathbb{R}^-) \times L^1(\mathbb{R}^-) \times L^1(\mathbb{R}^+),$$

the isometric isomorphism $I : \mathbb{X}_0 \to \mathbb{Y}$ being given by:

$$I(f_1, f_2) = (f_{1|\mathbb{R}^-}, f_{2|\mathbb{R}^-}, (f_1 + f_2)_{|\mathbb{R}^+}).$$

We have $I^{-1}(f_1, f_2, f_3) = (f_1 \cup \alpha' f_3, f_2 \cup \beta' f_3)$, where for $f \in L^1(\mathbb{R}^-)$ and $g \in L^1(\mathbb{R}^+)$, $f \cup g$ is a "union of graphs"-function:

$$(f \cup g)(x) = \begin{cases} f(x), & x < 0 \\ g(x), & x \geq 0. \end{cases}$$

Next, let B_0 be the operator in $\mathbb{Y}$ given by:

$$B_0(f_1, f_2, f_3) = (a_1 f_1'', a_2 f_2'', a_a f_3'')$$

on the domain composed of (f_1, f_2, f_3) such that $f_1, f_2 \in W^{2,1}(\mathbb{R}^-)$, $f_3 \in W^{2,1}(\mathbb{R}^+)$ satisfy the transmission conditions:

$$\begin{aligned} f_1(0-) = \alpha' f_3(0+), \, f_2(0-) = \beta' f_3(0+) \\ a_\text{a} f_3'(0+) = a_1 f_1'(0-) + a_2 f_2'(0-). \end{aligned} \tag{59.12}$$

These transmission conditions are dual to those of the previous chapter: to be more precise, they describe the same physical/biological phenomenon, yet in a different, "dual" space.

Now, the isomorphic copy of B_0 in $\mathbb{X}_0$ is given by:

$$B(f_1, f_2) = \left(\chi_1 f_1'', \chi_2 f_2''\right),$$

where $\chi_i = a_i 1_{(-\infty,0)} + a_\text{a} 1_{[0,\infty)}$, on the domain composed of (f_1, f_2) such that f_1 and f_2 are continuous on $\mathbb{R}$ (so that in particular:

$$f_i(0-) = f_i(0+); \tag{59.13}$$

this corresponds to the first two conditions in (59.12)),

$$f_{1|\mathbb{R}^-}, f_{2|\mathbb{R}^-} \in W^{2,1}(\mathbb{R}^-), \quad f_{1|\mathbb{R}^+} \in W^{2,1}(\mathbb{R}^+)$$

and:

$$a_\text{a} f_1'(0+) = \alpha' \left(a_1 f_1'(0-) + a_2 f_2'(0-)\right). \tag{59.14}$$

(It goes without saying that $(f_1, f_2) \in \mathbb{X}_0$.)

It is quite easy to solve the resolvent equation for B. Given $\lambda > 0$ and $(g_1, g_2) \in \mathbb{X}$, we have:

$$f_i(x) = \begin{cases} m_i \mathrm{e}^{\sqrt{\frac{\lambda}{a_i}} x} + \frac{1}{2\sqrt{\lambda a_i}} \int_{-\infty}^0 \mathrm{e}^{-\sqrt{\frac{\lambda}{a_i}}|y-x|} g_i(y)\,\mathrm{d}y, & x < 0, \\ n \mathrm{e}^{-\sqrt{\frac{\lambda}{a_\text{a}}} x} + \frac{1}{2\sqrt{\lambda a_\text{a}}} \int_0^\infty \mathrm{e}^{-\sqrt{\frac{\lambda}{a_\text{a}}}|x-y|} g_1(y)\,\mathrm{d}y, & x \ge 0, i = 1, \\ \frac{\beta}{\alpha} f_1(x) & x \ge 0, i = 2. \end{cases} \tag{59.15}$$

Here, the constants m_1, m_2, and n are chosen so that transmission conditions (59.13) and (59.14) are satisfied, that is,

$$\begin{aligned} m_1 + D_1 = n + C, \quad m_2 + D_2 = \frac{\beta}{\alpha}(n + C) \\ a_\text{a}\sqrt{\frac{\lambda}{a_\text{a}}}(C - n) = \alpha' \left[a_1 \sqrt{\frac{\lambda}{a_1}}(l_1 - D_1) + a_2 \sqrt{\frac{\lambda}{a_2}}(l_2 - D_2) \right], \end{aligned} \tag{59.16}$$

where $C = \frac{1}{2\sqrt{\lambda a_a}} \int_0^\infty e^{-\sqrt{\frac{\lambda}{a_a}}y} g_1(y)\,dy$ and, as previously,

$$D_i := \frac{1}{2\sqrt{\lambda a_i}} \int_{-\infty}^0 e^{\sqrt{\frac{\lambda}{a_i}}y} g_i(y)\,dy.$$

Finally, for $(g_1, g_2) \in \mathbb{X}_0$, $\alpha'(g_1 + g_2)_{|\mathbb{R}^+} = g_{1|\mathbb{R}^+}$ so that $C_{2,\infty} = C$, and (59.15) is the same as (59.10), except perhaps for the constants. Moreover, a bit of algebra shows that for k of (59.11), we have then:

$$\begin{aligned} k &= \frac{\left((\alpha a_1 + \beta a_2)\sqrt{\frac{\lambda}{a_a}} - \alpha a_1\sqrt{\frac{\lambda}{a_1}} - \beta a_2\sqrt{\frac{\lambda}{a_2}}\right) C + 2\alpha\sqrt{\lambda a_1} D_1 + 2\alpha\sqrt{\lambda a_2} D_2}{(\alpha a_1 + \beta a_2)\sqrt{\frac{\lambda}{a_a}} + \alpha a_1\sqrt{\frac{\lambda}{a_1}} + \beta a_2\sqrt{\frac{\lambda}{a_2}}} \\ &= \frac{(\sqrt{a_a} - \alpha'\sqrt{a_1} - \beta'\sqrt{a_2})C + 2\alpha'\sqrt{a_1} D_1 + 2\alpha'\sqrt{a_2} D_2}{\sqrt{a_a} + \alpha'\sqrt{a_1} + \beta'\sqrt{a_2}}. \end{aligned}$$

This, however, is just the assertion that k is the same as n of (59.16). It follows that $m_i = l_i$, $i = 1, 2$ and so:

$$R_\lambda(g_1, g_2) = (\lambda - B)^{-1}(g_1, g_2), \qquad (g_1, g_2) \in \mathbb{X}_0. \tag{59.17}$$

More generally,

$$R_\lambda = (\lambda - B)^{-1}\mathcal{P}, \tag{59.18}$$

where $\mathcal{P}$ is defined in (59.1).

Theorem 59.4 *We have*

$$\lim_{\kappa\to+\infty} e^{A_\kappa t} f = e^{tB}\mathcal{P}f, \qquad t > 0, f \in \mathbb{X} \tag{59.19}$$

strongly and almost uniformly in $t > 0$.

Proof Relation (59.18) (in fact, (59.17) suffices) shows that condition (a) of Kurtz's Singular Perturbation Theorem is satisfied. The rest of the argument is precisely the same as in the proof of Theorem 58.1. □

Exercise 59.1 Use the fact that t_is are roots of the quadratic to show that (compare (59.5)):

$$f_2 = \sum_{i=1}^{2} \left(\widetilde{k}_i e_{\sqrt{t_i}} + (-1)^i G_{t_i}\widetilde{h}_i\right),$$

where $\widetilde{k}_i = (\lambda + \beta - a_1 t_i)k_i$ and $\widetilde{h}_i = \frac{1}{2\sqrt{t_i}\Delta}(\beta g_1 + (\lambda + \beta - a_1 t_i)g_2)$.

Exercise 59.2

(a) Show that for G_t of (59.6), $\|G_t\| \le \frac{2}{\sqrt{t}}$.

(b) For $t > 0$, let $H_t \in \mathcal{L}(L^1(\mathbb{R}^-))$ be given by (see (59.2)):

$$H_t f(x) = \int_{-\infty}^{0} \mathrm{e}^{-\sqrt{t}|x-y|} f(y)\,\mathrm{d}y, \qquad x < 0.$$

The Banach spaces $L^1(\mathbb{R}^+)$ and $L^1(\mathbb{R}^-)$ are isometrically isomorphic with isomorphism $J : L^1(\mathbb{R}^+) \to L^1(\mathbb{R}^-)$ given by $Jf(x) = f(-x), x \le 0$. Check that $H_t = JG_tJ^{-1}$.

☞ Chapter's summary

We prove a "dual" version of the main result of the previous chapter. In contrast to the arguments presented there, the question of convergence of resolvents is reduced to that of a (singular) convergence of solutions of a system of linear equations in $\mathbb{R}^4$. Probabilistically, the main theorem of this chapter speaks of convergence of densities of the stochastic processes involved in L^1 norm. On the way to our theorem, we introduce a dual form of Freidlin–Wentzell transmission conditions (see (59.13) and (59.14)).

PART III

Convergence of cosine families

60

Regular Convergence of Cosine Families

$\dots$ Cos_{A_n}

$\dots$ Cos_{A_3} Cos_{A_2} Cos_{A_1}

Suppose that A_n are the generators of equibounded cosine operator functions $\{\mathrm{Cos}_{A_n}(t), t \in \mathbb{R}\}$ in a Banach space $\mathbb{X}$:

$$\|\mathrm{Cos}_{A_n}(t)\| \le M, \qquad t \in \mathbb{R}, n \ge 1.$$

Suppose furthermore that the strong limit:

$$\lim_{n\to\infty} (\lambda^2 - A_n)^{-1} =: R_{\lambda^2} \tag{60.1}$$

exists for all $\lambda > 0$. Then, by (1.3) and Lemma 2.1, the related **sine functions**: $\mathrm{Sin}_{A_n}(t) = \int_0^t \mathrm{Cos}_{A_n}(s)\,\mathrm{d}s$, $t \ge 0$ converge strongly as well, and for each member of $\mathbb{X}$, the limit is almost uniform in $t \in [0, \infty)$. Moreover, arguing as in Theorem 14.1, we obtain that for all $\phi \in L^1(\mathbb{R}^+)$ there exists the strong limit,

$$\lim_{n\to\infty} \int_0^\infty \phi(t)\mathrm{Cos}_{A_n}(t)\,\mathrm{d}t =: H_0(\phi). \tag{60.2}$$

In contrast to the case of semigroups, H_0 is *not* a representation of the convolution algebra $L^1(\mathbb{R}^+)$. However, introducing $L^1_e(\mathbb{R})$, the space of integrable even functions on $\mathbb{R}$, we check that the map $H(\varphi) = H_0(2\phi)$ where ϕ is the restriction of $\varphi \in L^1_e(\mathbb{R})$ to $\mathbb{R}^+$, is a representation of $L^1_e(\mathbb{R})$. The latter space is a convolution algebra with convolution $\varphi_1 * \varphi_2(t) = \int_{-\infty}^\infty \varphi_1(t-s)\varphi_2(s)\,\mathrm{d}s$, and the map $\varphi \to 2\phi$ is an isometric isomorphism of $L^1(\mathbb{R}^+)$ and $L^1_e(\mathbb{R})$ (as

Banach spaces, but not as Banach algebras). Then:

$$\lim_{n\to\infty}\int_{-\infty}^{\infty}\varphi(t)\mathrm{Cos}_{A_n}(t)\,\mathrm{d}t = H(\phi), \qquad \varphi\in L^1_{\mathrm{e}}(\mathbb{R}). \tag{60.3}$$

We note the formula:

$$\mathrm{Cos}_{A_n}(t)(\lambda^2-A_n)^{-1} = \int_0^t \mathrm{Sin}_{A_n}(s)(\lambda^2(\lambda^2-A_n)^{-1}-I)\,\mathrm{d}s + (\lambda^2-A_n)^{-1} \tag{60.4}$$

relating the cosine families with their resolvents and sine functions. For the proof of (60.4) we multiply both sides by $\mathrm{e}^{-\mu t}$, integrate over $\mathbb{R}^+$ and multiply the result by $\mu>0$, obtaining equality $\mu^2(\mu^2-A_n)^{-1}(\lambda^2-A_n)^{-1} = \lambda^2(\mu^2-A_n)^{-1}(\lambda^2-A_n)^{-1} - (\mu^2-A_n)^{-1} + (\lambda^2-A_n)^{-1}$. Since the latter equality is true by the Hilbert Equation, the Laplace transforms of both sides coincide, and (60.4) follows because its both sides are continuous in t.

Formula (60.4) implies that the limit $\lim_{n\to\infty}\mathrm{Cos}_{A_n}(t)R_{\lambda^2}x$ exists for all x, that is, that the limit $\lim_{n\to\infty}\mathrm{Cos}_{A_n}(t)x$ exists for all x in the range $\mathcal{R}$ of R_λ, $\lambda>0$, and is almost uniform in $t\in[0,\infty)$. Since the cosine families are equibounded, the same is true for x in the closure of the range, that is, in the regularity space of the limit pseudoresolvent R_λ, $\lambda>0$. On the other hand, the Weierstrass Formula makes it clear that convergence of cosine families implies convergence of the related semigroups of operators, and these do not converge almost uniformly outside of the regularity space.

This leads to the conclusion that, as in the case of semigroups, *the regularity space $\mathbb{X}_0$ where a sequence of equibounded cosine families converges strongly with the limit almost uniform in $t\in[0,\infty)$, is equal to the closure of the range of the limit pseudoresolvent:*

$$\mathbb{X}_0 = cl\,\mathcal{R}. \tag{60.5}$$

In particular, the regularity spaces for semigroups and cosine families coincide. This is a generalization of the Trotter–Kato Theorem analogue Thm. 8.6 in [163], p. 119 due, independently, to Goldstein [160] and Konishi [218]. The reader should convince herself that the generator of the limit cosine family $\{C(t), t\in\mathbb{R}\}$ is the part in $\mathbb{X}_0$ of the extended limit of A_n's.

But there is much more going on here than in the case of semigroups (see [47]). First of all, differentiating (60.4) with respect to t and arguing as earlier we obtain:

$$\lim_{n\to\infty}\frac{\mathrm{d}}{\mathrm{d}t}\mathrm{Cos}_{A_n}(t)(\lambda^2-A_n)^{-1}x = \frac{\mathrm{d}}{\mathrm{d}t}C(t)R_{\lambda^2}x.$$

This result is readily generalized; we have:

$$\lim_{n\to\infty} \frac{\mathrm{d}}{\mathrm{d}t}\mathrm{Cos}_{A_n}(t)x_n = \frac{\mathrm{d}}{\mathrm{d}t}C_0(t)x \tag{60.6}$$

(where $x = \lim_{n\to\infty} x_n$) for all $(x_n)_{n\geq 1}$ in the domain of the operator $\mathcal{A}$ defined by (14.1) and in its completion with respect to the norm $|||(x_n)_{n\geq 1}|||_1 = \sup_{n\in N} \|x_n\| + \sup_{t\in[0,1]} \sup_{n\in N} \|\frac{dC_n(t)x_n}{dt}\|$.

Put otherwise: $\mathcal{A}$ satisfies all assumptions of the Sova–DaPrato–Giusti generation theorem, but that of being densely defined. Hence, it generates a strongly continuous cosine family only on the closure of its domain: this is a restatement of (60.5). As is always the case, on the related Kisyński space, the cosine family is differentiable: this is a restatement of (60.6).

Finally, A_n's generate the semigroups $(\mathrm{e}^{tA_n})_{t\geq 0}, n \geq 1$ related to the cosine families by the Weierstrass Formula (1.4). Although, as we shall shortly see, for $x \notin \mathbb{X}_0$ the limit $\lim_{n\to\infty} \mathrm{Cos}_{A_n}(t)x$ need not exist (see next), the limit:

$$\lim_{n\to\infty} \mathrm{e}^{A_n t}x, \qquad x \in \mathbb{X}, t > 0 \tag{60.7}$$

always does! This follows from (60.3) with $\varphi(s) = \frac{1}{2\sqrt{\pi t}}\mathrm{e}^{-\frac{s^2}{4t}}$, and the Weierstrass Formula.

Let us turn to examples. First, consider a Banach space $\mathbb{X}'$ and a bounded cosine operator function $\{C(t), t \in \mathbb{R}\}$ in $\mathbb{X}'$ with the generator A. For all $n \geq 1$ the family:

$$C_n(t)\begin{pmatrix} x \\ a \end{pmatrix} = \begin{pmatrix} C(t)x \\ a\cos nt \end{pmatrix}, \qquad t \in \mathbb{R}, x \in \mathbb{X}', a \in \mathbb{R}, \tag{60.8}$$

of operators acting in $\mathbb{X} := \mathbb{X}' \times \mathbb{R}$, forms a cosine operator function. We have:

$$\lim_{n\to\infty} \int_0^\infty \mathrm{e}^{-\lambda t}C_n(t)\begin{pmatrix} x \\ a \end{pmatrix} \mathrm{d}t = \lim_{n\to\infty}\begin{pmatrix} \lambda(\lambda^2 - A)^{-1}x \\ \frac{\lambda a}{\lambda^2+n^2} \end{pmatrix} = \begin{pmatrix} \lambda(\lambda^2 - A)^{-1}x \\ 0 \end{pmatrix},$$

so that (60.1) is satisfied. By contrast, $\lim_{n\to\infty} C_n(t)\begin{pmatrix} x \\ a \end{pmatrix}$ does not exist unless $a = 0$ (in other words, $\mathbb{X}_0 = \mathbb{X}' \times \{0\}$). However, the related sine functions always converge:

$$\lim_{n\to\infty} \int_0^t C_n(s)\begin{pmatrix} x \\ a \end{pmatrix} \mathrm{d}s = \lim_{n\to\infty}\begin{pmatrix} \int_0^t C(s)x\,\mathrm{d}s \\ \frac{a}{n}\sin(nt) \end{pmatrix} = \begin{pmatrix} \int_0^t C(s)x\,\mathrm{d}s \\ 0 \end{pmatrix},$$

Furthermore, (60.3) takes the form:

$$\begin{aligned}\lim_{n\to\infty} \int_{-\infty}^\infty \phi(t)C_n(t)\begin{pmatrix} x \\ a \end{pmatrix} \mathrm{d}t &= \lim_{n\to\infty}\begin{pmatrix} \int_{-\infty}^\infty \phi(t)C(t)x\,\mathrm{d}t \\ a\int_{-\infty}^\infty \phi(t)\cos(nt)\,\mathrm{d}t \end{pmatrix} \\ &= \lim_{n\to\infty}\begin{pmatrix} \int_{-\infty}^\infty \phi(t)C(t)x\,\mathrm{d}t \\ 0 \end{pmatrix};\end{aligned}$$

the fact that $\lim_{n\to\infty}\int_{-\infty}^{\infty}\phi(t)\cos(nt)\,\mathrm{d}t = 0$ is known as a Riemann Theorem, or Riemann Lemma. Finally, in agreement with (60.7),

$$\lim_{n\to\infty} \mathrm{e}^{tA_n}\begin{pmatrix} x \\ a \end{pmatrix} = \lim_{n\to\infty}\begin{pmatrix} \mathrm{e}^{tA}x \\ \mathrm{e}^{-n^2t}a \end{pmatrix} = \begin{pmatrix} \mathrm{e}^{tA}x \\ 0 \end{pmatrix} \qquad \text{for } t > 0, x \in \mathbb{X}'.$$

A more elaborate example of this type is related to the elastic Brownian motion semigroups $(\mathrm{e}^{tA_\epsilon})_{t\ge0}$ of Chapters 10 and 31. In Chapter 31, we showed that the regularity space for these semigroups is $C_0(0,\infty]$, but they converge to the minimal Brownian motion semigroup on the entire $C[0,\infty]$. Here, we will show that A_ϵ generate cosine families $\{\mathrm{Cos}_{A_\epsilon}, t \in \mathbb{R}\}$ with:

$$\|\mathrm{Cos}_{A_\epsilon}(t)\| \le 3, \qquad t \in \mathbb{R}, \tag{60.9}$$

but in spite of this estimate and in spite of convergence of the semigroups, these families do not converge outside of $C_0(0,\infty] \subset C[0,\infty]$.

The method of constructing $\mathrm{Cos}_{A_\epsilon}(t)$ used next is **Lord Kelvin's method of images** [53, 54, 86]: we start with the key lemma (see [86] for an inspiration for this result, and [53] for its more general version). Let $C[-\infty,\infty]$ be the space of continuous functions with limits at $+\infty$ and $-\infty$, and let:

$$C(t)f(x) = \frac{1}{2}(f(x+t) + f(x-t)), \qquad x, t \in \mathbb{R} \tag{60.10}$$

be the **basic cosine family**. Certainly, $\{C(t), t \in \mathbb{R}\}$ is strongly continuous and its generator is $A = \frac{\mathrm{d}^2}{\mathrm{d}x^2}$ with (maximal) domain $D(A) = C^2[-\infty,\infty]$, composed of all twice differentiable functions with the second derivative in $C[-\infty,\infty]$.

Lemma 60.1 *For every $f \in D(A_\epsilon)$, there exists a unique function $f_\epsilon \in C[-\infty,\infty]$ such that:*

- $f_\epsilon(x) = f(x), x \ge 0,$
- $[C(t)f_\epsilon]_{|\mathbb{R}^+} \in D(A_\epsilon), t \in \mathbb{R}.$

Proof Our task is to find $g(x) := f_\epsilon(-x), x \ge 0$. Since $C(t) = C(-t), t \ge 0$, the second condition requires that:

$$\epsilon\frac{\mathrm{d}}{\mathrm{d}x}[f_\epsilon(x+t) + f_\epsilon(x-t)]_{|x=0} = f_\epsilon(t) + f_\epsilon(-t), \qquad t \ge 0.$$

Hence, it suffices to find the g satisfying:

$$\epsilon g'(t) + g(t) = \epsilon f'(t) - f(t) \tag{60.11}$$

with the initial condition $g(0) = f(0)$, where $f(t), t \geq 0$ is treated as given. It follows that the unique g satisfying our condition is:

$$g(t) = f(t) - 2\epsilon^{-1} e_\epsilon * f(t), \tag{60.12}$$

where $e_\epsilon(t) = \mathrm{e}^{-\epsilon^{-1}t}, t \geq 0$. The resulting f_ϵ is twice continuously differentiable since $g'(0) = -f'(0)$ (f is in $D(A_\epsilon)$!) and $g''(0) = f''(0)$. □

For $f \in C[0, \infty]$ and $\epsilon > 0$, the function $f_\epsilon \in C[-\infty, \infty]$ given by $f_\epsilon(x) = f(x), x \geq 0$ and $f_\epsilon(x) = g(-x), x < 0$, where g is given by (60.12), will be called the ϵ-extension of f.

Theorem 60.2 *Let $\mathbb{X}_\epsilon \subset C[-\infty, \infty]$ be composed of ϵ-extensions of members of $C[0, \infty]$. Then:*

- *$\mathbb{X}_\epsilon$ is isomorphic to $C[0, \infty]$,*
- *$\mathbb{X}_\epsilon$ is a closed invariant subspace for $\{C(t), t \in \mathbb{R}\}$,*
- *A_ϵ generates the cosine family $\{\mathrm{Cos}_{A_\epsilon}, t \in \mathbb{R}\}$ given by*

$$\mathrm{Cos}_{A_\epsilon} f(x) = C(t) f_\epsilon(x), \qquad x \geq 0. \tag{60.13}$$

Proof The isomorphism spoken of in the first point is $\mathcal{I} : C[0, \infty] \to \mathbb{X}_\epsilon$ given by $\mathcal{I} f = f_\epsilon$. By (60.12), $\|\mathcal{I}\| \leq 3$ and $\|\mathcal{I}^{-1}\| \leq 1$. The image of $D(A_\epsilon)$ via $\mathcal{I}$ is by Lemma 60.1 invariant for C. Hence, the same is true of its closure, which equals $\mathbb{X}_\epsilon$. The infinitesimal generator of the restriction of $\{C(t), t \in \mathbb{R}\}$ to $\mathbb{X}_\epsilon$ is the part A_p of A in $\mathbb{X}_\epsilon$. By (60.12), $f \in D(A_\epsilon)$ is a necessary condition for f_ϵ to be twice continuously differentiable. Thus $\mathcal{I}D(A_\epsilon) \supset D(A_p)$. The converse inclusion is immediate, and we have $\mathcal{I}^{-1}A_p\mathcal{I} = A_\epsilon$. Hence, A_p and A_ϵ are indistinguishable up to the isomorphism $\mathcal{I}$; this shows (60.13). □

Note that $\|\mathcal{I}\| \leq 3$ (i.e., $\|g\|_{C[0,\infty]} \leq 3\|f\|_{C[0,\infty]}$) implies (60.9).

The basic cosine family commutes with symmetries $P_\alpha, \alpha \in \mathbb{R}$ given by $P_\alpha f(x) = f(2\alpha - x), x \in \mathbb{R}$. Hence, it leaves the subspace of odd functions in $C[-\infty, \infty]$ invariant. Since this subspace is isometrically isomorphic to $C_0(0, \infty]$ composed of functions in $C[0, \infty]$ vanishing at zero, A_m defined on:

$$D(A_\mathrm{m}) = \{f \in C_0(0, \infty], f \text{ is twice differentiable with } f'' \in C_0(0, \infty]\}$$

generates the cosine family:

$$C_\mathrm{m}(t) f = C(t) f_\mathrm{m}, \qquad t \in \mathbb{R}, \tag{60.14}$$

where f_m is the odd extension of f.

Theorem 60.3 *For $f \in C[0,\infty]$ and $x, t \geq 0$,*

$$\lim_{\epsilon \to 0} \mathrm{Cos}_{A_\epsilon}(t) f(x) = \begin{cases} \frac{1}{2}[f(x+t) + f(x-t)], & x \geq t, \\ \frac{1}{2}[f(x+t) - f(t-x)], & 0 \leq x < t, \end{cases} \tag{60.15}$$

(pointwise in $x \in \mathbb{R}$). In particular, for $f \in C_0(0,\infty]$,

$$\lim_{\epsilon \to 0} \mathrm{Cos}_{A_\epsilon}(t) f = C_{\mathrm{m}}(t) f \tag{60.16}$$

strongly and uniformly in $t \in [0,\infty)$. Outside of $C_0(0,\infty]$, the limit (60.16) does not exist in the supremum norm sense.

Proof By (60.12) and (60.13),

$$\mathrm{Cos}_{A_\epsilon}(t) f(x) = \begin{cases} \frac{1}{2}[f(x+t) + f(x-t)], & x \geq t \\ \frac{1}{2}[f(x+t) + f(t-x) - 2\epsilon^{-1} e_\epsilon * f(t-x)], & 0 \leq x < t. \end{cases}$$

Since the measures on $\mathbb{R}^+$ with densities $\epsilon^{-1} e_\epsilon$ tend to the Dirac measure at 0, a standard argument shows $\lim_{\epsilon \to 0} \epsilon^{-1} e_\epsilon * f(x) = f(x)$, $x > 0$, and the limit is uniform in $x \in [x_0, \infty)$ for any $x_0 > 0$. This shows (60.15). Moreover, the limit considered here is uniform in $x \in [0,\infty]$ iff $f(0) = 0$. Hence, for $f \in C_0(0,\infty]$, the limit (60.15) is uniform in $x, t \in [0,\infty)$, establishing (60.16). For the same reason, for $f \notin C_0(0,\infty]$ the limit is not uniform. □

Relation (60.16) may be obtained from our Trotter–Kato Theorem analogue for cosine families, but this approach does not render uniform convergence in $t \in [0,\infty)$.

Since both cosine families defined by (60.8) and those of Theorem 60.3 converge, on their regularity space, uniformly in $t \in [0,\infty)$, we conclude this chapter by presenting cosine families converging merely almost uniformly in $t \in [0,\infty)$ (see [62] for more on this subject; see also Exercises 60.2 and 60.3). Our example is a modification of (60.8): in $\mathbb{X} = \mathbb{R}^2$ we consider the cosine families $C_n(t)\binom{x}{a} = \binom{x \cos \lambda_n t}{a \cos nt}$, $t \in \mathbb{R}$, $\binom{x}{a} \in \mathbb{R}^2$, where $(\lambda_n)_{n \geq 1}$ is a sequence of real, nonzero numbers converging to 0. Again, the regularity space here is $\mathbb{R} \times \{0\}$, and the limit cosine family is $C(t)\binom{x}{0} = \binom{x \cos 0t}{0} = \binom{x}{0}$, $x \in \mathbb{R}, t \in \mathbb{R}$: for any $T > 0$, $\sup_{t \in [-T,T]} \|C_n(t)\binom{x}{0} - \binom{x}{0}\| = |x| \sup_{t \in [-T,T]} |\cos \lambda_n t - \cos 0t| \leq 2T|x||\lambda_n|$ converges to 0, as $n \to \infty$. However, $\sup_{t \in \mathbb{R}} \|C_n(t)\binom{x}{0} - \binom{x}{0}\| = |x| \sup_{t \in \mathbb{R}} |\cos \lambda_n t - 1| = 2|x|$, proving that the limit is not uniform on the whole real line, unless $x = 0$.

Exercise 60.1 Assume that $\mathbb{X}$ is a complex Banach space. Show that the semigroups $(\mathrm{e}^{tA_n})_{t \geq 0}$ related to a sequence of equibounded cosine families are uniformly holomorphic and deduce (60.7) from Theorem 31.2.

Exercise 60.2 Let c_0 be the space of all (real or complex) sequences $\xi = (\xi_k)_{k\in\mathbb{N}}$ convergent to 0, equipped with the usual supremum norm. For each $n \in \mathbb{N}$, let e_n be the element of c_0 such that $(e_n)_k = 1$ if $k = n$ and $= 0$ otherwise. Then every $\xi = (\xi_k)_{k\in\mathbb{N}}$ in c_0 can conveniently be represented as the series $\xi = \sum_{k=1}^{\infty} \xi_k e_k$ convergent in the norm of c_0. For each $t \in \mathbb{R}$ and each $n \in \mathbb{N}$, let $C_n(t)$ be the linear operator on c_0 defined by $C_n(t)\xi = \sum_{k=1}^{n}(\cos kt)\xi_k e_k + \sum_{k=n+1}^{\infty} \xi_k e_k$, and, for each $t \in \mathbb{R}$, let $C(t)$ be the linear operator on c_0 given by $C(t)\xi = \sum_{k=1}^{\infty}(\cos kt)\xi_k e_k$, with $\xi = (\xi_k)_{k\in\mathbb{N}}$ being an arbitrary member of c_0. Show that, for each $n \in \mathbb{N}$, $C_n = \{\mathrm{Cos}_n(t)\}_{t\in\mathbb{R}}$ is a strongly continuous cosine family of contractions on c_0, and also $C = \{\mathrm{Cos}(t)\}_{t\in\mathbb{R}}$ is a strongly continuous cosine family of contractions on c_0. Prove that:

$$\sup_{t\in\mathbb{R}} \|C(t)\xi - C_n(t)\xi\| = 0.$$

Exercise 60.3 For the cosine family C_{m} defined in (60.14) and $\lambda_n \neq 1$ such that $\lim_{n\to\infty} \lambda_n = 1$, consider the cosine families $C_n(t) = C_{\mathrm{m}}(\lambda_n t)$, $t \in \mathbb{R}, n \geq 1$. Show that these families converge to C_{m} but the limit is not uniform on $\mathbb{R}$. *Hint:* show that for each $f \in C_0(0, \infty]$,

$$\|C_{\mathrm{m}}(t)f - C_n(t)f\| \geq \left| \frac{f(2t) - f((1+\lambda_n)t)}{2} - \frac{f_{\mathrm{m}}((1-\lambda_n)t)}{2} \right|, \quad t \geq 0.$$

Exercise 60.4 Use Lord Kelvin's method of images to find an explicit formula for the cosine family generated by the operator related to the stopped Brownian motion, introduced in Exercise 32.3. With the help of this formula give an independent proof of the convergence result in Exercise 32.3.

☞ Chapter's summary

In many respects, the theory of convergence of cosine families is analogous to that of semigroups. In both cases, convergence of resolvents is *a sine qua non* condition, and in both cases this condition guarantees convergence of the original family (be it semigroup or cosine family) on the regularity space, which for both families is the same and equals the closure of the range of the limit pseudoresolvent. However, in the case of cosine families, there are effects that are not visible for semigroups, for example, convergence of derivatives. More importantly, outside of the regularity space interesting things start to happen: while the cosine families *may or may not converge* here, the related semigroups (given by the Weierstrass Formula) always do. As we shall see in Chapter 61, outside of the regularity space, the cosine families *must not converge*.

61

Cosines Converge in a Regular Way

In the examples of the previous chapter, it is striking that the cosine families considered there do not converge outside of the regularity space in spite of the fact that the related semigroups do. Hence, a natural questions arises: can we find a sequence of cosine families that converges outside of its regularity space? Or, perhaps, cosine families *by nature* cannot converge in the irregular way?

To answer this question, we start by discussing the question of continuity of trajectories of cosine families. Let us recall that the main result of the paper [81] of Chander and Singh (see also [228]) says that a strongly measurable cosine family is strongly continuous in $t \in \mathbb{R} \setminus \{0\}$ (this is a cosine family analogue of [180], p. 305, Thm. 10.2.3). As shown by H. Fattorini even earlier (see [135, Lemma 5.2] and [139, pp. 24–26]), this result can be considerably strengthened: a strongly measurable cosine family on a Banach space is strongly continuous on the whole of $\mathbb{R}$. (See, e.g., [289, p. 529] and [331, Proposition 2.1.2] for versions of this theorem for other operator families.) This explains why there is no need for a cosine family analogue of Abel summable and related classes of semigroups (see [180]) – all measurable cosine families form just one class – and also marks the fundamental difference between the theories of semigroups and cosine families.

For the proof of Fattorini's Theorem we use the following argument of W. Chojnacki [62]:

Proposition 61.1 *Let $\mathbb{X}$ be a Banach space and* $\mathrm{Cos}\colon \mathbb{R} \to \mathcal{L}(\mathbb{X})$ *be a cosine function. If, for an $x \in \mathbb{X}$, the function $t \mapsto \mathrm{Cos}(t)x$, $x \in \mathbb{X}$, is continuous on (a, ∞) for some $a \in \mathbb{R}$, then it is continuous on $\mathbb{R}$.*

Proof To show continuity at a given $t \in \mathbb{R}$, we choose $h > 0$ so that $t + h > a + 1$. Then:

$$\mathrm{Cos}(\tau)x = 2\mathrm{Cos}(h)\mathrm{Cos}(\tau + h)x - \mathrm{Cos}(\tau + 2h)x$$

for each $\tau \in \mathbb{R}$. Noting that $\tau + h > a$ and $\tau + 2h > a$ whenever $\tau > t - 1$ and exploiting the assumption, we get:

$$\lim_{\tau \to t} \mathrm{Cos}(\tau + h)x = \mathrm{Cos}(t + h)x \quad \text{and} \quad \lim_{\tau \to t} \mathrm{Cos}(\tau + 2h)x = \mathrm{Cos}(t + 2h)x.$$

As $\mathrm{Cos}(h)$ is bounded, we also have:

$$\lim_{\tau \to t} \mathrm{Cos}(h)\mathrm{Cos}(\tau + h)x = \mathrm{Cos}(h)\mathrm{Cos}(t + h)x.$$

Therefore:

$$\lim_{\tau \to t} \mathrm{Cos}(\tau)x = 2\mathrm{Cos}(h)\mathrm{Cos}(t + h)x - \mathrm{Cos}(t + 2h)x = \mathrm{Cos}(t).$$

Since t was chosen arbitrarily, the result follows. □

Combining this observation with the main result of Chander and Singh [81], we immediately obtain the announced theorem.

Theorem 61.2 (Fattorini) *A strongly measurable cosine family of operators is strongly continuous in $t \in \mathbb{R}$.*

As it transpires, a stronger result holds: measurability of a single trajectory of a cosine function implies its continuity. This may be deduced from Fattorini's Theorem (see Exercise 61.1): a direct proof of this theorem can be found in Chapter 63.

Theorem 61.2 has an immediate bearing on the issue of convergence of cosine families, as we will explain now. Suppose that, as in the previous chapter, A_n are the generators of equibounded cosine operator functions $\{\mathrm{Cos}_{A_n}(t), t \in \mathbb{R}\}$ in a Banach space $\mathbb{X}$, and that limit in (60.1) exists. Then, the regularity space $\mathbb{X}_0$ where the cosine families converge almost uniformly in $t \in [0, \infty)$ is characterized by equation (60.5). Now, let $\mathbb{Y}$ be the subspace of $\mathbb{X}$ where the strong limit $\mathrm{Cos}(t) := \lim_{n\to\infty} \mathrm{Cos}_{A_n}(t)$ exists for all $t \in \mathbb{R}$. Since in the definition of $\mathbb{Y}$ we do not require the limit to be uniform in any way, $\mathbb{X}_0 \subset \mathbb{Y}$. What is surprising is that the inclusion may also be reversed, and the reader should note the key role of Theorem 61.2 in the proof presented here.

Theorem 61.3 *We have:*

$$\mathbb{Y} = \mathbb{X}_0.$$

Proof As mentioned earlier, it suffices to show $\mathbb{Y} \subset \mathbb{X}_0$. This may be done by following the argument presented in Chapter 28 (Proposition 28.3 and Theorem 28.4), but here is a much shorter proof based on (60.5).

Clearly, $\mathbb{Y}$ is a closed subspace of $\mathbb{X}$. Moreover, $\mathbb{Y}$ is left invariant by Cos, and Cos satisfies the cosine functional equation in $\mathbb{Y}$, since so do Cos_{A_n}. By Theorem 61.2, Cos being measurable as a strong limit of Cos_{A_n}, is strongly continuous in $t \in \mathbb{R}$. Let A be its generator. By assumption and the Lebesgue Dominated Convergence Theorem, $R_\lambda x = \lim_{n\to\infty}(\lambda - A_n)^{-1}x = (\lambda - A)^{-1}x, x \in \mathbb{Y}, \lambda > 0$. Now, any $x \in D(A)$ may be represented as $x = (\lambda - A)^{-1}y = R_\lambda y$ for some $\lambda > 0$ and $y \in \mathbb{Y}$. This implies $D(A) \subset \mathcal{R}$. Since $D(A)$ is dense in $\mathbb{Y}$ and both $\mathbb{Y}$ and $\mathbb{X}_0$ are closed, we are done. □

Theorem 61.3 may be phrased as follows: outside of the regularity space (60.5) the trajectories of cosine families must not converge; a statement not valid for semigroups. To put it yet another way: examples of the previous chapter seemed to show that cosine families *may* not converge outside of regularity space. Our theorem makes it clear that cosine families *must* not converge outside of this space. Theorem 61.2 shows that in contrast to semigroups there is no need for dividing cosine families into classes, such as Abel summable and strongly continuous cosine families, according to their regularity around origin (compare [180], p. 305): all measurable cosine families belong to the single class of strongly continuous families. Theorem 61.3, in turn, shows that there is no need to develop a theory of irregular convergence of cosine families (analogous to the existing theory of semigroups): if cosine families converge, they do so regularly. In particular, there is no singular perturbation theory for cosine families, analogous to the singular perturbation theory for semigroups.

☺ **And this seems like a perfect ending for this book.** ☺

Exercise 61.1 (See [62].) Let $\{\mathrm{Cos}(t)\}_{t\in\mathbb{R}}$ be a cosine family on a Banach space $\mathbb{X}$. Let $x \in \mathbb{X}$ be such that the function $t \mapsto \mathrm{Cos}(t)x$ is measurable on $\mathbb{R}$. Show that the function $t \mapsto \mathrm{Cos}(t)x$ is continuous on $\mathbb{R}$. *Hint:* let $\mathbb{X}_0$ be the set of all those $y \in \mathbb{X}_0$ for which $t \mapsto \mathrm{Cos}(t)y$ is measurable on $\mathbb{R}$. Show that $\mathbb{X}_0$ is a closed linear subspace of $\mathbb{X}_0$, left invariant by the cosine family, and deduce the result from Theorem 61.2.

Exercise 61.2 Suppose that $\{C(t), t \in \mathbb{R}\}$ is a cosine family in a Banach space $\mathbb{X}$ such that the strong limit:

$$\lim_{t\to\infty} C(t)x$$

exists for all $x \in \mathbb{X}$. Conclude that $C(t)x = x, t \in \mathbb{R}, x \in \mathbb{X}$. *Hint:* Consider the cosine families $C_n(t) = C(nt)$ and find their limit family.

Exercise 61.3 Prove the following individual version of the previous exercise: if $\{C(t), t \in \mathbb{R}\}$ is a bounded cosine family in a Banach space $\mathbb{X}$, and for some $x \in \mathbb{X}$ the limit $\lim_{t\to\infty} C(t)x$ exists, then for all $t \in \mathbb{R}$, we have $C(t)x = x$. *Hint:* Consider the subspace of $\mathbb{X}$ where the limit specified earlier exists, and prove that this subspace is closed and invariant for the cosine family. (Note that this is in contrast to the case of semigroups; see Chapter 32.)

Exercise 61.4 Let $(e^{tA})_{t\geq 0}$ be a strongly continuous semigroup in a Banach space $\mathbb{X}$ such that the strong limit:

$$Px = \lim_{t\to\infty} e^{tA}x$$

exists for all $x \in \mathbb{X}$. Show that either the kernel of P is nontrivial or $A = 0$. State and prove an analogous result in the case where the strong limit $\lim_{\lambda\to 0} \lambda(\lambda - A)^{-1}x$ exists.

☞ Chapter's summary

Trajectories of (equibounded, strongly continuous) cosine families either converge almost uniformly in $t \in \mathbb{R}$ or do not converge at all. In particular, there is no singular perturbation theory for cosine families, like the one presented in this book for semigroups. Hence, do not anticipate the second volume of this book! ☹

PART IV

Appendixes

62

Appendix A: Representation Theorem for the Laplace Transform

The Hennig–Neubrander Representation Theorem

Suppose $h : \mathbb{R}^+ \to \mathbb{R}$ (or $\mathbb{C}$) is of exponential growth, that is, there exists $M \geq 0$ and $\omega \in \mathbb{R}$ such that $|h(t)| \leq M\mathrm{e}^{\omega t}$ for all $t \geq 0$. Then, the Laplace transform:

$$f(\lambda) = \int_0^\infty \mathrm{e}^{-\lambda t} h(t)\,\mathrm{d}t$$

of h exists for all $\lambda > \omega$ and, differentiating under the integral sign (which is legitimate by the exponential growth assumption) we obtain:

$$|f^{(n)}(\lambda)| \leq \frac{Mn!}{(\lambda - \omega)^{n+1}}, \qquad \lambda > \omega, n \in N. \tag{62.1}$$

The classical theorem of Widder ([345], p. 315, Theorem 16a) states that the converse is also true: an infinitely differentiable function satisfying (62.1) is a Laplace transform of a function h satisfying $|h(t)| \leq M\mathrm{e}^{\omega t}, t \geq 0$.

If we turn our attention to functions with values in a Banach space $\mathbb{X}$, exponential growth condition implies estimate (62.2) in the same way as above, but (62.2) does not force f to be the Laplace transform of a function of exponential growth. The typical example here is $f(\lambda) = e_\lambda \in L^1_\omega(\mathbb{R}^+)$, where $L^1_\omega(\mathbb{R}^+)$ is the space of equivalence classes of integrable with weight $\mathbb{R}^+ \ni t \mapsto \mathrm{e}^{\omega t}$, and e_λ is the equivalence class of the function $\mathbb{R}^+ \ni t \mapsto \mathrm{e}^{-\lambda t}$ (see Example 62.2, later on). The Hennig–Neubrander Representation Theorem [176] shows that this example is archetypical: all functions satisfying (62.2) are images of e_λ via a linear map from $L^1_\omega(\mathbb{R}^+)$ to $\mathbb{X}$.

Theorem 62.1 (Hennig–Neubrander) *Fix $\omega \in R$ and let f be an $\mathbb{X}-$valued function. The following statements are equivalent.*

(1) f is infinitely differentiable and there exists a constant $M > 0$ such that:

$$\|f^{(n)}(\lambda)\| \leq \frac{Mn!}{(\lambda-\omega)^{n+1}}, \qquad \lambda > \omega, n \in N. \tag{62.2}$$

(2) There exists a bounded linear operator $\mathcal{T} : L^1_\omega(\mathbb{R}^+) \to \mathbb{X}$, such that $f(\lambda) = \mathcal{T}(e_\lambda)$ for $\lambda > \omega$.

Since in this book we need this result merely in the case $\omega = 0$, we present the proof in this particular case. This is done without loss of generality, since the general proof is the same modulo technicalities [46]. Other proofs may be found, for example, in [9, 176].

Let $L^1(\mathbb{R}^+)$ be the convolution algebra of (equivalence classes of) Lebesgue integrable functions $\psi : R^+ \to R$. We will use the following notation: $L^1(\mathbb{R}^+) \ni \psi = \{\psi(t)\}$ means that ψ is the equivalence class of the function $R^+ \ni t \to \psi(t) \in R$. In such cases, to avoid misunderstandings, we will always use t as an independent variable of the function ψ. Other variables should be considered as parameters. We recall that $L^1(\mathbb{R}^+)$ is also a Banach algebra, multiplication being the standard convolution:

$$(\psi * \varphi)(t) = \int_0^t \psi(t-s)\varphi(s)\,\mathrm{d}s.$$

The function $(0,\infty) \ni \lambda \mapsto e_\lambda = \{e^{-\lambda t}\} \in L^1(\mathbb{R}^+)$ satisfies the Hilbert Equation:

$$(\lambda - \mu)e_\lambda * e_\mu = e_\mu - e_\lambda,$$

and we have $\|e_\lambda\|_{L^1(\mathbb{R}^+)} = \frac{1}{\lambda}$. Therefore, it is infinitely differentiable, and:

$$\begin{aligned}\left\|\frac{\mathrm{d}^n}{\mathrm{d}\lambda^n}e_\lambda\right\|_{L^1(\mathbb{R}^+)} &= \left\|n!e_\lambda^{*(n+1)}\right\|_{L^1(\mathbb{R}^+)} = \|\{e^{-\lambda t}t^n\}\|_{L^1(\mathbb{R}^+)} \\ &= \int_0^\infty \mathrm{e}^{-\lambda t}t^n\,\mathrm{d}t = \frac{n!}{\lambda^{n+1}}.\end{aligned} \tag{62.3}$$

This shows sufficiency of (2) in the Henning–Neubrander Theorem; the converse will be shown in the next section.

Example 62.2 To see that there is no function h such that $e_\lambda = \int_0^\infty \mathrm{e}^{-\lambda t}h(t)\,\mathrm{d}t$ consider $u(t) = 1_{[0,t)} \in L^1(\mathbb{R}^+)$, $t \geq 0$. $u(\cdot)$ is continuous and

$$\lambda \int_0^t e^{-\lambda s}u(s)ds = [e_\lambda - e^{-\lambda t}]u(t). \tag{62.4}$$

Hence $\lambda\int_0^\infty e^{-\lambda t}u(t)\,\mathrm{d}t = e_\lambda$. Now, if $\int_0^\infty \mathrm{e}^{-\lambda t}h(t)\,\mathrm{d}t = e_\lambda$ holds for a bounded measurable function h, one gets, by invertibility of the Laplace transform for continuous functions, $u(t) = \int_0^t h(s)ds$, which is a contradiction, since $u(t)$ is not differentiable. Indeed, for any $t \geq 0$, $q(h) = \frac{1}{h}(u(t+h) - u(t)) = \frac{1}{h}1_{[t,t+h)}$ does not converge in $L^1(\mathbb{R}^+)$, as $h \to 0$, since $\|q(2h) - q(h)\| = 1$, $h > 0$. □

A Proof via Yosida Approximation

Throughout this section $(0, \infty) \ni \lambda \to f(\lambda)$ is an infinitely differentiable function with values in a Banach space $\mathbb{X}$, satisfying:

$$\|f^{(n)}(\lambda)\| \leq \frac{Mn!}{\lambda^{n+1}}, \tag{62.5}$$

where $M > 0$ is a constant. We put:

$$g_\mu(t) = \mathrm{e}^{-\mu t} t\mu^2 \sum_{n=0}^{\infty} \frac{t^n \mu^{2n}(-1)^n f^{(n)}(\mu)}{n!(n+1)!}, \qquad \text{for } \mu > 0. \tag{62.6}$$

$g_\mu(t)$ will be called the Yosida approximation of f. This approximation in (almost) this form has been studied first by R. Phillips [180, 286]. We follow closely [46]. (See also [36] for further developments.)

This definition agrees well with the definition introduced in Chapter 8 in the sense that if $f(\lambda) = (\lambda - A)^{-1}$ is the resolvent of an operator A satisfying (8.3) then, by $\frac{\mathrm{d}^n}{\mathrm{d}\lambda^n}(\lambda - A)^{-n} = (-1)^n n!(\lambda - A)^{n+1}$, we have $\mathrm{e}^{-\mu t} + g_\mu(t) = \mathrm{e}^{-\mu t} \sum_{n=0}^{\infty} \frac{t^n \mu^{2n}(\mu - A)^{-n}}{n!} = \mathrm{e}^{tA_\mu}$. The absence of the term $\mathrm{e}^{-\mu t}$ in our definition is due to the fact that instead of working in the Banach algebra of operators we want to deal with an arbitrary Banach space, where it does not make sense to speak of a unit element.

Lemma 62.3 *g_μ, $\mu > 0$ are continuous and bounded by M.*

Proof Fix $\mu > 0$. The estimate:

$$\sum_{n=0}^{\infty} \left\| \frac{t^{n+1}\mu^{2n+2}(-1)^n f^{(n)}(\mu)}{n!(n+1)!} \right\| \leq M \sum_{n=0}^{\infty} \frac{t^{n+1}\mu^{(n+1)}}{(n+1)!} = M(\mathrm{e}^{\mu t} - 1),$$

which is valid by (62.5), proves that the series appearing in (62.6) is convergent absolutely and almost uniformly. Its sum is therefore a continuous function. Furthermore, $\|g_\mu(t)\| \leq M(1 - \mathrm{e}^{-\mu t}) \leq M$, as claimed. □

Lemma 62.4 *The Laplace transform of g_μ is:*

$$\int_0^\infty \mathrm{e}^{-\lambda t} g_\mu(t)\,\mathrm{d}t = \frac{\mu^2}{(\lambda + \mu)^2} f\left(\frac{\lambda\mu}{\lambda + \mu}\right), \qquad \mu, \lambda > 0.$$

Proof For $\nu \in (0, \mu)$ and $n \geq 1$ (see (7.4)),

$$f(\nu) = \sum_{k=0}^{n} f^{(k)}(\mu)\frac{(\nu-\mu)^k}{k!} + \int_{\mu}^{\nu} f^{(n+1)}(\xi)\frac{(\nu-\xi)^n}{n!}\,\mathrm{d}\xi.$$

By (62.5), the norm of the integral reminder here does not exceed:

$$M(n+1)\int_{\nu}^{\mu}\left(\frac{\xi-\nu}{\xi}\right)^n \frac{1}{\xi^2}\,\mathrm{d}\xi = \frac{M(n+1)}{\nu}\int_0^{\frac{\mu-\nu}{\mu}} \alpha^n\,\mathrm{d}\alpha$$
$$= \frac{M}{\nu}\left(\frac{\mu-\nu}{\mu}\right)^{(n+1)},$$

and so it converges to zero. Thus $f(\nu) = \sum_{n=0}^{\infty} f^{(n)}(\mu)\frac{(\nu-\mu)^n}{n!}$. In particular, $f(\frac{\lambda\mu}{\lambda+\mu}) = \sum_{n=0}^{\infty}\frac{f^{(n)}(\mu)}{n!}(\frac{\lambda\mu}{\lambda+\mu}-\mu)^n$, $\lambda, \mu > 0$ since $\mu > \frac{\lambda\mu}{\lambda+\mu} > 0$.

Therefore,

$$\int_0^{\infty} \mathrm{e}^{-\lambda t} g_\mu(t)\,\mathrm{d}t = \sum_{n=0}^{\infty}\frac{\mu^{2n+2}(-1)^n f^{(n)}(\mu)}{n!(n+1)!}\int_0^{\infty}\mathrm{e}^{-(\lambda+\mu)t}t^{n+1}\,\mathrm{d}t$$
$$= \left(\frac{\mu}{\lambda+\mu}\right)^2\sum_{n=0}^{\infty}\left(\frac{-\mu^2}{\lambda+\mu}\right)^n\frac{f^{(n)}(\mu)}{n!}$$
$$= \left(\frac{\mu}{\lambda+\mu}\right)^2\sum_{n=0}^{\infty}\left(\frac{\lambda\mu}{\lambda+\mu}-\mu\right)^n\frac{f^{(n)}(\mu)}{n!}$$
$$= \left(\frac{\mu}{\lambda+\mu}\right)^2 f\left(\frac{\lambda\mu}{\lambda+\mu}\right).$$

□

Lemma 62.5 *Suppose f satisfies (62.5). For $\phi \in L^1(\mathbb{R}^+)$ the limit:*

$$\mathcal{T}(\phi) = \lim_{\mu\to\infty}\int_0^{\infty}\phi(t)g_\mu(t)\,\mathrm{d}t, \tag{62.7}$$

exists and $\mathcal{T}: L^1(\mathbb{R}^+) \to \mathbb{X}$ is a bounded linear operator with $\|\mathcal{T}\| \leq M$.

Proof By (62.4), the operators $\mathcal{T}_\mu : L^1(\mathbb{R}^+) \to \mathbb{X}$ defined as $\mathcal{T}_\mu(\psi) = \int_0^{\infty}\psi(t)g_\mu(t)\,\mathrm{d}t$ satisfy $\|\mathcal{T}_\mu\| \leq M$. By Lemma 62.4,

$$\mathcal{T}_\mu(e_\lambda) = \frac{\mu^2}{(\lambda+\mu)^2} f\left(\frac{\lambda\mu}{\lambda+\mu}\right)$$

converges to $f(\lambda)$, as $\mu \to \infty$. Since the e_λ, $\lambda > 0$ form a total set in $L^1(\mathbb{R}^+)$, the proof is complete. □

Lemma 62.5 proves in particular the necessity of (2) in the Henning–Neubrander Theorem for, as we have seen, $\mathcal{T}$ defined there satisfies $\mathcal{T}(e_\lambda) = f(\lambda)$.

Applications: The Kisyński, the Widder, and the Arendt Theorems

Let us consider some applications of this theorem. If f has values in a Banach algebra and satisfies the Hilbert Equation, the corresponding $\mathcal{T}$ is a homomorphism of the convolution algebra $L^1(\mathbb{R}^+)$; this is the **Kisyński Homomorphism Theorem**. Indeed, it is enough to show $\mathcal{T}(e_\lambda)\mathcal{T}(e_\mu) = \mathcal{T}(e_\lambda * e_\mu)$, the set e_λ, $\lambda > \omega$ being total in $L^1_\omega(\mathbb{R}^+)$. This formula, however, follows from the Hilbert Equation:

$$\begin{aligned}\mathcal{T}(e_\lambda)\mathcal{T}(e_\mu) = R_\lambda R_\mu &= \frac{1}{\mu - \lambda}[R_\lambda - R_\mu] = \mathcal{T}\left(\frac{1}{\mu-\lambda}(e_\lambda - e_\mu)\right)\\ &= \mathcal{T}(e_\lambda * e_\mu).\end{aligned}$$

Corollary 62.6 *If f satisfies (62.5), then the limit:*

$$\lim_{\mu\to\infty} \int_0^t g_\mu(s)\,\mathrm{d}s =: u(t) \tag{62.8}$$

exists (almost uniformly in $t \in [0, \infty)$) and is a Lipschitz continuous function with the Lipschitz constant M. Furthermore,

$$\int_0^\infty \mathrm{e}^{-\lambda t} u(t)\,\mathrm{d}t = \frac{f(\lambda)}{\lambda}, \qquad \text{for } \lambda > \omega. \tag{62.9}$$

Proof By Lemma 62.5 there exists the limit $\lim_{\mu\to\infty} \int_0^t g_\mu(s)\,\mathrm{d}s := u(t)$, and since the integrands here are bounded by M, the limit is almost uniform in $[0, \infty)$ and u is Lipschitz continuous. Formula (62.9) follows by $\mathcal{T}(e_\lambda) = f(\lambda)$ and integration by parts. □

Theorem 62.7 (Widder) *A necessary and suffficient condition for an infinitely differentiable complex-valued function f to be the Laplace transform of a bounded measurable function is that there exist a constant $M > 0$ such that:*

$$|f^{(n)}(\lambda)| \leq \frac{Mn!}{\lambda^{n+1}}. \tag{62.10}$$

Proof We have seen that (62.10) is necessary. Conversely, if (62.10) holds, then, by Corollary 1, there exists a Lipschitz continuous function $u(t)$, $|u(t) - u(s)| \leq M|t - s|$ satisfying (62.9). Since $u(t)$ is complex-valued, then, by the Lebesgue Theorem, it is differentiable a.e., and, by (62.9), $u'(t)$ is a function we were looking for. □

As a by-product of our proof we also obtain an inversion formula for the Laplace transform, valid if original function is bounded:

$$\mathcal{L}^{-1}(f) = \frac{d}{dt} \lim_{\mu \to \infty} \int_0^t g_\mu(s)\, ds.$$

Theorem 62.8 **(Arendt)** *Suppose* $(0, \infty) \ni \lambda \to f(\lambda)$ *is an infinitely differentiable vector-valued function, then the following two conditions are equivalent:*

a) there exists an $M \geq 0$ *such that* (62.5) *holds,*
b) there exists a function u *with values in the same Banach space, Lipschitz continuous with the Lipschitz constant* M*, such that* (62.9) *holds.*

Proof The proof of b) ⇒ a) is straightforward. To prove the converse, we use u from Corollary 1. □

Generation Theorem for Solution Families

The theory presented here allows unifying approaches to Hille–Yosida–Feller–Phillips–Miyadera and Da Prato–Giusti–Sova theorems and in a straightforward way explains the striking parallelism and similarity between them. It is remarkable that the following theorem, which includes both theorems mentioned earlier as special cases $n = 1, n = 2$, respectively, is so easily derived from Theorem 62.8 (for more about this subject see the excellent "Green Book" [9]). It is necessary to add here, however, that it is only these two cases that are of importance in the theory, since higher order Cauchy problems are well posed (in a certain, reasonable sense) iff the operator appearing on the right-hand side of the involved equation $x^{(n)}(t) = Ax(t)$ is bounded ([163, p. 125], [135, 136], and [179]) We present the proof in the general case, since it is equally simple for all $n \in \mathbf{N}$.

Theorem 62.9 *Let* $n \in \mathbf{N}$ *and let* A *be a densely defined closed linear operator acting in a Banach space* L*. The following two conditions are equivalent:*

1) There exists a strongly continuous family $\{T(t), t \geq 0\}$ *of bounded linear operators, such that:*
   a) $T(0)x = x$,
   b) there exist constants $M \geq 1, \omega \geq 0$ *satisfying:*

$$\|T(t)\| \leq Me^{\omega t}, \quad \textit{for all} \quad t \geq 0,$$

c) for $\lambda > \omega$, $\lambda^n \in \rho(A)$ and:

$$\lambda^{n-1}(\lambda^n - A)^{-1}x = \int_0^\infty \mathrm{e}^{-\lambda t} T(t)x\,\mathrm{d}t,$$

2) there exist constants $M \geq 1$, $\omega \geq 0$ such that for all $\lambda > \omega$:

$$\left\| \frac{d^k}{d\lambda^k}\left(\lambda^{n-1}(\lambda^n - A)^{-1}x\right)\right\| \leq \frac{Mk!}{(\lambda-\omega)^{k+1}}, \qquad k \in \mathbf{N}.$$

Furthermore, if one of the above assumptions is satisfied, then:

(i) the operators $T(t), t \geq 0$ commute,
(ii) x belongs to $D(A)$ iff there exists the limit $\lim_{h\to 0} \frac{n!(T(h)x-x)}{h^n}$ (and equals Ax),
(iii) the function $t \to x(t) = T(t)x$ is a solution to the Cauchy problem

$$\frac{d^n x(t)}{dt^n} = Ax, \quad x(0) = x \in D(A), \quad x'(0) = \cdots = x^{(n-1)}(0) = 0.$$

Proof 1) $\Rightarrow$ 2)

$$\left\| \frac{\mathrm{d}^k}{\mathrm{d}\lambda^k}(\lambda^{n-1}(\lambda^n - A)^{-1}x)\right\| = \left\| \int_0^\infty \mathrm{e}^{-\lambda t}(-t)^k T(t)x\,\mathrm{d}t\right\|$$
$$\leq M \int_0^\infty \mathrm{e}^{-\lambda t} e^{\omega t} t^k\,\mathrm{d}t = \frac{Mk!}{(\lambda-\omega)^{k+1}}.$$

2) $\Rightarrow$ 1). By Arendt's Theorem, there exists a family $\{U(t), t \geq 0\}$ of bounded linear operators such that:

$$\lambda^{n-2}(\lambda^n - A)^{-1}x = \int_0^\infty \mathrm{e}^{-\lambda t} U(t)x\,\mathrm{d}t, \tag{62.11}$$

$$\|U(t) - U(s)\| \leq M \int_s^t e^{\omega u}\,\mathrm{d}u. \tag{62.12}$$

Furthermore, the Yosida approximation gives:

$$\begin{aligned} U(t) &= \mathcal{H}(1_{[0,t)}) \\ &= \lim_{\lambda\to\infty} \int_0^t e^{-\lambda s} s\lambda^2 \sum_{k=0}^{\infty} \frac{s^k \lambda^{2k}(-1)^k \frac{\mathrm{d}^k}{\mathrm{d}\lambda^k}\lambda^{n-1}(\lambda^n - A)^{-1}}{k!(k+1)!}\,\mathrm{d}s, \end{aligned} \tag{62.13}$$

where the limit is taken in the uniform operator topology. This formula shows that $U(t)$ commute with A.

Let $x \in D(A)$ be given. Define $\{T(t)x,\ t \geq 0\}$ by:

$$T(t)x := \frac{1}{(n-2)!} \int_0^t (t-s)^{(n-2)} U(s)Ax\,\mathrm{d}s + x.$$

(For $n = 1$ put $T(t)x = U(t)Ax + x$). For $\lambda > \omega$,

$$\begin{aligned}\int_0^\infty e^{-\lambda t} T(t)x\,dt &= \frac{1}{\lambda^{n-1}}\int_0^\infty e^{-\lambda t} U(t)Ax\,dt + \frac{1}{\lambda}x \\ &= \frac{1}{\lambda^{n-1}} A\lambda^{n-2}(\lambda^n - A)^{-1}x + \frac{1}{\lambda}x \\ &= (\lambda - x)^{-1}\,. \end{aligned} \tag{62.14}$$

Combining (62.14) and (62.11),

$$\int_0^\infty e^{-\lambda t}\int_0^t T(s)x\,ds\,dt = \lambda^{n-2}(\lambda^n - A)^{-1}x = \int_0^\infty e^{-\lambda t} U(t)x\,dt.$$

Consequently, $U(t)x = \int_0^t T(s)x\,ds$. It follows that:

$$T(t)x = U'(t)x, \qquad x \in D(A). \tag{62.15}$$

Using (62.12) we get $\|T(t)x\| \leq Me^{\omega t}\|x\|$. Therefore, the operators $T(t)$ may be extended to the whole of the space $\mathbb{X}$ by continuity, and formulae (62.14)–(62.15), remain true for every $x \in \mathbb{X}$.

Condition (i) follows from (62.13) and (62.15). To prove (ii) consider $x \in D(A)$. By definition,

$$\begin{aligned} T(t)x - x &= \frac{1}{(n-2)!}\int_0^t (t-s)^{(n-2)} U(s)Ax\,ds \\ &= \frac{1}{(n-1)!}\int_0^t (t-s)^{(n-1)} T(s)Ax\,ds. \end{aligned} \tag{62.16}$$

Thus,

$$\lim_{t\to 0}\frac{n!(T(t)x - x)}{t^n} = \lim_{t\to 0}\frac{n!}{t^n}\frac{1}{(n-1)!}\int_0^t (t-s)^{(n-1)} T(s)Ax\,ds = Ax.$$

Conversely, if $\lim_{h\to 0}\frac{n!(T(h)x-x)}{h^n} = y$, we can apply the already proven result to obtain:

$$\begin{aligned} x &= (\lambda^n - A)(\lambda^n - A)^{-1}x = \frac{1}{\lambda^{n-1}}(\lambda^n - A)\int_0^\infty e^{-\lambda t} T(t)x\,dt \\ &= \lambda\int_0^\infty e^{-\lambda t} T(t)x\,dt - \frac{1}{\lambda^{n-1}}\lim_{h\to 0}\frac{n!(T(h) - I)}{h^n}\int_0^\infty e^{-\lambda t} T(t)x\,dt \\ &= \lambda^n(\lambda^n - A)^{-1}x - \frac{1}{\lambda^{n-1}}\int_0^\infty e^{-\lambda t} T(t)\lim_{h\to 0}\frac{n!(T(h)x - x)}{h^n}\,dt \\ &= \lambda^n(\lambda^n - A)^{-1}x - (\lambda^n - A)^{-1}y \\ &= (\lambda^n - A)^{-1}[\lambda^n x - y] \in D(A). \end{aligned}$$

Condition (iii) is a direct consequence of (62.16). □

63

Appendix B: Measurable Cosine Functions Are Continuous

This chapter is devoted to the proof of the individual version of 'measurability implies continuity theorem' for cosine families, announced in Chapter 61. The entire direct argument presented here is due to W. Chojnacki, whom the author is grateful for granting permission to include this proof in this book. The main result reads:

Theorem 63.1 *Let $\mathbb{X}$ be a Banach space and let $\{\mathrm{Cos}(t)\}_{t\in\mathbb{R}}$ be a cosine family on $\mathbb{X}$. Let $x \in \mathbb{X}$. If the function $t \mapsto \mathrm{Cos}(t)x$ is measurable on $\mathbb{R}$, then it is continuous on $\mathbb{R}$.*

Theorem 63.1 will be deduced from the following result:

Theorem 63.2 *Let $\mathbb{X}$ be a Banach space, let $\{\mathrm{Cos}(t)\}_{t\in\mathbb{R}}$ be a cosine family on $\mathbb{X}$, and let:*

$$\mathbb{Y} = \{x \in \mathbb{X} \mid t \mapsto \|\mathrm{Cos}(t)x\| \text{ is measurable on } \mathbb{R}\}.$$

If $\mathbb{Y} \neq \{0\}$, then the function $d\colon \mathbb{R} \to [0, \infty)$ defined by:

$$d(t) = \sup\{\|x\|^{-1}\|C(t)x\| \mid x \in \mathbb{Y} \setminus \{0\}\} \quad (t \in \mathbb{R})$$

is locally bounded.

Background

We start by laying out some background material. Denote by m the Lebesgue measure and by m^* the outer Lebesgue measure, both on $\mathbb{R}$. First, we present a minor modification of a result of Kurepa [227, Lemma 1], which in turn is akin to the famous theorem of Steinhaus [49, 225].

Lemma 63.3 *Let $K \subset \mathbb{R}$ be a measurable set such that $0 < m(K) < \infty$. Then there exists $a > 0$ with the property that for every $-a < s < a$ there is a measurable set $W_s \subset K$ of positive measure such that $t, t + \frac{1}{2}s, t + s \in K$ whenever $t \in W_s$.*

Proof Define the function $u \colon \mathbb{R} \to [0, \infty)$ by:

$$u(s) = m\left(K \cap \left(K - \frac{1}{2}s\right) \cap (K - s)\right) \quad (s \in \mathbb{R}).$$

We have:

$$\begin{aligned} |u(s) - u(0)| &= \left| \int_{\mathbb{R}} 1_K(t) \left[1_K\left(t + \frac{1}{2}s\right) 1_K(t + s) - 1_K(t) 1_K(t + s) \right. \right. \\ &\qquad \left. \left. + 1_K(t) 1_K(t + s) - 1_K(t) \right] \mathrm{d}t \right| \\ &\le \int_{\mathbb{R}} \left| 1_K\left(t + \frac{1}{2}s\right) - 1_K(t) \right| \mathrm{d}t + \int_{\mathbb{R}} |1_K(t + s) - 1_K(t)| \, \mathrm{d}t, \end{aligned}$$

where 1_A denotes the characteristic function of the set A. As the two right-most integrals tend to zero as $s \to 0$, u is continuous at $s = 0$. Since $u(0) = m(K) \neq 0$, there exists $a > 0$ such that $u(s) \neq 0$ for each $-a < s < a$. But $u(s) \neq 0$ implies that the set:

$$W_s := K \cap \left(K - \frac{1}{2}s\right) \cap (K - s)$$

has positive measure. Moreover, it is immediately seen that $t, t + \frac{1}{2}s, t + s \in K$ whenever $t \in W_s$. The proof is complete. □

In what follows we shall exploit the concept of the *measurable boundaries* of an arbitrary real-valued function on $\mathbb{R}$, due to Blumberg [42]. For a set $E \subset \mathbb{R}$ and a point $x \in \mathbb{R}$, the *exterior density* of E at x, $D_e(E, x)$, is defined by:

$$D_e(E, x) = \lim_{m(I) \to 0} \frac{m^*(E \cap I)}{m(I)},$$

with I denoting a closed interval that contains x, whenever the limit exists. Let $f \colon \mathbb{R} \to \mathbb{R}$ be an arbitrary function. For any $t \in \mathbb{R}$, let:

$$\begin{aligned} A_t &= \{x \in \mathbb{R} \mid f(x) > t\}, \\ B_t &= \{x \in \mathbb{R} \mid f(x) < t\}. \end{aligned}$$

The *upper* and *lower approximate limits* of f at a point $x \in \mathbb{R}$ are defined by:

$$\text{ap}\limsup_{\xi \to x} f(\xi) = \inf\{t \in \mathbb{R} \mid D_e(A_t, x) = 0\},$$
$$\text{ap}\liminf_{\xi \to x} f(\xi) = \sup\{t \in \mathbb{R} \mid D_e(B_t, x) = 0\},$$

with the proviso that $\inf \emptyset = \infty$ and $\sup \emptyset = -\infty$. Blumberg calls the extended real-valued functions:

$$x \mapsto \text{ap}\limsup_{\xi \to x} f(\xi) \quad \text{and} \quad x \mapsto \text{ap}\liminf_{\xi \to x} f(\xi)$$

the *upper* and *lower measurable boundaries* of f; we shall denote these functions as $u_B(f)$ and $l_B(f)$, respectively. This terminology and notation allude to the fact, established by Blumberg, that $u_B(f)$ and $l_B(f)$ are both measurable and satisfy

$$l_B(f)(x) \leq f(x) \leq u_B(f)(x) \quad \text{for almost all } x \in \mathbb{R}.$$

Goffman and Zink [157] have provided an additional substantiation to the above nomenclature by proving the following result:

- if $g\colon \mathbb{R} \to \mathbb{R}$ is measurable and $f(x) \leq g(x)$ for almost all $x \in \mathbb{R}$, then $u_B(f)(x) \leq g(x)$ for almost all $x \in \mathbb{R}$.
- if $g\colon \mathbb{R} \to \mathbb{R}$ is measurable and $g(x) \leq f(x)$ for almost all $x \in \mathbb{R}$, then $g(x) \leq l_B(f)(x)$ for almost all $x \in \mathbb{R}$.

When we choose not to distinguish between classes of equivalent measurable functions and measurable functions themselves, the combined result of Blumberg, Goffman, and Zink can be phrased as saying that $u_B(f)$ and $l_B(f)$ are the least measurable majorant and the greatest measurable minorant of f, respectively.

We note the following simple property of the mappings $f \mapsto u_B(f)$ and $f \mapsto l_B(f)$, which will be of relevance later on: If $f\colon \mathbb{R} \to \mathbb{R}$ is an arbitrary function, then:

$$u_B(T_t f) = T_t(u_B(f)) \quad \text{and} \quad l_B(T_t f) = T_t(l_B(f))$$

for each $t \in \mathbb{R}$. Here, given $g \in \mathbb{R} \to \mathbb{R}$ and $t \in \mathbb{R}$, $T_t g$ denotes the *translate* of g by t, defined by:

$$T_t g(x) = g(x + t) \quad (x \in \mathbb{R}).$$

The two constituent parts of this property will be referred to as the *translational equivariance* of the upper and lower measurable boundary, respectively.

Proof of Theorem 63.2

We are now in position to establish Theorem 63.2. Our proof will essentially be a refinement of an argument due to Kurepa (see the proof of [227, Thm. 1]). We begin by noting that, by the cosine functional equation, for any $s, t \in \mathbb{R}$,

$$\begin{aligned} C(s) &= 2C(s+t)C(t) - C(s+2t) \\ &= 2C(s+t)C(t) + I_{\mathbb{X}} - 2C^2\left(\frac{1}{2}s+t\right), \end{aligned}$$

where $I_{\mathbb{X}}$ denotes the identity operator on $\mathbb{X}$, and this implies:

$$\|C(s)\| \leq 2\|C(s+t)\|\,\|C(t)\| + 2\left\|C\left(\frac{1}{2}s+t\right)\right\|^2 + 1.$$

Our next lemma establishes an analogous inequality to this one in which the function $t \mapsto \|C(t)\|$ on the right-hand side is replaced by the lower measurable boundary $k \colon \mathbb{R} \to \mathbb{R} \cup \{-\infty\}$ of $t \mapsto \|C(t)\|$:

$$k(t) = l_B\,(\tau \mapsto \|\mathrm{Cos}(\tau)\|)\,(t) \quad (t \in \mathbb{R}).$$

Lemma 63.4 *Suppose that $\mathbb{Y} \neq \{0\}$ and let $x \in \mathbb{Y} \setminus \{0\}$. Fix $s \in \mathbb{R}$. Then:*

$$\|\mathrm{Cos}(s)x\| \leq \left(2k(s+t)k(t) + 2\left[k\left(\frac{1}{2}s+t\right)\right]^2 + 1\right)\|x\| \tag{63.1}$$

for almost all $t \in \mathbb{R}$.

Proof As $t \mapsto \|x\|^{-1}\|\mathrm{Cos}(t)x\|$ is a measurable minorant of $t \mapsto \|C(t)\|$, we have:

$$\|x\|^{-1}\|\mathrm{Cos}(t)x\| \leq k(t) \quad \text{for almost all } t \in \mathbb{R}$$

and further:

$$\|\mathrm{Cos}(t)x\| \leq k(t)\|x\| \quad \text{for almost all } t \in \mathbb{R}. \tag{63.2}$$

Fix $s \in \mathbb{R}$ arbitrarily. Note that, for each $t \in \mathbb{R}$, the identity:

$$\mathrm{Cos}(s)x + \mathrm{Cos}(s+2t)x = 2\mathrm{Cos}(s+t)\mathrm{Cos}(t)x$$

implies:

$$\|\mathrm{Cos}(s)x\| - \|\mathrm{Cos}(s+2t)x\| \leq 2\|\mathrm{Cos}(s+t)\mathrm{Cos}(t)x\|.$$

Since the function $t \mapsto \|\mathrm{Cos}(s)x\| - \|\mathrm{Cos}(s+2t)x\|$ is measurable, it follows that:

$$\|\mathrm{Cos}(s)x\| - \|\mathrm{Cos}(s+2t)x\| \leq 2l_B\,(\tau \mapsto \|\mathrm{Cos}(s+\tau)\mathrm{Cos}(\tau)x\|)\,(t) \tag{63.3}$$

for almost all $t \in \mathbb{R}$. On the other hand,

$$\begin{aligned} l_B\,(\tau \mapsto \|\mathrm{Cos}(s+\tau)\mathrm{Cos}(\tau)x\|)\,(t) &\leq \|\mathrm{Cos}(s+t)\mathrm{Cos}(t)x\| \\ &\leq \|\mathrm{Cos}(s+t)\|\,\|\mathrm{Cos}(t)x\| \end{aligned} \tag{63.4}$$

for almost all $t \in \mathbb{R}$. Define the function $m\colon \mathbb{R} \mapsto [0, \infty) \cup \{-\infty\}$ by:

$$m(t) = \begin{cases} \|\mathrm{Cos}(t)x\|^{-1} l_B\,(\tau \mapsto \|\mathrm{Cos}(s+\tau)\mathrm{Cos}(\tau)x\|)\,(t), & \text{if } \mathrm{Cos}(t)x \neq 0; \\ 0, & \text{otherwise.} \end{cases}$$

Clearly, m is measurable and, by (63.4), it is a minorant of $t \mapsto \|\mathrm{Cos}(s+t)\|$ almost all on $\mathbb{R}$. Hence, by the translational equivariance of the lower measurable boundary, $m(t) \leq k(s+t)$ for almost all $t \in \mathbb{R}$. Moreover, again by (63.4), $l_B(\tau \mapsto \|\mathrm{Cos}(s+\tau)\mathrm{Cos}(\tau)x\|)(t) = 0$ for almost every t such that $\mathrm{Cos}(t)x = 0$. Consequently,

$$l_B\,(\tau \mapsto \|\mathrm{Cos}(s+\tau)\mathrm{Cos}(\tau)x\|)\,(t) \leq k(s+t)\|\mathrm{Cos}(t)x\|$$

for almost all $t \in \mathbb{R}$. Combining this estimate with (63.2) and (63.3), we get:

$$\|\mathrm{Cos}(s)x\| \leq \|\mathrm{Cos}(s+2t)x\| + 2k(s+t)k(t)\|x\| \tag{63.5}$$

for almost all $t \in \mathbb{R}$.

Observe that, for each $t \in \mathbb{R}$, the identity:

$$\mathrm{Cos}(s+2t)x = 2\mathrm{Cos}^2\left(\frac{1}{2}s+t\right)x - x$$

implies:

$$\|\mathrm{Cos}(s+2t)x\| - \|x\| \leq 2\left\|\mathrm{Cos}^2\left(\frac{1}{2}s+t\right)x\right\|.$$

Since the function $t \mapsto \|\mathrm{Cos}(s+2t)x\| - \|x\|$ is measurable, it follows that:

$$\|\mathrm{Cos}(s+2t)x\| - \|x\| \leq 2l_B\left(\tau \mapsto \left\|\mathrm{Cos}^2\left(\frac{1}{2}s+\tau\right)x\right\|\right)(t) \tag{63.6}$$

for almost all $t \in \mathbb{R}$. On the other hand,

$$\begin{aligned} l_B\left(\tau \mapsto \left\|\mathrm{Cos}^2\left(\frac{1}{2}s+\tau\right)x\right\|\right)(t) &\leq \left\|\mathrm{Cos}^2\left(\frac{1}{2}s+t\right)x\right\| \\ &\leq \left\|\mathrm{Cos}\left(\frac{1}{2}s+t\right)\right\|^2 \|x\| \end{aligned}$$

for almost all $t \in \mathbb{R}$. Thus:

$$t \mapsto \left(\|x\|^{-1} l_B\left(\tau \mapsto \left\|\mathrm{Cos}^2\left(\frac{1}{2}s+\tau\right)x\right\|\right)(t)\right)^{1/2}$$

is a measurable minorant of $t \mapsto \|\mathrm{Cos}(\frac{1}{2}s + t)\|$ almost all on $\mathbb{R}$. Consequently, by the translational equivariance of the lower measurable boundary,

$$\left(\|x\|^{-1} l_B \left(\tau \mapsto \left\|\mathrm{Cos}^2\left(\frac{1}{2}s + \tau\right)x\right\|\right)(t)\right)^{1/2} \leq k\left(\frac{1}{2}s + t\right)$$

for almost all $t \in \mathbb{R}$, whence:

$$l_B \left(\tau \mapsto \left\|\mathrm{Cos}^2\left(\frac{1}{2}s + \tau\right)x\right\|\right)(t) \leq \left[k\left(\frac{1}{2}s + t\right)\right]^2 \|x\|$$

for almost all $t \in \mathbb{R}$, and further, on account of (63.6),

$$\|\mathrm{Cos}(s + 2t)x\| - \|x\| \leq 2\left[k\left(\frac{1}{2}s + t\right)\right]^2 \|x\|$$

for almost all $t \in \mathbb{R}$. Combining this inequality with (63.5), we complete the proof. □

Proof of Theorem 63.2 As k is measurable and attains only finite values from above, there exists a measurable set $K \subset \mathbb{R}$ of finite positive measure such that:

$$L := \sup_{t \in K} k(t) < \infty.$$

By Lemma 63.3, there exists $a > 0$ with the property that for each $0 \leq s \leq a$ there is a measurable set $W_s \subset K$ of positive measure such that $t, t + \frac{1}{2}s, t + s \in K$ whenever $t \in W_s$. Fix $0 \leq s \leq a$ and $x \in \mathbb{Y} - \{0\}$. In view of Lemma 63.4, the set W_s' of those $t \in W_s$ for which (63.1) holds is of full measure in W_s, and, given that W_s has positive measure, W_s' has positive measure and in particular is nonvoid. Select any $t \in W_s'$. Invoking (63.1) with this t, we see that:

$$\|\mathrm{Cos}(s)x\| \leq (4L^2 + 1)\|x\|.$$

In view of the arbitrariness of x, we have:

$$d(s) \leq (4L^2 + 1). \tag{63.7}$$

For any $b > 0$, let:

$$D(b) = \sup_{s \in [0,b]} d(s).$$

As s in (63.7) is arbitrary within $[0, a]$, it follows that:

$$D(a) \leq (4L^2 + 1). \tag{63.8}$$

Now, using:

$$\|\mathrm{Cos}(b + h)x\| \leq 2\|\mathrm{Cos}(b)\| \, \|\mathrm{Cos}(h)x\| + \|\mathrm{Cos}(b - h)x\|$$

with $0 \le h \le b$, we find that, for every $b > 0$,

$$\sup_{s\in[b,2b]} d(s) \le 2\|\mathrm{Cos}(b)\|D(b) + D(b)$$

and further:

$$\begin{aligned} D(2b) &= \max\big(\sup_{s\in[0,b]} d(s),\ \sup_{s\in[b,2b]} d(s)\big) \\ &\le \max(D(b),\ 2\|\mathrm{Cos}(b)\|D(b) + D(b)) \\ &= 2\|\mathrm{Cos}(b)\|D(b) + D(b). \end{aligned}$$

As an immediate consequence, if $D(b) < \infty$, then $D(2b) < \infty$ for every $b > 0$. This together with (63.8) implies that $t \mapsto d(t)$ is bounded on $[0, b]$ for each $0 < b < \infty$. Since $t \mapsto d(t)$ is even, it follows that $t \mapsto d(t)$ is bounded on every finite interval of $\mathbb{R}$. □

Proof of Theorem 63.1

We now proceed to establish Theorem 63.1.

Proof of Theorem 63.1 Let:

$$\mathbb{Z} = \{x \in \mathbb{X} \mid t \mapsto \mathrm{Cos}(t)x \text{ is measurable on } \mathbb{R}\}.$$

It is clear that $\mathbb{Z}$ is a linear subspace of $\mathbb{X}$. Since the pointwise limit of a sequence of measurable functions is a measurable function, $\mathbb{Z}$ is also closed. Moreover, $\mathbb{Z}$ is invariant for each operator $\mathrm{Cos}(s)$, $s \in \mathbb{R}$, that is,

$$\mathrm{Cos}(s)\mathbb{Z} \subset \mathbb{Z} \tag{63.9}$$

for every $s \in \mathbb{R}$. For if $s \in \mathbb{R}$ and $x \in \mathbb{Z}$, then $t \mapsto \mathrm{Cos}(t + s)x$ and $t \mapsto \mathrm{Cos}(t - s)x$ are both measurable, and consequently the function

$$t \mapsto C(t)C(s)x = \frac{1}{2}(\mathrm{Cos}(t + s)x + \mathrm{Cos}(t - s)x)$$

is measurable too. Retaining the notation from the statement of Theorem 63.2, we clearly have $\mathbb{Z} \subset \mathbb{Y}$, and if only $\mathbb{Z}$ contains a nonzero element, then it follows from Theorem 63.2 that, for each $t \in \mathbb{R}$,

$$\|\mathrm{Cos}(t)|_{\mathbb{Z}}\| \le d(t),$$

where $\mathrm{Cos}(t)|_{\mathbb{Z}}$ denotes the restriction of $\mathrm{Cos}(t)$ to Z.

Let $x \in \mathbb{Z}$. If $x = 0$, then the assertion of the theorem is trivial. Suppose then that x is nonzero. Let $a, b \in \mathbb{R}$ be such that $a < b$. Then, for any $s, t, \tau \in \mathbb{R}$,

$$\begin{aligned}\operatorname{Cos}(\tau) &= 2\operatorname{Cos}(s)\operatorname{Cos}(\tau - s) - \operatorname{Cos}(\tau - 2s)\\ \operatorname{Cos}(t) &= 2\operatorname{Cos}(s)\operatorname{Cos}(t - s) - \operatorname{Cos}(t - 2s)\end{aligned}$$

so that:

$$\begin{aligned}\operatorname{Cos}(\tau)x - \operatorname{Cos}(t)x = {} & 2\operatorname{Cos}(s)(\operatorname{Cos}(\tau - s)x - \operatorname{Cos}(t - s)x)\\ & + (\operatorname{Cos}(t - 2s)x - \operatorname{Cos}(\tau - 2s)x). \qquad (63.10)\end{aligned}$$

The functions $s \mapsto \operatorname{Cos}(t - 2s)x$ and $s \mapsto \operatorname{Cos}(\tau - 2s)x$ are measurable, and, by Theorem 63.2, they also are locally bounded. Hence both functions are locally integrable. Eq. (63.10) now implies that the function:

$$s \mapsto \operatorname{Cos}(s)(\operatorname{Cos}(\tau - s)x - \operatorname{Cos}(t - s)x)$$

is locally integrable, too, and we have:

$$\begin{aligned}(b - a)(\operatorname{Cos}(\tau)x - C(t)x) = {} & 2\int_a^b \operatorname{Cos}(s)(\operatorname{Cos}(\tau - s)x - \operatorname{Cos}(t - s)x)\,\mathrm{d}s\\ & + \int_a^b (\operatorname{Cos}(t - 2s)x - \operatorname{Cos}(\tau - 2s)x)\,\mathrm{d}s.\end{aligned}$$

By the invariance property of $\mathbb{Z}$ (as per (63.9)), $\operatorname{Cos}(\tau - s)x$ and $\operatorname{Cos}(t - s)x$ are in $\mathbb{Z}$, and so, for $s \in [a, b]$,

$$\begin{aligned}&\|\operatorname{Cos}(s)(\operatorname{Cos}(\tau - s)x - \operatorname{Cos}(t - s)x)\|\\ &\quad \le \|\operatorname{Cos}(s)|_{\mathbb{Z}}\|\,\|\operatorname{Cos}(\tau - s)x - \operatorname{Cos}(t - s)x\|\\ &\quad \le M\|\operatorname{Cos}(\tau - s)x - \operatorname{Cos}(t - s)x\|\end{aligned}$$

where $M := \sup_{s\in[a,b]} d(s)$ is, by Theorem 63.2, a finite positive number. Consequently,

$$\begin{aligned}(b - a)\|\operatorname{Cos}(\tau)x - C(t)x\| \le {} & 2M\int_a^b \|\operatorname{Cos}(\tau - s)x - \operatorname{Cos}(t - s)x\|\,\mathrm{d}s\\ & + \int_a^b \|\operatorname{Cos}(t - 2s)x - \operatorname{Cos}(\tau - 2s)x\|\,\mathrm{d}s.\end{aligned}$$

But the integrals on the right-hand side tend to zero as $\tau \to t$ (see, e.g., [180, Thm. 3.8.3]), so $\lim_{\tau\to t} \|\operatorname{Cos}(\tau)x - C(t)x\| = 0$, as required. □

We finally remark that while the space $\mathbb{Z}$ introduced in the proof of Theorem 63.1 is contained in the set $\mathbb{Y}$ from the statement of Theorem 63.2, the two sets are generally different. To see this, let $l^1(\mathbb{R})$ be the space of all absolutely summable functions $f\colon \mathbb{R} \to \mathbb{C}$ with the norm:

$$\|f\|_1 := \sum_{x \in \mathbb{R}} |f(x)|.$$

For each $t \in \mathbb{R}$, let:

$$\mathrm{Cos}(t) = \frac{1}{2}(T_t + T_{-t}),$$

where T_t denotes the translation operator by t on $l^1(\mathbb{R})$. The family $\mathrm{Cos} = \{\mathrm{Cos}(t)\}_{t \in \mathbb{R}}$ is a cosine family of contractions on $l^1(\mathbb{R})$. Denote by e_t the element of $l^1(\mathbb{R})$ such that:

$$e_t(x) = \begin{cases} 1, & \text{if } x = t; \\ 0, & \text{otherwise.} \end{cases}$$

It is readily seen that:

$$\mathrm{Cos}(t)e_0 = \frac{1}{2}(e_t + e_{-t})$$

for each $t \in \mathbb{R}$. Consequently, $\|\mathrm{Cos}(t)e_0\|_1 = 1$ for each $t \in \mathbb{R}$, and we see that e_0 is a member of the corresponding space $\mathbb{Y}$ for Cos. On the other hand,

$$\|\mathrm{Cos}(s)e_0 - \mathrm{Cos}(t)e_0\|_1 = 2$$

whenever $s \neq t$. This implies that for every uncountable set $A \subset \mathbb{R}$, the set $\{\mathrm{Cos}(t)e_0 \mid t \in A\}$ is nonseparable. In particular, $\{\mathrm{Cos}(t)e_0 \mid t \in A\}$ is nonseparable for every measurable set $A \subset \mathbb{R}$ of positive measure. By the Pettis Measurability Theorem (see, e.g., [180, Thm. 3.5.3]), the function $t \mapsto \mathrm{Cos}(t)e_0$ is not measurable, meaning that e_0 does not belong to the space $\mathbb{Z}$ for Cos. We have thus identified e_0 as a member of $\mathbb{Y} \setminus \mathbb{Z}$.

References

[1] A. D. Aczel, *Fermat's Last Theorem: Unlocking the Secret of an Ancient Mathematical Problem*, A Delta Book, New York, Dell Publishing, 1997. (Cited on page 183.)

[2] R. A. Adams and J. J. F. Fournier, *Sobolev Spaces*, second ed., Pure and Applied Mathematics, vol. 140, Elsevier/Academic Press, Amsterdam, 2003. (Cited on page 228.)

[3] U. Alon, *An Introduction to Systems Biology. Design Principles of Biological Circuts*, Chapman and Hall/CRC, Boca Raton, 2007. (Cited on pages 150, 216, and 286.)

[4] A. Alonso and B. Simon, The Birman–Kreĭn–Vishik theory of self-adjoint extensions of semibounded operators, *J. Oper. Th.* **4** (1980), no. 2, 251–270. (Cited on page 36.)

[5] S. S. Andrews, Accurate particle-based simulation of adsorption, desorption, and partial transmission, *Phys. Biol.* (2010), no. 6, 046015. (Cited on page 66.)

[6] W. Arendt, Vector-valued Laplace transforms and Cauchy problems, *Israel J. Math.* **59** (1987), 327–352. (Cited on page 82.)

[7] W. Arendt, *Semigroups and Evolution Equations: Functional Calculus, Regularity, and Kernel Estimates*, Evolutionary Equations Vol. 1 (C. M. Dafermos and E. Feireisl, eds.), Handbook of Differential Equations, Elsevier B.V., 2004, pp. 1–85. (Cited on pages 179 and 180.)

[8] W. Arendt, *Heat Kernels – Manuscript of the* 9th *Internet Seminar*, 2006, Freely available at http://www.uni-ulm.de/fileadmin/website_uni_ulm/mawi.inst.020/arendt/downloads/internetseminar.pdf. (Cited on pages 32 and 35.)

[9] W. Arendt, C. J. K. Batty, M. Hieber, and F. Neubrander, *Vector-Valued Laplace Transforms and Cauchy Problems*, Birkhäuser, Basel, 2001. (Cited on pages xi, 1, 3, 4, 29, 87, 89, 141, 157, 159, 179, 180, 250, 398, and 402.)

[10] W. Arendt, A. Driouich, and O. El-Mennaoui, On the infinite product of C_0-semigroups, *J. Functional Analysis* **160** (1998), 524–542. (Cited on pages 86, 87, and 91.)

[11] O. Arino, E. Sanchez, R. Bravo de la Parra, and P. Auger, A singular perturbation in an age-structured population, *SIAM J. Appl. Math.* **60** (1999–2000), no. 2, 408–436. (Cited on pages 304, 305, and 345.)

[12] F. Aristizabal and M. I. Glavinovic, Simulation and parameter estimation of dynamics of synaptic depression, *Biol. Cybern.* **90** (2004), 3–18. (Cited on pages 189, 190, and 192.)

[13] R. Bafico and P. Baldi, Small random perturbations of Peano phenomena, *Stochastics* **6** (1982), 279–292. (Cited on page xiii.)

[14] J. Banasiak, Singularly perturbed linear and semilinear hyperbolic systems: Kinetic theory approach to some folks theorems, *Acta Applicandae Mathematica* **49** (1997), no. 2, 199–228. (Cited on page 73.)

[15] J. Banasiak, Global classical solutions of coagulation fragmentation equations with unbounded coagulation rates, *Nonlinear Analysis: Real World Applications* **13** (2012), no. 1, 91–105. (Cited on page 221.)

[16] J. Banasiak and L. Arlotti, *Perturbations of Positive Semigroups with Applications*, Springer, 2006. (Cited on pages 77, 80, 219, 227, and 301.)

[17] J. Banasiak and A. Bobrowski, Interplay between degenerate convergence of semigroups and asymptotic analysis: a study of a singularly perturbed abstract telegraph system, *J. Evol. Equ.* **9** (2009), no. 2, 293–314. (Cited on pages 54, 70, 73, 137, 341, 342, and 344.)

[18] J. Banasiak and A. Bobrowski, A semigroup related to a convex combination of boundary conditions obtained as a result of averaging other semigroups, *J. Evol. Equ.* **15** (2015), no. 1, 223–237. (Cited on pages 345 and 352.)

[19] J. Banasiak, A. Falkiewicz, and P. Namayanja, Asymptotic state lumping in transport and diffusion problems on networks, *ArXiv* e-prints (2015), 1–32. (Cited on page 203.)

[20] J. Banasiak, A. Falkiewicz, and P. Namayanja, Semigroup approach to diffusion and transport problems on networks, *ArXiv* e-prints (2015), 1–16. (Cited on page 203.)

[21] J. Banasiak and A. Goswami, Singularly perturbed population models with reducible migration matrix. 1. Sova–Kurtz theorem and the convergence to the aggregated model, *DCDS-A* **35** (2015), 617–635. (Cited on pages 258, 345, and 359.)

[22] J. Banasiak, A. Goswami, and S. Shindin, Aggregation in age and space structured population models: an asymptotic analysis approach, *J. Evol. Equ.* **11** (2011), 121–154. (Cited on pages 304, 342, 345, and 359.)

[23] J. Banasiak, A. Goswami, and S. Shindin, Singularly perturbed population models with reducible migration matrix: 2. Asymptotic analysis and numerical simulations, *Mediterr. J. Math.* **11** (2014), no. 2, 533–559. (Cited on pages 304, 345, and 359.)

[24] J. Banasiak and M. Lachowicz, Around the Kato generation theorem for semigroups, *Studia Mathematica* **179** (2007), no. 3, 217–238. (Cited on pages 75 and 148.)

[25] J. Banasiak and M. Lachowicz, *Methods of Small Parameter in Mathematical Biology*, Modeling and Simulation in Science, Engineering, and Technology, Birkhäuser, 2014. (Cited on pages xiii, 344, 345, and 359.)

[26] J. Banasiak and W. Lamb, Analytic fragmentation semigroups and continuous coagulation-fragmentation equations with unbounded rates, *J. Math. Anal. Appl.* **391** (2012), no. 1, 312–322. (Cited on page 221.)

[27] J. Banasiak and J. R. Mika, Singularly perturbed telegraph equations with applications in the random walk theory, *J. Appl. Math. Stochastic Anal.* **11** (1998), no. 1, 9–28. (Cited on page 73.)

[28] A. Bátkai, P. Csomós, B. Farkas, and G. Nickel, Operator splitting for non-autonomous evolution equations, *J. Functional Analysis* **260** (2011), 2163–2190. (Cited on page 123.)

[29] A. Bátkai, I. Kiss, E. Sikolya, and P. L. Simon, Differential equation approximations of stochastic network processes: an operator semigroup approach, *Networks and Heterogeneous Media* **7** (2012), no. 1, 43–58. (Cited on page 337.)

[30] C. J. K. Batty, Derivations on compact spaces, *Proc. London Math. Soc. (3)* **42** (1981), no. 2, 299–330. (Cited on page 302.)

[31] C. J. K. Batty, Derivations on the line and flows along orbits, *Pacific J. Math.* **126** (1987), no. 2, 209–225. (Cited on page 302.)

[32] C. J. K. Batty and A. F. M. ter. Elst, On series of sectorial forms, *J. Evol. Equ.* **14** (2014), no. 1, 29–47. (Cited on page 365.)

[33] F. A. Berezin, Covariant and contravariant symbols of operators, *Izv. Akad. Nauk SSSR Ser. Mat.* **36** (1972), 1134–1167. (Cited on page xiii.)

[34] A. Bielecki, Une remarque sur la méthode de Banach–Cacciopoli–Tikhonov, *Bull. Polish Acad. Sci.* **4** (1956), 261–268. (Cited on pages 114, 122, 211, and 225.)

[35] A. Bielecki and P. Kalita, Model of neurotransmitter fast transport in axon terminal of presynaptic neuron, *J. Math. Biol.* **56** (2008), 559–576. (Cited on pages 190, 191, and 192.)

[36] S. Bjon, On an integral transform of R. S. Phillips, *Cent. Eur. J. Math.* **8** (2010), no. (1), 98–113. (Cited on page 399.)

[37] D. Blackwell, Another countable Markov process with only instantaneous states, *Ann. Math. Statist.* **29** (1958), 313–316. (Cited on page 90.)

[38] Ph. Blanchard, Ph. Combe, M. Sirugue, and M. Sirugue-Collin, Jump processes related to the two-dimensional Dirac equation, *Stochastic Processes–Mathematics and Physics, II*, (Bielefeld, 1985), Lecture Notes in Mathematics 1250, Springer, 1985, pp. 1–13. (Cited on page 130.)

[39] Ph. Blanchard, Ph. Combe, M. Sirugue, and M. Sirugue-Collin, Path integral representation of the Dirac equation in presence of an external electromagnetic field, *Path Integrals from meV to MeV* (Bielefeld, 1985), Bielefeld Encount. Phys. Math., VII, World Scientific, 1985, pp. 396–413. (Cited on page 130.)

[40] Ph. Blanchard, Ph. Combe, M. Sirugue, and M. Sirugue-Collin, Stochastic jump processes associated with Dirac equation, *Stochastic Processes in Classical and Quantum System* (Ascona, 1985), Lecture Notes in Physics 262, Springer, 1986, pp. 87–104. (Cited on page 130.)

[41] Ph. Blanchard, Ph. Combe, M. Sirugue, and M. Sirugue-Collin, Jump processes. An introduction and some applications to quantum physics, *Rend. Circ. Mat. Palermo* (2) Suppl. (1987), no. 17, 47–104 (1988), Functional integration with emphasis on the Feynman integral (Sherbrooke, PQ, 1986). (Cited on page 130.)

[42] H. Blumberg, The measurable boundaries of an arbitrary function, *Acta Math.* **65** (1935), no. 1, 263–282. (Cited on page 406.)

[43] A. Bobrowski, Some remarks on the two-dimensional Dirac equation, *Semigroup Forum* **45** (1992), 77–91. (Cited on pages 121, 123, and 130.)

[44] A. Bobrowski, Degenerate convergence of semigroups, *Semigroup Forum* **49** (1994), no. 3, 303–327. (Cited on pages 60, 70, 71, and 171.)

[45] A. Bobrowski, Integrated semigroups and the Trotter–Kato theorem, *Bull. Polish Acad. Sci.* **41** (1994), 297–304. (Cited on page 82.)

[46] A. Bobrowski, On the Yosida approximation and the Widder–Arendt theorem, *Studia Mathematica* **124** (1997), no. 3, 281–290. (Cited on pages 83, 398, and 399.)

[47] A. Bobrowski, The Widder–Arendt theorem on inverting of the Laplace transform, and its relationships with the theory of semigroups of operators, *Methods Funct. Anal. Topology* **3** (1997), no. 4, 1–39. (Cited on pages 83, 109, and 384.)

[48] A. Bobrowski, A note on convergence of semigroups, *Ann. Polon. Math.* **69** (1998), no. 2, 107–127. (Cited on pages 157, 171, and 250.)

[49] A. Bobrowski, *Functional Analysis for Probability and Stochastic Processes*, Cambridge University Press, Cambridge, 2005, An introduction. (Cited on pages xiv, 1, 17, 29, 38, 39, 71, 73, 83, 89, 93, 96, 98, 100, 112, 116, 152, 199, 264, 331, 365, and 405.)

[50] A. Bobrowski, Degenerate convergence of semigroups related to a model of stochastic gene expression, *Semigroup Forum* **73** (2006), no. 3, 345–366. (Cited on pages 283 and 335.)

[51] A. Bobrowski, On a semigroup generated by a convex combination of two Feller generators, *J. Evol. Equ.* **7** (2007), no. 3, 555–565. (Cited on pages 293 and 299.)

[52] A. Bobrowski, On limitations and insufficiency of the Trotter–Kato theorem, *Semigroup Forum* **75** (2007), no. 2, 317–336. (Cited on pages 145 and 335.)

[53] A. Bobrowski, Generation of cosine families via Lord Kelvin's method of images, *J. Evol. Equ.* **10** (2010), no. 3, 663–675. (Cited on page 386.)

[54] A. Bobrowski, Lord Kelvin's method of images in the semigroup theory, *Semigroup Forum* **81** (2010), 435–445. (Cited on pages 186, 195, 248, 349, 358, and 386.)

[55] A. Bobrowski, From diffusions on graphs to Markov chains via asymptotic state lumping, *Ann. Henri Poincare* **13** (2012), 1501–1510. (Cited on pages 203, 204, 205, and 208.)

[56] A. Bobrowski, Emergence of Freidlin–Wentzell's transmission conditions as a result of a singular perturbation of a semigroup, *Semigroup Forum* **92** (2016), no. 1, 1–22. (Cited on pages 362 and 370.)

[57] A. Bobrowski, Families of operators describing diffusion through permeable membranes, *Operator Semigroups Meet Complex Analysis, Harmonic Analysis and Mathematical Physics*, Arendt, W., Chill, R., Tomilov, Y. eds. Birkhauser, 2015, pp. 87–105. (Cited on pages 24 and 65.)

[58] A. Bobrowski, On a somewhat forgotten condition of Hasegawa and on Blackwell's example, *Archiv der Mathematik* **104** (2015), no. 3, 237–246. (Cited on page 95.)

[59] A. Bobrowski, Singular perturbations involving fast diffusion, *J. Math. Anal. Appl.* **427** (2015), no. 2, 1004–1026. (Cited on pages 236, 243, 327, and 328.)

[60] A. Bobrowski and R. Bogucki, Semigroups generated by convex combinations of several Feller generators in models of mathematical biology, *Studia Mathematica* **189** (2008), 287–300. (Cited on pages 283 and 293.)

[61] A. Bobrowski and R. Bogucki, Two theorems on singularly perturbed semigroups with applications to models of applied mathematics, *Discr. Cont. Dyn. Syst. B* **17** (2012), no. 3, 735–757. (Cited on pages 263, 271, and 287.)

[62] A. Bobrowski and W. Chojnacki, Cosine families and semigroups really differ, *J. Evol. Equ.* **13** (2013), no. 4, 897–916. (Cited on pages 4, 388, 390, and 392.)

[63] A. Bobrowski and W. Chojnacki, Isolated points of the set of bounded cosine families on a Banach space, *Studia Mathematica* **217** (2013), no. 3, 219–241. (Cited on pages 41, 42, 43, 44, and 109.)

[64] A. Bobrowski, W. Chojnacki, and A. Gregosiewicz, On close-to-scalar one-parameter cosine families, *J. Math. Anal. Appl.* **429** (2015), no. 1, 383–394. (Cited on page 43.)

[65] A. Bobrowski and M. Kimmel, A random evolution related to a Fisher–Wright–Moran model with mutation, recombination, and drift, *Mathematical Methods in the Applied Sciences* **2003** (2003), no. 26, 1587–1599. (Cited on pages 311, 313, 315, and 317.)

[66] A. Bobrowski and M. Kimmel, *An Operator Semigroup in Mathematical Genetics*, Springer Briefs in Applied Sciences and Technology. Mathematical Methods, Springer, 2015. (Cited on pages 119 and 152.)

[67] A. Bobrowski, M. Kimmel, R. Chakraborty, and O. Arino, A semigroup representation and asymptotic behavior of the Fisher–Wright–Moran coalescent, Chapter 8 in *Handbook of Statistics 19: Stochastic Processes: Theory and Methods*, C. R. Rao and D. N. Shanbhag, eds., Elsevier Science, Amsterdam, 2001. (Cited on page 152.)

[68] A. Bobrowski, M. Kimmel, and T. Wojdyła, Asymptotic behavior of a Moran model with mutations, drift, and recombinations among multiple loci, *J. Math. Biol.* **61** (2010), 455–473. (Cited on pages 318 and 326.)

[69] A. Bobrowski, T. Lipniacki, K. Pichór, and R. Rudnicki, Asymptotic behavior of distributions of mRNA and protein levels in a model of stochastic gene expression, *J. Math. Anal. Appl.* **333** (2007), 753–769. (Cited on page 283.)

[70] A. Bobrowski and K. Morawska, From a PDE model to an ODE model of dynamics of synaptic depression, *Discr. Cont. Dyn. Syst. B* **17** (2012), no. 7, 2313–2327. (Cited on pages 190, 193, and 194.)

[71] A. Bobrowski and D. Mugnolo, On moments-preserving cosine families and semigroups in $C[0,1]$, *J. Evol. Equ.* **13** (2013), no. 4, 715–735. (Cited on pages 37, 184, and 186.)

[72] R. Bogucki, *Convergence of semigroups of operators related to piece-wise deterministic Markov processes, with applications*, PhD thesis (in Polish), Institute of Mathematics, Polish Academy of Sciences, 2014, pp. 1–103. (Cited on pages 293 and 310.)

[73] M. Boulanouar, A mathematical study for a Rotenberg model, *J. Math. Anal. Appl.* **265** (2002), no. 2, 371–394. (Cited on page 30.)

[74] B. M. Brown, D. Elliot, and D. F. Paget, Lipschitz constants for the Bernstein polynomials of a Lipschitz continuous function, *J. Approx. Theor.* **49** (1987), 196–199. (Cited on page 264.)

[75] G. C. Brown and B. N. Kholodenko, Spatial gradients of cellular phosphoproteins, *FEBS Letters* **457** (1999), no. 3, 452–454. (Cited on pages 241 and 242.)

[76] P. Cannarsa and G. Da Prato, On a functional analysis approach to parabolic equations in infinite dimensions, *J. Functional Analysis* **118** (1993), 22–42. (Cited on page 86.)

[77] C. Canzi and G. Guerra, A simple counterexample related to the Lie–Trotter product formula, *Semigroup Forum* **84** (2012), no. 3, 499–504. (Cited on page 103.)

[78] N. L. Carothers, *A Short Course on Banach Space Theory*, London Mathematical Society Student Texts, vol. 64, Cambridge University Press, 2004. (Cited on pages 90, 139, and 222.)

[79] J. A. Carrillo, S. Cuadrado, and B. Perthame, Adaptive dynamics via Hamilton–Jacobi approach and entropy methods for a juvenile-adult model, *Math. Biosci.* **205** (2007), 137–161. (Cited on page xiii.)

[80] V. Casarino, K.-J. Engel, R. Nagel, and G. Nickel, A semigroup approach to boundary feedback systems, *Integral Equations Operator Theory* **47** (2003), no. 3, 289–306. (Cited on page 249.)

[81] R. Chander and H. Singh, On the measurability and continuity properties of the cosine operator, *Indian J. Pure Appl. Math.* **12** (1981), no. 1, 81–83. (Cited on pages 390 and 391.)

[82] P. R. Chernoff, Elements of a normed algebra whose 2^nth powers lie close to the identity, *Proc. Amer. Math. Soc.* **23** (1969), no. 2, 386–387. (Cited on page 44.)

[83] P. R. Chernoff, Universally commutatable operators are scalars, *Michigan Math. J.* **20** (1973), 101–107. (Cited on page 104.)

[84] R. Chill and E. Fašangová, *Gradient Systems. 13th International Internet Seminar*, Matfyzpress, Prague, 2010. (Cited on page 228.)

[85] R. Chill and A. Haraux, An optimal estimate for the time singular limit of an abstract wave equation, *Funkcial. Ekvac.* **47** (2004), no. 2, 277–290. (Cited on page 73.)

[86] R. Chill, V. Keyantuo, and M. Warma, Generation of cosine families on $L^p(0, 1)$ by elliptic operators with Robin boundary conditions, *Functional Analysis and Evolution Equations. The Günter Lumer Volume*, Birkhäuser, Basel, 2007, Amann, H. et al. (eds.), pp. 113–130. (Cited on page 386.)

[87] R. Chill and Y. Tomilov, Operators $L_1(R^+) \to X$ and the norm continuity problem for semigroups, *J. Functional Analysis* **256** (2009), no. 2, 352–384. (Cited on page 253.)

[88] W. Chojnacki, On the equivalence of a theorem of Kisyński and the Hille–Yosida generation theorem, *Proc. Amer. Math. Soc.* **126** (1998), no. 2, 491–497. (Cited on page 83.)

[89] W. Chojnacki, On cosine families close to scalar cosine families, *Journal of the Australian Mathematical Society* **FirstView** (2015), 1–9. (Cited on page 43.)

[90] K. L. Chung, *Markov Chains with Stationary Transition Probabilities*, Springer, 1960. (Cited on page 258.)

[91] K. L. Chung, On the exponential formulas of semigroup theory, *Math. Scand.* **10** (1962), 153–162. (Cited on page 110.)

[92] G. M. Coclite, A. Favini, C. G. Gal, G. R. Goldstein, J. A. Goldstein, E. Obrecht, and S. Romanelli, The role of Wentzell boundary conditions in linear and nonlinear analysis, *Advances in Nonlinear Analysis: Theory Methods and Applications*, Math. Probl. Eng. Aerosp. Sci., vol. 3, Camb. Sci. Publ., Cambridge, 2009, pp. 277–289. (Cited on page 18.)

[93] G. M. Coclite, A. Favini, G. R. Goldstein, J. A. Goldstein, and S. Romanelli, Continuous dependence on the boundary conditions for the Wentzell Laplacian, *Studia Mathematica* **77** (2008), 101–108. (Cited on page 18.)

[94] G. M. Coclite, A. Favini, G. R. Goldstein, J. A. Goldstein, and S. Romanelli, Continuous dependence in hyperbolic problems with Wentzell boundary conditions, *Commun. Pure Appl. Anal.* **13** (2014), no. 1, 419–433. (Cited on page 18.)

[95] E. Conway, D. Hoff, and J. Smoller, Large time behavior of solutions of systems of nonlinear reaction-diffusion equations, *SIAM J. Appl. Math.* **35** (1978), 1–16. (Cited on page 228.)

[96] R. H. Cox, Matrices all of whose powers lie close to the identity (Abstract), *Amer. Math. Monthly* **73** (1966), 813. (Cited on page 44.)

[97] E. Crooks, B. Kaźmierczak, and T. Lipniacki, A spatially extended model of kinase-receptor interaction, *SIAM J. Appl. Math.* **73** (2013), no. 1, 374–400. (Cited on page 241.)

[98] D. J. Daley and D. Vere-Jones, *An Introduction to the Theory of Point Processes*, Springer, 1998. (Cited on page 271.)

[99] R. Dautray and J.-L. Lions, *Mathematical Analysis and Numerical Methods for Science and Technology. Vol. 2*, Springer-Verlag, Berlin, 1988, Functional and variational methods, with the collaboration of Michel Artola, Marc Authier, Philippe Bénilan, Michel Cessenat, Jean Michel Combes, Hélène Lanchon, Bertrand Mercier, Claude Wild and Claude Zuily, Translated from the French by Ian N. Sneddon. (Cited on page 32.)

[100] R. Dautray and J.-L. Lions, *Mathematical Analysis and Numerical Methods for Science and Technology. Vol. 3*, Springer-Verlag, Berlin, 1990, Spectral theory and applications, with the collaboration of Michel Artola and Michel Cessenat, Translated from the French by John C. Amson. (Cited on page 190.)

[101] R. Dautray and J.-L. Lions, *Mathematical Analysis and Numerical Methods for Science and Technology. Vol. 5*, Springer-Verlag, Berlin, 1992, Evolution problems. I, with the collaboration of Michel Artola, Michel Cessenat, and Hélène Lanchon, Translated from the French by Alan Craig. (Cited on pages xi, 1, and 95.)

[102] E. B. Davies, A model for absorption or decay, *Helv. Phys. Acta* **48** (1975), no. 3, 365–382. (Cited on page 365.)

[103] E. B. Davies, *One-parameter Semigroups*, Academic Press, London, 1980. (Cited on page 1.)

[104] M. H. A. Davis, *Lectures on Stochastic Control and Nonlinear Filtering*, Springer, 1984. (Cited on page 271.)

[105] M. H. A. Davis, Piece-wise deterministic Markov processes, *J. Royal Statistical Soc., Ser. B.* **46** (1984), 353–388. (Cited on page 271.)

[106] M. H. A. Davis, *Markov Processes and Optimization*, Chapman and Hall, 1993. (Cited on page 271.)

[107] L. Debnath and P. Mikusiński, *Introduction to Hilbert Spaces with Applications*, 3rd ed., Elsevier, Amsterdam, 2005. (Cited on pages 8, 32, and 39.)

[108] A. Defant and K. Floret, *Tensor Norms and Operator Ideals*, North Holland, Amsterdam, 1993. (Cited on pages 153, 318, 327, and 330.)

[109] R. deLaubenfels, Well-behaved derivations on $C[0, 1]$, *Pacific J. Math.* **115** (1984), no. 1, 73–80. (Cited on page 302.)

[110] R. deLaubenfels, Correction to: "Well-behaved derivations on $C[0,1]$" [*Pacific J. Math.* **115** (1984), no. 1, 73–80], *Pacific J. Math.* **130** (1987), no. 2, 395–396. (Cited on page 302.)

[111] R. deLaubenfels, *Existence Families, Functional Calculi, and Evolution Equations*, Lecture Notes in Mathematics, vol. 1570, Springer Verlag, Berlin, 1994. (Cited on page 139.)

[112] N. Deo, *Graph Theory with Applications to Engineering and Computer Science*, Prentice-Hall, Inc., Englewood Cliffs., 1974. (Cited on page 203.)

[113] O. Diekmann, P.-E. Jabin, S. Mischler, and B. Perthame, The dynamics of adaptation: an illuminating example and a Hamilton-Jacobi approach, *Theor. Popul. Biol.* **67** (2005), no. 4, 257–271. (Cited on page xiii.)

[114] R. L. Dobrushin, An example of a countable homogeneous Markov process all states of which are instantaneous, *Teor. Veroyatnost. i Primenen.* **1** (1956), 481–485. (Cited on page 90.)

[115] J. L. Doob, *Stochastic Processes*, Wiley, 1953. (Cited on page xi.)

[116] B. Dorroh, Contraction semigroups in a function space, *Pacific J. Math.* **19** (1966), 35–38. (Cited on pages 299 and 300.)

[117] R. M. Dudley, *Real Analysis and Probability*, Cambridge University Press, Cambridge, 2003, Second edition. (Cited on pages 215 and 263.)

[118] N. Dunford and J. T. Schwartz, *Linear Operators. Part I*, Wiley Classics Library, John Wiley & Sons, Inc., New York, 1988, General theory, with the assistance of William G. Bade and Robert G. Bartle, Reprint of the 1958 original, A Wiley-Interscience Publication. (Cited on page 1.)

[119] R. Durret, *Probability Models for DNA Sequence Evolution*, Springer, New York, 2002. (Cited on page 311.)

[120] R. Durrett and S. Kruglyak, A new stochastic model of microsatellite evolution, *J. Appl. Prob.* (1999), no. 36, 621–631. (Cited on page 152.)

[121] E. B. Dynkin, Infinitesimal operators of Markov random processes, *Dokl. Akad. Nauk SSSR (N.S.)* **105** (1955), 206–209. (Cited on page 20.)

[122] E. B. Dynkin, *Markov Processes. Vols. I, II*, Translated with the authorization and assistance of the author by J. Fabius, V. Greenberg, A. Maitra, G. Majone. Die Grundlehren der Mathematischen Wissenschaften, Bände 121, vol. 122, Academic Press Inc., Publishers, New York; Springer-Verlag, Berlin-Göttingen-Heidelberg, 1965. (Cited on pages 1 and 20.)

[123] E. B. Dynkin, *Selected papers of E. B. Dynkin with Commentary*, American Mathematical Society, Providence, RI; International Press, Cambridge, MA, 2000, Edited by A. A. Yushkevich, G. M. Seitz, and A. L. Onishchik. (Cited on page 20.)

[124] E. B. Dynkin and A. A. Yushkevich, *Markov Processes. Theorems and Problems*, Plenum Press, New York, 1969. (Cited on pages 20, 22, and 183.)

[125] R. E. Edwards, *Functional Analysis. Theory and Applications*, Dover Publications, 1995. (Cited on pages 114, 122, 211, and 225.)

[126] T. Eisner, *Stability of Operators and Operator Semigroups*, Birkhäuser, 2010. (Cited on page 177.)

[127] E. Yu. Emel′yanov, *Non-Spectral Asymptotic Analysis of One-Parameter Operator Semigroups*, Operator Theory: Advances and Applications, vol. 173, Birkhäuser Verlag, Basel, 2007. (Cited on pages 1, 80, 154, and 179.)

[128] K.-J. Engel and R. Nagel, *One-Parameter Semigroups for Linear Evolution Equations*, Springer, New York, 2000. (Cited on pages xi, 1, 29, 87, 89, 98, 114, 123, 137, 159, 178, 179, 201, 250, 299, 300, 349, and 358.)

[129] K.-J. Engel and R. Nagel, *A Short Course on Operator Semigroups*, Springer, New York, 2006. (Cited on pages 1, 29, 199, and 349.)

[130] R. Erban and S. J. Chapman, Reactive boundary conditions for stochastic simulation of reaction-diffusion processes, *Phys. Biol.* (2007), no. 4, 16–28. (Cited on page 66.)

[131] J. Esterle, Bounded cosine functions close to continuous scalar-bounded cosine functions, *ArXiv* e-prints (2015), 1–11. (Cited on page 43.)

[132] S. N. Ethier and T. G. Kurtz, *Markov Processes. Characterization and Convergence*, Wiley, New York, 1986. (Cited on pages xi, 1, 15, 17, 53, 73, 89, 112, 114, 133, 253, 255, 266, 301, and 331.)

[133] S. N. Evans and R. B. Sowers, Pinching and twisting Markov processes, *Ann. Probab.* **31** (2003), no. 1, 486–527. (Cited on page xiii.)

[134] W. J. Ewens, *Mathematical Population Genetics*, Springer, New York, 2004, Second edition. (Cited on pages 20, 96, 119, and 195.)

[135] H. O. Fattorini, Ordinary differential equations in linear topological spaces. I, *J. Differential Equations* **5** (1969), 72–105. (Cited on pages 390 and 402.)

[136] H. O. Fattorini, Ordinary differential equations in linear topological spaces. II, *J. Differential Equations* **6** (1969), 50–70. (Cited on page 402.)

[137] H. O. Fattorini, Singular perturbation and boundary layer for an abstract Cauchy problem, *J. Math. Anal. Appl.* **97** (1983), no. 2, 529–571. (Cited on page 73.)

[138] H. O. Fattorini, On the Schrödinger singular perturbation problem, *SIAM J. Math. Anal.* **16** (1985), no. 5, 1000–1019. (Cited on page 73.)

[139] H. O. Fattorini, *Second Order Linear Differential Equations in Banach Spaces*, North-Holland, Amsterdam, 1985. (Cited on pages 4, 70, 73, and 390.)

[140] H. O. Fattorini, The hyperbolic singular perturbation problem: an operator theoretic approach, *J. Differential Equations* **70** (1987), no. 1, 1–41. (Cited on page 73.)

[141] W. Feller, *An Introduction to Probability Theory and Its Applications*, vol. 1, Wiley, New York, 1950, Third edition, 1970. (Cited on page 96.)

[142] W. Feller, Diffusion processes in genetics, *Proceedings of the Second Berkeley Symposium on Mathematical Statistics and Probability*, 1950 (Berkeley and Los Angeles), University of California Press, 1951, pp. 227–246. (Cited on page 20.)

[143] W. Feller, Two singular diffusion problems, *Ann. Math.* **54** (1951), 173–182. (Cited on page 195.)

[144] W. Feller, Diffusion processes in one dimension, *Trans. Amer. Math. Soc.* **77** (1952), no. 1, 468–519. (Cited on pages 17 and 20.)

[145] W. Feller, The parabolic differential equations and the associated semigroups of transformations, *Ann. Math.* **55** (1952), 468–519. (Cited on pages 20 and 359.)

[146] W. Feller, *An Introduction to Probability Theory and Its Applications*, vol. 2, Wiley, New York, 1966, Second edition, 1971. (Cited on pages 166, 181, 183, 187, and 297.)

[147] W. Feller and H. P. McKean, Jr., A diffusion equivalent to a countable Markov chain, *Proc. Nat. Acad. Sci. U.S.A.* **42** (1956), 351–354. (Cited on page 90.)

[148] E. Fieremans, D. S Novikov, J. H. Jensen, and J. A. Helpern, Monte Carlo study of a two-compartment exchange model of diffusion, *NMR in Biomedicine* (2010), no. 23, 711–724. (Cited on page 66.)

[149] W. H. Fleming and Q. Zhang, Risk-sensitive production planning of a stochastic manufacturing system, *SIAM J. Control Optim.* **36** (1998), 1147–1170. (Cited on page 287.)

[150] D. Freedman, *Markov Chains*, Holden-Day, Inc., 1971. (Cited on page 90.)

[151] M. I. Freidlin and A. D. Wentzell, Diffusion processes on graphs and the averaging principle, *Ann. Math.* **21** (1993), 2215–2245. (Cited on pages 327 and 361.)

[152] M. I. Freidlin and A. D. Wentzell, Diffusion processes on an open book and the averaging principle, *Stochastic Processes and their Applications* **113** (2004), no. 1, 101–126. (Cited on page xiii.)

[153] M. I. Freidlin and A. D. Wentzell, *Random Perturbations of Dynamical Systems*, third ed., Grundlehren der Mathematischen Wissenschaften [Fundamental Principles of Mathematical Sciences], vol. 260, Springer, Heidelberg, 2012, Translated from the 1979 Russian original by Joseph Szücs. (Cited on pages 327 and 361.)

[154] Y. Fujita, A probabilistic approach to Volterra equations in Banach spaces, *Diff. Int. Eqs.* **5** (1992), 769–776. (Cited on page 4.)

[155] F. R. Gantmacher, *Applications of the Theory of Matrices*, Interscience Publishers, Inc. New York, 1959. (Cited on page 305.)

[156] J. H. Gillespie, *The Causes of Molecular Evolution*, Oxford University Press, New York, 1991. (Cited on page 311.)

[157] C. Goffman and R. E. Zink, Concerning the measurable boundaries of a real function, *Fund. Math.* **48** (1959/1960), 105–111. (Cited on page 407.)

[158] G. R. Goldstein, Derivation and physical interpretation of general boundary conditions, *Adv. Differential Equations* **11** (2006), no. 4, 457–480. (Cited on page 20.)

[159] J. A. Goldstein, A Lie product formula for one parameter groups of isometries on Banach spaces, *Math. Ann.* **186** (1970), 299–306. (Cited on page 104.)

[160] J. A. Goldstein, On the convergence and approximation of cosine functions, *Aequationes Math.* **11** (1974), 201–205. (Cited on page 384.)

[161] J. A. Goldstein, Semigroup-theoretic proofs of the central limit theorem and other theorems of analysis, *Semigroup Forum* **12** (1976), no. 3, 189–206. (Cited on page 110.)

[162] J. A. Goldstein, Cosine functions and the Feynman–Kac formula, *Quart. J. Math. Oxford Ser. (2)* **33** (1982), no. 131, 303–307. (Cited on page 104.)

[163] J. A. Goldstein, *Semigroups of Linear Operators and Applications*, Oxford University Press, New York, 1985. (Cited on pages xi, 1, 125, 127, 137, 337, 384, and 402.)

[164] A. Gomilko and Y. Tomilov, On convergence rates in approximation theory for operator semigroups, *J. Functional Analysis* **266** (2014), no. 5, 3040–3082. (Cited on pages 44 and 110.)

[165] R. L. Graham, D. E. Knuth, and O. Patashnik, *Concrete Mathematics*, Addison-Wesley, 1994, second edition. (Cited on pages 116 and 325.)

[166] A. Gregosiewicz, Asymptotic behaviour of diffusions on graphs, *Probability in Action*, Banek, T., Kozlowski, E., eds., Lublin University of Technology, 2014, pp. 83–96. (Cited on page 208.)

[167] G. Greiner, Perturbing the boundary conditions of a generator, *Houston J. Math.* **13** (1987), no. 2, 213–229. (Cited on pages 345 and 346.)

[168] R. J. Griego and R. Hersh, Random evolutions, Markov chains, and systems of partial differential equations, *Proc. Nat. Acad. Sci. U.S.A.* **62** (1969), 305–308. (Cited on page 271.)

[169] R. J. Griego and R. Hersh, Theory of random evolutions with applications to partial differential equations, *Trans. Amer. Math. Soc.* **156** (1971), 405–418. (Cited on page 271.)

[170] K. Gustafson, A note on left multiplication of semigroup generators, *Pacific J. Math.* **24** (1968), 463–465. (Cited on page 299.)

[171] M. Hasegawa, A note on convergence of semigroups of operators, *Proc. Japan Acad.* **40** (1964), 262–266. (Cited on page 85.)

[172] B. Hat, P. Paszek, M. Kimmel, K. Piechór, and T. Lipniacki, How the number of alleles influences gene expression, *J. Statist. Phys.* **128** (2007), no. 1/2, 511–533. (Cited on page 283.)

[173] A. Haurie, A two-timescale stochastic game framework for climate change policy assessment, *Dynamic Games: Theory and Applications* (ed. A. Haurie, G. Zaccour), Springer, 2005. (Cited on page 289.)

[174] A. Haurie and F. Moresino, Singularly perturbed piecewise deterministic games, *SIAM J. Control Optim.* **47** (2008), no. 1, 73–91. (Cited on pages 288 and 289.)

[175] J. Hein, M. H. Schierup, and C. Wiuf, *Gene Genealogies, Variation, and Evolution*, Oxford University Press, Oxford, 2006. (Cited on page 311.)

[176] B. Hennig and F. Neubrander, On representations, inversions, and approximations of Laplace transform in Banach spaces, *Appl. Anal.* **49** (1993), 151–170. (Cited on pages 397 and 398.)

[177] D. Henry, *Geometric Theory of Semilinear Parabolic Equations*, Springer, Berlin, 1981. (Cited on page 210.)

[178] H. Heyer, *Probability Measures on Locally Compact Groups*, Springer-Verlag, Berlin-New York, 1977, Ergebnisse der Mathematik und ihrer Grenzgebiete, Band 94. (Cited on pages 95 and 297.)

[179] E. Hille, Une généralisation du problème de Cauchy, *Ann. Inst. Fourier (Grenoble)* **4** (1952), 31–48. (Cited on page 402.)

[180] E. Hille and R. S. Phillips, *Functional Analysis and Semi-Groups*, Amer. Math. Soc. Colloq. Publ. 31, Amer. Math. Soc., Providence, R. I., 1957. (Cited on pages xi, 1, 75, 152, 157, 160, 297, 390, 392, 399, 412, and 413.)

[181] R. A. Hirschfeld, On semigroups in Banach algebras close to the identity, *Proc. Japan Acad.* **44** (1968), 755. (Cited on page 44.)

[182] F. Hoppensteadt, Stability in systems with parameter, *J. Math. Anal. Appl.* **18** (1967), 129–134. (Cited on page 268.)

[183] G. A. Hunt, Semigroups of measures on Lie groups, *Trans. Amer. Math. Soc.* **81** (1956), 264–293. (Cited on pages 95 and 297.)

[184] M. Iannelli, *Mathematical Theory of Age-Structured Population Dynamics*, Applied Mathematical Monographs 7, Giardini Editori E Stampatori, Pisa, 1995. (Cited on pages 304, 308, 349, and 358.)

[185] A. M. Il'in, R. Z. Khasminskii, and G. Yin, Asymptotic expansions of solutions of integro-differential equations for transition densities of singularly perturbed switching diffusions: rapid switchings, *J. Math. Anal. Appl.* **238** (1999), 516–539. (Cited on page 298.)

[186] M. Iosifescu, *Finite Markov Processes and Their Applications*, Wiley, New York, 1980. (Cited on page 323.)

[187] A. Irle and J. Kauschke, *Diffusion approximation of birth and death process with applications in financial market herding models*, working paper, August 2008, pp. 1–13. (Cited on pages 133 and 135.)

[188] K. Itô and McKean, Jr. H. P., *Diffusion Processes and Their Sample Paths*, Springer, Berlin, 1996, Repr. of the 1974 ed. (Cited on pages 20 and 333.)

[189] J. Janas, An extension of Berezin's approximation method, *J. Oper. Th.* **29** (1993), no. 1, 43–56. (Cited on page xiii.)

[190] N. L. Johnson, A. W. Kemp, and S. Kotz, *Univariate Discrete Distributions*, Wiley, 1992. (Cited on page 114.)

[191] P. E. T. Jørgensen, Monotone convergence of operator semigroups and the dynamics of infinite particle systems, *J. Approx. Theor.* **43** (1985), 205–230. (Cited on page 15.)

[192] M. Kac, *Some Stochastic Problems in Physics and Mechanics*, Colloq. Lect. 2, Magnolia Petrolum Co., 1956. (Cited on pages 71 and 73.)

[193] O. Kallenberg, *Foundations of Modern Probability*, Springer, 1997. (Cited on pages 1, 20, 112, 297, 301, and 331.)

[194] N. Kalton, S. Montgomery-Smith, K. Oleszkiewicz, and Y. Tomilov, Power-bounded operators and related norm estimates, *J. London Math. Sci.* (2) (2004), no. 70, 463–478. (Cited on page 44.)

[195] I. Karatzas and S. E. Shreve, *Brownian Motion and Stochastic Calculus*, Springer, New York, 1991. (Cited on page 20.)

[196] S. Karlin and H. M. Taylor, *A Second Course in Stochastic Processes*, Academic Press Inc. [Harcourt Brace Jovanovich Publishers], New York, 1981. (Cited on pages 20, 114, and 266.)

[197] T. Kato, On the semigroups generated by Kolmogoroff's differential equations, *J. Math. Soc. Japan* **6** (1954), 1–15. (Cited on page 75.)

[198] T. Kato, On the Trotter–Lie product formula, *Proc. Japan Acad.* **50** (1974), 694–698. (Cited on page 149.)

[199] T. Kato, Singular perturbation and semigroup theory, *Turbulence and Navier–Stokes Equations* (Proc. Conf., Univ. Paris-Sud, Orsay, 1975), Springer, Berlin, 1976, pp. 104–112. Lecture Notes in Math., Vol. 565. (Cited on pages xii and 15.)

[200] T. Kato, Trotter's product formula for an arbitrary pair of self-adjoint contraction semigroups, *Topics in Functional Analysis*, Adv. Math. Suppl. Stud., vol. 3, Academic Press, New York, 1978, Essays dedicated to M. G. Krein on the occasion of his 70th birthday, pp. 185–195. (Cited on page 149.)

[201] T. Kato, *Perturbation Theory for Linear Operators*, Classics in Mathematics Series, Springer, 1995, reprint of the 1980 edition. (Cited on pages 1, 32, 34, 36, 37, 38, 39, 62, 87, 159, 250, and 365.)

[202] B. Kaźmierczak and T. Lipniacki, Regulation of kinase activity by diffusion and feedback, *J. Theor. Biol.* **259** (2009), 291–296. (Cited on pages 241, 242, 243, and 248.)

[203] J. P. Keener, Activators and inhibitors in pattern formation, *Stud. Appl. Math.* **59** (1978), no. 1, 1–23. (Cited on page 236.)

[204] T. B. Kepler and T. C. Elston, Stochasticity in transcriptional regulation: origins, consequences, and mathematical representations, *Biophysical J.* **81** (2001), 3116–3136. (Cited on pages 281, 282, and 296.)

[205] J. Kevorkian and J. D. Cole, *Multiple Scale and Singular Perturbation Methods*, Applied Mathematical Sciences, vol. 114, Springer-Verlag, New York, 1996. (Cited on page xiii.)

[206] R. Z. Khasminskii, Principle of averaging for parabolic and elliptic equations for Markov processes with small parameter, *Teor. Veroyatnost. i Primenen.* **8** (1963), 1–21. (Cited on page 327.)

[207] J. F. C. Kingman, The coalescent, *Stoch. Proc. Appl.* **13** (1982), 235–248. (Cited on page 118.)

[208] J. F. C. Kingman, Exchangeability and the evolution of large populations, *Exchangeability in Probability and Statistics*, Koch, G., Spizzichino, F., eds., North-Holland, 1982, pp. 97–112. (Cited on page 119.)

[209] J. F. C. Kingman, On the genealogy of large populations, *J. Appl. Prob.* **19A** (1982), 27–43. (Cited on page 118.)

[210] J. Kisyński, Sur les équations hyperboliques avec petit paramètre, *Colloq. Math.* **10** (1963), no. 2, 331–343. (Cited on page 73.)

[211] J. Kisyński, A proof of the Trotter–Kato theorem on approximation of semigroups, *Colloq. Math.* **18** (1967), 181–184. (Cited on pages xi and 82.)

[212] J. Kisyński, On second order Cauchy's problem in a Banach space, *Bull. Acad. Polon. Sci. Ser. des. Sci. Math. Astr. et Phys.* **18** (1970), no. 7, 371–374. (Cited on pages 71 and 73.)

[213] J. Kisyński, On cosine operator functions and one-parameter groups of operators, *Studia Mathematica* **44** (1972), 93–105. (Cited on page 4.)

[214] J. Kisyński, On M. Kac's probabilistic formula for the solutions of the telegraphist's equation, *Ann. Polon. Math.* **29** (1974), 259–272. (Cited on page 71.)

[215] J. Kisyński, *Semigroups of Operators and Some of Their Applications to Partial Differential Equations*, Control Theory and Topics in Functional Analysis (Internat. Sem., Internat. Centre Theoret. Phys., Trieste, 1974), vol. III, International Atomic Energy Agency, Vienna, 1976, pp. 305–405. (Cited on pages 1, 32, 35, and 102.)

[216] J. Kisyński, Around Widder's characterization of the Laplace transform of an element of $L^\infty(\mathbb{R}^+)$, *Ann. Polon. Math.* **74** (2000), 161–200, Dedicated to the memory of Bogdan Ziemian. (Cited on page 83.)

[217] J. Kisyński, On Fourier transforms of distribution semigroups, *J. Functional Analysis* **242** (2007), 400–441. (Cited on page 50.)

[218] Y. Konishi, Cosine functions of operators in locally convex spaces, *J. Fac. Sci. Univ. Tokyo Sect. IA Math.* **18** (1971/1972), 443–463. (Cited on page 384.)

[219] V. S. Korolyuk and Turbin A. F., *Mathematical Foundations of the State Space Lumping of Large Systems*, Kluwer Academic, 1993. (Cited on page 298.)

[220] V. S. Korolyuk and Limnios N., *Stochastic Systems in Merging Phase Space*, World Scientific, 2005. (Cited on page 298.)

[221] V. S. Korolyuk and Korolyuk V. V., *Stochastic Models of Systems*, Kluwer Academic, 1999. (Cited on page 298.)

[222] M. Krein, The theory of self-adjoint extensions of semi-bounded Hermitian transformations and its applications. I, *Rec. Math. [Mat. Sbornik] N.S.* **20(62)** (1947), 431–495. (Cited on page 36.)

[223] S. G. Krein, *Linear Differential Equations in Banach Space*, Transl. of Math. Monographs, vol 29, Amer. Math. Soc., Providence, 1971. (Cited on page 122.)

[224] C. S. Kubrusly, *The Elements of Operator Theory*, second ed., Birkhäuser/ Springer, New York, 2011. (Cited on page 366.)

[225] M. Kuczma, *An Introduction to the Theory of Functional Equations and Inequalities: Cauchy's Equation and Jensen's Inequality*, Birkhäuser, Basel, 2009, Edited by A. Gilányi. (Cited on page 405.)

[226] F. Kühnemund and M. Wacker, The Lie–Trotter product formula does not hold for arbitrary sums of generators, *Semigroup Forum* **60** (1999), no. 3, 478–485. (Cited on page 103.)

[227] S. Kurepa, A cosine functional equation in Hilbert space, *Canad. J. Math.* **12** (1960), 45–50. (Cited on pages 405 and 408.)

[228] S. Kurepa, Semigroups and cosine functions, *Functional Analysis (Dubrovnik, 1981)*, Lecture Notes in Math., vol. 948, Springer, Berlin, 1982, pp. 47–72. (Cited on page 390.)

[229] T. G. Kurtz, Extensions of Trotter's operator semigroup approximation theorems, *J. Functional Analysis* **3** (1969), 354–375. (Cited on pages xi, 15, 45, and 137.)

[230] T. G. Kurtz, A general theorem on the convergence of operator semigroups, *Trans. Amer. Math. Soc.* **148** (1970), 23–32. (Cited on pages 82 and 86.)

[231] T. G. Kurtz, Limit theorems for sequences of jump Markov processes approximating ordinary differential processes, *J. Appl. Prob.* **8** (1971), no. 2, 344–356. (Cited on page 338.)

[232] T. G. Kurtz, A limit theorem for perturbed operator semigroups with applications to random evolutions, *J. Functional Analysis* **12** (1973), 55–67. (Cited on pages 54 and 253.)

[233] T. G. Kurtz, Applications of an abstract perturbation theorem to ordinary differential equations, *Houston J. Math.* **3** (1977), no. 1, 67–82. (Cited on pages 253 and 263.)

[234] A. Lasota and M. C. Mackey, *Chaos, Fractals, and Noise. Stochastic Aspects of Dynamics*, Springer, 1994. (Cited on pages 24 and 80.)

[235] P. D. Lax, *Functional Analysis*, Pure and Applied Mathematics (New York), Wiley-Interscience [John Wiley & Sons], New York, 2002. (Cited on page 366.)

[236] P. D. Lax, *Linear Algebra and Its Applications*, Wiley, 2007. (Cited on pages 150, 277, 279, and 280.)

[237] P. D. Lax and Phillips R. S., *Scattering Theory*, Academic Press, 1967. (Cited on page 5.)

[238] J. P. Lehoczky, Real-time queueing theory, *Proceedings of the IEEE Real-Time Systems Symposium*, IEEE, New York, 1996, pp. 186–195. (Cited on page 57.)

[239] A. Lejay, *The snapping out Brownian motion*, hal–00781447, December 2013. (Cited on page 66.)

[240] T. M. Liggett, *Continuous Time Markov Processes. An Introduction*, Amer. Math. Soc., 2010. (Cited on pages 90, 98, and 121.)

[241] T. Lipniacki, P. Paszek, A. R. Brasier, B. Luxon, and M. Kimmel, Stochastic regulation in early immune response, *Biophys. J.* **90** (2006), 725–742. (Cited on page 283.)

[242] T. Lipniacki, P. Paszek, A. Marciniak-Czochra, A. R. Brasier, and M. Kimmel, Transcriptional stochasticity in gene expression, *J. Theor. Biol.* **238** (2006), 348–367. (Cited on pages 283 and 284.)

[243] J. H. Liu, Singular perturbations of integro-differential equations in Banach space, *Proc. Amer. Math. Soc.* **122** (1994), no. 3, pp. 791–799. (Cited on page 73.)

[244] C. Lizama and H. Prado, Singular perturbations of integro-differential equations, *Applied Mathematics and Computation* **175** (2006), no. 2, 1582–1595. (Cited on page 73.)

[245] G. G. Lorentz, *Bernstein Polynomials*, University of Toronto Press, 1953, Second edition, New York 1986. (Cited on page 264.)

[246] A. Lunardi, *Analytic Semigroups and Optimal Regularity in Parabolic Problems*, Birkhäuser, 1995. (Cited on page 159.)

[247] P. Mandl, *Analytical Treatment of One-Dimensional Markov Processes*, Springer, 1968. (Cited on pages 20, 21, 106, 297, and 301.)

[248] A. Marciniak-Czochra, S. Harting, G. Karch, and K. Suzuki, Dynamical spike solutions in a nonlocal model of pattern formation, *arXiv:*1307.6236 (2013). (Cited on page 240.)

[249] A. Marciniak-Czochra and M. Kimmel, Modeling of early lung cancer progression: influence of growth factor production and cooperation between partially transformed cells, *Math. Mod. Meth. Appl. Sci.* **17** (2007), 1693–1719. (Cited on page 216.)

[250] A. Marciniak-Czochra and M. Kimmel, Reaction-diffusion model of early carcinogenesis: the effects of influx of mutated cells, *Math. Model. Nat. Phenom.* **3** (2008), no. 7, 90–114. (Cited on page 216.)

[251] A. Marciniak-Czochra and M. Ptashnyk, Derivation of a macroscopic receptor-based model using homogenization techniques, *SIAM J. Math. Anal.* **40** (2008), 215–237. (Cited on page 55.)

[252] J. L. Massera, On Liapounoff's conditions of stability, *Ann. Math.* **50** (1949), 705–721. (Cited on page 269.)

[253] T. Mátrai, Resolvent norm decay does not characterize norm continuity, *Israel J. Math.* **168** (2008), 1–28. (Cited on page 253.)

[254] K. Maurin, *Methods of Hilbert Spaces*, Translated from the Polish by Andrzej Alexiewicz and Waclaw Zawadowski. Monografie Matematyczne, Tom 45, Państwowe Wydawnictwo Naukowe, Warsaw, 1967. (Cited on page 366.)

[255] S. McAllister, F. Neubrander, A. Reiser, and Y. Zhuang, Stabilizations of the Trotter–Kato theorem and the Chernoff product formula, *Semigroup Forum* **86** (2013), 511–524. (Cited on pages 139 and 140.)

[256] A. C. McBride, *Semigroups of Linear operators: An Introduction*, Pitman Research Notes in Mathematics Series, vol. 156, Longman Scientific & Technical, Harlow; John Wiley & Sons, Inc., New York, 1987. (Cited on pages 1 and 95.)

[257] A. C. McBride, A. L. Smith, and W. Lamb, Strongly differentiable solutions of the discrete coagulation fragmentation equation, *Physica D: Nonlinear Phenomena* **239** (2010), no. 15, 1436–1445, Evolution Equations in Pure and Applied Sciences. (Cited on pages 219, 225, and 227.)

[258] H. P. McKean, Jr., Chapman–Enskog–Hilbert expansion for a class of solutions of the telegraph equation, *J. Mathematical Phys.* **8** (1967), 547–552. (Cited on page 73.)

[259] M. A. McKibben, *Discovering Evolution Equations with Applications. Volume 1: Deterministic Equations*, CRC Press, 2011. (Cited on page 1.)

[260] R. McVinish and P. K. Pollett, The deterministic limit of a stochastic logistic model with individual variation, *Math. Biosci.* **241** (2013), 109–114. (Cited on page 338.)

[261] R. E. Megginson, *An Introduction to Banach Space Theory*, Springer, 1998. (Cited on page 279.)

[262] C. Meyer, *Matrix Analysis and Applied Linear Algebra*, SIAM, 2000. (Cited on page 274.)

[263] J. R. Mika and J. Banasiak, *Singularly Perturbed Equations with Applications to Kinetic Theory*, Series on Advances in Mathematics for Applied Sciences, vol. 34, World Scientific, Singapore, 1995. (Cited on pages xiii, 341, 342, and 344.)

[264] W. Mlak, *Hilbert Spaces and Operator Theory*, Mathematics and Its Applications (East European Series), vol. 51, Kluwer Academic Publishers Group, Dordrecht; PWN—Polish Scientific Publishers, Warsaw, 1991, Translated from the fourth Polish edition by Marcin E. Kuczma. (Cited on pages 38 and 365.)

[265] D. Mugnolo, *Semigroup Methods for Evolution Equations on Networks*, Springer, 2014. (Cited on page 203.)

[266] D. Mugnolo, Some remarks on the Krein–von Neumann extension of different Laplacians, *Semigroups of Operators – Theory and Applications* (J. Banasiak, A. Bobrowski, and M. Lachowicz, eds.), Springer Proceedings in Mathematics & Statistics, vol. 113, Springer International Publishing, 2015, pp. 69–87 (English). (Cited on page 36.)

[267] D. Mugnolo and R. Nittka, Convergence of operator semigroups associated with generalized elliptic forms, *J. Evol. Equ.* **12** (2012), 593–619. (Cited on pages 18 and 39.)

[268] J. D. Murray, *Mathematical Biology*, Springer, 1993, Second, corrected edition. (Cited on page 304.)

[269] R. Nagel (ed.), *One-parameter Semigroups of Positive Operators*, Lecture Notes in Mathematics, vol. 1184, Springer, 1986. (Cited on pages 42, 80, 87, 146, 154, 179, and 312.)

[270] M. Nagisa and S. Wada, Averages of operators and their positivity, *Proc. Amer. Math. Soc.* **126** (1998), no. 2, 499–506. (Cited on page 44.)

[271] M. Nakamura and M. Yoshida, On a generalization of a theorem of Cox, *Proc. Japan Acad.* **43** (1967), 108–110. (Cited on page 44.)

[272] J. Nedoma, Cauchy problem for the hyperbolic equation with the small parameter, *Časopis pro pěstování matematiky* **092** (1967), no. 4, 392–417 (In Czech). (Cited on page 73.)

[273] E. Neher and R. S. Zucker, Multiple calcium-dependent process related to secretion in bovine chromaffin cells, *Neuron* **10** (1993), 2–30. (Cited on page 189.)

[274] E. Nelson, *Topics in Dynamics. I: Flows*, Mathematical Notes, Princeton University Press, Princeton, N.J.; University of Tokyo Press, Tokyo, 1969. (Cited on page 104.)

[275] G. Nickel, Evolution semigroups and product formulas for nonautonomous Cauchy problems, *Math. Nachr.* **212** (2000), 101–116. (Cited on page 123.)

[276] G. Nickel, A semigroup approach to dynamic boundary value problems, *Semigroup Forum* **69** (2004), no. 2, 159–183. (Cited on page 249.)

[277] R. Nittka, Approximation of the semigroup generated by the Robin Laplacian in terms of the Gaussian semigroup, *J. Functional Analysis* **257** (2009), no. 5, 1429–1444. (Cited on pages 108, 149, and 150.)

[278] J. R. Norris, *Markov Chains*, Cambridge University Press, Cambridge, 1997. (Cited on pages 152 and 258.)

[279] A. Ochab-Marcinek and M. Tabaka, Bimodal gene expression in noncooperative regulatory systems, *Proc. Nat. Acad. Sci. U.S.A.* **107** (2010), no. 51, 22096–22101. (Cited on page 28.)

[280] R. E. O'Malley, *Historical Developments in Singular Perturbations*, Springer, Cham, 2014. (Cited on page xiii.)

[281] H. G. Othmer and T. Hillen, The diffusion limit of transport equations derived from velocity-jump processes, *SIAM J. Appl. Math.* **61** (2000), no. 3, 751–775. (Cited on page 73.)

[282] E. M. Ouhabaz, *Analysis of Heat Equations on Domains*, Lond. Math. Soc. Monograph Series, vol. 30, Princeton Univ. Press, Princeton, 2005. (Cited on pages 32 and 35.)

[283] E. Pardoux and Yu. Veretennikov, Averaging of backward stochastic differential equations with applications to semi-linear PDEs, *Stochastics and Stochastics Reports* **60** (1997), 255–270. (Cited on page 235.)

[284] A. Pazy, *Semigroups of Linear Operators and Applications to Partial Differential Equations*, Springer, 1983. (Cited on pages xi, 1, 137, 159, 210, 211, 250, and 301.)

[285] B. Perthame, *Transport Equations in Biology*, Frontiers in Mathematics, Birkhäuser Verlag, Basel, 2007. (Cited on page xiii.)

[286] R. S. Phillips, An inversion formula for the Laplace transform and semigroups of linear operators, *Ann. Math.* **59** (1954), 325–356. (Cited on page 399.)

[287] V. Q. Phóng, Stability of semigroups commuting with a compact operator, *Proc. Amer. Math. Soc.* **124** (1996), no. 10, 3207–3209. (Cited on page 154.)

[288] M. A. Pinsky, *Lectures on Random Evolutions*, World Scientific, Singapore, 1991. (Cited on page 73.)

[289] S. Piskarev and S.-Y. Shaw, On certain operator families related to cosine operator functions, *Taiwanese J. Math.* **1** (1997), no. 4, 527–546. (Cited on page 390.)

[290] G. Da Prato and F. Giusti, Una carracterizzazione dei generatori di funzioni coseno astratte, *Bull. Un. Mat. Ital.* **22** (1967), 357–362. (Cited on page 4.)

[291] G. Da Prato and E. Sinestrari, Differential operators with nondense domain, *Ann. Scuola Norm. Sup. Pisa Cl. Sci.* **14** (1987), no. 4, 285–344. (Cited on pages 52 and 159.)

[292] J. Prüss, *Evolutionary Integral Equations and Applications*, Birkhäuser, 1993. (Cited on page 4.)

[293] M. Rausand and A. Høyland, *System Reliability Theory: Models and Statistical Methods*, Wiley, 2004. (Cited on page 287.)

[294] M. Reed and B. Simon, *Methods of Modern Mathematical Physics. I*, second ed., Academic Press, Inc. [Harcourt Brace Jovanovich, Publishers], New York, 1980, Functional analysis. (Cited on pages 15, 38, and 39.)

[295] M. I. Reiman, Some diffusion approximations with state-space collapse, *Proceedings of International Seminar on Modeling and Performance Evaluation*

Methodology, Lecture Notes in Control and Inform. Sci. 60, Springer, Berlin, 1983, pp. 209–240. (Cited on pages xiii and 57.)

[296] D. Revuz and M. Yor, *Continuous Martingales and Brownian Motion*, Springer, 1999, Third edition. (Cited on pages 17, 20, 195, 242, 299, 300, and 333.)

[297] D. W. Robinson, *The Thermodynamic Pressure in Quantum Statistical Mechanics*, Springer-Verlag, Berlin-New York, 1971, Lecture Notes in Physics, Vol. 9. (Cited on page 365.)

[298] L. C. G. Rogers and D. Williams, *Diffusions, Markov Processes, and Martingales, Vol. 1, Foundations*, Cambridge University Press, Cambridge, 2000. (Cited on pages 121 and 299.)

[299] M. Rotenberg, Transport theory for growing cell populations, *J. Theor. Biol.* **103** (1983), no. 2, 181–199. (Cited on page 30.)

[300] F. Rothe, *Global Solutions of Reaction-Diffusion Systems*, Lecture Notes in Mathematics, vol. 1072, Springer-Verlag, Berlin, 1984. (Cited on page 240.)

[301] R. Rudnicki, *Models of Population Dynamics and Their Applications in Genetics*, From Genetics to Mathematics. M. Lachowicz and J. Miekisz (eds.), Ser. Adv. Math. Appl. Sci., vol. 79, World Sci. Publ., Hackensack, NJ, 2009, pp. 103–147. (Cited on page 308.)

[302] R. Rudnicki, *Models and Methods of Mathematical Biology*, Institute of Mathematics of the Polish Academy of Sciences, 2014, (in Polish). (Cited on pages 304 and 308.)

[303] R. Rudnicki and M. Tyran-Kamińska, Piecewise deterministic Markov processes in biological models, *Semigroups of Operators – Theory and Applications* (J. Banasiak, A. Bobrowski, and M. Lachowicz, eds.), Springer Proceedings in Mathematics & Statistics, vol. 113, Springer International Publishing, 2015, pp. 235–255 (English). (Cited on page 166.)

[304] R. A. Ryan, *Introduction to Tensor Products of Banach Spaces*, Springer, 2002. (Cited on pages 153, 318, 327, 330, and 331.)

[305] L. Saloff-Coste, *Lectures on Finite Markov Chains*, Lectures on Probability Theory and Statistics, Lecture Notes in Mathematics 1665, Springer, Berlin, 1997. (Cited on pages 278 and 280.)

[306] E. Sanchez, R. Bravo de la Parra, P. Auger, and P. Gomez-Mourelo, Time scales in linear delayed differential equations, *J. Math. Anal. Appl.* **323** (2006), 680–699. (Cited on page 304.)

[307] V. V. Sarafyan and A. V. Skorohod, On fast switching dynamical systems, *Theory Probab. Appl.* **32** (1987), 595–607. (Cited on page 298.)

[308] F. Schwenninger and H. Zwart, Less than one, implies zero, *arXiv:*1310.6202 (2013), 1–12. (Cited on pages 43 and 44.)

[309] F. Schwenninger and H. Zwart, Zero-two law for cosine families, *J. Evol. Equ.* **15** (2015), no. 3, 559–569. (Cited on page 43.)

[310] S. Sethi, H. Zhang, and Q. Zhang, *Average-Cost Control of Stochastic Manufacturing Systems*, Springer, 2005. (Cited on page 287.)

[311] E. Shchepakina, V. Sobolev, and M. P. Mortell, *Singular Perturbations*, Lecture Notes in Mathematics, vol. 2114, Springer, Cham, 2014, Introduction to system order reduction methods with applications, with a foreword by Robert O'Malley. (Cited on page xiii.)

[312] W. Sierpiński, *Infinite Operations*, Mathematical Monographs, Czytelnik, 1948. (In Polish.) (Cited on page 92.)
[313] B. Simon, A canonical decomposition for quadratic forms with applications to monotone convergence theorems, *J. Functional Analysis* **28** (1978), no. 3, 377–385. (Cited on page 365.)
[314] E. Sinestrari, On the abstract Cauchy problem of parabolic type in spaces of continuous functions, *J. Math. Anal. Appl.* **107** (1985), 16–66. (Cited on page 159.)
[315] P. F. Slade, Simulation of 'hitch-hiking' genealogies, *J. Math. Biol.* **42** (2001), 41–70. (Cited on page 313.)
[316] D. R. Smith, *Singular-Perturbation Theory*, Cambridge University Press, Cambridge, 1985, An introduction with applications. (Cited on page xiii.)
[317] J. A. Smoller, Singular perturbations of Cauchy's problem, *Comm. Pure Appl. Math.* **18** (1965), 665–677. (Cited on page 73.)
[318] J. A. Smoller, *Shock-Waves and Reaction-Diffusion Equations*, Springer, 1994. (Cited on pages 210, 215, and 228.)
[319] M. Sova, Cosine operator functions, *Rozprawy Mat.* **49** (1966), 47. (Cited on page 4.)
[320] M. Sova, Convergence d'opérations linéaires non bornées, *Rev. Roumaine Math. Pures Appl.* **12** (1967), 373–389. (Cited on pages 15 and 45.)
[321] M. Sova, Équations hyperboliques avec petit paramètre dans les éspaces de Banach généraux, *Colloq. Math.* **21** (1970), no. 2, 303–320. (Cited on page 73.)
[322] R. B. Stein, A theoretical analysis of neuronal variability, *Biophys. J.* **5** (1965), 173–194. (Cited on page 166.)
[323] R. B. Stein, Some models of neuronal variability, *Biophys. J.* **7(1)** (1967), 37–68. (Cited on page 166.)
[324] K. Taira, *Semigroups, Boundary Value Problems, and Markov Processes*, Springer Monographs in Mathematics, Springer-Verlag, Berlin, 2004. (Cited on pages 1 and 21.)
[325] F. Tajima, Evolutionary relationship of DNA sequences in finite populations, *Genetics* **105** (1983), 437–460. (Cited on page 118.)
[326] H. Tanabe, *Equations of Evolution*, Pitman, 1979. (Cited on page 122.)
[327] M. E. Taylor, *Partial Differential Equations I. Basic Theory*, second ed., Applied Mathematical Sciences, vol. 115, Springer, New York, 2011. (Cited on page 338.)
[328] A. N. Tikhonov, Systems of differential equations containing small parameters in the derivatives, *Mat. Sb.* **31** (1952), 575–586. (Cited on pages 267 and 268.)
[329] E. C. Titchmarsh, *The Theory of Functions*, Oxford Scientific Publications, 1939. (Cited on page 175.)
[330] J. van Neerven, *The Asymptotic Behaviour of Semigroups of Linear Operators*, Operator Theory: Advances and Applications, vol. 88, Birkhäuser Verlag, Basel, 1996. (Cited on page 179.)
[331] V. V. Vasil′ev and S. I. Piskarev, Differential equations in Banach spaces. II. Theory of cosine operator functions, *J. Math. Sci. (N. Y.)* **122** (2004), no. 2, 3055–3174. (Cited on page 390.)
[332] A. B. Vasil′eva and B. F. Butuzov, *Asymptotic Expansions of Solutions of Singularly Perturbed Equations*, Nauka, Moscow, 1973. (Cited on pages 267 and 268.)
[333] F. Verhulst, *Methods and Applications of Singular Perturbations*, Texts in Applied Mathematics, vol. 50, Springer, New York, 2005, Boundary layers and multiple timescale dynamics. (Cited on page xiii.)

[334] I. I. Vrabie, *C_0-semigroups and Applications*, North-Holland Mathematics Studies, vol. 191, North-Holland Publishing Co., Amsterdam, 2003. (Cited on pages xi and 1.)

[335] J. Wakeley, *Coalescent Theory*, Ben Roberts Publishing, 2008. (Cited on page 311.)

[336] L. J. Wallen, On the magnitude of $x^n - 1$ in a normed algebra, *Proc. Amer. Math. Soc.* **18** (1967), no. 5, 956. (Cited on pages 44 and 84.)

[337] W. Walter, *Differential and Integral Inequalities*, Ergebnise d. Mathematik u. ihrer Granzgebiete, vol. 55, Springer, New York, 1970. (Cited on pages 214 and 263.)

[338] W. Walter, Differential inequalities and maximum principles: theory, new methods, and applications, *Proceedings of the Second World Congress of Nonlinear Analysts*, Part 8 (Athens, 1996), vol. 30, 1997, pp. 4695–4711. (Cited on pages 214 and 263.)

[339] W. Walter, *Ordinary Differential Equations*, Springer, 1998. (Cited on pages 263 and 277.)

[340] G. F. Webb, A representation formula for strongly continuous cosine families, *Aequationes Math.* **21** (1980), no. 1, 251–256. (Cited on page 103.)

[341] G. F. Webb, *Theory of Nonlinear Age-Dependent Population Dynamics*, Monographs and Textbooks in Pure and Applied Mathematics, vol. 89, Marcel Dekker Inc., New York, 1985. (Cited on page 304.)

[342] G. F. Webb, Structured population dynamics, *Mathematical Modeling of Population Dynamics*, R. Rudnicki ed., Banach Center Publ., vol. 63, Polish Acad. Sci., Warsaw, 2004, pp. 123–163. (Cited on page 304.)

[343] A. D. Wentzell, *A Course in the Theory of Stochastic Processes*, McGraw-Hill International Book Co., New York, 1981, Translated from the Russian by S. Chomet, with a foreword by K. L. Chung. (Cited on pages 1 and 95.)

[344] W. Wharton, *Birdy*, Knopf, 1978. (Cited on page 262.)

[345] D. V. Widder, *The Laplace Transform*, Princeton Univ. Press, 1946. (Cited on page 397.)

[346] W. J. Wilbur and J. Rinzel, An analysis of Stein's model for stochastic neuronal excitation, *Biol. Cybern.* **45** (1982), no. 2, 107–114. (Cited on pages 166 and 167.)

[347] D. Williams, *Probability with Martingales*, Cambridge Mathematical Textbooks, Cambridge University Press, Cambridge, 1991. (Cited on page 93.)

[348] W. Wils, On semigroups near the identity, *Proc. Amer. Math. Soc.* **21** (1969), 762–763. (Cited on page 44.)

[349] M. Winkler, Aggregation vs. global diffusive behavior in the higher-dimensional Keller-Segel model, *J. Differential Equations* **248** (2010), no. 12, 2889–2905. (Cited on page 240.)

[350] G. Yin, On limit results for a class of singularly perturbed switching diffusions, *J. Theor. Probab.* **14** (2001), 673–697. (Cited on page 298.)

[351] G. Yin and M. Kniazeva, Singularly perturbed multidimensional switching diffusions with fast and slow switchings, *J. Math. Anal. Appl.* **229** (1999), 605–630. (Cited on page 298.)

[352] G. G. Yin and Q. Zhang, *Continuous-Time Markov Chains and Applications. A Singular Perturbation Approach*, Springer, New York, 1998. (Cited on pages 258, 262, and 287.)

[353] K. Yosida, *Functional Analysis*, Springer, 1965. (Cited on pages xi, 1, 87, and 301.)

[354] T. Zastawniak, Path integrals for the telegraphers and Dirac equations: the analytic family of measures and the underlying Poisson process, *Bull. Polish Acad. Sci. Math.* **36** (1988), 341–356. (Cited on page 130.)

[355] M. Zlámal, Sur l'équation des télégraphistes avec un petit paramètre, *Atti Accad. Naz. Lincei. Rend. Cl. Sci. Fis. Mat. Nat.* **27** (1959), no. 8, 324–332. (Cited on page 73.)

[356] P. J. Żuk, M. Kochańczyk, J. Jaruszewicz, W. Bednorz, and T. Lipniacki, Dynamics of a stochastic spatially extended system predicted by comparing deterministic and stochastic attractors of the corresponding birth-death process, *Phys. Biol.* **9** (2012), no. 5, 055002 (12pp). (Cited on page 233.)

[357] H. Zwart, *Three line proof: less than one law*, Ulmer Seminar (2014), p. 341. (Cited on page 42.)

Index

CPSIA information can be obtained
at www.ICGtesting.com
Printed in the USA
LVOW02*0141240816
501573LV00001B/2/P

9 781107 137431